METHODS IN MOLECULAR BIOLOGY

Series Editor
John M. Walker
School of Life and Medical Sciences
University of Hertfordshire
Hatfield, Hertfordshire, AL10 9AB, UK

For further volumes:
http://www.springer.com/series/7651

Natural Products From Marine Algae

Methods and Protocols

Edited by

Dagmar B. Stengel

Botany and Plant Science, School of Natural Sciences, Ryan Institute for Environmental, Marine and Energy Research, National University of Ireland Galway, Galway, Ireland

Solène Connan

Photobiotechnology, INTECHMER, Conservatoire National des Arts et Métiers, Cherbourg, Cedex, France; CNRS, GEPEA, UMR6144, Boulevard de l'Université, Saint Nazaire, Cedex, France

Humana Press

Editors
Dagmar B. Stengel
Botany and Plant Science
School of Natural Sciences
Ryan Institute for Environmental
Marine and Energy Research
National University of Ireland Galway
Galway, Ireland

Solène Connan
Photobiotechnology, INTECHMER
Conservatoire National des Arts et Métiers
Cherbourg, Cedex, France

CNRS, GEPEA, UMR6144
Boulevard de l'Université
Saint Nazaire, Cedex, France

ISSN 1064-3745 ISSN 1940-6029 (electronic)
Methods in Molecular Biology
ISBN 978-1-4939-2683-1 ISBN 978-1-4939-2684-8 (eBook)
DOI 10.1007/978-1-4939-2684-8

Library of Congress Control Number: 2015940760

Springer New York Heidelberg Dordrecht London

Printed on acid-free paper

Humana Press is a brand of Springer
Springer Science+Business Media LLC New York is part of Springer Science+Business Media (www.springer.com)

Preface

Over the last decade or so, there has been an explosion in the global interest in marine algae including both seaweeds (macroalgae) and microalgae. This has commonly focused on their application as a source of bioenergy but also, more recently, on their potential as an "untapped" resource of natural products. In tandem with scientific and technological developments, public awareness of algae has increased considerably and their inclusion in our daily lives does not appear as alien anymore as it might have been, at least in the western world, a few years ago. Numerous algae-based products are on offer to the consumer, ranging from agri-horticultural to food and cosmetic products. Despite this enhanced presence, the general understanding of the diversity and complexity of what this unfortunate all-encompassing term "algae" entails is usually still underestimated by the public as well as some non-phycological researchers; however many scientists globally are currently engaged in unraveling the chemical and taxonomic richness of this diverse group of organisms. In parallel to working towards a better understanding of the basic biology of the many algal groups and their strategies to survive in the marine environments, considerable research efforts have resulted in significant advances in algal biotechnology. There has also been excellent progress in the field of chemical and structural identification of bioactive compounds as promising (marine-derived) natural products with potential in drug development in the long term. On the other hand, algal products also have the capability to be integrated in our daily lives as consumers for example as health-promoting foods.

Valuable compounds from marine algae include pigments, lipids, and fatty acids and sterols, polysaccharides, proteins and peptides, as well as many secondary metabolites such as mycosporine-like amino acids, phenolic compounds, and terpenes, all of which are highly specific to different algal groups and even to species within these. Bioactivities of algal compounds described to date range from antioxidant, anti-inflammatory, antidiabetic, anticancer, antiviral, antimicrobial, antifungal to anti-obesity and antidiabetic; more recently, high potencies of natural algal products against specific parasites have also been discovered.

Whilst of traditional and current economic value, and with high social acceptance as commodities in Asia, the western world is lagging in its appreciation of this marine resource, but this is about to change. For example, algae are anticipated to play an important role in the future within the European bioeconomy, with climate change, food security, and an aging population presenting global challenges; industry, researchers, politicians, and, increasingly, the public currently look towards the oceans as a source of novel and sustainable source of biomass to supply human food and support health and well-being. The provision of sustainable and safe biomass, together with more effective extraction of novel valuable compounds, is thus a growing concern and at the forefront of many national and international multidisciplinary research programs.

Regardless of the ultimate application, assuming that suitable algal biomass can be provided sustainably, the vast diversity and complexity of algal biomass demands reliable, fast, and efficient extraction techniques that allow safe provision of the target compound(s) and accurate but affordable analytical techniques. Also, rapid and reliable tests for bioactivities are required for the manifold applications that are known—and those yet to be discovered.

This volume aims to provide examples of the recent advances in extraction methodologies, analytical techniques, and commonly used bioactive assays currently applied to marine algae. Chapters include protocols for a suite of both routinely used standard procedures and newly developed, highly advanced and specialized techniques, which display currently available tools for characterizing algal chemical composition for the vast array of applications. Whilst the book cannot attempt to be complete (both due to the diversity and complexity of algal compounds, as well as ongoing technological developments), protocols were chosen to represent a range of extraction and analytical methods currently applied to both marine macro- and microalgae. They also cover a range of different compounds families that are of current and potential future interest, concentrating on high-value (i.e., non-bioenergy) applications in the food, agricultural, cosmetics, and pharmaceutical sectors.

Specifically, a review of sources of available algal biomass, current applications of marine algae in different industries, and the recent trends in algal biotechnology is presented at the beginning of this volume (Chapter 1). This is followed by a description of secondary metabolites (structure and function) produced by macroalgae (Chapter 2); then different extraction techniques are outlined, ranging from the traditional Solid-Liquid Extraction (SLE; Chapters 4–7, 10, 11, 13–18, 21) to the use of enzymes (Chapter 8) or Microwave Assisted Extraction (MAE; Chapter 9). The following chapters detail several analytical methods: Spectrophotometry (Chapters 3, 5, 7, 20, 21), Thin Layer Chromatography (TLC; Chapters 11, 13, 14), Electrophoresis (Chapter 5), Liquid Chromatography with or without Mass Spectrometer(s) (LC with/without MS; Chapters 6, 10, 15–18), Gas Chromatography associated with different detectors or Mass Spectrometer (GC; Chapters 11, 14, 21), Mass Spectrometer (Chapters 19, 21), liquid or solid state Nuclear Magnetic Resonance Spectroscopy (RMN; Chapters 7, 12–14, 20–22), Infra-red Spectroscopy (IF; Chapters 21, 22), and Raman Spectroscopy (Chapter 23). Also methodologies to highlight different bioactivity of compounds or extracts are described: antioxidant (Chapters 7 and 24), antimicrobial (Chapter 25), antifungal (Chapter 26), and antifouling (Chapter 27). In each case, these techniques are applied to primary or secondary algal metabolites: proteins (Chapters 3, 4), polysaccharides (Chapters 3, 9, 19–22), lipids (Chapter 11), pigments (Chapters 3, 5, 15), mycosporine-like amino acids (MAAs; Chapter 6), phenolic compounds (Chapters 3, 7, 16), oxylipins (Chapter 10), terpenes (Chapters 13, 14), betaines (Chapter 17), and different biotoxins (Chapter 18).

Active in both ecophysiological and applied biotechnological algal research, we continue to be intrigued by the newly described diverse adaptations of algae to their extreme and fluctuating environments and the resultant "twists" in chemical structures that are discovered. Algal responses to their physical and chemical habitats and microhabitats, chemical ecology and its application in natural products research, are likely to continue to be a fascinating and rich field that will yield new chemicals of value to humans. However the exploitation and commercial application of algae will need to ascertain, perhaps with the aid of novel cultivation methods and further biotechnological developments, that natural resources, their biological and chemical diversity, and their surrounding marine ecosystems will be protected, in particular, under globally increased environmental pressures.

We are grateful to our colleagues who have provided support and encouragement throughout our careers, including our Ph.D. supervisors, colleagues at NUI Galway and other institutions in Ireland, France, and abroad, and especially our colleagues in the field of marine biotechnology for stimulating and challenging discussions. A special "thank you" goes to Dr Zoë Popper. We would like to thank our families for their continued support and patience, and the valuable advice and support from Springer during the preparation of this volume.

Galway, Ireland
Cherbourg, France
Saint Nazaire, France

Dagmar B. Stengel
Solène Connan

Contents

Preface . . . . *v*
Contributors . . . . *xi*

1 Marine Algae: a Source of Biomass for Biotechnological Applications 1
Dagmar B. Stengel and Solène Connan

2 Structure and Function of Macroalgal Natural Products 39
Ryan M. Young, Kathryn M. Schoenrock, Jacqueline L. von Salm, Charles D. Amsler, and Bill J. Baker

3 Spectrophotometric Assays of Major Compounds Extracted from Algae 75
Solène Connan

4 Extraction and Enrichment of Protein from Red and Green Macroalgae 103
Pádraigín A. Harnedy and Richard J. FitzGerald

5 Extraction and Purification of R-phycoerythrin from Marine Red Algae. 109
Justine Dumay, Michèle Morançais, Huu Phuo Trang Nguyen, and Joël Fleurence

6 Extraction and Analysis of Mycosporine-Like Amino Acids in Marine Algae. 119
Nedeljka N. Rosic, Christoph Braun, and David Kvaskoff

7 Extraction and Purification of Phlorotannins from Brown Algae 131
Erwan Ar Gall, Florian Lelchat, Mélanie Hupel, Camille Jégou, and Valérie Stiger-Pouvreau

8 Enzyme-Enhanced Extraction of Antioxidant Ingredients from Algae 145
Björn V. Adalbjörnsson and Rósa Jónsdóttir

9 Microwave-Assisted Extraction of Fucoidan from Marine Algae. 151
Solange I. Mussatto

10 Extraction and Analysis of Oxylipins from Macroalgae Illustrated on the Example *Gracilaria vermiculophylla* 159
Dominique Jacquemoud and Georg Pohnert

11 Lipids and Fatty Acids in Algae: Extraction, Fractionation into Lipid Classes, and Analysis by Gas Chromatography Coupled with Flame Ionization Detector (GC-FID) 173
Freddy Guihéneuf, Matthias Schmid, and Dagmar B. Stengel

12 HRMAS NMR Analysis of Algae and Identification of Molecules of Interest via Conventional 1D and 2D NMR: Sample Preparation and Optimization of Experimental Conditions 191
Gaëlle Simon, Nelly Kervarec, and Stéphane Cérantola

13 Extraction, Purification, and NMR Analysis of Terpenes from Brown Algae. 207
Marc Gaysinski, Annick Ortalo-Magné, Olivier P. Thomas, and Gérald Culioli

14 Extraction, Isolation, and Identification of Sesquiterpenes from *Laurencia* Species 225
Angélica Ribeiro Soares

15 The Use of HPLC for the Characterization of Phytoplankton Pigments 241
José L. Garrido and Suzanne Roy

16 Characterization of Phlorotannins from Brown Algae by LC-HRMS 253
Jeremy E. Melanson and Shawna L. MacKinnon

17 Analysis of Betaines from Marine Algae Using LC-MS-MS 267
Shawna L. MacKinnon and Cheryl Craft

18 Analysis of Marine Biotoxins Using LC-MS/MS 277
Bernd Luckas, Katrin Erler, and Bernd Krock

19 Fucoidan Analysis by Tandem MALDI-TOF and ESI Mass Spectrometry 299
Stanislav D. Anastyuk, Natalia M. Shevchenko, and Vladimir I. Gorbach

20 Determination of Substitution Patterns of Galactans from Green Seaweeds of the Bryopsidales 313
Paula Ximena Arata, Paula Virginia Fernández, and Marina Ciancia

21 Structural Characterization of a Hybrid Carrageenan-Like Sulfated Galactan from a Marine Red Alga *Furcellaria lumbricalis* 325
Youjing Lv, Bo Yang, Xia Zhao, Junzeng Zhang, and Guangli Yu

22 Characterization of Alginates by Nuclear Magnetic Resonance (NMR) and Vibrational Spectroscopy (IR, NIR, Raman) in Combination with Chemometrics 347
Henrik Max Jensen, Flemming Hofmann Larsen, and Søren Balling Engelsen

23 Imaging and Identification of Marine Algal Bioactive Compounds by Surface Enhanced Raman Spectroscopy (SERS) 365
Mats Josefson, Alexandra Walsh, and Katarina Abrahamsson

24 In Vitro Protocols for Measuring the Antioxidant Capacity of Algal Extracts 375
Owen Kenny, Nigel P. Brunton, and Thomas J. Smyth

25 Disk Diffusion Assay to Assess the Antimicrobial Activity of Marine Algal Extracts 403
Andrew P. Desbois and Valerie J. Smith

26 Screening of a Marine Algal Extract for Antifungal Activities 411
Graciliana Lopes, Paula B. Andrade, and Patrícia Valentão

27 Protocol for Assessing Antifouling Activities of Macroalgal Extracts 421
Claire Hellio, Rozenn Trepos, R. Noemí Aguila-Ramírez, and Claudia J. Hernández-Guerrero

Index *437*

Contributors

KATARINA ABRAHAMSSON • *Department of Chemistry and Molecular Biology, University of Gothenburg, Gothenburg, Sweden*

BJÖRN V. ADALBJÖRNSSON • *Biotechnology and Biomolecules/Faculty of Food Science and Nutrition, Matis Ltd. Icelandic Food and Biotech R&D/University of Iceland, Reykjavík, Iceland*

R. NOEMÍ AGUILA-RAMÍREZ • *Instituto Politecnico Nacional-Centro Interdisciplinario de Ciencias Marinas, La Paz, Mexico*

CHARLES D. AMSLER • *Department of Biology, University of Alabama at Birmingham, Birmingham, AL, USA*

STANISLAV D. ANASTYUK • *G.B. Elyakov Pacific Institute of Bioorganic Chemistry, Far Eastern Branch, Russian Academy of Sciences, Vladivostok, Russian Federation*

PAULA B. ANDRADE • *REQUIMTE/Laboratório de Farmacognosia, Departamento de Química, Faculdade de Farmácia, Universidade do Porto, Porto, Portugal*

PAULA XIMENA ARATA • *Cátedra de Química de Biomoléculas, Departamento de Biología Aplicada y Alimentos, Facultad de Agronomía, Universidad de Buenos Aires, Buenos Aires, Argentina*

ERWAN AR GALL • *LEMAR – UMR6539, IUEM, Plouzane, Brittany, France; University of Brest (University of Western Brittany – UEB), Brest, France*

BILL J. BAKER • *Department of Chemistry, Center for Drug Discovery and Innovation, University of South Florida, Tampa, FL, USA*

CHRISTOPH BRAUN • *School of Biological Sciences, University of Queensland, St. Lucia, QLD, Australia*

NIGEL P. BRUNTON • *Department of Agriculture and Food Science, UCD, Dublin, Ireland*

STÉPHANE CÉRANTOLA • *Plateforme technologique de Résonance Magnétique Nucléaire, Résonance Paramagnétique Électronique et Spectrométrie de Masse, Brest, France*

MARINA CIANCIA • *Cátedra de Química de Biomoléculas, Departamento de Biología Aplicada y Alimentos, Facultad de Agronomía, Universidad de Buenos Aires, Buenos Aires, Argentina; CIHIDECAR-CONICET, Departamento de Química Orgánica, Facultad de Ciencias Exactas y Naturales, Universidad de Buenos Aires, Ciudad Universitaria, Buenos Aires, Argentina*

SOLÈNE CONNAN • *Photobiotechnology, INTECHMER, Conservatoire National des Arts et Métiers, Cherbourg, France; CNRS, GEPEA, UMR6144, Boulevard de l'Université, Saint Nazaire, France*

CHERYL CRAFT • *Aquatic and Crop Resource Development, National Research Council of Canada, Halifax, NS, Canada*

GÉRALD CULIOLI • *Nice Institute of Chemistry – PCRE and PFTC, UMR 7272 CNRS, Université de Nice Sophia-Antipolis (UNS), Nice, France; MAPIEM, EA 4323, Université de Toulon (UTLN), La Garde, France*

ANDREW P. DESBOIS • *Marine Biotechnology Research Group, Institute of Aquaculture, School of Natural Sciences, University of Stirling, Stirlingshire, United Kingdom*

JUSTINE DUMAY • *LUNAM Université, Université de Nantes, MMS, Nantes, France*

Søren Balling Engelsen • *Department of Food Science, Faculty of Science, University of Copenhagen, Frederiksberg C, Denmark*
Katrin Erler • *Institute of Nutrition, Faculty of Biology & Pharmacy, Friedrich-Schiller University, Jena, Germany*
Paula Virginia Fernández • *Cátedra de Química de Biomoléculas, Departamento de Biología Aplicada y Alimentos, Facultad de Agronomía, Universidad de Buenos Aires, Buenos Aires, Argentina; CIHIDECAR-CONICET, Departamento de Química Orgánica, Facultad de Ciencias Exactas y Naturales, Universidad de Buenos Aires, Ciudad Universitaria, Buenos Aires, Argentina*
Richard J. FitzGerald • *Department of Life Sciences, University of Limerick, Limerick, Ireland*
Joël Fleurence • *LUNAM Université, Université de Nantes, MMS, Nantes, France*
José L. Garrido • *Instituto de Investigaciones Marinas (CSIC), Vigo, Pontevedra, Spain*
Marc Gaysinski • *Nice Institute of Chemistry – PCRE and PFTC, UMR 7272 CNRS, Université de Nice Sophia-Antipolis (UNS), Nice, France*
Vladimir I. Gorbach • *G.B. Elyakov Pacific Institute of Bioorganic Chemistry, Far Eastern Branch, Russian Academy of Sciences, Vladivostok, Russian Federation*
Freddy Guihéneuf • *Botany and Plant Science, School of Natural Sciences, Ryan Institute for Environment, Marine and Energy Research, National University of Ireland Galway, Galway, Ireland*
Pádraigín A. Harnedy • *Department of Life Sciences, University of Limerick, Limerick, Ireland*
Claire Hellio • *LEMAR UMR 6539 IUEM, Université de Bretagne Occidentale, Plouzané, France*
Claudia J. Hernández-Guerrero • *Instituto Politecnico Nacional-Centro Interdisciplinario de Ciencias Marinas, Av. IPN S/N, La Paz, Mexico*
Mélanie Hupel • *Salipouss, Fouesnant, Brittany, France*
Dominique Jacquemoud • *Institute for Inorganic and Analytical Chemistry, Bioorganic Analytics, Friedrich Schiller University Jena, Jena, Germany*
Camille Jégou • *Lubem Quimper – EA3882, Quimper, Brittany, France; University of Brest (University of Western Brittany – UEB), Brest, France*
Henrik Max Jensen • *DuPont Nutrition & Health, DuPont Nutrition Biosciences ApS, Braband, Denmark*
Rósa Jónsdóttir • *Biotechnology and Biomolecules, Matis Ltd., Icelandic Food and Biotech R&D, Reykjavík, Iceland*
Mats Josefson • *Pharmaceutical Development, AstraZeneca R&D, Mölndal, Sweden*
Owen Kenny • *Department of Food Biosciences, Teagasc Food Research Centre, Dublin, Ireland*
Nelly Kervarec • *Plateforme technologique de Résonance Magnétique Nucléaire, Résonance Paramagnétique Électronique et Spectrométrie de Masse, Brest, France*
Bernd Krock • *Ecological Chemistry, Alfred-Wegener Institut Helmholtz-Zentrum für Polar- und Meeresforschung, Bremerhaven, Germany*
David Kvaskoff • *University of Queensland Centre for Clinical Research Herston, University of Queensland, St. Lucia, QLD, Australia*
Flemming Hofmann Larsen • *Department of Food Science, Faculty of Science, University of Copenhagen, Frederiksberg C, Denmark*
Florian Lelchat • *BMM Laboratory, Ifremer-Brest, ZI Began Diaoul, Plouzane, Brittany, France*

Graciliana Lopes • *REQUIMTE/Laboratório de Farmacognosia, Departamento de Química, Faculdade de Farmácia, Universidade do Porto, Porto, Portugal*
Bernd Luckas • *Institute of Botany and Plant Physiology, Faculty of Biology & Pharmacy, Friedrich-Schiller University, Jena, Germany*
Youjing Lv • *Shandong Provincial Key Laboratory of Glycoscience and Glycotechnology, Ocean University of China, Qingdao, People's Republic of China; Key Laboratory of Marine Drugs, Ministry of Education, School of Medicine and Pharmacy, Ocean University of China, Qingdao, People's Republic of China*
Shawna L. MacKinnon • *Aquatic and Crop Resource Development, National Research Council of Canada, Halifax, NS, Canada*
Jeremy E. Melanson • *Measurement Science and Standards, National Research Council of Canada, Ottawa, ON, Canada*
Michèle Morançais • *LUNAM Université, Université de Nantes, MMS, Nantes, France*
Solange I. Mussatto • *Department of Biotechnology, Delft University of Technology, Delft, The Netherlands*
Huu Phuo Trang Nguyen • *LUNAM Université, Université de Nantes, MMS, Nantes, France*
Annick Ortalo-Magné • *MAPIEM, EA 4323, Université de Toulon (UTLN), La Garde, France*
Georg Pohnert • *Institute for Inorganic and Analytical Chemistry, Bioorganic Analytics, Friedrich Schiller University Jena, Jena, Germany*
Nedeljka N. Rosic • *School of Biological Sciences, University of Queensland, St. Lucia, QLD, Australia*
Suzanne Roy • *ISMER, Université du Québec à Rimouski, Rimouski, QC, Canada*
Jacqueline L. von Salm • *Department of Chemistry, Center for Drug Discovery and Innovation, University of South Florida, Tampa, FL, USA*
Matthias Schmid • *Botany and Plant Science, School of Natural Sciences, Ryan Institute for Environment, Marine and Energy Research, National University of Ireland Galway, Galway, Ireland*
Kathryn M. Schoenrock • *Department of Biology, University of Alabama at Birmingham, Birmingham, AL, USA*
Natalia M. Shevchenko • *G.B. Elyakov Pacific Institute of Bioorganic Chemistry, Far Eastern Branch, Russian Academy of Sciences, Vladivostok, Russian Federation*
Gaëlle Simon • *Plateforme technologique de Résonance Magnétique Nucléaire, Résonance Paramagnétique Électronique et Spectrométrie de Masse, Brest, France*
Valerie J. Smith • *Scottish Oceans Institute, School of Biology, University of St Andrews, St Andrews, Fife, United Kingdom*
Thomas J. Smyth • *Department of Life Sciences, Institute of Technology Sligo, Sligo, Ireland*
Angélica Ribeiro Soares • *Núcleo em Ecologia e Desenvolvimento Sócioambiental de Macaé, Grupo de Produtos Naturais de Organismos Aquáticos (GPNOA), Universidade Federal do Rio de Janeiro, Rio de Janeiro, Brazil*
Dagmar B. Stengel • *Botany and Plant Science, School of Natural Sciences, Ryan Institute for Environmental, Marine and Energy Research, National University of Ireland Galway, Galway, Ireland*
Valérie Stiger-Pouvreau • *LEMAR – UMR6539, IUEM, Plouzané, Brittany, France; University of Brest (University of Western Brittany – UEB), Brest, France*
Olivier P. Thomas • *Nice Institute of Chemistry – PCRE and PFTC, UMR 7272 CNRS, Université de Nice Sophia-Antipolis (UNS), Nice, France*

ROZENN TREPOS • *School of Biological Sciences, University of Portsmouth, Portsmouth, UK*
PATRÍCIA VALENTÃO • *REQUIMTE/Laboratório de Farmacognosia, Departamento de Química, Faculdade de Farmácia, Universidade do Porto, Porto, Portugal*
ALEXANDRA WALSH • *Department of Chemistry and Molecular Biology, University of Gothenburg, Gothenburg, Sweden*
BO YANG • *Shandong Provincial Key Laboratory of Glycoscience and Glycotechnology, Ocean University of China, Qingdao, People's Republic of China; Key Laboratory of Marine Drugs, Ministry of Education, School of Medicine and Pharmacy, Ocean University of China, Qingdao, People's Republic of China*
RYAN M. YOUNG • *Department of Chemistry, Center for Drug Discovery and Innovation, University of South Florida, Tampa, FL, USA*
GUANGLI YU • *Shandong Provincial Key Laboratory of Glycoscience and Glycotechnology, Ocean University of China, Qingdao, People's Republic of China; Key Laboratory of Marine Drugs, Ministry of Education, School of Medicine and Pharmacy, Ocean University of China, Qingdao, People's Republic of China*
JUNZENG ZHANG • *Natural Products Chemistry, Aquatic and Crop Resource Development, National Research Council of Canada, Halifax, NS, Canada*
XIA ZHAO • *Shandong Provincial Key Laboratory of Glycoscience and Glycotechnology, Ocean University of China, Qingdao, People's Republic of China; Key Laboratory of Marine Drugs, Ministry of Education, School of Medicine and Pharmacy, Ocean University of China, Qingdao, People's Republic of China*

Chapter 1

Marine Algae: a Source of Biomass for Biotechnological Applications

Dagmar B. Stengel and Solène Connan

Abstract

Biomass derived from marine microalgae and macroalgae is globally recognized as a source of valuable chemical constituents with applications in the agri-horticultural sector (including animal feeds and health and plant stimulants), as human food and food ingredients as well as in the nutraceutical, cosmeceutical, and pharmaceutical industries. Algal biomass supply of sufficient quality and quantity however remains a concern with increasing environmental pressures conflicting with the growing demand. Recent attempts in supplying consistent, safe and environmentally acceptable biomass through cultivation of (macro- and micro-) algal biomass have concentrated on characterizing natural variability in bioactives, and optimizing cultivated materials through strain selection and hybridization, as well as breeding and, more recently, genetic improvements of biomass. Biotechnological tools including metabolomics, transcriptomics, and genomics have recently been extended to algae but, in comparison to microbial or plant biomass, still remain underdeveloped. Current progress in algal biotechnology is driven by an increased demand for new sources of biomass due to several global challenges, new discoveries and technologies available as well as an increased global awareness of the many applications of algae. Algal diversity and complexity provides significant potential provided that shortages in suitable and safe biomass can be met, and consumer demands are matched by commercial investment in product development.

Key words Algal biomass, Biotechnology, Cultivation, Commercial applications, Diversity, Genetic improvement, Natural products

1 Marine Algae: An "Untapped Resource" of Natural Products?

The vast and largely unexplored taxonomic, genetic, and chemical diversity and complexity of marine organisms including algae is considered an "untapped source" of valuable products and is currently investigated at many levels. Recent research has focused on a range of novel uses of algae in a wide range of industrial sectors, from agricultural, health care and cosmetics, to pharmaceutical uses, supported by advances in the chemical characterization and analytical techniques of new compounds and the ability to screen for different bioactivities in high-throughput systems.

Dagmar B. Stengel and Solène Connan (eds.), *Natural Products From Marine Algae: Methods and Protocols*, Methods in Molecular Biology, vol. 1308, DOI 10.1007/978-1-4939-2684-8_1, © Springer Science+Business Media New York 2015

Research efforts are largely driven by an appreciation that many cultures have traditionally used natural marine resources as foods and medicines, and the fact that 70 % of the Earth's surface is covered by the oceans which may provide additional opportunities where terrestrial resources are limited. More recently there has been a globally increasing search to meet demands in biomass and address the many challenges of an expanding human population, including climate change, aging populations, and food security resulting in the urgent need for sustainable biomass production from non-traditional sources.

Consequently, there has been a surge of investment globally in marine biotechnology, with research primarily targeting sponges, invertebrates, microorganisms, and algae. Algae are of particular interest because of their apparent high productivity, unequalled diversity, and because of the many new discoveries in bioactivities and compound diversity, and technical developments in cultivation.

The term "algae" refers to a large diversity of unrelated phylogenetic entities, ranging from picoplanktonic cells to macroalgal kelps. Existing taxonomic and chemical diversity between and within algal groups and its implication for commercial application is discussed in more detail by Stengel et al. [1]. Current systematic screening of algae is commonly based on systematic screening protocols, and disconnected from traditional phycological research, despite the high likelihood that the specific chemical adaptations of algal populations to extreme environments to yield discovery of interesting novel compounds. Algal biotechnology has gained significant importance in agri-horticulture [2, 3], marine functional foods [4], cosmetics [5], and human health (e.g., 6) applications, including the provision of ingredients for functional foods.

The development of new products from bioresources including algae involves many steps (Fig. 1) that are driven by new markets and consumer demands, but requires significant investment. A further obstacle is the lengthy approval process of the relevant safety agencies ensuring consumer safety and product and trading standards. The current dominant markets for high value products of algae are in the field of agri-horticultural products, including animal feeds and plant stimulants, cosmeceuticals, and nutraceuticals. There is also potential for new drug development although this requires longer development times and larger investments. The "time-to-market" is likely to be shortest for products in the agri-horticultural sector and intermediate in nutraceuticals and cosmeceuticals.

Algal biomass can be derived from natural or cultivated resources, which may or may not be predefined by strain selection and characterization, or hybridization, or by targeted wild-harvesting of specific phenotypes or materials which are likely to vary according to their origin [1]. Once compounds have been

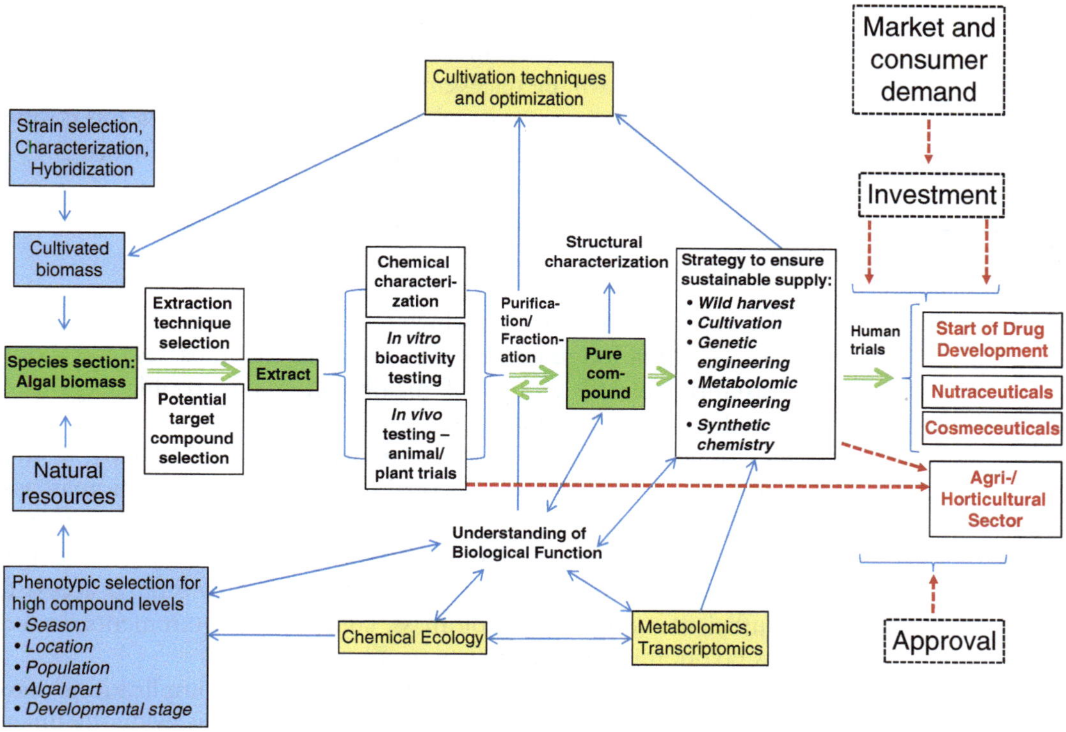

Fig. 1 Critical components and processes in the development of marine algal products

chosen for product development, extraction protocols can be selected accordingly, with the choice being influenced by the anticipated end-product, e.g., for food products GRAS technologies should be selected. Purified or crude extracts containing mixtures of compounds are screened for bioactivities of interest. Following a positive result, extracts are fractionated and the purified active ingredient chemically and structurally analyzed. One or several avenues to ensure a sustainable and consistent supply of the bioactive compound may be pursued; if the source organism can be cultivated to provide sufficient biomass and compounds, cultivation techniques can be applied and potentially optimized to achieve higher target compound yields. Knowledge of the biological function of the compound, and the wider field of chemical ecology, will make an invaluable contribution here with such biological insight additionally likely to give insight into the biochemical mechanisms that will downstream support, for example, drug development. Depending on the organism, metabolic or genetic engineering approaches may be possible, if sufficient genomic information for the organism is available, as well as the metabolomics and transcriptomics of the compound biosynthesis. Where the cultivation of the source organisms or genetic approaches are not feasible, chemical synthesis may be an avenue for supplying sufficient quantities for further testing and eventual product development. Considerable financial investment and long-term commitment from industry will be necessary at this point to conduct human trials for the downstream development of nutraceuticals and cosmeceuticals, and in particular for the development of new drugs.

2 Algal Diversity

In contrast to land plants which all shared a single common ancestor ~470 million years ago, algal diversity encompasses several only distantly related groups of mainly photoautotrophic organisms predominantly inhabiting aquatic environments.

Recent focus has been on marine macroalgal groups, as they comprise a large (known) diversity, with conspicuous biomass visible along coastlines and traditionally attracting attention from coastal communities. The vast taxonomic diversity of microalgae has received comparatively less attention due to limited access to sufficient mono-specific biomass, cultivation requirements, as well as taxonomic complexities (e.g., 7).

The chemical diversity of algae and its implications have recently been reviewed [1], and algal chemistry is evidently linked to evolution but phenotypic modifications also result from environmental and biological pressures. Local conditions at the collection site such as light, nutrient, and temperature regimes considerably impact on metabolite levels and thus bioactive composition. Additionally, the biological status of algae including life cycle- and developmental-stage, as well as macroalgal thallus structure, influences biochemical composition and the eventual value of the algal source material. The major algal groups were traditionally classified according to their chemical composition, but recent technological advances have allowed more detailed differentiation within compound families. These have revealed an even greater chemical richness, suggesting complex biochemical pathways and adaptive mechanisms, which promise to exceed the anticipated commercial potential for natural product development. However, more precise analytical techniques providing added definition also demand more careful sampling as to algal origin, time of collection, or culture conditions as well as, for macroalgae, thallus parts which display critical chemical distinctions [1].

The richness and diversity of algal compounds with conceivable commercial interest may be explained by complexities including a vast taxonomy of algal species groups and species within groups, significant chemical diversity correlated with the genetic diversity of these groups and the geographical distribution (large and small scale). Biological stimuli enhance the chemical diversity as individual populations and individuals exhibit considerable phenotypic plasticity and adaption to their habitats. Specific limitations of individual species fall within the physiological restrictions that delimit their biogeographic boundaries. However, because they live in stressful and highly fluctuating environments, most algae have evolved complex chemical mechanisms to facilitate short- or medium-term acclimations to stressors (e.g., UV radiation, salinity and temperature stress) as well as to defend themselves from biological pressures such as competitors, grazers, epiphytes, or parasites. Chemical ecology is a fascinating and rapidly expanding field at the interface of algal biology and natural product chemistry. It encompasses the study of chemicals involved in defining species interactions, either between different algal species, or between algae and herbivores, or the secondary metabolites (e.g., phenolic compounds in brown algae, or mycosporine-like amino acids (MAAs) in red algae, some Cyanobacteria, and dinoflagellates) produced by algae to protect themselves against environmental stressors such as UV radiation. The next chapter (Chapter 2) provides an overview of the major classes of chemical compounds that have important biological functions in defense and signalling but also have been identified as having potentially valuable applications as natural products.

3 Current Applications of Algae

Table 1 summarizes some of the most commonly used algal compounds, species from which they are derived including cultivated or wild stocks, and geographic distributions of major exploited species.

3.1 Algae and Food

Algae have traditionally been used as human food; famously Nori (*Pyropia/Porphyra*) and kelp (*Laminaria*, *Saccharina*, *Undaria*) species have been consumed in the Far East, but there is also evidence that the Cyanobacteria, *Nostoc*, *Arthrospira/Spirulina*, and *Aphanizomenon*, have served as human food for thousands of years in China [18, 25, 26], Chad and Mexico [26]. In the West, algal consumption has predominantly been restricted to red seaweeds; particularly *Palmaria palmata* (Dulse), *Chondrus cripsus* and *Mastocarpus stellatus* (Irish Moss) in Ireland, Scotland, Brittany, and North America. The main microalgal genera used today as food or food ingredients include mainly freshwater representatives of the genera *Chlorella*, *Arthrospira/Spirulina*, *Nostoc*, *Aphanizomenon*, *Haematococcus* and *Dunaliella* [19]. Whilst chemical composition can vary according to species and growth environment (e.g., 27), microalgae contain high levels of nutritional constituents such as proteins [28], vitamins [29], and minerals [30], as well as bioactive compounds (e.g., anticancer, antibacterial, antiviral etc., *see* 25 for review) and considerable interest and investment has focused on microalgal biomass

Table 1
Examples of algal extracts and compounds: algal species producing these compounds from wild or cultivated stocks, their origin, and some applications of these compounds

Compound/extract	Algal species	Source	Continent (country of production)	Applications	Example reference
Seaweed manure and extract	*Ascophyllum nodosum* *Ecklonia maxima* *Laminaria/Saccharina* sp. *Fucus* sp. *Sargassum* sp. Maerl (*Phymatolithon calcareum* + *Lithothamnion corallioides*)	Wild stock	Europe (Ireland, Norway, France, UK, Ireland) Asia-Pacific (Australia; Japan, China) Africa (South Africa) America (USA, Mexico, Chile)	Agri-horticulture (e.g., Kelpak®, Goëmar BM86®) + Animal Feed	[3, 8]
Polysaccharides					
Carbohydrates	*Mastocarpus stellatus*	Wild stock	Europe (France)	Cosmetics (Isopeptinoside® Science et Mer)	[9]
Polysaccharides	*Porphyridium cruentum* *Rhodella* sp. Cyanobacteria	Culture	Europe (France)	Cosmetics	[10]
Agar (hydrocolloid)	*Gelidium* sp. *Gracilaria* sp.	Wild stock Except *Gracilaria*: wild stock + culture	Europe (Spain, Portugal) Asia (Japan, Korea; Indonesia, China) America (Chile) Africa (Morocco; South Africa, Namibia)	Biotechnology (Bacteriological Agar) Human food (Agar-agar) Health Horticulture	[11, 12]

Alginates (hydrocolloid)	*Ascophyllum nodosum* *Durvillaea potatorum* *Ecklonia maxima* *Laminaria/Saccharina* sp. *Lessonia nigrescens* *Macrocystis pyrifera*	Wild stock Except *Laminaria/Saccharina* sp.	Europe (Ireland, Norway, France, UK, Ireland) Asia-Pacific (Australia; Japan, China) Africa (South Africa) America (USA, Mexico, Chile)	Pharmaceutic and medical industry Cosmetics Human food Immobilized biocatalyst Paper industry Animal feed Algogel™ Satialgine™ (Cargill) Protanal®; Protacid® (FMC BioPolymer) Textile printing	[11, 12]
Carrageenan (hydrocolloid)	*Eucheuma denticulatum* *Kappaphycus alvarezii* *Gigartina skottsbergii* *Sarcothalia crispata* *Chondrus crispus*	Wild stock Except *Eucheuma* sp. and *Kappaphycus* sp.	Asia (Indonesia, Philippines) Africa (Tanzania) America (Chile, Mexico; Canada) Europe (France)	Human food Pharmaceutical and medical industry (vaginal suppositories with Tenofovir against HIV) Immobilized biocatalyst Home care (Satiagel™, Cargill)	[11, 12]
Ulvans	*Ulva* sp.	Culture + Wild stock	Europe (France)	Animal feed (Amadeite®; Olmix)	[13]
Fucanes (fucoidan)	*Ascophyllum nodosum*	Wild stock	Europe	Cosmetics (Ascophyscient® Algue et Mer) Human food Nutraceutical Animal feed (Leilinature Fucoidan, AlgaNovo) Pharmaceutical (Maritech®; Marinova)	[14]
Laminarin	Laminariales			Horticulture (Vacciplant® Goëmar)	[15]

(continued)

Table 1
(continued)

Compound/extract	Algal species	Source	Continent (country of production)	Applications	Example reference
Fatty acids					
Fatty acids	*Chlorella* sp. *Synechocystis* sp. *Chlamydomonas* sp. *Isochrysis* sp. (20 strains tested)	Culture		Biofuel	[16]
Docosahexaenoic acid (DHA) Eicosapentaenoic acid (EPA)	*Crypthecodinium cohnii* *Schizochytrium* sp. *Nannochloropsis* sp. *Spirulina/Arthrospira* sp. *Ulkenia* sp.	Culture (heterotrophy mainly) Genetic engineering	Europe (Germany) America (USA) Asia-Pacific (Australia)	Nutraceutical (UltraClean Algal DHA and Omega 7®, BioCeuticals; life'sDHA™, DSM) Pharmaceutical Animal feed (ReNew™, Cellana)	[10, 17, 18]
Pigments					
β-Carotene	*Dunaliella salina* *Haematococcus pluvialis*	Culture	Asia (Israel) Asia-Pacific(Australia, India, China, Japan) America (USA)	Animal feed Human food Pharmaceutical Nutraceutical (Colorant + Pro-vitamin A + antioxidant)	[10, 17–20]
Astaxanthin	*Haematococcus pluvialis* *Chlorella* sp.	Culture	Asia (Israel) Asia-Pacific(India) America (USA)	Animal feed (aquaculture) Nutraceutical (BioAstin®; Cyanotech) Pharmaceutical	[10, 17, 18, 20]
Canthaxanthin	*Chlorella* sp. *Haematococcus pluvialis* Green algae	Culture	Asia (Israel) Asia-Pacific(India) America (USA)	Animal feed (colorant: aquaculture, poultry, food)	[10, 17]

Echineone	*Botryococcus braunii* Cyanobacteria	Culture		Animal feed (colorant + antioxidant)	[10]
Fucoxanthin	*Phaeodactylum tricornutum*	Culture		Nutraceutical Pharmaceutical	[10, 21]
Lutein	*Scenedesmus* sp. *Muriellopsis* sp. Green algae	Culture		Nutraceutical Pharmaceutical	[10, 20]
Zeaxanthin	*Chlorella ellipsoidea* *Dunaliella salina* (mutant)	Culture		Animal feed (Antioxidant + colorant)	[10]
Phycobiliprotein (PC, PE, APC)	*Spirulina/Arthrospira* sp. *Pophyridium* sp. *Rhodella* sp. *Galdieria* sp. Cyanobacteria, Rhodophyta, Cryptophyta, Glaucophyta	Culture	Asia-Pacific(Chine, Japan, Myanmar) America (USA)	Human food (Colorant) Cosmetics (colorant + antioxidant) Histochemistry, fluorescence microscopy, flow cytometry (fluorescent conjugates/ probes; PhycoPro™, Prozyme) Nutraceutical Medical (photodynamic therapy in cancer)	[10, 17, 19, 22]
Other compounds					
Mycosporine-like amino acids	*Porphyra umbilicalis* (porphyra-334 + shinorine) Cyanobacteria	Culture		Cosmetics (sunscreens; Helioguard 365; Mibelle group)	[10]
Organohalogenated compounds	*Asparagopsis armata*	Culture	Europe (France)	Cosmetics (Ysaline®; Algue et Mer)	[23]
Phlorotannins (phenolics)	Fucales	Wild stock		Pharmaceutical	[24]

(continued)

Table 1
(continued)

Compound/extract	Algal species	Source	Continent (country of production)	Applications	Example reference
Phytoene, phytofluene	*Dunaliella* sp.	Culture		Cosmetics	[10]
Polyhydroxyalkanoates (P3HB; PHA)	*Synechocystis* sp. *Nostoc* sp. Cyanobacteria	Culture (mixotrophy)		Bioplastics (Cereplast Algae Bioplastics™)	[10]
Squalene	*Aurantiochytrium* sp.	Culture		Cosmetics	[10]

production for human consumption. Several reviews have outlined the feasibility of macroalgae (e.g., 31) and microalgae as a source of proteins (e.g., 28) and nutraceuticals (e.g., 17, 19), but algal chemical composition and associated bioactivity levels are species- and sample-specific (e.g., 28, 31, 32).

A search for new sources of polyunsaturated fatty acids (PUFAs) has led to intensive screening of the natural fatty acid composition of macroalgae and microalgae; e.g., detailed studies of macroalgal species demonstrated that most are rich in polyunsaturated fatty acids (PUFAs) and have an advantageous n6/n3 ratio (0.61–5.15:1). However, fatty acid composition and proportions of C18 and C20 PUFAs vary within and between different phyla [33–35]. Algal pigments including carotenoids, such as β-carotene, fucoxanthin and astaxanthin, and phycobilins, as well as UV-absorbing MAAs are of interest to the food industry as dyes and for their strong antioxidant and health promoting properties (e.g., 36–38).

Algae absorb minerals from the surrounding water but can also take up heavy metals, if present, which means that water quality and good management of cultivation procedures are critical. Marine algae can hyperaccumulate iodine which has associated health risks, but iodine content and bioavailability depend on species, with generally higher levels in kelps [39], and also vary seasonally and geographically (for review *see* 40).

Besides Cyanobacteria-derived microalgal food extracts, *Chlorella* products are commonly used in powder or tablet form to supplement dietary requirements. However, a recent analytical study of commercial *Chlorella* products revealed that not only were many products not monospecific (as claimed), containing several bacterial contaminants, but the biochemical composition also differed containing breakdown products rather than the intended compounds [41]. This highlights that to ensure consumer safety there is a necessity for greater taxonomic awareness of industry, as some microalgal species (such as *Chlorella*) are difficult to differentiate, as well as careful reinforcement of environmental controls in cultivation and manufacturing to prevent bacterial contamination and processing-derived degradation of algal products.

3.2 Algae and Agri-horticulture

The potential uses of algae, in particular of microalgae, as animal feed (e.g., in aquaculture) have long been recognized [42] and, alongside traditional uses as human food, probably represent the most established uses. Principally diatoms and green algae are grown in intense cultivation for animal feeds. Species selection for particular application and animals (e.g., molluscs, crustaceans) has traditionally been based on their chemical composition, the associated production costs, and local expertise. Borowitzka [42] pointed out that the choice of species grown for animal feed was limited to a few species with the result that the large diversity of algae available was not fully exploited. Macroalgae are also commonly cultivated, sometimes in multi-trophic and integrated systems aiming to maximize biomass potential and minimize production cost (e.g., 43). The application of macroalgae as animal feed for aquaculture is expanding (e.g., 43) but it remains to be seen whether it can compete with the more established [44] cultivation of microalgae. Microalgae commercially are cultivated to provide a basic source of protein, carbohydrates, and minerals and fatty acids [45, 46] but also commonly as a colorant (e.g., astaxanthin [44]) to improve a products visual property and thus add value. The incorporation of both microalgae and macroalgae into feeds for (land-based) domestic animals including pigs, lambs and poultry is developing due to their nutritional and potential bioactive properties [45, 47, 48].

Seaweeds have traditionally been used as fertilizers and soil conditioners, but quality of early products was sometimes poor [3]. Recent progress has been made in developing algal extracts based on an improved analytical understanding of the biochemical composition, variability and stability in macroalgal (e.g., derived from fucoids, kelps, and maerl) but also microalgal (e.g., 49) products. Algal constituents of particular value as plant stimulants include hormones but also hormone-like substances such as betaines, sterols, polysaccharides, and peptides (recently reviewed by Khan et al. [2] and Craigie et al. [50]). Critical new knowledge has allowed improved analytical characterization of extract components with activities for specific horticultural applications (e.g., 51), more reliable bioassays (e.g., 52) and the targeted application of bioactives such as specific types of brown algal phlorotannins. Further improvement has been made in recognizing the need for standardized seasonal and site-specific algal sampling to obtain optimum activity of compounds of interest to achieve product quality and consistency (e.g., 50).

3.3 Algae and Cosmetics

In addition to a wealth of algae-based body care products utilizing the intrinsic properties of phycocolloids such alginates, agars and carrageenans, the potential algal applications in cosmetics are based on three properties [5, 22, 53–55]: (1) their use as emulsifying and thickening agents (based on polysaccharides such as carrageenan and alginates), (2) as UV-protective agents due to the UV-absorbing properties of extracts including phenolics, carotenoids, and MAAs; (3) antioxidant, anti-inflammatory, and anti-tumor properties of algal extracts. Whilst reference is commonly made to "potential applications" of algal extracts in the cosmetics industries, actual scientific and medical evidence for the skin care properties of algal compounds in products currently on the market is limited and a major attraction appears to be the "organic" label commonly awarded to algae-containing products. However, recent studies have clearly outlined the potential of algal extracts in cosmetic applications; specific activities of extracts are photoprotection, antioxidant, moisturizing, and whitening activity, but also anti-cellulite and slimming activity which have been attributed to a range of algal components including pigments, polysaccharides, terpenoids, and halogenated compounds, MAAs and phenolics, as well as proteins and amino acids [5]. Further studies into the bioactivity and composition stability of algal compounds from known and scientifically proven sources are required.

3.4 Algae and Health

Perhaps the greatest current activity in algal biotechnological research is the search for new compounds for human well-being based on health-promoting activities of algal extracts. Natural products chemistry is one of the fastest developing fields in biotechnology, and discovery of new compounds from microalgae and macroalgae is commonly supported by analytical methods including NMR and spectroscopy. Many recent reviews have highlighted the wealth of compounds that have potential pharmaceutical applications (e.g., 6, 36), or in turn outlined the many chemicals that have applications in both food and health, such as antioxidant activities (e.g., 56). Whilst commonly screening for new chemicals is not specific to algae but involves a wider range of marine organisms including microbes and sponges (e.g., 57), new bioassays and screening techniques applied to algae have accelerated discoveries (e.g., 17, 58, 59). The screening stretches across all algal groups (e.g., 49, 55, 60, 61) and the areas of biomedical applications include general antioxidant, anticancer, antimicrobial, antiviral, antiobesity, antifouling, and biomaterials (e.g., 62, 63). Compounds of interest include phycocolloids [64, 65], phenolics (e.g., 55, 61), pigments (e.g., 36), peptides and proteins [66], lipids [58], halogenated compounds (e.g., 67), and terpenoids [68].

4 Sources of Algal Biomass

4.1 Harvesting of Natural Resources

According the Food and Agriculture Organization of United Nations (FAO; 12), 1.1 million tonnes of wild algal stocks are harvested annually, with the majority of biomass used in the phycocolloid and food industries [11]. Sustainable exploitation of natural resources needs to take into account the impact of harvesting which is known and monitored in only very few cases; for example long-term data on kelp populations have been invaluable to assess effects of commercial exploitation of kelps (e.g., in Brittany, France [69]) and the highly complex structure of kelp forests and potential ecological impacts of mechanical harvesting have repeatedly been demonstrated (e.g., 70, 71). Even though detailed studies have been undertaken in some locations, data on productivity, recruitment, population structure and demography cannot be extrapolated to other populations [72]. On the other hand, there are several examples where overharvesting reduced natural populations (e.g., 73), and where biomass supply by cultivation was pursued initially by land-based and, subsequently, open-shore cultivation to meet demands [74].

Demands on seaweed biomass have multiplied driven by a globally increasing awareness, as well as a larger pool of scientific evidence, of their potential benefits and the many applications within various sectors, but also involve shifts in algal sources supplying existing markets. For example, the harvested kelp biomass in Europe primarily for the production of alginates increased from about 5,000 tonnes to 30,500 tonnes between 1999 and 2009, whilst harvesting of the Pacific kelp *Macrocystis* decreased from 35,000 to 5,000 over the same period [11]. Prudent resource management is thus critical not only to support local industries and related socioeconomic benefits of coastal communities in the long-term but also to avoid habitat loss and damage to marine ecosystems as the macroalgae as primary producers and habitat providers are essential functional constituents of coastal systems (e.g., 75).

An additional consideration when utilizing natural resources is the potential variability in chemical composition of natural stocks. This may be advantageous in some instances where the most suitable sites and harvesting times can be selected. However, the natural variation in biomass composition and quality [50] may not be appropriate for all applications such as high value food products (e.g., 76).

Suitability of the available harvesting technologies, i.e., various hand- and mechanical methods, as well as harvesting regimes and intensities needs to be tested and adapted to individual seaweed species and locations (e.g., 77). Effects of multiple and interacting impacts of recently observed indicators of global change, including eutrophication, increased water temperatures, changes in CO_2 levels and associated processes leading to ocean acidification, on commercial (and other) seaweed stocks and their productivity are yet to be quantified. Environmental impacts of harvesting natural populations also include effects on non-target species. Also, the risk of contaminants in the water that may be absorbed or adsorbed by macroalgae reducing their value or limiting their application should thus be of concern not only from a safety but also economic perspective [78].

4.2 Algal Cultivation

According to FAO [12] the total global production of algae (including seaweeds and microalgae) amounts to ca. 25 million tonnes for various applications including human food and agricultural products such as animal feed and fertilizers, as well as health care products. Despite the large algal diversity potentially available, algal cultivation is limited to around 37 species, including both freshwater and marine (macro- and micro-) algae [12]; seaweed production is dominated by a handful of species such as the carrageenophytes *Eucheuma* spp. and *Kappaphycus* spp., followed by *Saccharina japonica*, agarophytes (mainly *Gracilaria* spp.), Wakame (*Undaria pinnatifida*—also classified as highly invasive species and listed on the Global Register for invasives), and Nori species (*Porphyra* and *Pyropia*). Based on figures from FAO [12], seaweed biomass derived from cultivation doubled between 2010 and 2012. Production is concentrated in Asia; China is a major producer with an estimated annual yield of 13.5 million tonnes (mainly kelps); substantial increases in carragenophyte production in Indonesia have also been recorded [12]. On the other hand, Japanese Nori production was significantly reduced possibly due to the fallout of the Fukushima nuclear incident in 2011.

Traditionally the main purpose of microalgal production has been the production of animals (fish, molluscs) in aquaculture [42], with more recent objectives being the production of biomass for human food stimulated by the high protein contents of commonly grown genera such as *Chlorella* and *Arthrospira/Spirulina* [79], and the production of third generation biofuel [80]. Microalgae are now more commonly accepted as human food supplements and widely available commercially [17, 81].

The opportunity to harness species-specific responses to the environmental conditions that control algal metabolism, and by extension composition, in cultivation systems offers the potential to maximize the production of compounds of interest. There is a long tradition of microalgal biomass production for specific applications, and virtually all biomass is derived from cultivated sources with various degrees of intensities. By contrast, only recently, a few commercial ventures have developed effective but well-controlled cultivation systems allowing production of macroalgal biomass for high value applications [76]. Whilst it is critical, and perhaps obvious, to match taxonomic groups to particular applications based on their natural chemical composition, the scope to harness the responses of different algal groups to environmental stimuli, which can be artificially imposed in cultivation, is enormous but remains a challenge. The objective is to modify not only levels of compounds but also their composition which is related to their bioactive potential and thus final application.

4.3 Cultivation Systems

4.3.1 Macroalgal Cultivation Systems

There are two principle approaches to macroalgal cultivation; (1) seaweeds can be grown on land in tanks or ponds or, (2), more commonly, in open-sea coastal or off-shore environments. Apart from careful site selection, off-shore production by default offers only limited opportunities for optimizing algal growth and composition, and other factors including water quality, access, and conflicting interests of coastal usage impose constraints.

Most basic cultivation methods have involved an artificial increase of substrates that might attract algal settlement in convenient locations, by placing rocks in sheltered intertidal regions to encourage fucoid recruitment and increase harvestable algal biomass (e.g., in Ireland). Other simple cultivation systems avail of existing aquaculture infrastructure (e.g., disused and modified fish or shrimp ponds), but more usually macroalgae are farmed on floating or bottom-fastened net or long-line structures in coastal or off-shore waters. More recently the integration of macroalgal cultivation into existing aquaculture systems (e.g., fish or mussel farms) has gained interest, but only few fully commercial farms of such integrated multi-trophic aquaculture (IMTA) systems exist due to the complexity of the different components and most farms are at pilot stage.

Once suitable functional components of algae are identified for the development of a particular product, demand needs to be secured and this must allow traceability of materials, sustainability, and quality assurance. Whilst on-land cultivation fits most of these criteria [76] it is associated with high costs due to energy inputs and labor requirements. Although based on long-standing traditions and significant research efforts, the large-scale cultivation of food species underway in South-East Asia has recently suffered from environmental pressures such as the consequences of eutrophication, as well as industrial incidents such as the 2011 Fukushima nuclear disaster. Consequently, the demand for edible seaweeds produced for export to markets globally which are usually supplied by Japanese seaweed production has increased and raised interest in seaweed production in regions considered safe and non-contaminated.

Although the cultivation of macroalgae is regarded as a tool for water quality improvement through the nutrient capacity of some species (including *Porphyra* species), good cultivation and management practises are essential. A possible connection between large-scale seaweed cultivation and the occurrence of green tides has been suggested, where the green tide forming *Ulva* [*Enteromorpha*] *prolifera* discarded from *Pyropia yezoensis* nets appeared to be the origin of the world's largest (1,200 km^2) green tide just before the 2008 Beijing Olympics sailing event at Quindao [82]. Also, whilst cultivation of macroalgae such as *Gracilaria* and *Gelidium* for agar production has developed to make a major economic and social contribution to local communities (e.g., in Chile), improved management techniques to combat pathogens and pests, as well as issues with changed sedimentation due to reduced water movement, need to be developed further [83, 84].

As generally effective nutrient scrubbers, algae also absorb trace elements and heavy metals from their surrounding seawater which potentially include toxic contaminants. Seaweed also have a complex and largely unexplored epiflora [85, 86] which needs to be taken into consideration, and should be carefully monitored, as it could reduce the quality of open-shore cultivated macroalgae.

4.3.2 Microalgal Cultivation Systems

Despite the vast potential due to the diversity of species available, the multiple applications of microalgae which are recognized, as well as biotechnological advances made, only a handful of microalgal species are produced at commercial scale. Cultivation is limited by the cost of production balanced against the value of the end-product, and technological issues in maximizing algal productivity. Geographic location and associated production cost due to energy demand is also a concern, and with few exceptions, most commercial microalgal production is limited to low latitudes (e.g., 87).

Clearly, end-uses, product value, and overall economic constraints drive the scope for algal production. Principle differences between open or closed cultivation systems, and the many types of bioreactor design available and under development, determine production and running costs, product quality and value, and ultimately the economic viability of these cultivation systems. Besides obvious advantages and disadvantages of indoor and outdoor cultivation with regard to the ability to control environmental conditions, energy costs, and factors such as space limitation and local climate, product choice also places constraints on microalgal production [88]. For example, the production of food ingredients requires ideally sterile, axenic cultures and thus closed vessels to avoid contaminants but energy input requirements are higher [19]. The principle argument for open systems is the potentially cheaper production of organisms utilizing natural solar energy. However, efficiency and thus productivity may be reduced and production is restricted to locations with adequate space, water and energy provisions and to species with wider environmental tolerances (e.g., 89); contamination risks pose and additional threat. Eriksen [22] reviewed different tank and photobioreactor designs (tubular, flat panel, column) with mixing and light distribution being critical factors limiting algal productivity. Applications of the physiological responses of microalgae to light include the opportunity to apply intermittent high-light–low-light cycles to stimulate algal productivity in photobioreactors by reducing the risk of photoinhibition (e.g., 90). The design of bioreactors for optimum algal productivity is an area of significant research efforts which have not only focused on impacts of environmental parameters such as light, temperature, salinity, and nutrient regimes (and combinations thereof) but also on optimizing flow-rates, turbulence, and light exposure through specific bioreactor design. A number of reviews have outlined the many outstanding issues regarding the link between optimum light absorption, turbulence, algal biomass, and compound production for different purposes (including the production of biofuels) using different groups of microalgae [22, 91–98], with critical factors identified as economics, independency of external climatic factors, end-products, and species choice.

4.4 Adapting Algal Cultivation Systems to Specific Biomass Applications

With increasing interest in a wider range of algal products, cultivation techniques particularly have made significant advances by optimizing culture conditions for specific groups, species, or strains, recognizing the scope to fine-tune conditions according to organism requirements, and identifying the environmental parameters controlling compound production. The need to adjust culture conditions to particular requirements is directly linked to the basic biological function(s) of the desired compound, as well as the ability of different taxonomic groups or subgroups to regulate compound levels and composition.

In contrast to macroalgal cultivation, the main objective of microalgal production has been for animal feed enrichment (or replacement of other food components). Subsequently, demand for carotenoid or astaxanthin production has driven the production of these compounds in commonly grown species such as *Dunaliella* sp. or *Haematococcus* sp. for a particular purpose in aquaculture from an early stage of research. Despite considerable advances there is still a lack of expertise in identifying optimum conditions for synthesis of novel compounds in strains (or species) and in identifying particularly suitable strains for industrial application. New data of ecophysiological impacts are essential to achieve economically viable algal production. Whilst some generalities on the environmental impacts on groups of algae are possible with regard to levels of metabolite production (e.g., in response to light and nutrients), species-specific requirements are critical, and reports on impacts for individual species cannot necessarily be applied to species (or strains) within the same group. Although in principle the same approach could be applied to macroalgal cultivation, most progress has been made by adopting specific culture conditions for the production of microalgae biomass with specific, mainly high-value, applications such as bioactive production [99].

4.4.1 Light

The basic ecophysiological principle of light limitation of photosynthesis, up to a saturating irradiance, which should result in increased growth rates under otherwise non-limiting conditions obviously also applies to microalgae. Here, compared to multicellular macroalgae, the physiological relationships are often easier to control in culture. Simple light–growth relationships have been established for different microalgal strains adapted to different light climates [100–102]. High irradiance regimes at low nutrient levels have been applied to cultivated algae to achieve high carbohydrate contents (e.g., 103), with carbohydrate composition being more directly related to light levels in microalgae than they are in macroalgae which have a more complex distribution of storage vs. structural carbohydrates (e.g., 104).

For both macroalgae and microalgae, responses to light can be categorized in terms of light quantity or quality, and classic irradiance responses include changes in pigment levels and composition [105, 106]. Optimum light levels vary considerably according to type of cultivation (indoor vs. outdoor using natural sunlight), the rationale for cultivation (e.g., where low-light conditions are to be achieved to simulate light limitation or where sun-screen compounds such as carotenoids or phenolics are desired), and the target species. For example, commercial carotenoid production has been successfully achieved in *Dunaliella* grown under high irradiances [107], although the impact of multiple stresses (high salinity, low temperature, nutrient deprive, heavy metal stress) can further enhance carotenoid production (in *Dunaliella*) [108].

As synthesis of some desired products, such as carotenoids, is further controlled by spectral qualities, these can be manipulated in culture to optimize production. Whilst most research on UV protective mechanisms to date has focused on climate change impacts and thus not necessarily involved study of commercial species, a wealth of knowledge regarding processes that control compositional changes in response to UV exposure has been established (e.g., 54, 109). There is ample evidence for physiological and biochemical mechanisms to provide protection for microalgae from mild UV radiation although high levels in particular of UV-A result in DNA damage and/or loss of viability. A more novel approach is the deliberate exposure of microalgae to UV-B to stimulate the production of compounds of interest, including pigments and tocopherol levels (e.g., *Synechocystis*; [106]). It should be noted that UV responses differ according to the natural ability of different algal groups, species, and strains, to adjust their photosynthetic processes. This is commonly linked to the natural fluctuations in UV that the different species are adapted to.

By manipulating the culture conditions such as growing some microalgal species under heterotrophic or mixotrophic conditions, production of some compounds of interest can be enhanced; synthesis of a compound not synthesized under normal conditions can even be stimulated, e.g., *Galdieria sulphuraria* cultivated under heterotrophic conditions, i.e., in the absence of light and photosynthesis produces higher levels of phycocyanin than under phototrophic conditions (reviewed by Eriksen [22]). Heterotrophic conditions are not limited by incident light as such, so this regime can lead to higher production potentials [22].

4.4.2 Nutrients

Nutrient levels are often assumed to be directly related to algal biomass production, as eutrophic waters are usually characterized by higher algal biomass, to the extent that oceanic or coastal nutrient levels are used as a proxy for algal productivity (e.g., 110). However, relationships between nutrients, chlorophyll, cell number, cell size, and actual algal biomass production are not necessarily directly correlated. Although several specific growth media are suitable for growth of different algal groups under laboratory conditions [111], economic constraints usually do not allow their application to commercial scale.

Early studies have demonstrated the potential for enhancement of total protein levels in *Chlorella* grown for potential food production by increasing nitrogen levels [112]. Experiments on microalgal growth responses to nutrient addition were also conducted to establish simple ways to achieve cheap algal biomass at large volumes, e.g., by testing impacts of commercial fertilizers [113]. Very high nitrogen concentrations result in nitrite accumulation and overall growth reduction in a bloom-forming *Microcystis* species. CO_2-enrichment under these conditions enhanced growth, suggesting that intracellular nitrite accumulation is linked to high external nitrate levels [114]. However, the direct observations from natural environments that high nutrient concentrations generally result in high productivity, have led to over-simplified assumptions regarding algal cultures, without considering species-specific impacts of types of nutrients, nutrient ratios, or impacts of different nutrients on biochemical composition.

High nitrogen levels, in principle, achieve increases in levels of proteins or other N-containing compounds, which, in red algae include proteinaceous phycobilins. Thus simple nutrient adjustments and balances may be used to increase phycocyanin and phycoerythrin levels; both are compounds of commercial interest (e.g., 37, 115). Alternatively, some responses to nutrient enhancement and interactions with other environmental parameters can be utilized to optimize protein levels are species- and group-specific (e.g., 116). Total protein levels can be directly and significantly (up to 9.2-fold) enhanced by growing microalgae, e.g., *Chlorella* sp., under different nutrient regimes [112], and the composition and levels of high-value polysaccharides of *Arthrospira/Spirulina* in batch culture were increased under nitrogen limitation [117].

Lipid and fatty acid levels and composition may also be also modified under environmental control including nutrient levels. Although most studies have concentrated on the impact of nitrogen on lipid and fatty acid levels (e.g., 27, 118), they were shown to be modified by different phosphorous regimes in the diatom *Phaeodactylum tricornutum* and the green alga *Dunaliella tertiolecta* [119]. Also fatty acid and lipid composition of some green algae were altered during N-starvation but such effects were not seen for Cyanobacteria (e.g., 120).

Different nutrient combinations, including different levels and ratios of nitrogen and phosphate, as well as different nitrogen sources, were the subject of a detailed study to improve growth, life cycle development, and astaxanthin production in *Haematococcus pluvialis* [44] for optimized large-scale production.

In this instance, carotenoid production was stimulated by nitrogen starvation which, in turn, was enhanced by lower phosphate levels. Whilst overall findings were comparable to reported positive impacts of nutrients on carotene formation for other green algae (e.g., *Dunaliella* sp. [121]), results also indicated important strain-specific preferences for nitrogen source. Silica deficiency also influenced lipid composition in some diatom species [122] but such effects would not be expected for other algal groups with a lower demand for silica. Overall these examples illustrate the importance of nutrient source-, species- and even strain-specific studies with regard to optimized compound production.

4.4.3 Temperature

Besides the obvious impacts of temperature on microalgal growth and productivity, with different microalgal species or strains exhibiting temperature optima according to their origin, growth temperatures induce physiological changes in a range of species; e.g., reduced temperature resulted in changes in commercially value components, including fatty acids and tocopherols, in *Porphyridium* [123]. Similarly, combined impacts of temperature and light (not unexpectedly) have been observed in Cyanobacteria (e.g., 102), but both extreme high and low temperatures had negative effects on phycobilin levels, irrespective of growth irradiance.

As for other compounds of interest, the production of proteins, lipids, and phenolic compounds by *Arthrospira/Spirulina platensis* did not occur at their optimum growth rates, but at higher temperatures, suggesting a compromise between optimum production of biomass and that of desired compounds needs to be achieved under commercial conditions [27, 118].

4.4.4 Salinity

Impacts of salinity on the biochemical composition are particularly important in open pond or tank cultivation where evaporation can cause fluctuations in salinity, for example where halophilic species such as *Dunaliella salina* are grown, and where compounds of interest are specifically stimulated under salinity stress (e.g., fatty acids [124]; pigments [101]). The genus *Dunaliella* is probably one of the most intensely studied microalgal groups with respect to impacts of salinity (for review *see* 125); this is due to the ability of some species to grow at high salinities and produce high levels of β-carotene (up to 10 % dry weight), glycerol, and proteins under nitrogen deficiency, high light, and high salinity, making them ideal candidates for large outdoor cultivation in hot climates. Salinity fluctuations can induce specific proteins and enzymes [126], and some *Dunaliella* species accumulate glycerol to counteract osmotic stress [101]. Impacts of salinity on biochemical composition, in combination with other environmental factors, are also reported for other green algae [127] and diatoms [128].

4.4.5 Carbon Source and Level

Whilst photoautotrophic microalgae largely rely on an external supply of inorganic carbon for photosynthesis, in some diatoms including *Nitzschia inconspicuosa* biomass production was enhanced when algae were provided with an exogenous carbon source as well as CO_2, with the resultant acetate and fatty acid composition dependent on the carbon source [129]. The potential of using algae to sequester CO_2 from the atmosphere in large-scale cultivation to mitigate enhanced atmospheric CO_2 concentrations has received a lot of attention (e.g., 103, 130). It is viable only where large-scale production is still economically viable, thus demanding that either little additional energy input is needed, or that other high value products, ideally other than biofuel, can be produced. Similar to other current biotechnological constraints, algal species that efficiently uptake carbon do not necessarily fulfil the other criteria outlined above. Whilst the prospects for bioengineering are promising in this field, significant progress has also been made using naturally rather than genetically modified species which, theoretically, may have a greater variety of future applications. For example, using extremely high CO_2 concentrations under otherwise naturally high light photosynthetic changes in *Chlorella minutissima* have been induced resulting in higher biomass production [131]. Such research highlights that the physiological acclimation to high CO_2 levels would be advantageous for CO_2 sequestration under otherwise natural conditions: such system could be applied as a truly environmentally friendly carbon trap provided that a use can be found for the biomass produced. Another example is *Dunaliella* sp.: production at 5–10 % CO_2 supply resulted in increased biomass and lipid content and a different lipid composition [132].

4.4.6 Interactive Effects of Culture Conditions

No environmental parameter affects algal growth and chemical composition in isolation; although light, temperature, and nutrient regimes influence growth and productivity by directly affecting photosynthesis through light absorption and enzymatic processes, their combined impacts on biochemical pathways and the resulting bioactive composition are less well defined. Until recently, combinatorial effects have received less attention since frequently only algal cell size, cell number, or biomass were measured with biomass, rather than compound production, being the central aim.

Commonly interactive effects of essential parameters have been investigated to identify growth limiting factors or differentiate between adaptations of different species. A wealth of research has also been conducted to study impacts of light, nutrient, temperature, and salinity on selected algal species or strains of commercial value. However, given existing algal diversity, intraspecific and strain-specific responses, as well as the ability of cultures to adapt to different conditions, the available knowledge for commercial exploitation is more limited than might be anticipated.

Consideration of interactive effects on the biochemical composition of different microalgae is particularly important where multiple uses or end-products are sought. As an example, whilst CO_2 levels and temperature interact positively to affect growth of the diatom *Chaetoceros* cf. *wighamii* in culture, lipid and carbohydrate contents were higher at lower temperatures, and there was no impact on proteins; on the other hand, elevated CO_2 levels increased protein but not carbohydrate contents, and salinity enhanced carbohydrates, but not lipids or proteins [128]. Also, specific lipid composition and fatty acid accumulation could be enhanced by a specific nitrogen and carbon enrichment regime [133] indicating the need for careful consideration of chosen end products with regard to optimum culture conditions, rather than maximum biomass production. Interactive effects of light, salinity, and nutrients on β-carotene production of different *Dunaliella* spp. have received significant attention in light of the economic interest in the genus [101]. Another example is the definition of optimum growth and compound production by different dinoflagellates under different light and nutrients; as might be anticipated, UV-related responses varied significantly between species [134].

Although fundamental relationships between for example irradiance and carbohydrate vs. nitrogen and protein levels have been established, in principle, for decades, assessment of in-depth composition requires further attention, particularly with regard to species- or strain-specific responses and physiological adaptations under different culture conditions. When multiple compound production is the objective, some compromises may be unavoidable; for example, carbohydrate, exopolysaccharide, and chlorophyll concentrations in *Chroomonas* sp. responded positively to nutrient enrichment, but carotenoid and protein levels and cell numbers were unaffected [135]. Commonly, overall culture productivity in terms of cell number or algal biomass is suboptimal at increased compound production due to the biological functions of the different compounds. Interactive impacts of combinations of environmental parameters (e.g., temperature, salinity, and CO_2 regimes on growth rates vs. protein, carbohydrate, and lipid contents [128]), need to be taken into account but can also be utilized for targeted production.

More fundamental research into establishing basic relationships between biomass to compound ratios is still needed. The selected growth conditions for algal production need to take into account the strain-specific relationship between biomass (or cell number) and compound production, e.g., when growth rates are inversely related to desired compound production (e.g., [103]). Such knowledge is required for some species that are currently cultivated but also when considering the new opportunities for future utilization based on the wider algal biodiversity available. Recent advances in microalgal cultivation demonstrate that a better understanding of processes and functions of different compounds in different species and strains will allow more efficient choice of culture parameters.

The current increase in interest in novel products from algae in biotechnology demands sustainable supplies of algal resources. Recent advances in cultivation methods for maximum biomass production have been made but research also highlights the potential for exploiting specific algal physiological tolerances and biochemical responses to environmental stimuli to optimize compound production for high value products; the fact that such responses are group-, species-, or even strain-specific can be considered either a hindrance, or an opportunity.

5 Current Trends in Algal Biotechnology: Tools Available

Applications of molecular biology to support the mining of algal biomass for valuable compounds may be based on a range of processes, but generally involves genetic characterization of the source materials (including both wild and cultivated stocks). While the genomes of only few (in comparison to the diversity available) algal species have been fully sequenced, the identification of specific genes and the elucidation of metabolic pathways involved in the biosynthesis of the desired product seem mandatory. Proteomic and transcriptomic approaches thus represent essential components of the development of different genetic engineering approaches for natural products, although in comparison to microbial, microalgal or plant biomass, such options remain limited for marine macroalgae.

5.1 Breeding and "Domestication" of Algae?

Different approaches to genetic improvement of algal biomass for commercial applications, whether for high value products or bioenergy production, are of increasing interest to provide sustainable, high value, consistent, and safe biomass for the multiple uses that algae offer. In addition to facilitating product improvement by growing algae under specific environmental conditions to induce the production of the desired compound/s, several levels of genetic improvement can also, at least theoretically, be accomplished by modifying their genome. Such methods are commonly used in domestic animals and food crops. However, the feasibility, practicality and safety of applying such methods to algae requires further development, and careful scrutiny applied to different organisms and cultivation systems. Different levels of genetic modifications can be realized, ranging from cloning and strain selection, hybridization, to genetic engineering and targeted genetic modification. Whilst strain selection followed by cloning is a commonly adopted method for both on-land [76] and open-shore cultivation [136] for high value food species and carrageenophytes, respectively, compared to plants, few algal species (both macro- and micro-) have been successfully transformed.

Traditionally strain selection and hybridization have been used as tools for developing more resistant and valuable seaweed crops. Through long-term breeding programs, particular strains with favorable traits have been created; these include resistance to pathogens as well as environmental extremes (e.g., temperature and salinity), and food quality [137]. Thus, in contrast to the manifold examples of animal- and plant-domestication for human food applications, and despite the long tradition of algal uses as food, in particular in south East Asia, progress on the domestication of algae has been restricted to a small number of species, including *Saccharina/Laminaria japonica*, *Porphyra*, and *Pyropia* spp. The complexity of the life cycle of commercially valuable macroalgae, particularly that of Nori species, has hindered development of large-scale cultivation in the west, but also resulted in a large range of genetically differing strains and varieties now commonly used in the Far East. Significant progress has been made in the breeding and hybridization of *Pyropia/Porphyra* and *Saccharina/Laminaria* strains which are grown commercially in China [138, 139].

Molecular techniques such as random amplified polymorphic DNA (RAPD) and amplified fragment length polymorphism (AFLP) have revealed, and enabled differentiation between, the many commercially grown *Porphyra* strains [138]. Detailed DNA fingerprinting of currently exploited strains is also underway and will allow identification and protection of pure *Pyropia/Porphyra* lines [140]. Investigation of species currently cultivated in Japan has revealed that intensive breeding of strains that were originally considered a mixture of *Pyropia tenera* and *P. yezoenzis* reduced genetic diversity [141]. In several studies, Chinese *Saccharina/Laminaria* populations have been examined by molecular techniques (e.g., 142) and related to their environmental tolerance characteristics with a view to identifying new sources for genetic breeding of environmentally tolerant hybrids. Many new varieties of *Saccharina* spp. and *Undaria pinnatifida* have been breed over the last 50 years and are now grown in the wild, and some "elite" crops currently grown in China [139, 143] are interspecifc hybrids between *S. japonica* and *S. longissima* artificially bred in culture.

5.2 Genetic Modification

While the genomes of just over 20 algal (mostly microalgal) species have been fully sequenced, the phylogenetic diversity of algal taxa is far greater than that of vascular plants, and also encompasses wide genotypic plasticity and the occurrence of many strains within single species, as well as their complex life cycles which usually involve several generations. Several attempts have been made to genetically modify seaweeds (largely *Saccharina/Laminaria* and *Pyropia/Porphyra* species) and microalgae. However, in comparison to terrestrial plant biotechnology including genetic engineering, limited progress has been made in applying genetic modification (GM) technologies to algae. Considering the vast taxonomic and genetic diversity, especially of microalgae, compared to plant diversity, the number of species for which methods for GM have successfully been developed is minute. However, exponentially increasing research outputs can be anticipated given the technological developments.

Future successes of algal genetic engineering, and particularly that of seaweeds with complex multi-stage life cycles, will require the development of new vectors and specific transformation methods suitable for seaweeds [144, 145]. In contrast genetic transformation of microalgae which has been successful for about 30 strains [146], many belonging to the "model" organisms *Chlamydomonas, Chlorella*, *Dunaliella salina* (Chlorophyta; e.g., 147) and diatoms such as *Thallasiosira* (Bacillariophyceae 148). Genomes of seaweeds are large and complications arise from complex life cycles, and large diversity of strains within species, and the presence of bacterial symbionts (e.g., 149). Transformation methods for commonly used plants have yielded limited success in seaweeds. Techniques based on particle bombardment, electroporation, and use of glass beads have largely been successful. However, other commonly used methods for transforming plants for example using *Agrobacterium* or viruses are less easily applied to algae; the latter (amongst other constraints) due to more complex interactions between algae and viruses and commonly lethal effects of infections in algae. It may be anticipated that the application of transposons routinely used in food plants, may represent a route forward [144].

Gene regulation for key enzymes involved in biosynthesis of high value compounds is poorly understood despite significant research on processes to enhance biofuel production by microalgae [145, 146] by greater and specific lipid accumulation suitable for applications as biodiesel, biohydrogen production, and more efficient photosynthesis for higher productivity rates (e.g., 80, 150). Compared to macroalgae, transcriptomic analyses have been more commonly applied to microalgae over several decades (*see* 151), and are currently subject of a large, integrated multi-institutional scale investigation via the Marine Microbial Eukaryote Transcriptome Sequencing Project [152].

With further developments and progress on transformed algae, cultivation methods for genetically modified algae need to be reconsidered since they pose a potential threat to aquatic ecosystem biodiversity. For example the open-water cultivation of GM algae, as may be anticipated if successful for food species such as *Porphyra* or kelps, raises concerns regarding biosafety due to the far less controllable aquatic systems where potential escapes and their subsequent dispersal will be difficult to manage. Additionally, the complex life cycles of many species which have resulted already in the interbreeding of cultivated strains in open water, again increase the risk of cross-contamination and spread of GM materials in the environment.

Amongst the few macroalgal species that have been fully sequenced, knowledge of the genome of the filamentous brown macroalga *Ectocarpus siliculosus* [153] has triggered significant detailed genetic insights so far unprecedented for other macroalgae (e.g., 154, 155). Advances in sequencing capabilities coupled with decreasing sequencing costs extended to algae have led to recent promising progress in algal transcriptomics, when assessing algal responses to external cues; for gene expression patterns under environmental or biotic stress scenarios, light or grazing stress in kelps have been described. For example, Konotchik et al. [156] assessed expressed sequence tags in *Macrocystis pyrifera* blades grown under different light regimes which suggests that light is an important regulator for gene expression that allows small-scale and temporal acclimations of the blade to changing regimes. As further examples of an increasing research activity in this field, Leblanc et al. [157] identified defense-related gene transcripts for molecular responses of kelps to grazing, and Cosse et al. [158] reported defenses against pathogens. Several further studies have examined gene expression in economically valuable crop species (e.g., 159, 160). Such information, in addition to the knowledge of the biological function of valuable compounds will underpin potential genetic or metabolic engineering of natural products in future commercial development.

5.3 Synthetic Biology and Molecular Pharming

The concept of synthetic biology has lately been extended to algae, and has been proposed as a way forward for example by using algal cells as factories for desired compounds after targeted genetic modification [146, 161], in particular in biofuel research, where current algal production capabilities are below the threshold of economic feasibility.

Following recent developments in plant and microbial biotechnology (e.g., 162, 163), the concept of "molecular pharming" which uses genetically modified plants or microbes to produce valuable proteins such as mammalian antibodies and vaccines and other therapeutic substances has recently been incorporated into algal biotechnology (e.g., 164–166) as eukaryotic algae might be considered as safe as plants since they are generally free from human pathogens. So far progress has been hindered largely by a lack of available technologies to achieve algal transformation [145]. However, genetic tools for algal transformation are in development and likely to make an important contribution to algal biotechnology.

5.4 Synthetic Chemistry

Another approach to increase the yield of valuable bioactive compounds that are naturally produced by algae only in small quantities is synthetic engineering of interesting molecules identified from natural sources, including marine algae. Clearly such an approach is expensive and time-consuming; however, it has potential after structural elucidation of promising bioactive compounds that are of interest to industry (in particular to pharmaceutical industries), and where biomass production and related compound yields are limiting and failing to meet market demand. Structural identification of potent compounds is essential where commercial exploitation, e.g., in pharmaceutics, is likely. So far most new drugs under development, based on compounds isolated from marine sources, are derived from bacteria or invertebrates [167], even though the field of algal biodiscovery is highly active and productive. To overcome the low yields of compounds which naturally occur at low levels in natural marine resources including algae, synthetic chemistry has created interest in particular where it is either too expensive or not feasible to cultivate and mass produce the organism of interest, as an alternative approach to synthetic biology. Total synthesis of algal compounds has been achieved on a number of instances following significant research efforts, in particular focussing on compounds isolated from the genus *Laurencia* (e.g., 168). On the other hand, progress in analytical technologies has also revealed that a range of compounds described over the last few decades were mistakenly structurally assigned [169] suggesting that more reliable results can be expected in the future.

6 Conclusions

Despite the significant global progress which has further outlined the potential of algal biomass as a source of natural products, some challenges remain and further research needs can be recognized. Significant gaps exist concerning biomass supply but also in the taming, enhancement, and simulation of compounds biosynthesis applying some of the biotechnological approaches described above.

The commonly used metaphoric description of algal biomass as an "untapped resource" implies that once the appropriate valve is identified and unlocked, natural products and associated benefits to the economy and human well-being, as well as financial gain, will flow. Staying with this metaphor, however, there is no bottomless tank filled with algal biomass or useful chemicals ripe for exploitation once the appropriate products have been developed. Rather, the sustainability of algal biomass, whether from wild stocks or cultivated, and the production of valuable metabolites through novel biotechnological processes, including metabolic engineering, requires significant further attention. While algae are playing an increasing important role in our lives, there will be no quick-fix for the current global challenges.

Demand for biomass for commercial applications is unlikely to be met, in the short-term for any, and in the long-term not for all, applications by either wild or cultivated sources, so that a major bottleneck remains regarding the sustainable supply for safe applications. Successful exploitation of the many opportunities that exist for the multiple, currently emerging applications of algae will require a critical assessment of biomass supply and will need to reckon with algal complexity which is so far only poorly described.

Further research needs to focus on sustainable harvesting and economic cultivation procedures:

- Critical assessment of harvesting regimes and a better understanding natural local populations dynamics of macroalgae
- Novel and product-specific cultivation techniques including land-based cultivation
- Cost and energy-efficiency of cultivation and harvesting processes to make it economically viable with regard to end-product value and consumer safety

Developments are required for new extraction and purification techniques, detection methods to better characterize algal resources. Continued technological progress will achieve new discoveries and an improved understanding of functional implications of the observed chemical diversity. Natural compound variability within algal biomass and stability during processing are likely to remain a major focus. In parallel, more detailed fundamental biology is required to decipher the metabolic processes that control the levels and composition of compounds, and to link these to observed bioactivities within different groups of algae.

Some developments in high-tech processing for precise and efficient extraction of valuable compounds are currently not considered economically viable and are unlikely to be up-scaled commercially in the near future. On the other hand, some processes currently used by algal industries are recognized as unsuitable for yielding high-value products. Additionally source traceability and quality control of processes, including addressing environmental concerns, will be essential to ensure a high quality product that meets regulatory requirements and public and environmental safety.

Further research will need to explore the level of bioavailability of extracted compounds and their efficacy, which may, in turn, be supported by chemical ecological knowledge. The many animal trials and human intervention studies, including clinical trials required, will demand long-term industrial investment to match the current, largely academic research, efforts funded from public monies. Recently increased public awareness and consumer acceptance of algal products may encourage private investment in further market and product development.

Given the complexity and diversity of algal biomass in terms of supply chain (including the issues outlined above such as algal chemodiversity, biomass supply and stability), in the face of growing international interest in the valorization of algae, biotechnology will need to respond by addressing the many existing challenges in parallel approaches; research will need to cater for the development of several different pipelines that are likely to progress at different speeds. The intense biodiscovery investigations currently underway, with many new compounds and bioactivities having been described, in parallel with technological developments in screening (High Throughput Screening, new and faster assays, smaller sample quantities required for structural analysis), suggest a bright future for the discovery of new natural products with drug-development potential. However, the "valley of death" outlined by Gerwick and Moore [167] highlighting the dramatic reduction from a vast number of promising leads to a handful of commercial drugs, primarily resulting from lack of long-term investment, needs to be addressed. Other available "shorter routes" to commercialization in the agri-horticultural, cosmetics and food markets may be more promising in the short- to medium-term. At the same time, research to ensure sustainable biomass supply is critical to support the many emerging branches of algal biotechnology and secure long-term prospects of the sectors. Consequently, public and private investment in algal biotechnology must address the multitude of research requirements regarding the complexity of sustainable algal biomass supply and facilitate an integrated development of several application pipelines.

References

1. Stengel DB, Connan S, Popper ZA (2011) Algal chemodiversity and bioactivity: sources of natural variability and implications for commercial application. Biotechnol Adv 29(5):483–501. doi:10.1016/j.biotechadv.2011.05.016
2. Khan W, Rayir UP, Subramanian S et al (2009) Seaweed extracts as biostimulants of plant growth and development. J Plant Growth Regul 28:386–399
3. Craigie JS (2011) Seaweed extract stimuli in plant science and agriculture. J Appl Phycol 23:371–393
4. Freitas AC, Rodrigues D, Rocha-Santos TA et al (2012) Marine biotechnology advances towards applications in new functional foods. Biotechnol Adv 30:1506–1515
5. Bedoux G, Hardouin K, Burlot AS et al (2014) Bioactive components from seaweeds: cosmetic applications and future development. Adv Bot Res 71:345–378
6. Ioannou E, Roussis V (2009) Natural products from seaweeds. In: Osbourn AE, Lanzotti V (eds) Plant-derived natural products: synthesis, function, and application. Springer, New York, pp 51–81
7. Stern RF, Horak A, Andrew RL et al (2010) Environmental barcoding reveals massive dinoflagellate diversity in marine environments. PLoS One 5:e13991. doi:10.1371/journal.pone.0013991
8. Evans FD, Critchley AT (2014) Seaweeds for animal production use. J Appl Phycol 26: 891–899
9. Bodeau C, Deslandes E (2007) Composition cosmétique comprenant une association de floridoside et d'acide iséthionique particulièrement un extrait d'algue rouge. French Patent No. FR2888504. Paris, INPI
10. Borowitzka MA (2013) High-value products from microalgae—their development and

commercialisation. J Appl Phycol 25: 743–756

11. Bixler HJ, Porse H (2011) A decade of change in the seaweed hydrocolloids industry. J Appl Phycol 23:321–335
12. FAO (2014) The state of world fisheries and aquaculture 2014. Rome
13. Paradossi G, Cavalieri F, Chiessi E (2002) A conformational study on the algal polysaccharide ulvan. Macromolecules 35:6404–6411
14. Berteau O, Mulloy B (2003) Sulfated fucans, fresh perspectives: structures, functions, and biological properties of sulfated fucans and an overview of enzymes active toward this class of polysaccharide. Glycobiology 13:29R–40R
15. Mery AB, Joubert JM (2012) Laminarin (Vacciplant®) against apple scab (*Venturia inaequalis*) and gloeosporium on apple (*Gloeosporium album et Perenans*). In: AFPP (ed). Association Française de Protection des Plantes. 10ème Conférence Internationale sur les Maladies des Plantes, Tours, France, 3–5 December 2012. pp. 630–639
16. Harun R, Singh M, Forde GM et al (2010) Bioprocess engineering of microalgae to produce a variety of consumer products. Ren Sust Energ Rev 14:1037–1047
17. Plaza M, Herrero M, Cifuentes A et al (2009) Innovative natural functional ingredients from microalgae. J Agric Food Chem 57: 7159–7170
18. Spolaore P, Joannis-Cassan C, Duran E et al (2006) Commercial applications of microalgae. J Biosci Bioeng 101:87–96
19. Chacón-Lee TL, Gonzáles-Mariño GE (2010) Microalgae for 'healthy' foods—possibilities and challenges. Compr Rev Food Sci Food Safety 9:655–675
20. Guedes AC, Amaro HM, Malcata FX (2011) Microalgae as sources of carotenoids. Mar Drugs 9:625–644
21. Cho Y-S, Kim S-K (2011) Pharmaceutical compositions containing fucoidan for stimulating and activating osteogenesis. U.S. Patent No. US20110301119 A1 Washington, DC, U.S. Patent and Trademark Office
22. Eriksen NT (2008) Production of phycocyanin—a pigment with applications in biology, biotechnology, foods and medicine. Appl Microbiol Biotechnol 80:1–14
23. Moigne JY (1998) Use of algae extract as antibacterial and/or antifungal agent and composition containing same. World Patent No. WO1998010656 A1Washington, DC, U.S. Patent and Trademark Office
24. Kim S-K, Kim Y-M, Lee M-S, et al (2013) Antibiotic composition including phlorotannin compound derived from *Eisenia bicyclis* as effective component. U.S. Patent No. US20130338217 A1 Washington, DC, U.S. Patent and Trademark Office
25. Singh S, Kate BN, Banerjee UC (2005) Bioactive compounds from Cyanobacteria and microalgae: an overview. Crit Rev Biotechnol 25:73–95
26. Chisti Y (2006) Microalgae as sustainable cell factories. Environ Eng Manag J 5:261–274
27. Colla LM, Bertolin TE, Vieira Costa JA (2004) Fatty acids profile of *Spirulina platensis* grown under different temperatures and nitrogen concentrations. Z Naturforsch 59c: 55–59
28. Becker EW (2007) Micro-algae as a source of protein. Biotechnol Adv 25:207–210. doi:10.1016/j.biotechadv.2006.11.002
29. Fabregas J, Herrero C (1990) Vitamin content of four marine microalgae. Potential use as source of vitamin in nutrition. J Ind Microbiol 5:259–263
30. Gouveia L, Batista AP, Sousa I et al (2009) Microalgae in novel food products. In: Hagen KN (ed) Algae: nutrition, pollution control and energy sources. Nova Science Publishers, New York, pp 265–300
31. Chandini SK, Ganesan P, Suresh PV et al (2008) Seaweeds as a source of nutritionally beneficial compounds—a review. J Food Sci Technol 45:1–13
32. Kumari P, Kumar M, Gupta V et al (2010) Tropical marine macroalgae as potential sources of nutritionally important PUFAs. Food Chem 120:749–757
33. Kumar M, Kumari P, Trivedi N et al (2011) Minerals, PUFAs and antioxidant properties of some tropical seaweeds from Saurashtra coast of India. J Appl Phycol 23:797–810. doi:10.1007/s10811-010-9578-7
34. Schmid M, Guihéneuf F, Stengel DB (2014) Fatty acid contents and profiles of 16 macroalgae collected from the Irish coast at two seasons. J Appl Phycol 26:451–463. doi:10.1007/s10811-013-0132-2
35. Schmid M, Stengel DB (2015) Intra-thallus differentiation of fatty acid and pigment profiles in some temperate Fucales and Laminariales. J Phycol (in press)
36. Nishino H, Murakoshi M, Ii T et al (2002) Carotenoids in cancer chemoprevention. Cancer Metast Rev 21:257–264
37. Prasanna R, Sood A, Suresh A et al (2007) Potentials and applications of algal pigments in biology and industry. Acta Bot Hung 49:131–156. doi:10.1556/ABot. 49.2007.1-2.14

38. de la Coba F, Aguilera J, Figueroa FL et al (2009) Antioxidant activity of mycosporine-like amino acids isolated from three red macroalgae and one marine lichen. J Appl Phycol 21:161–169
39. Nitschke U, Stengel DB (2014) A new HPLC method for the detection of iodine applied to natural samples of edible seaweeds and commercial seaweed food products. Food Chem 172:326–334
40. Müssig M (2009) Iodine-induced toxic effects due to seaweed consumption. In: Preedy VR, Burrow GN, Watson R (eds) Comprehensive handbook of iodine. Elsevier Academic Press, Oxford, pp 897–908
41. Görs M, Schumann R, Hepperle D et al (2010) Quality analysis of commercial *Chlorella* products used as dietary supplement in human nutrition. J Appl Phycol 22: 265–276
42. Borowitzka MA (1997) Microalgae for aquaculture: opportunities and constraints. J Appl Phycol 9:393–401
43. Mao Y, Yang H, Zhou Y et al (2009) Potential of the seaweed *Gracilaria lemaneiformis* for integrated multi-trophic aquaculture with scallop *Chlamys farreri* in North China. J Appl Phycol 21:649–656
44. Borowitzka MA, Huisman JM, Osborn A (1991) Culture of the astaxanthin-producing green alga *Haematococcus pluvialis* 1. Effects of nutrients on growth and cell type. J Appl Phycol 3:295–304
45. Belay A, Kato T, Ota Y (1996) *Spirulina* (*Arthrospira*): potential application as an animal feed supplement. J Appl Phycol 8: 303–311
46. Patil V, Källqvist T, Olsen E et al (2007) Fatty acid composition of 12 microalgae for possible use in aquaculture feed. Aquacult Int 15:1–9
47. Archer GS, Friend TH, Caldwell D et al (2008) Impacts of feeding several components of the seaweed *Ascophyllum nodosum* on transported lambs. Animal Feed Sci Technol 140:258–271
48. O'Sullivan L, Murphy B, McLoughlin P et al (2010) Prebiotics from marine macroalgae for human and animal health applications. Mar Drugs 8:2038–2064
49. Ördög V, Stirk WA, Lenobel R et al (2004) Screening microalgae for some potentially useful agricultural and pharmaceutical secondary metabolites. J Appl Phycol 16: 309–314
50. Craigie JS, MacKinnon SL, Walter JA (2008) Liquid seaweed extracts identified using ^{1}H NMR profiles. J Appl Phycol 20:665–671
51. MacKinnon SL, Hiltz D, Ugarte R et al (2010) Improved methods of analysis for betaines in *Ascophyllum nodosum* and its commercial seaweed extracts. J Appl Phycol 22:489–494
52. Rayorath P, Jithesh MN, Farid A et al (2008) Rapid bioassays to evaluate the plant growth promoting activity of *Ascophyllum nodosum* (L.) Le Jol. using a model plant, *Arabidopsis thaliana* (L.) Heynh. J Appl Phycol 20:423–429
53. de Nys R, Steinberg PD (2002) Linking marine biology and biotechnology. Curr Opin Biotechnol 13:244–248
54. Häder DP, Kumar HD, Smith RC et al (1998) Effects on aquatic ecosystems. J Photochem Photobiol B 46:53–68
55. Zubia M, Payri C, Deslandes E (2008) Alginate, mannitol, phenolic compounds and biological activities of two range-extending brown algae, *Sargassum mangarevense* and *Turbinaria ornata* (Phaeophyta: Fucales), from Tahiti (French Polynesia). J Appl Phycol 20:1033–1043
56. Cornish ML, Garbary DJ (2010) Antioxidants from macroalgae: potential applications in human health and nutrition. Algae 25: 155–171
57. Blunt JW, Copp BR, Keyzers RA et al (2014) Marine natural products. Nat Prod Rep 31:160–258. doi:10.1039/c3np70117d
58. Plaza M, Santoyo S, Jaime L et al (2010) Screening for bioactive compounds from algae. J Pharmaceut Biomed 51:450–455
59. de la Mare JA, Sterrenberg JN, Sukhthankar MG et al (2013) Assessment of potential anti-cancer stem cell activity of marine algal compounds using an in vitro mammosphere assay. Cancer Cell Int 13:39. doi:10.1186/1475-2867-13-39
60. Rastogi RP, Sinha RS (2009) Biotechnological and industrial significance of cyanobacterial secondary metabolites. Biotechnol Adv 27: 521–539
61. Zubia M, Fabre MS, Kerjean V et al (2009) Antioxidant and antitumoural activities of some Phaeophyta from Brittany coasts. Food Chem 116:693–701
62. Felício-Fernandes G, Laranjeira M (2000) Calcium phosphate biomaterials from marine algae. Hydrothermal synthesis and characterisation. Quím Nova 23:441–446. doi:10.1590/S0100-40422000000400002
63. Dimartino S, Lir I, Haber M et al (2013) Characterization of biomimetic adhesives from the red alga *Gracilaria conferta* for biomedical applications. In: Flammang P, Santos R, Aldred N, Gorb S (eds) Biological

and biomimetic adhesives: challenges and opportunities. Royal Society of Chemistry, UK, pp 117–131

64. Rinaudo M (2007) Seaweed polysaccharides. In: Kalmerling JP (ed) Comprehensive glycoscience from chemistry to systems biology, vol 2. Elsevier, London, pp 691–735
65. Nakayasu S, Soegima R, Yamaguchi K et al (2009) Biological activities of fucose-containing polysaccharide ascophyllan isolated from the brown alga *Ascophyllum nodosum*. Biosci Biotechnol Biochem 73: 961–964
66. Harnedy PA, FitzGerald RJ (2011) Bioactive protein, peptides and amino acids from macroalgae. J Phycol 47:218–232
67. Cabrita MT, Vale C, Rauter AP (2010) Halogenated compounds from marine algae. Mar Drugs 8:2301–2317
68. Plouguerné E, Hellio C, Cesconetto C et al (2010) Antifouling activity as a function of population variation in *Sargassum vulgare* from the littoral of Rio de Janeiro (Brazil). J Appl Phycol 22:717–724. doi:10.1007/s10811-010-9511-0
69. Davoult D, Engel CR, Arzel P et al (2011) Environmental factors and commercial harvesting: exploring possible links behind the decline of the kelp *Laminaria digitata* in Brittany, France. Cah Biol Mar 52:429–435
70. Sivertsen K (1997) Geographic and environmental factors affecting the distribution of kelp beds and barren grounds and changes in biota associated with kelp reduction at sites along the Norwegian coast. Can J Fish Aquat Sci 54:2872–2887
71. Christie H, Fredriksen S, Rinde E (1998) Regrowth of kelp and colonization of epiphyte and fauna community after kelp trawling at the coast of Norway. In: Baden S, Phil L, Rosenberg R et al (eds) Recruitment, colonization and physical-chemical forcing in marine biological systems. Springer, Netherlands, pp 49–58
72. Valero M, Destombe C, Mauger S et al (2011) Using genetic tools for sustainable management of kelps: a literature review and the example of *Laminaria digitata*. Cah Biol Mar 52:467–483
73. Sharp G, Semple R, Wilson M et al (2008) A survey of the distribution and abundance of Irish Moss (*Chondrus crispus*) on the south shore of Nova Scotia. Port Medway, Shelburne Co. to Pennant Point, Halifax Co. Cano Manuser. Rep Fish Aquat Sci 2856:Iii+34
74. Chopin T, Sharp G, Belyea E et al (1999) Open-water aquaculture of the red alga *Chondrus crispus* in Prince Edward Island, Canada. Hydrobiologia 398:417–425
75. Stagnol D, Renaud M, Davoult D (2013) Effects of commercial harvesting of intertidal macroalgae on ecosystem biodiversity and functioning. Estuar Coast Shelf Sci 130: 99–110
76. Hafting JT, Critchley AT, Cornish ML et al (2012) On-land cultivation of functional seaweed products for human usage. J Appl Phycol 24:385–392
77. Stengel DB, Dring MJ (2000) Copper and iron concentrations in *Ascophyllum nodosum* (Fucales, Phaeophyta) from different sites in Ireland and after culture experiments in relation to thallus age and epiphytism. J Exp Mar Biol Ecol 246:145–161
78. Gellenbeck KW (2012) Utilization of algal materials for nutraceutical and cosmeceutical applications—what do manufacturers need to know? J Appl Phycol 24:309–313
79. Jassby A (1988) *Spirulina*: a model for microalgae as human food. In: Lembi CA, Waaland JR (eds) Algae and human affairs. Cambridge University Press, Cambridge, pp 149–179
80. Daroch M, Geng S, Wang G (2013) Recent advances in liquid biofuel production from algal feedstocks. Appl Energ 102:1371–1381
81. Pulz O, Gross W (2004) Valuable products from biotechnology of microalgae. Appl Microbiol Biotechnol 65:635–648
82. Liu D, Keesing JK, Xing Q et al (2009) World's largest macroalgal bloom caused by expansion of seaweed aquaculture in China. Mar Poll Bull 58:888–895
83. Buschmann AH, Correa JA, Westermeier R (2001) Red algal farming in Chile: a review. Aquaculture 194:203–220
84. Buschmann AH, Hernández-González MDC, Varela D (2008) Seaweed future cultivation in Chile: perspectives and challenges. Int J Environ Pollut 33:432–456. doi:10.1504/IJEP.2008.020571
85. Longford SR, Tujula NA, Crocetti GR et al (2007) Comparisons of diversity of bacterial communities associated with three sessile marine eukaryotes. Aquat Microb Ecol 48: 217–229
86. Tujula NA, Crocetti GR, Burke C et al (2010) Variability and abundance of the epiphytic bacterial community associated with a green marine Ulvacean alga. ISME J 4:301–311
87. Moody JW, McGinty CM, Quinn JC (2014) Global evaluation of biofuel potential from microalgae. Proc Natl Acad Sci U S A 111:8691–8696. doi:10.1073/pnas.1321652111

88. Mata TM, Martins AA, Caetano NS (2010) Microalgae for biodiesel production and other applications: a review. Renew Sust Energ Rev 14:217–232
89. Luque R (2010) Algal biofuels: the eternal promise? Energy Environ Sci 3:254–257
90. Ugwu CU, Ogbonna JC, Tanaka H (2005) Light/dark cyclic movement of algal culture (*Synechocystis aquatilis*) in outdoor inclined tubular photobioreactor equipped with static mixers for efficient production of biomass. Biotechnol Lett 27:75–78
91. Pulz O, Scheibenbogen K (1998) Photobioreactors: design and performance with respect to light energy input. Adv Biochem Eng/Biotechnol 59:123–152
92. Ugwu CU, Aoyagi H, Uchiyama H (2008) Photobioreactors for mass cultivation of algae. Bioresour Technol 99:4021–4028
93. Ugwu CU, Ogbonna JC, Tanaka H (2002) Improvement of mass transfer characteristics and productivities of inclined tubular photobioreactors by installation of internal static mixers. Appl Microbiol Biotechnol 58: 600–607
94. Del Campo JA, Garcia-Gonzáles M, Guerrero MG (2007) Outdoor cultivation of microalgae for carotenoid production: current state and perspectives. Appl Microbiol Biotechnol 74:1163–1174
95. Doucha J, Lívanský K (2006) Productivity, CO_2/O_2 exchange and hydraulics in outdoor open high density microalgal (*Chlorella* sp.) photobioreactors operated in a middle and southern European climate. J Appl Phycol 18:811–826
96. Doucha J, Lívanský K (2009) Outdoor open thin-layer microalgal photobioreactor: potential productivity. J Appl Phycol 21:111–117
97. Grobbelaar JU (2009) Factors governing algal growth in bioreactors. The 'open' versus 'closed' debate. J Appl Phycol 21:489–492
98. Chen CY, Yeh KL, Aisyah R et al (2011) Cultivation, photobioreactor design and harvesting of microalgae for biodiesel production: a critical review. Bioresour Technol 102: 71–81
99. Gacheva GV, Gigova LG (2014) Biological activity of microalgae can be enhanced by manipulating the cultivation temperature and irradiance. Cent Eur J Biol 9:1168–1181
100. Chrismadha T, Borowitzka MA (1994) Effect of cell density and irradiance on growth, proximate composition and eicosapentaenoic acid production of *Phaeodactylum tricornutum* grown in a tubular photobioreactor. J Appl Phycol 6:67–74
101. Çelekli A, Dönmez G (2006) Effect of pH, light intensity, salt and nitrogen concentrations on growth and β-carotene accumulation by a new isolate of *Dunaliella* sp. World J Microbiol Biotechnol 22:183–189
102. Chaneva G, Furnadzhieva S, Minkova K et al (2007) Effect of light and temperature on the cyanobacterium *Arthronema africanum* - a prospective phycobiliprotein-producing strain. J Appl Phycol 9:37–544
103. Wang B, Li Y, Wu N et al (2008) CO_2 bio-mitigation using microalgae. Appl Microbiol Biotechnol 79:707–718
104. Bondu S, Kervarec N, Deslandes E et al (2008) The use of HRMAS NMR spectroscopy to study the in vivo intra-cellular carbon/nitrogen ratio of *Solieria chordalis* (Rhodophyta). J Appl Phycol 20:673–679
105. Venugopal V, Prasanna R, Sood A et al (2006) Stimulation of pigment accumulation in *Anabaena azollae* strains: effects of light intensity and sugars. Folia Microbiol 51: 50–56
106. Marxen K, Vanselow KH, Hintze R et al (2010) Comparison of two different modes of UV-B irradiation on synthesis of some cellular substances in the cyanobacterium *Synechocystis* sp. PCC6803. J Appl Phycol 22:677–690
107. Raja R, Hemaiswarya S, Rengasamy R (2007) Exploitation of *Dunaliella* for β-carotene production. Appl Microbiol Biotechnol 74: 517–523
108. Ye ZW, Jiang JG, Wu GH (2008) Biosynthesis and regulation of carotenoids in *Dunaliella*: progresses and prospects. Biotechnol Adv 26:352–360
109. Bancroft BA, Baker NJ, Blaustein AR (2007) Effects of UVB radiation on marine and freshwater organisms: a synthesis through meta-analysis. Ecol Lett 10:332–345
110. Behrenfeld MJ, Falkowski PG (1997) Photosynthetic rates derived from satellite-based chlorophyll concentration. Limnol Oceanogr 42:1–20
111. Andersen RA (2005) Algal culturing techniques. Elsevier Academic Press, New York
112. Fabregas J, Herrero C, Abalde J et al (1986) The marine microalga *Chlorella stigmatophora* as a potential source of single cell protein: enhancement of the protein content in response to nutrient enrichment. J Ind Microbiol 1:251–257
113. Corsini M, Karydis M (1990) An algal medium based on fertilizers and its evaluation in mariculture. J Appl Phycol 2:333–339
114. Chen W, Zhang Q, Dai S (2009) Effects of nitrate on intracellular nitrite and growth of

Microcystis aeruginosa. J Appl Phycol 21: 701–706

115. Prasanna R, Sood A, Jaiswal P et al (2010) Rediscovering cyanobacteria as valuable sources of bioactive compounds (Review). Appl Biochem Microbiol 46:119–134
116. Fatma T (2009) Screening of cyanobacteria for phycobiliproteins and effect of different environmental stress on its yield. Bull Environ Contam Toxicol 83:509–515
117. Nie ZY, Xia JL, Levert JM (2002) Fractionation and characterization of polysaccharides from cyanobacterium *Spirulina* (*Arthrospira*) *maxima* in nitrogen-limited batch culture. J Cent South Univ Technol 9:81–86
118. Colla LM, Reinehr CO, Reichert C et al (2007) Production of biomass and nutraceutical compounds by *Spirulina platensis* under different temperature and nitrogen regimes. Bioresour Technol 98:1489–1493
119. Siron R, Giusti G, Berland B (1989) Changes in the fatty acid composition of *Phaeodactylum tricornutum* and *Dunaliella tertiolecta* during growth and under phosphorus deficiency. Mar Ecol Prog Ser 55:95–100
120. Piorreck M, Baasch KH, Pohl P (1984) Biomass production, total protein, chlorophylls, lipids and fatty acids of freshwater green and blue-green algae under different nitrogen regimes. Phytochemistry 23: 207–216
121. Ben-Amotz A (1987) Effect of irradiance and nutrient deficiency on the chemical composition of *Dunaliella bardawil* Ben-Amotz and Avron (Volvocales, Chlorophyta). J Plant Physiol 131:479–487
122. Roessler PG (1988) Effects of silicon deficiency on lipid composition and metabolism in the diatom *Cyclotella cryptica*. J Phycol 24:394–400
123. Durmaz Y, Monteiro M, Bandarra N et al (2007) The effect of low temperature on fatty acid composition and tocopherols of the red microalga, *Porphyridium cruentum*. J Appl Phycol 19:223–227
124. Xu XQ, Beardall J (1997) Effect of salinity on fatty acid composition of a green microalga from an antarctic hypersaline lake. Phytochemistry 45:655–658
125. Oren A (2005) A hundred years of *Dunaliella* research: 1905–2005. Saline Syst 1:1–14. doi:10.1186/1746-1448-1-2
126. Chen H, Jiang JG (2009) Osmotic responses of *Dunaliella* to the changes of salinity. J Cell Physiol 219:251–258
127. Rao AR, Dayananda C, Sarada R et al (2007) Effect of salinity on growth of green alga *Botryococcus braunii* and its constituents. Bioresour Technol 98:560–564
128. de Castro Araújo S, Garcia VMT (2005) Growth and biochemical composition of the diatom *Chaetoceros* cf. *wighamii* brightwell under different temperature, salinity and carbon dioxide levels. I. Protein, carbohydrates and lipids. Aquaculture 246:405–412
129. Chu WL, Phang SM, Goh SH (1996) Environmental effects on growth and biochemical composition of *Nitzschia inconspicua* Grunow. J Appl Phycol 8:389–396
130. Huntley ME, Redalje DG (2007) CO_2 mitigation and renewable oil from photosynthetic microbes: a new appraisal. Mitig Adapt Strat GC 12:573–608
131. Papazi A, Makridis P, Divanach P et al (2008) Bioenergetic changes in the microalgal photosynthetic apparatus by extremely high CO_2 concentrations induce an intense biomass production. Physiol Plant 132:338–349
132. Jeon H, Lee Y, Chang KS et al (2013) Enhanced production of biomass and lipids by supplying CO_2 in marine microalga *Dunaliella* sp. J Microbiol 51:773–776
133. Guihéneuf F, Stengel DB (2013) Microalgae LC-PUFA enriched oil production: lipid and triacylglycerols containing n-3 LC-PUFA accumulation are triggered by nitrogen-limitation and inorganic carbon availability in the marine microalga *Pavlova lutheri*. Mar Drugs 11:4246–4266. doi:10.3390/md11114246
134. Marcoval MA, Villafane VE, Helbling EW (2007) Interactive effects of ultraviolet radiation and nutrient addition on growth and photosynthesis performance of four species of marine phytoplankton. J Photochem Photobiol B 89:78–87
135. Bermúdez J, Rosales N, Loreto C et al (2004) Exopolysaccharide, pigment and protein production by the marine microalga *Chroomonas* sp. in semicontinuous cultures. World J Microbiol Biotechnol 20:179–183
136. Halling C, Wikström S, Lilliesköld-Sjöö G et al (2013) Introduction of Asian strains and low genetic variation in farmed seaweeds: indications for new management practices. J Appl Phycol 25:89–95
137. Uppalapati SR, Fujita Y (2000) Red rot resistance in interspecific protoplast fusion product progeny of *Porphyra yezoensis* and *P. tenuipedalis* (Bangiales, Rhodophyta). Phycol Res 48:281–289

138. Xu P, Yang L, Zhu J et al (2011) Analysis of hybridization strains of *Porphyra* based on rbc L gene sequences. J Appl Phycol 23:235–241
139. Zhang QS, Tang XX, Cong YZ et al (2007) Breeding of an elite *Laminaria* variety 90-1 through inter-specific gametophyte crossing. J Appl Phycol 19:303–311
140. Weng M, Liu B, Jin D et al (2005) Identification of 27 *Porphyra* lines (Rhodophyta) by DNA fingerprinting and molecular markers. J Appl Phycol 17:91–97
141. Niwa K, Aruga Y (2006) Identification of currently cultivated *Porphyra* species by PCR-RFLP analysis. Fish Sci 72:143–148
142. Bi Y, Hu Y, Zhou Z (2011) Genetic variation of *Laminaria japonica* (Phaeophyta) populations in China as revealed by RAPD markers. Acta Oceanol Sin 30:103–112
143. Wu C, Guangheng L (1987) Progress in the genetics and breeding of economic seaweeds in China. Hydrobiologia 151:57–61
144. Lin H, Qin S (2014) Tipping points in seaweed genetic engineering: scaling up opportunities in the next decade. Mar Drugs 12:3025–3045. doi:10.3390/md12053025
145. Rasala BA, Chao SS, Pier M et al (2014) Enhanced genetic tools for engineering multigene traits into green algae. PLoS One 9:e94028. doi:10.1371/journal.pone.0094028
146. Radakovits R, Jinkerson RE, Darzins A et al (2010) Genetic engineering of algae for enhanced biofuel production. Eukaryot Cell 9:486–501
147. Kim S, Lee YC, Cho DH et al (2014) A simple and non-invasive method for nuclear transformation of intact-walled *Chlamydomonas reinhardtii*. PLoS One 9:e101018. doi:10.1371/journal.pone.0101018
148. Poulsen N, Chesley PM, Kröger N (2006) Molecular genetic manipulation of the diatom *Thalassiosira pseudonana* (Bacillariophyceae). J Phycol 42:1059–1065
149. Nakamura Y, Sasaki N, Kobayashi M et al (2013) The first symbiont-free genome sequence of marine red alga, Susabi-nori (*Pyropia yezoensis*). PLoS One 8:e57122. doi:10.1371/journal.pone.0057122
150. Lü J, Sheahan C, Fu P (2011) Metabolic engineering of algae for fourth generation biofuels production. Energy Environ Sci 4:2451–2466. doi:10.1039/1754-5706/2008#Link
151. Cadoret JP, Garnier M, Saint-Jean B (2012) Microalgae, functional genomics and biotechnology. Adv Bot Res 64:285–341
152. Keeling PJ, Burki F, Wilcox HM et al (2014) The marine microbial eukaryote transcriptome sequencing project (MMETSP): illuminating the functional diversity of eukaryotic life in the oceans through transcriptome sequencing. PLoS Biol 12:e1001889. doi:10.1371/journal.pbio.1001889
153. Cock JM, Sterck L, Rouzé P et al (2010) The *Ectocarpus* genome and the independent evolution of multicellularity in brown algae. Nature 465:617–621
154. Ritter A, Dittami SM, Goulitquer S et al (2014) Transcriptomic and metabolomic analysis of copper stress acclimation in *Ectocarpus siliculosus* highlights signaling and tolerance mechanisms in brown algae. BMC Plant Biol 14:116. doi:10.1186/1471-2229-14-116
155. Dittami SM, Scornet D, Petit JL et al (2009) Global expression analysis of the brown alga *Ectocarpus siliculosus* (Phaeophyceae) reveals large-scale reprogramming of the transcriptome in response to abiotic stress. Genome Biol 10:R66
156. Konotchick T, Dupont CL, Valas RE et al (2013) Transcriptomic analysis of metabolic function in the giant kelp, *Macrocystis pyrifera*, across depth and season. New Phytol 198:398–407
157. Leblanc C, Schaal G, Cosse A et al (2011) Trophic and biotic interactions in *Laminaria digitata* beds: which factors could influence the persistence of the marine kelp forests in northern Brittany? Cah Biol Mar 52:415–427
158. Cosse A, Leblanc C, Potin P (2007) Dynamic defense of marine macroalgae against pathogens: from early activated to gene-regulated responses. Adv Bot Res 46:221–266
159. Kitade Y, Asamizu E, Fukuda S et al (2008) Identification of genes preferentially expressed during asexual sporulation in *Porphyra yezoensis* gametophytes (Bangiales, Rhodophyta). J Phycol 44:113–123
160. Wu S, Sun J, Chi S et al (2014) Transcriptome sequencing of essential marine brown and red algal species in China and its significance in algal biology and phylogeny. Acta Oceanol Sin 33:1–12
161. Georgianna DR, Mayfield SP (2012) Exploiting diversity and synthetic biology for the production of algal biofuels. Nature 488:329–335
162. Goldstein DA, Thomas JA (2004) Biopharmaceuticals derived from genetically modified plants. Q J Med 97:705–716

163. Paul M, van Dolleweerd C, Drake PMW et al (2011) Molecular pharming—future targets and aspirations. Hum Vaccin 7:375–382
164. Geng D, Wang Y, Wang P et al (2003) Stable expression of Hepatitis B surface antigen gene in *Dunaliella salina* (Chlorophyta). J Appl Phycol 15:451–456
165. Jones AC, Gu L, Sorrels CM et al (2009) New tricks from ancient algae: natural products biosynthesis in marine Cyanobacteria. Curr Opin Chem Biol 13: 216–223
166. Hempel F, Lau J, Klingl A et al (2011) Algae as protein factories: expression of a human antibody and the respective antigen in the diatom *Phaeodactylum tricornutum*. PLoS One 6:e28424. doi:10.1371/journal.pone.0028424
167. Gerwick WH, Moore BS (2012) Lessons from the past and charting the future of marine natural products drug discovery and chemical biology. Chem Biol 19:85–98
168. Saitoh T, Suzuki T, Sugimoto M et al (2003) Total synthesis of (+) –laurallene. Tetrahedr Lett 44:3175–3178
169. Nicolaou KC, Snyder SA (2005) Chasing molecules that were never there: misassigned natural products and the role of chemical synthesis in modern structure elucidation. Angew Chem Int Ed 44:1012–1044

Chapter 2

Structure and Function of Macroalgal Natural Products

Ryan M. Young, Kathryn M. Schoenrock, Jacqueline L. von Salm, Charles D. Amsler, and Bill J. Baker

Abstract

Since the initial discovery of marine phyco-derived secondary metabolites in the 1950s there has been a rapid increase in the description of new algal natural products. These metabolites have multiple ecological roles as well as commercial value as potential drugs or lead compounds. With the emergence of resistance to our current arsenal of drugs as well as the development of new chemotherapies for currently untreatable diseases, new compounds must be sourced. As outlined in this chapter algae produce a diverse range of chemicals many of which have potential for the treatment of human afflictions.

In this chapter we outline the classes of metabolites produced by this chemically rich group of organisms as well as their respective ecological roles in the environment. Algae are found in nearly every environment on earth, with many of these organisms possessing the ability to shape the ecosystem they inhabit. With current challenges to climate stability, understanding how these important organisms interact with their environment as well as one another might afford better insight into how they respond to a changing climate.

Key words Chemical ecology, Macroalgae, Natural products, Secondary metabolites

1 Introduction

Fun fact: amongst more than 24,000 marine natural products represented in MarinLit [1], the industry standard database for such compounds, those with the greatest proportion of halogenation by weight are derived from algae. Marine algae are well recognized as producers of polyhalogenated compounds, though some of the most biologically significant algal products lack halogenation. Nonetheless, with several algal metabolites bearing 90 % or more of their mass in halogens, others with six or more halogen atoms, and one alga producing the only marine product bearing bromine, chlorine and iodine on the same structure, this reputation is not undeserved.

Halogenated compounds and others present in algal tissues are presumably biosynthesized by the producing organism to increase its fitness in its environment. Many algal compounds have been

Dagmar B. Stengel and Solène Connan (eds.), *Natural Products From Marine Algae: Methods and Protocols*, Methods in Molecular Biology, vol. 1308, DOI 10.1007/978-1-4939-2684-8_2, © Springer Science+Business Media New York 2015

found, for example, to deter feeding by predators, to reduce competition for space or as settlement cues. Herein we present a brief review of secondary metabolites produced by eukaryotic macroalgae, specifically belonging to Rhodophyta, Chlorophyta, and Phaeophyceae. Following the chemical synopsis is a discussion of the function of secondary metabolites in the ecology of the producing organism.

2 Algal Secondary Metabolites

2.1 Overview

A recent review recognized nearly 3,200 natural products from marine macroalgae, representing 13 % of compounds reported from marine organisms [2]. While marine algae continue to be a good resource for new compounds, overall reports of new metabolites have recently decreased [2, 3]. Each algal group tends to produce its own unique chemistry. Greater than 50 % of all algal secondary metabolites are isoprenoids (e.g., terpenes, steroids, carotenoids, prenylated quinones and hydroquinones), followed by polyketides and shikimates, the two of which contribute a significant number of aromatic natural products such as prenylated hydroquinones and tannins [2, 4]. Nitrogen-containing metabolites, including non-ribosomal peptides and alkaloids, are the vanishingly rare in natural products of eukaryotic macroalgae. For simplicity in conceptualizing structural aspects of compounds discussed in this review as they relate to ecological function, it is organized into Isoprenoid Metabolites, Aliphatic Acetogenic Metabolites, Aromatic Metabolites, and Nitrogenous Macroalgal Natural Products.

The Rhodophyta are perhaps the most chemically rich group of all the algae in terms of both abundance and diversity, with nearly 1,700 compounds reported to date [1]. Nearly all chemical classes of natural products are found in red algae; most remarkable is the number of halogenated secondary metabolites. Over 70 % of reported red algal natural products contain bromine, chlorine, or iodine, while only 11 % and 4 % of green and brown algal secondary metabolites contain these halogens, respectively [5]. The majority (57 %) of Rhodophyte chemistry is found in the Family Rhodomelaceae with approximately 85 % originating from the genus *Laurencia* [6]. Brown algae are well known for their production of polyphenolic phlorotannins. Currently there are over 1,200 compounds reported from brown algae (Phaeophyceae) with over a third coming from a single genus, *Dictyota*. The Chlorophyta are perhaps the least prolific producers of secondary metabolites amongst the macroalgae (ca. 270 compounds). Green algae have been reported to produce chemicals similar to those of red algae but lack the same degree of halogenation. Just short of half of all the reported green algal chemistry comes from the order

Bryopsidales, with over 85 % of that chemistry being terpenoid. Between 2011 and 2013, there were only a few reports of new compounds from green algae, all of which lack significant structural diversity from known metabolites [3]. However, fungi associated with green algae had a large increase in new compound isolation.

2.2 Isoprenoid Metabolites

For the purposes of this review, isoprenoids comprise the terpenoids and steroids. A significant portion of algal chemistry derives from prenylated phenyl rings, but those compounds will be discussed below in Subheading 2.4.

Terpenoids are a large class of organic compounds common to a wide range of organisms. Except for the rare class of hemiterpenoids, which derive biosynthetically from a single isoprene (C_5) unit, terpenoids are produced from condensation of multiple units of isoprene, building the carbon scaffold five carbons at a time. Two, three, and four isoprene units, representing monoterpenes, sesquiterpenes, and diterpenes, respectively, are the most common. Almost 60 % of all marine natural products reported from algae are terpenoids, with the greatest number (nearly 50 %) of them found in Rhodophytes, though Phaeophytes contain approximately 36 % of all reported algal terpenoids [2].

Steroids are derived from cholesterol via oxidation and other derivatization metabolism. They generally bear a C_{27} tetracyclic structure, reflecting three demethylation steps from a C_{30} triterpenoid such as lanosterol. Side-chain methylation can increase the carbon count, but steroids remain distinct from triterpenes based on lack of the 4,4-gem dimethyl and 14-methyl group found on triterpenes [6].

2.2.1 Monoterpenes

Common algal terpenoids are monoterpene (comprising two isoprene units, for a C_{10} carbon backbone), sesquiterpene (three isoprene units, C_{15} carbon backbone), and diterpene (four isoprene units, C_{20} carbon backbone). Of these, monoterpenes are the rarest amongst macroalgae. Examples of monoterpenes include linear and cyclic halogenated monoterpenes, such as anverene [**1**], *epi-plocamene* D [**2**], and pyranoid [**3**] from the red alga *Plocamium cartilagineum*, described with some interesting ecological implications in Subheading 3.5. Amongst this group of monoterpenes, halomon [**4**], from *Portieria hornemannii*, enjoyed considerable enthusiasm within the biomedical community as a selective antitumor agent [7]. Interestingly, *P. hornemannii* was shown to produce a suite of related compounds many of which have comparable antitumor activity [8]. Because harvesting algal biomass from nature is not sustainable, several research groups have developed synthetic pathways for the production of **4** [9, 10]. While reports of monoterpenes from red algae number well over 200, 95 % of which are halogenated, green and brown algae have very limited monoterpene diversity.

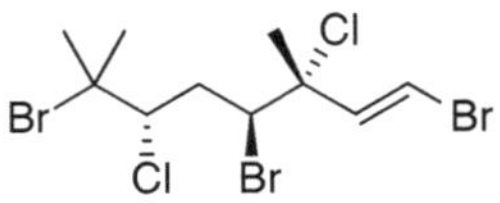

1

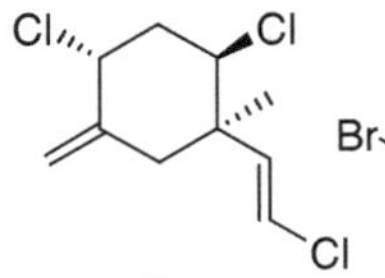

2

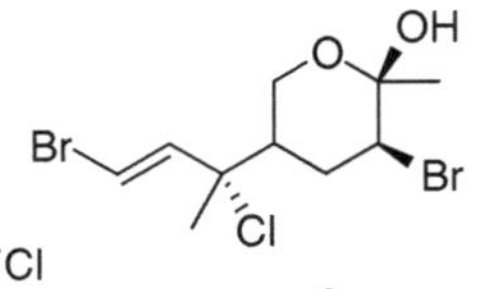

3

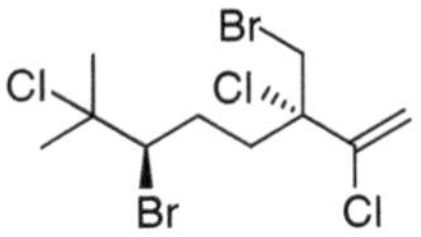

4

2.2.2 Sesquiterpenes

Algal terpenes diversify significantly amongst sesquiterpene and diterpene metabolites. Interestingly, red algae dominate the sesquiterpene class and brown algae the diterpene class. In both cases, the increased molecular size introduces increased complexity. Consider red algae sesquiterpenes with high oxygenation and polycyclization, such as hydroxyaplysistatin [**5**], reported from *Laurencia filiformis* and the intriguing tricycle perforatone [**6**], with four, five, and six membered fused rings, from *Palisada* (*Laurencia*) *perforata*. Spirocycle elatol [**7**], which is reported from a number of *Laurencia* species [11–16] showed potent antifouling activity against the larvae of the barnacle *Balanus amphitrite* [17, 18]. The brown algal sesquiterpenoids 11-epispathulenol [**8**], from *Taonia atomaria*, is a rare example of a brown algal sesquiterpene. And while Chlorophytes lack significant numbers of sesquiterpenes, those they produce are reminiscent of halogenated sesquiterpenes from Rhodophytes, such as neomeranol [**9**] from the green alga *Neomeris annulata*.

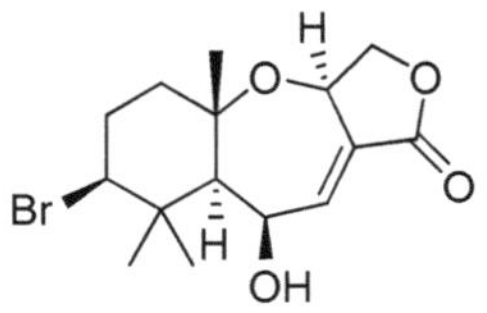

5

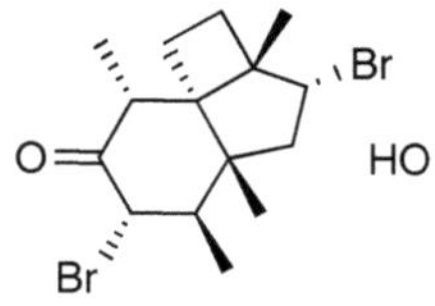

6

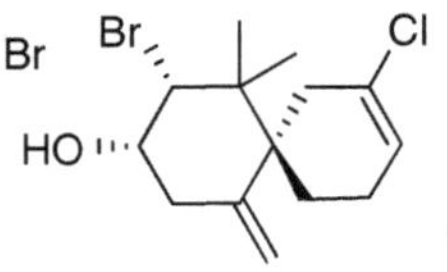

7

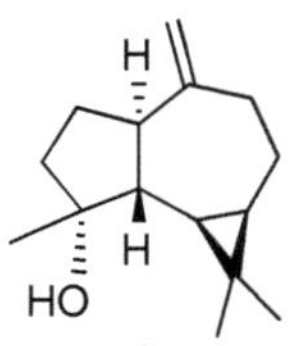

8

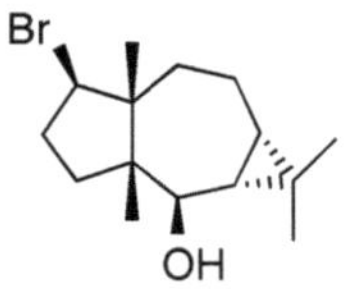

9

2.2.3 Diterpenes

Brown algae dominate the diterpenoid class of terpenes, comprising the largest single group of macroalgal natural products with nearly 500 representatives. Red algal diterpenoids are approximately one fourth as numerous, and are less common than either red algal monoterpenes or sesquiterpenes. Examples of red algal diterpenes include tetracyclic coronopifoliol [**10**] and oxatane prevezol B [**11**], from *Sphaerococcus coronopifolius* and *Laurencia obtusa*, respectively. The brown algal genus *Dictyota* is the major source of diterpenoids in this class, representatives of which include dictyolactone [**12**] and dictyoxetane [**13**]. Spatol [**14**], from *Spatoglossum schmittii*, and dilospirane B [**15**], from *Dilophus spiralis*, illustrate examples characteristic of other phaeophyte genera. Both red and brown algae elaborate innumerable examples of diterpenoid–shikimate hybrids, or meroterpenoids, with chromene, hydroquinone, quinone, and similar aromatic functionalities, not to mention cyclic and polycyclic diterpene moieties. These are discussed further in Subheading 2.4.1 based on their structural similarities to polyketide–shikimate hybrids. A key feature of both sesquiterpene and diterpene green algal metabolites is the 1,4-dialdehyde function, often with the aldehyde function masked as acetals and/or enol acetates, as illustrated in chlorodesmin [**16**], from *Chlorodesmis fastigiata*. Subheading 3.3 includes a discussion of such masked dialdehydes as activated defenses.

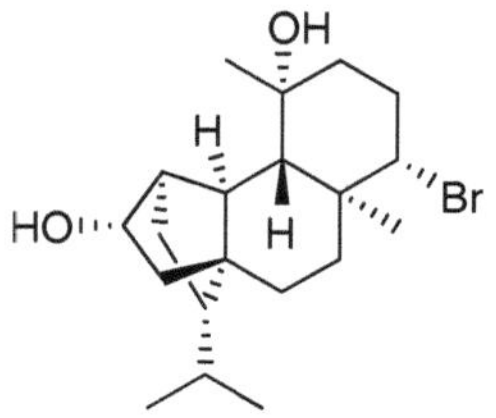

10

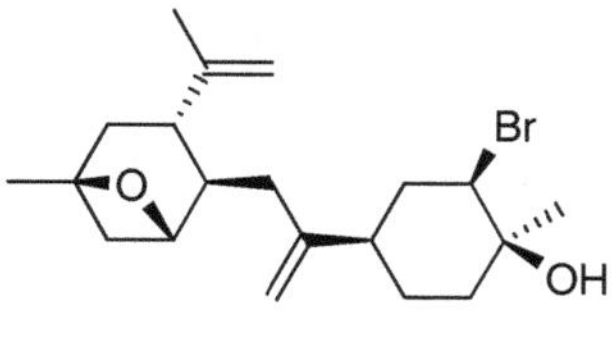

11

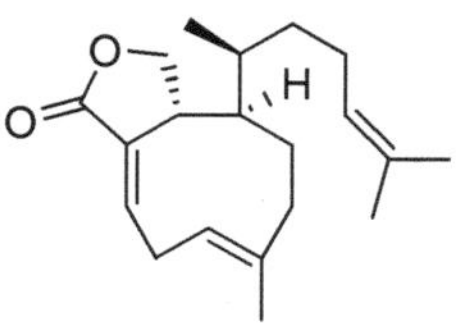

12

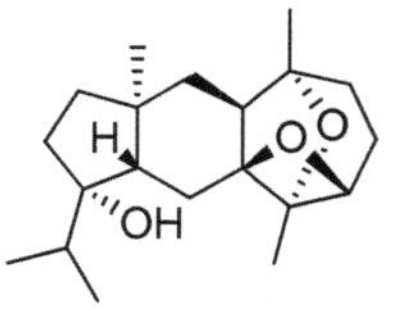

13

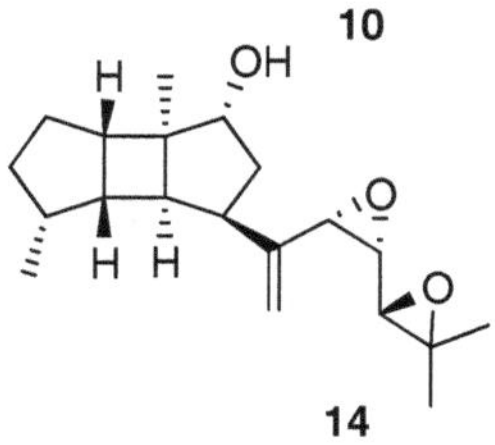

14

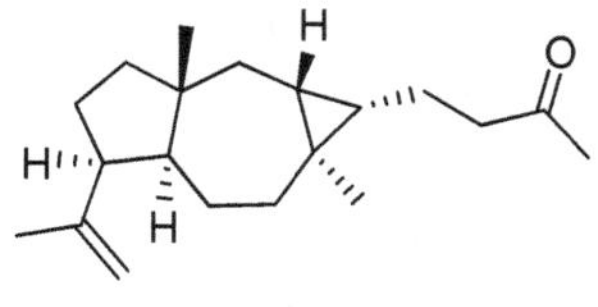

15

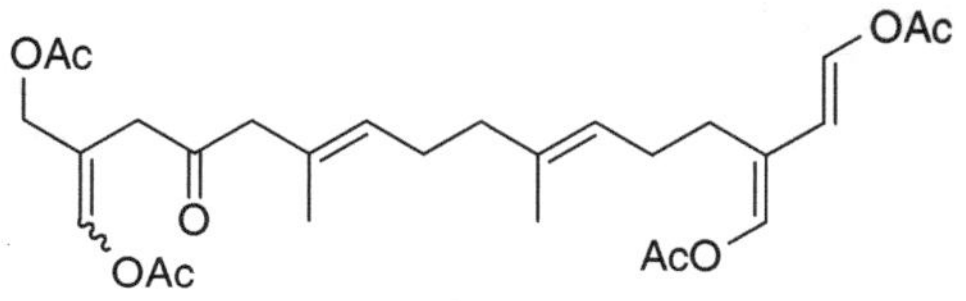

16

2.2.4 Higher Terpenes

Higher terpenes, including sesterterpenes, triterpenes, and even larger terpenes, are rare in algae. Nonetheless, there are some structurally interesting higher terpenes. Consider, for example, the series of more than four dozen triterpene polyethers from *Laurencia* spp. and *Chondria armata*, such as thyrsiferol [**17**] and amatol E [**18**]. The C_{40} dictyotadimer [19] is certainly among the largest of the algal terpenoids, along with innumerable C_{40} carotenoids found in brown and green algae [1].

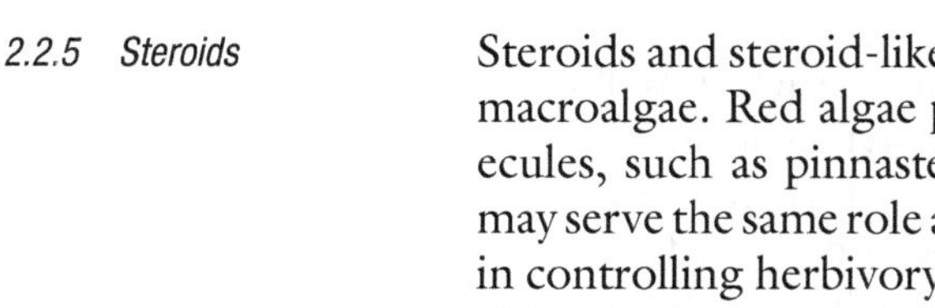

17

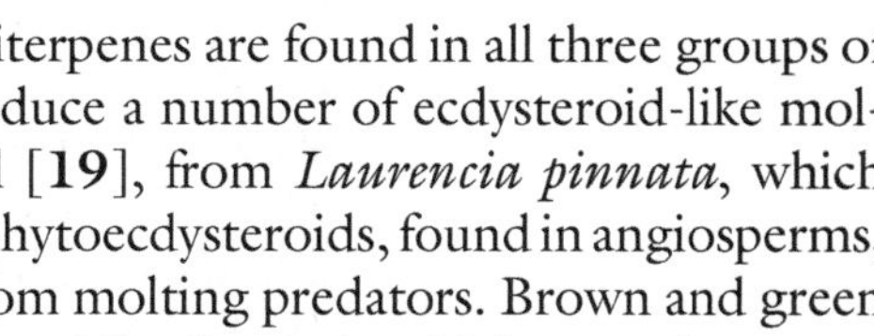

18

2.2.5 Steroids

Steroids and steroid-like triterpenes are found in all three groups of macroalgae. Red algae produce a number of ecdysteroid-like molecules, such as pinnasterol [**19**], from *Laurencia pinnata*, which may serve the same role as phytoecdysteroids, found in angiosperms, in controlling herbivory from molting predators. Brown and green algae produce very similar oxidized sterols which may also act as phytoecdysteroids, e.g., the cathasterone [**20**] from the brown alga *Polycladia* (*Cystoseira*) *myrica*. Red and green algae both elaborate lanosterol-based sulfonated triterpenoids, such as **21**, from the green alga *Tydemania expeditionis* [20].

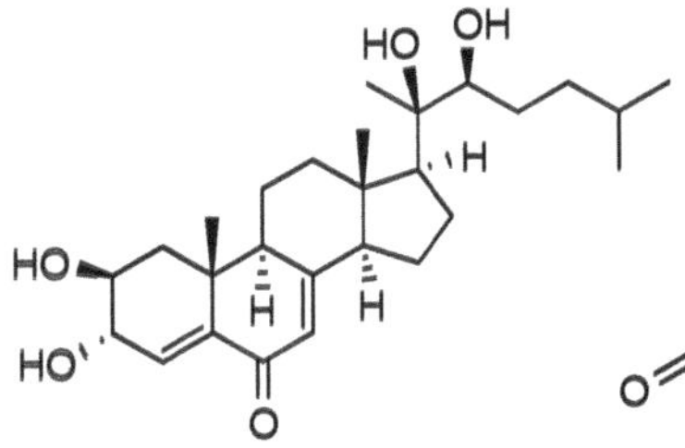

19

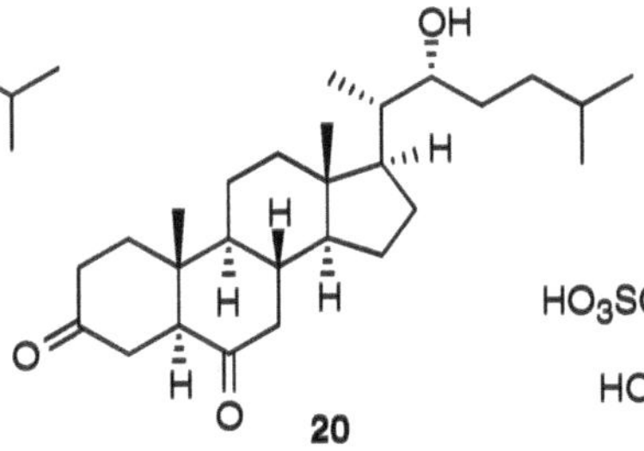

20

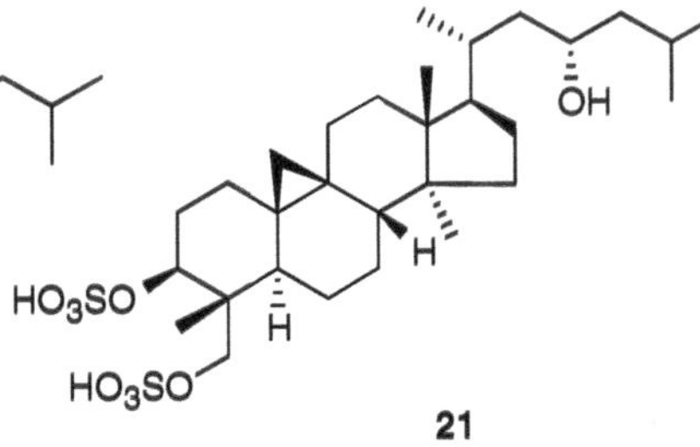

21

2.3 Aliphatic Acetogenic Metabolites

The second largest group of natural products from macroalgae are based on repeating C_2 acetate units (=acetogenin), Taken with isoprenoids, which also have acetate in their genesis, nearly 80 % of macroalgal natural products derive from acetate. In this section, we will address aliphatic acetogenins in three groups, reflecting structural and functional similarities: the C_{15} acetogenins from the Rhodophyta, C_{11} oxylipin pheromones from the Phaeophyceae, and other aliphatic polyketides including macrolides and larger oxylipins from all three macroalgal groups. Polyketides, strictly, are biosynthesized in polyketide synthase modules. However, many secondary metabolites from macroalgae that bear polyketide-like features undoubtedly originate from fatty acid starting materials, which are biosynthesized by fatty acid synthase enzymes. This review does not distinguish between the two since the focus is on their structure and function.

2.3.1 C_{15} Acetogenins from the Rhodophyta

Deriving from the corresponding polyunsaturated C_{16} fatty acid via decarboxylation, laurencenyne [**22**], originally discovered in *Laurencia okamurae* [21], may serve as the precursor to all of the red algal C_{15} acetogenins [22–24]. Examples of the diversity in this class of compounds include those containing allene moieties, such as okamurallene and obtusallene IV [**23**, **24**] from *Laurencia intricata* and *L. obtusa*, respectively [25–27], the polyether obtusin [**25**] from *L. obtusa* [28], as well as the more common enyne, such as maneonene [**26**] from *L. obtusa* [29]. A hypothesis for the biosynthetic origin of the obtusallene family of terpenes was proposed by Braddock [30] based on a series of electrophilic bromination events. Algal predators, such as molluscs [31] and amphipods [32], are known to sequester algal chemistry for their own defense [33]. The anaspidean herbivorous opisthobranch mollusc *Aplysia dactylomela*, has been proposed to sequester obtusallene IV from its algal diet and used by the mollusc to deter predation [25].

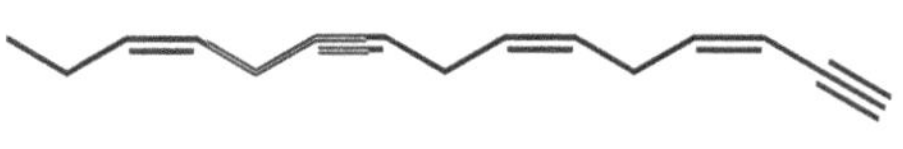

22

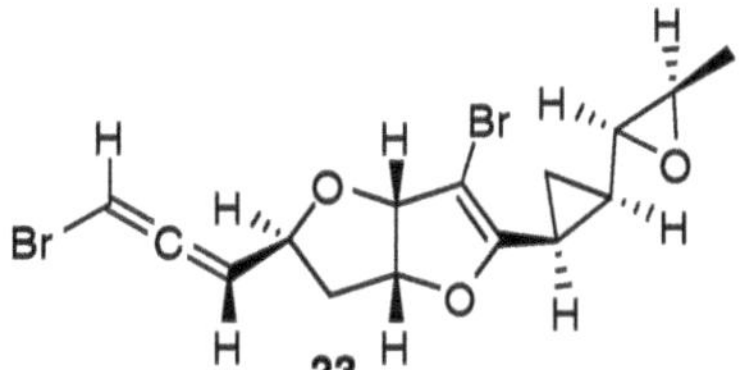

23

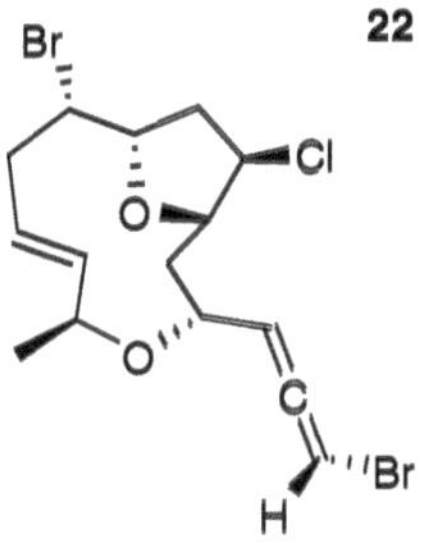

24

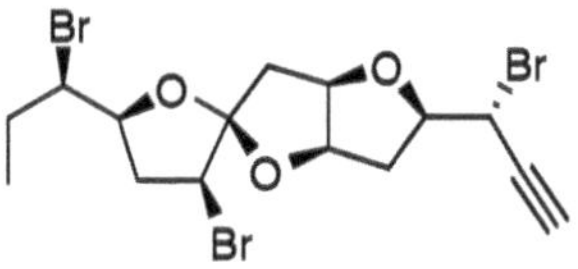

25

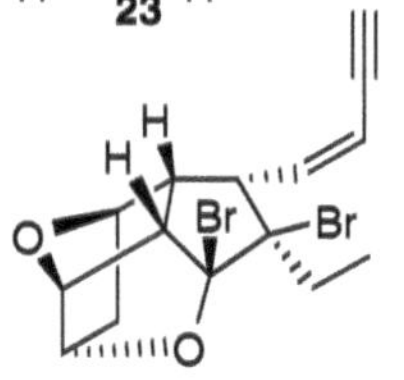

26

2.3.2 C_{11} Oxylipin Pheromones from the Phaeophyceae

Gametes of brown algae disperse pheromones to attract sperm to egg. To date, approximately a dozen pheromone structures have been identified, including a number of isomeric forms. With few exceptions (e.g., Fucales), these metabolites are degraded fatty acids bearing 11 carbon linear or mono-carbocyclic structures. The first such pheromone to be characterized was ectocarpene [**27**] [34], from *Ectocarpus siliculosus*, though that was later identified as a thermal rearrangement product of pre-ectocarpene [**28**] [35]. Other than several examples of epoxides, as observed in lamoxirene [**29**] from *Laminaria digitata* [36] and other species [37], structural diversity of the pheromones is largely captured in these three structures. These compounds are discussed in more detail in Subheading 3.2.

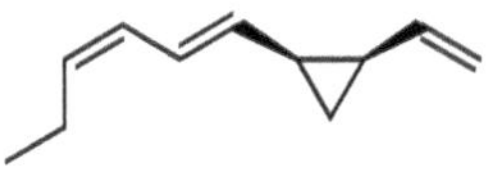

27

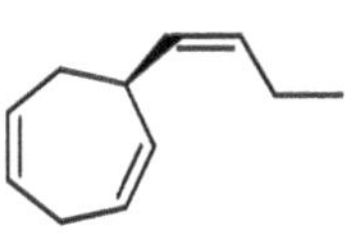

28

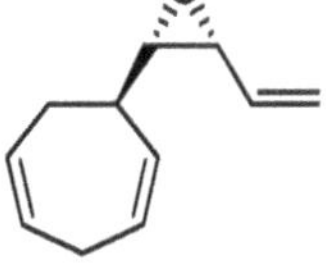

29

2.3.3 Other Polyketide or Fatty Acid C_2 Metabolites

Aliphatic polyketides in the traditional sense of macrolides, polyethers and related extended carbon backbones are rare in macroalgae, though many examples of smaller, simpler, polyketides have been reported. Two examples from the macrolide family of polyketides are noteworthy. Lobophorolide [**30**], from the brown alga *Lobophora variegata*, appears to protect the alga from attack by pathogenic and saprophytic marine fungi, displaying submicromolar activity against two strains of sympatric marine fungi, not to mention human HCT-116 colon cancer cells [38]. Polycavernoside, e.g., **31**, is a potent toxin derived from the red alga *Gracilaria edulis* [39]. Despite repeated examples of human toxicity (e.g., [40]), *Gracilaria* remains a target of biotech development for nutraceuticals, pharmaceuticals, and other human-use applications [41].

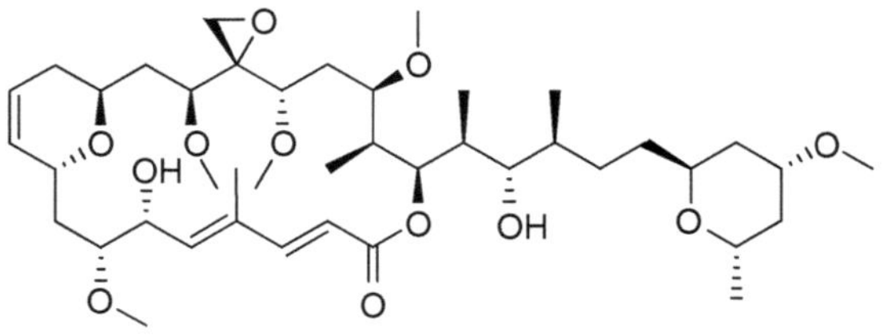

30

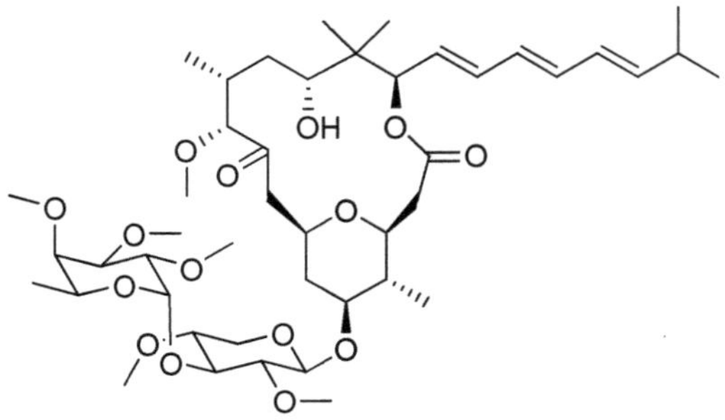

31

Among the smaller, simpler, polyketides are some of the most halogenated natural products known, as alluded to in the introduction. The red alga *Asparagopsis taxiformis* was the first macroalga known to produce a suite of one- to four-carbon polyhalogenated hydrocarbons [42], including CHBrClI. Recent research shows the phenomenon of volatile halogenated organic compound (VHOC) production to be widespread [43, 44]. The per-brominated acetoxy propene [**32**], from *A. taxiformis* [45], is a good example of the degree of halogenation (89 % by weight) achieved in this series of compounds. Halogenated furanones, such as pulchralide [**33**] from the red alga *Delisea pulchra*, serve as acylated homoserine lactone mimics that regulate biofilm production through quorum sensing, reducing biofouling of surfaces [46, 47]. Pyranosylmagellanicus C [**34**], from the red alga *Ptilonia magellanica*, is also thought to play a role in bacterial communication [48]. Hybridalactone [**35**] from the red alga *Osmundea* (*Laurencia*) *hybrida* [49], and constanolactone A [**36**] from the red alga *Constantinea simplex* [50], are examples of oxylipins from macroalgae. Hybridalactone was isolated from the antimicrobial extract of the algae and was the first marine oxylipin found with cyclopropane and macrolactone functionality. A biosynthetic pathway originating from eicosapentaenoic acid has been proposed [51]. Due to the importance of postanoids in mammalian systems for homeostasis this class of compounds is of great pharmaceutical interest [52].

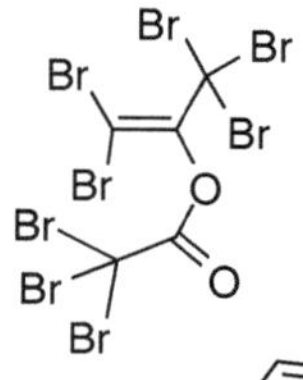

32

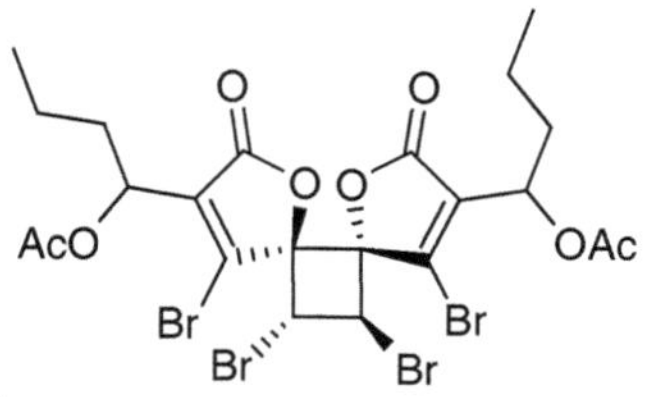

33

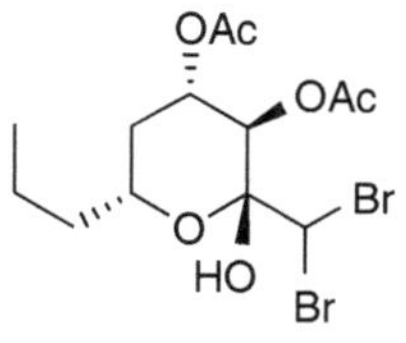

34

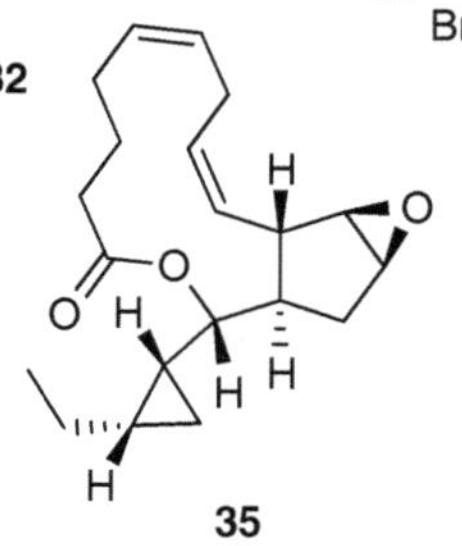

35

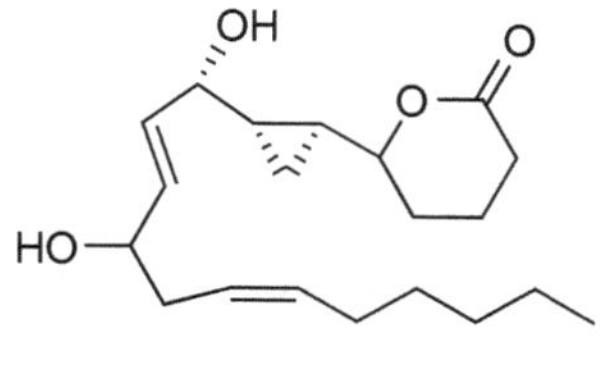

36

2.4 Aromatic Metabolites

Three distinct groups of aromatic metabolites are found in macroalgae. The largest group, with over 300 metabolites, contains a phenyl or quinone ring with a pendent alkyl chain comprising isoprene or acetate units. While brown algae dominate as the source of these hybrid metabolites, green algae have been reported with the occasional terpene–acetogenin hybrid [1]. While both of the other two groups of aromatic metabolites are similar in number, at approximately 150 compounds each, the phlorotannins are limited to phaeophytes while the bromophenol group predominantly occurs in rhodophytes. Like acetogenins discussed above, chlorophytes have very few representatives amongst the aromatic metabolites.

2.4.1 Alkylated Phenyl and Quinone Metabolites

Three structural motifs in this group of macroalgal compounds are similar based on chemical structure but differ biosynthetically. The two terpene-bearing motifs are meroterpenoids, mixing shikimate and/or acetate biosynthesis with terpene biosynthesis. Sargachromanol B [**37**], from the brown alga *Sargassum siliquastrum* [53], is an example of a chromene derived from a terpene-bearing shikimate metabolite. Terpene–shikimate hybrids are largely restricted to brown algae. A structurally related yet biosynthetically distinct group of compounds can be represented by 7-hydroxycymopochromanone [**38**] from the green alga *Cymopolia barbata* [54], which bears a terpene side chain attached to an acetogenic phenyl ring rather than the shikimate-derived phenyl ring of **37**. These terpene–acetogenin hybrids are distributed in both brown and green algae. The third motif that emerges, in brown algae again, is represented by spiralisone D [**39**], a resorcinol characteristic of *Zonaria* spp. [55], in which both the phenyl ring and side chain are acetogenic. Resorcinols are known to protect the algae from microbial, epiphytic algae, and larval settling as well as predation [56].

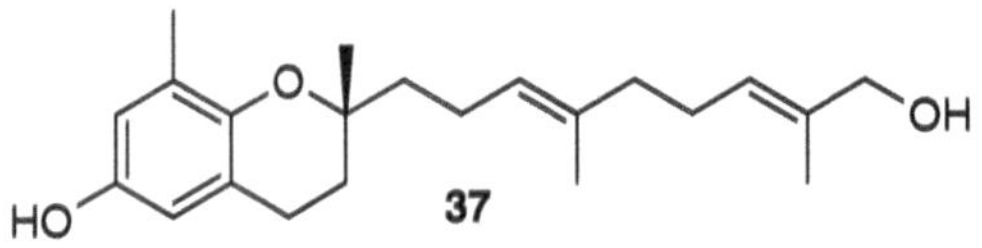

37

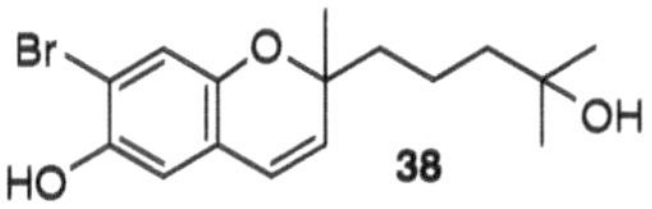

38

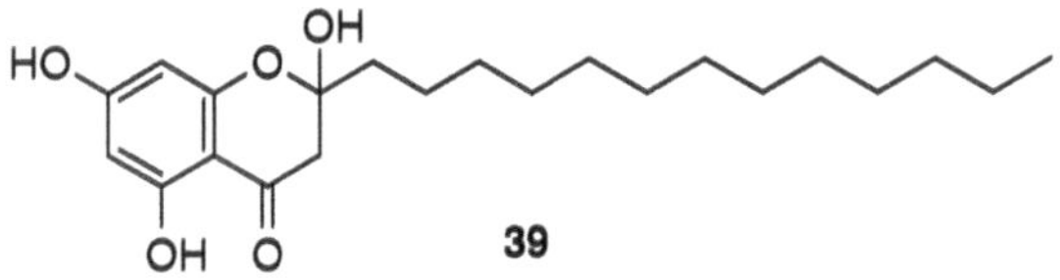

39

Compounds **37–39** can be thought of as "type specimens"; however, with the exception of **39**, the structural diversity of alkylated phenyl and quinone meroterpenoids is much broader. For example, cyclic and polycyclic diterpenes can be found appended to the shikimate phenyl ring depicted in **37**. This is demonstrated in cystoseirol [**40**] [57], from the brown alga *Cystoseira mediterranea*, or belearone [**41**], from *C. brachycarpa* (*C. balaerica*) [58]. In the absence of alkylation of the hydroquinone function (e.g., **37–41**), oxidation readily takes place to produce quinones, exemplified by stypoquinoic acid [**42**], from *Stypopodium zonale* [59].

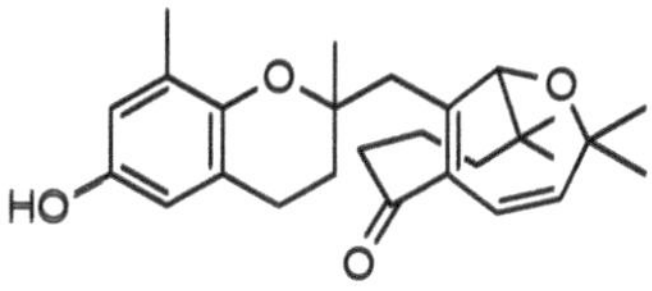

40

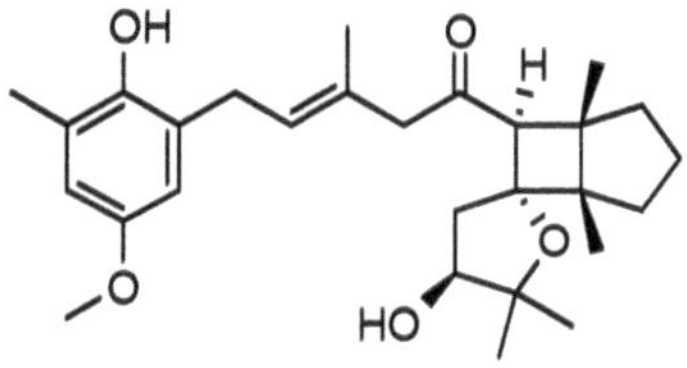

41

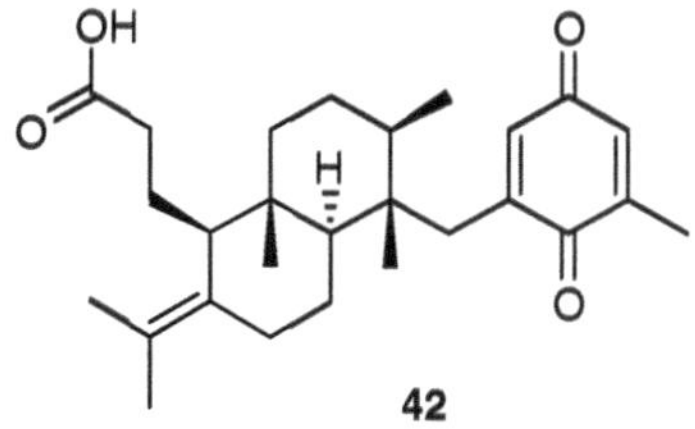

42

2.4.2 Phlorotannins

Phlorotannins are polyphenols based on polymerization of phloroglucinol (1,3,5-trihydroxybenzene) which are exclusively found in brown algae. These compounds are predominantly stored within special vesicles called physodes and have been reported to aid in wound healing, herbivore deterrence, defense against microbial infection, metal ion chelation, UV protection and allelopathic activities [60–64]. Phlorotannins can be divided into six distinct groups based on the polymerization patterns of the phloroglucinol subunits. These include fucols (e.g., trifucol **43**; [65]), phlorethols (e.g., triphlorethol **44**; [66]), eckols (e.g., eckol **45**; [67]), fuhalols (e.g., trifuhalol B **46**; [68]), isofuhalols, and fucophlorethols (e.g., fucophlorethol B **47**; [69]), though different polymerization patterns may be found in some of the larger polymers [1]. As with many marine compounds, halogenation is not uncommon (e.g., chlorodifucol **48**; [70, 71]). Unlike other natural products produced by this macroalgal group, polyphenolic compounds are found in higher concentrations in algae from temperate or polar waters than in tropical water [72, 73]. Ecological functions of phlorotannins are discussed in more detail in Subheading 3.

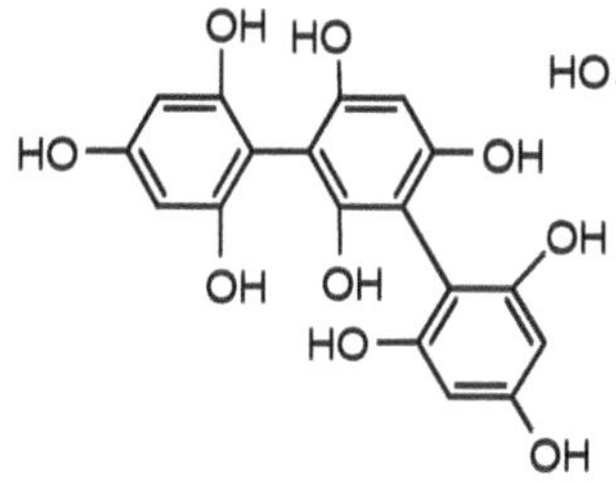

43

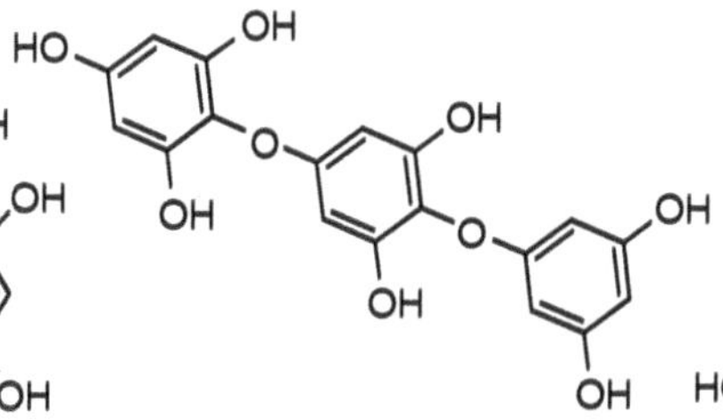

44

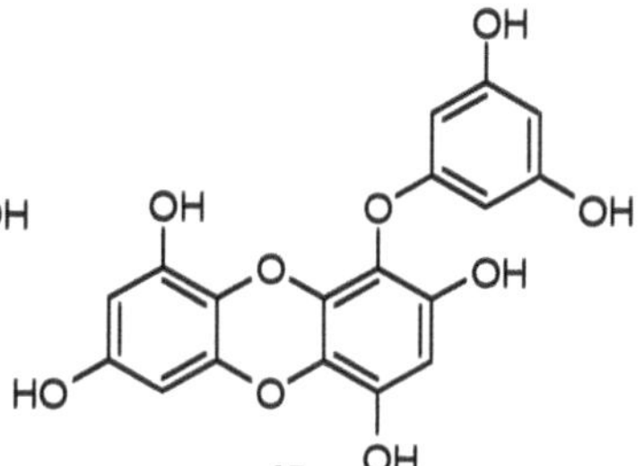

45

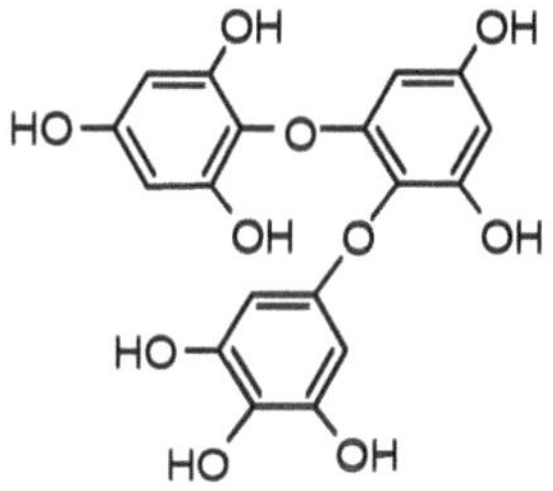

46

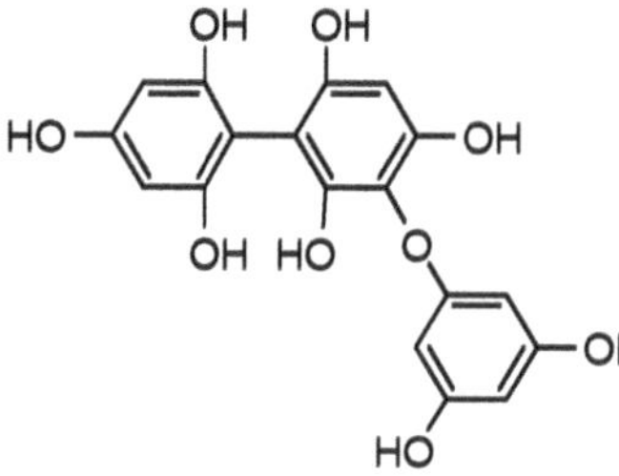

47

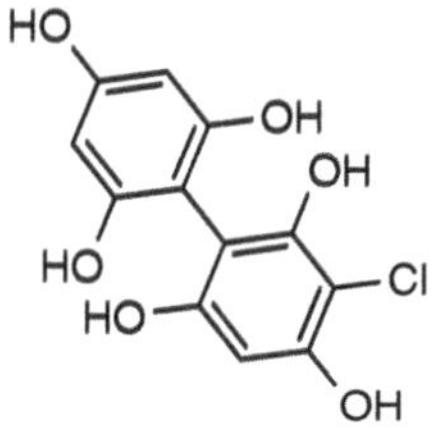

48

2.4.3 Bromophenols

A recent review described over 60 bromophenols, along with their biomedical activity [74]. This group of shikimate-derived metabolites is found in all three macroalgal groups. Examples include diphenyl methylene [**49**] [75] and the structurally similar dibenzyl ether [**50**] [76]. While most structures are simple variations on mono- or di-phenyl ring systems, occasional highly oxidized and/or cyclized members, such as colensolide A [**51**] [77], reflect the diversity of macroalgal chemistry described in other sections discussed in this review.

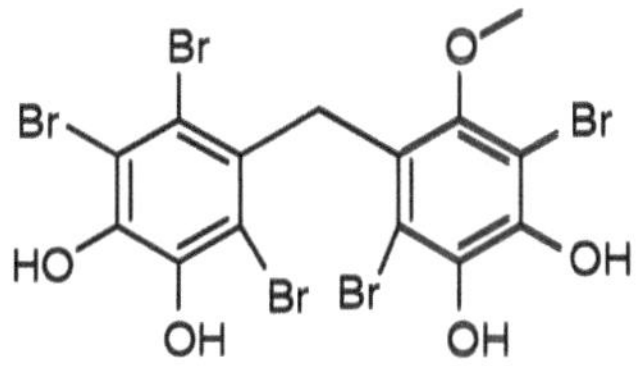

49

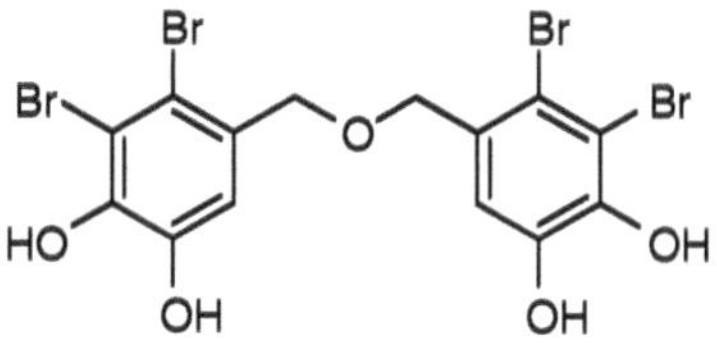

50

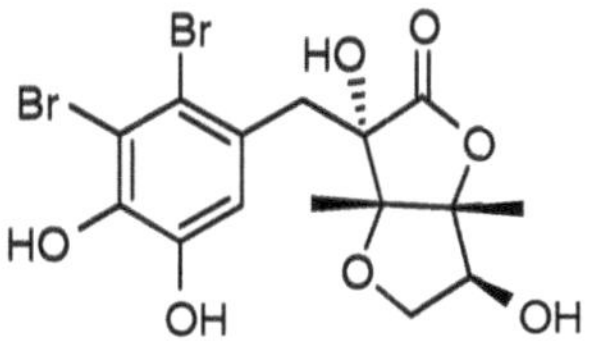

51

2.5 Nitrogenous Macroalgal Natural Products

Nitrogen-bearing metabolites are rare in eukaryotic macroalgae: perhaps 3–5 % of all macroalgal natural products. Peptide natural products are oligomers based on condensation of amino acids. Such oligomers often bear features not found amongst traditional peptides and proteins, such as incorporation of D-amino acids, hydroxylations, dehydrations, and ester linkages [1]. These compounds were found to be uncoupled from ribosomal biosynthesis and lack requirement for mRNA. Their biosynthetic enzymes, non-ribosomal peptide synthetases (NRPS), resemble polyketide and terpene synthases in their modular arrangement and lack of fidelity to substrate, which often results in mixed-biosynthetic products, such as NRPS–PKS hybrids [78]. Only two macroalgae, possibly three, have been reported with peptide chemistry.

Alkaloids and mycosporin-like amino acids (MAAs) are the other noteworthy groups of nitrogen-bearing natural products. Alkaloids are about as rare in macroalgae as peptides. What distinguishes alkaloids is a basic nitrogen functional group, and in a strict classical sense they are derived from shikimate pathway amino acids (tryptophan, tyrosine, phenylalanine), though the term is now more often used loosely to describe natural products with amine functional groups [6]. MAAs are widely distributed in the marine realm [79]. They are characterized by a cyclohexenimine ring, often bearing an amino acid as the enamine portion of the molecule, and approximately 20 different MAAs have been characterized [80]. They are widely regarded as serving the producing organisms as UV protectants, efficiently absorbing biologically detrimental UV light from 310 to 360 nm (e.g., [81]). They are discussed in more detail in Subheading 3.3.

2.5.1 Macroalgal Peptides

Amongst the three groups of macroalgal peptides, the kahalalide series of depsipeptides, e.g., kahalalide F [**52**] [82, 83], were originally found from the mollusc *Elysia rufescens*. Further investigation has shown that the molluscs sequester the kahalalides from their green algal diet *Bryopsis* sp. Kahalalide F especially has potent activity against a panel of human prostate and breast cancer [84]. After the successful completion of phase I clinical trial [85] **52** was entered in phase II trials [86]. Galaxamide [**53**], from the green alga *Galaxaura filamentosa*, also displays impressive cytotoxicity, with in vitro antiproliferative activity toward renal and liver lines [87]. The third peptide presented here, **54**, is a member of the ceratospongamide series of cyclic peptides which originate from a red alga–sponge symbiosis [88]. It is not clear whether the sponge or the alga is the source of the peptide.

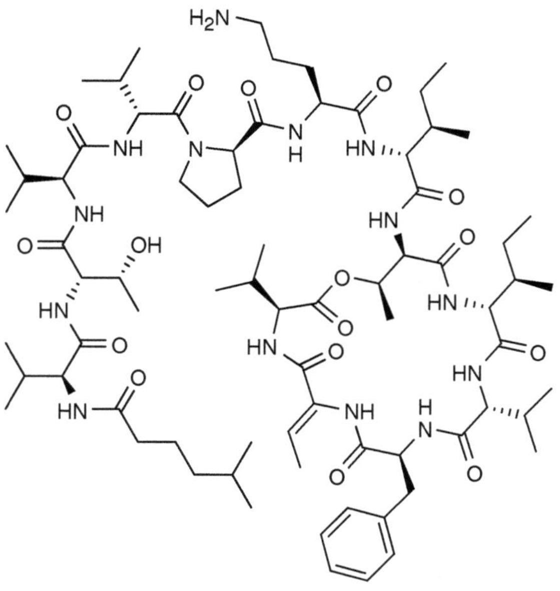

52

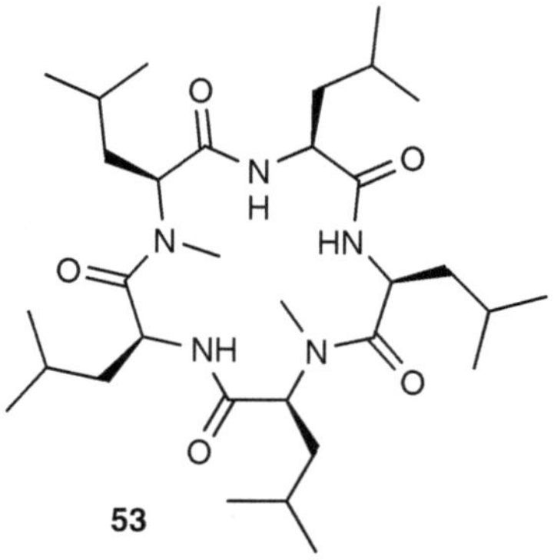

53

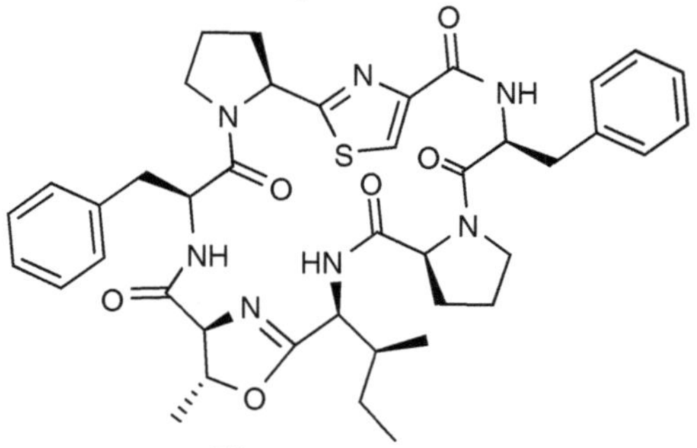

54

2.5.2 Alkaloids and MAAs

Red and green macroalgae have been the source of several alkaloids which remain nonexistent among brown algae [1]. The oxazolylindole ring system of martefragin A [**55**] is rare in nature and most commonly associated with *Streptomycetes* spp. [89]. In [**55**], from the red alga *Martensia fragilis*, inhibition of lipid peroxidation may well have a protective effect on the producing alga [90]. The oxazoyl ring itself is unusual for its 2,5-disubstitution pattern, likely arising from oxidation of a dipeptide not unlike almazole C [**56**] from the red alga *Haraldiophyllum* sp. [91]. Neurotoxins from the kainic and domoic acid families of alkaloids, e.g., domoilactone A [**57**] [92], are broadly distributed amongst Rhodophytes, though some site specificity has been noted [93]. A family of red pigments has been obtained from the green alga *Caulerpa* spp., including caulerpin [**58**] [94] and racemosins [95].

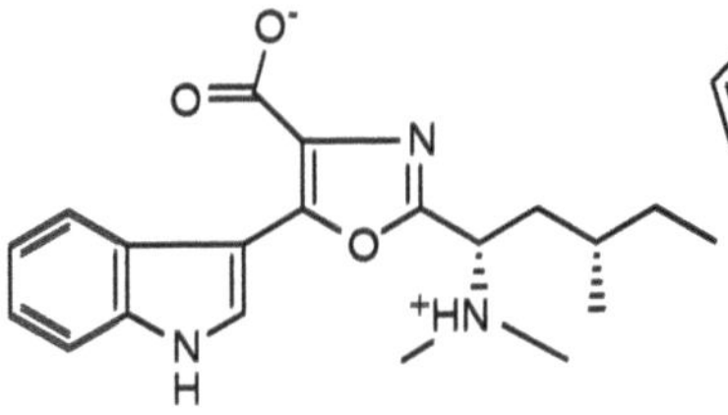

55

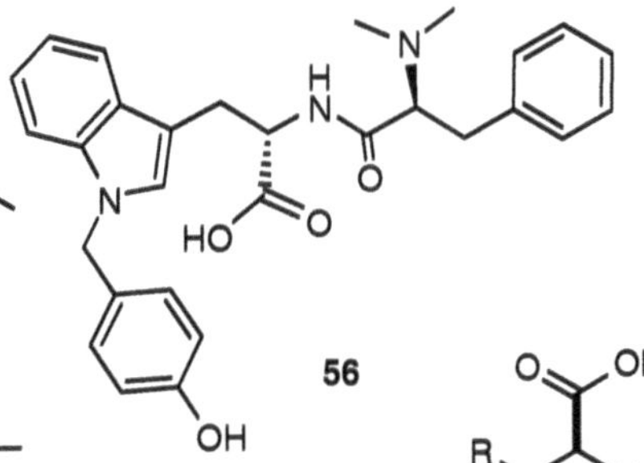

56

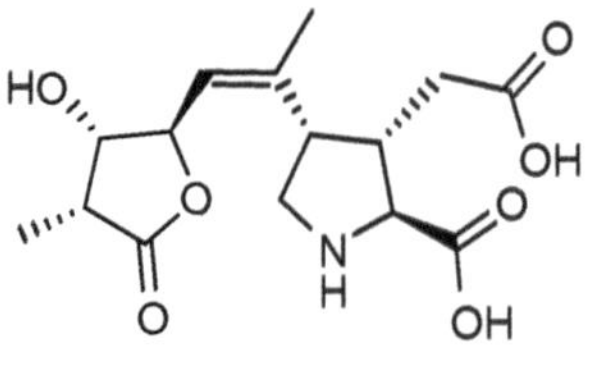

57

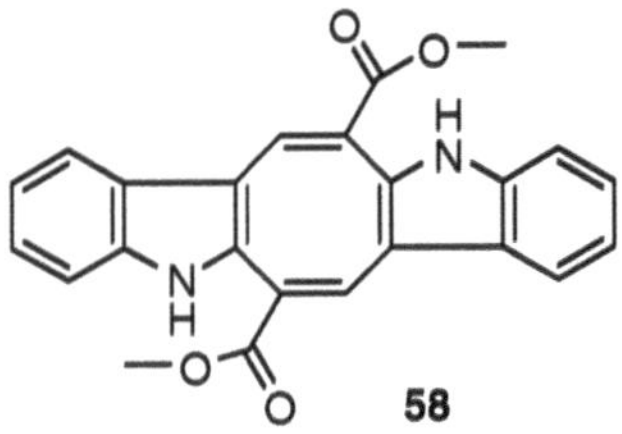

58

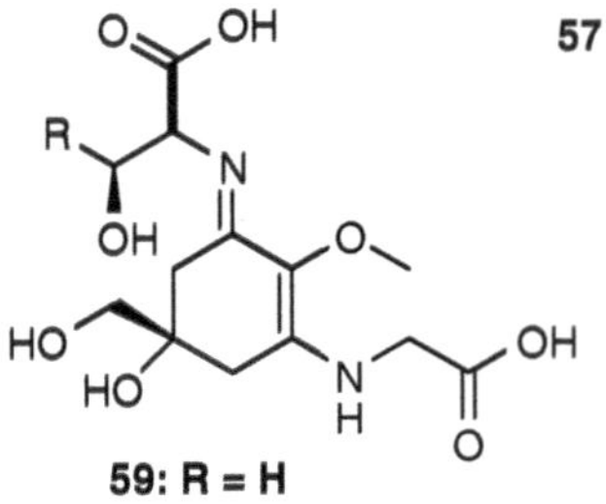

59: R = H
60: R = CH_3

Red algae were amongst the first organisms from which MAAs were reported, though now 87 % of marine taxa analyzed for MAAs have shown detectable levels [96]. Porphyra-334 [**59**] [97] and shinorine [**60**] [98], are examples of MAAs first described from macroalgae (rhodophytes *Porphyra tenera* and *Chondrus yendoi*, respectively). Garcia-Pichel et al. [99] estimated that MAAs absorb up to 30 % of photons from UVR, thus preventing this radiation from reaching vulnerable cytoplasmic elements. In two related studies, the photostability of this class of compounds was shown under a variety of conditions [100, 101]. *See* Subheading 3.3 for discussion of MAAs as biological sunscreens.

3 Ecological Roles of Algal Metabolites

3.1 Geographical Variation in Algal Chemical Ecology

Variability in algal defenses may be ascribed to many factors including geographic differences in the presence or importance of biotic and abiotic factors which act as selective forces on the evolution of secondary metabolites. The bioactivity of a secondary metabolite in ecological interactions depends on its structure, which is shaped at least in part by the environmental influences affecting metabolism [102]. In this manner each environment uniquely molds the chemical diversity within the local organisms. Algae with broad geographic ranges may change concentration or composition of secondary metabolites in different habitats (e.g., the brown alga *Stypopodium zonale*) [103–105]. More commonly, large scale patterns are illusive due to the number of local factors that are likely to differentially select for secondary metabolite variation within and between algal classes [102, 106]. Latitudinal clines have been identified in some compounds. For example dimethylsulfonioproprionate (DMSP) concentration in green algae decreases with lower latitudes [107].

Phlorotannin concentrations also vary with latitude in Indo-Pacific brown algae, but this is not observed in other regions [63]. Tropical species often have greater concentrations and more diversity of lipophilic compounds compared to temperate seaweeds, but temperate species have the largest diversity of compounds overall [108, 109].

The flora of tropical reefs is often dominated by green and red algae. While diversity in marine fauna reaches a maximum on these reefs, floral diversity does not [109]. However, the chemical ecology of tropical seaweeds has been well studied [109]. Faunal assemblages have a large impact on distribution and abundance of algae on tropical reefs [110–113], and many herbivore species are pantropical [109]. Perhaps because of broad species distributions, there are few specific herbivore–algae relationships on tropical reefs but herbivorous fishes may drive selection for secondary metabolites in reef algae [114]. There is some thought that tropical algae are better defended than their temperate counterparts [115]. Compounds from this region are structurally and biochemically diverse; predominantly terpenoids, acetogenins, and metabolites of mixed biosynthetic origin (meroterpenoids) [3]. A meta-analysis of metabolite diversity shows it peaks on temperate reefs, corresponding with algal diversity, but relatively fewer studies have examined algal secondary metabolite roles in tropical seagrass beds, mangrove forests, and rocky shores, which may be rich sources of additional compounds [109].

The large diversity of algae on temperate reefs is generated by small scale heterogeneity of habitat, seasonal growing conditions, depth gradients, regulation through trophic interaction and frequent disturbances which can counteract competitive exclusion [116]. In some parts of the world these reefs are dominated by brown algae in the Order Laminariales (commonly called kelps) which can form large forests along coastlines. Defenses in temperate algae may shape community structure, as herbivory in these reefs can wipe out local populations when unregulated [117]. One focus of algal chemical ecology research on these reefs is studies of the patterns of algal defenses, often focused on phlorotannin [**43–48**] concentrations [116].

Algal diversity decreases in polar habitats, but biomass levels are still high, and in some parts of Antarctica they can mirror those seen in temperate kelp forests [118, 119]. Far more secondary metabolites have been identified from Antarctic macroalgae [120] than from the Arctic [102], although 18 additional secondary metabolites are known from subarctic regions of the North Atlantic and North Pacific Oceans [3, 121]. Chemical defenses may play a role in observed unpalatability in only 2 of 17 Arctic macroalgae assayed at Spitsbergen, Norway [122]. By contrast, a majority of the macroalgal species tested in Antarctica are unpalatable to sympatric herbivores and chemical defenses have been shown to be widespread [118, 123–127]. Known secondary metabolites from

Antarctic macroalgae include MAAs (e.g., **59**, **60**) which defend against UV radiation [128] and a variety of volatile halogenated organic compounds (VHOCs; [129, 130]) for which ecological roles in Antarctica have not been established. Secondary metabolites responsible for the widespread unpalatability of Antarctic macroalgae include the halogenated monoterpenes anverene [**1**] and epi-plocamene D [**2**] which deter sympatric amphipod feeding [131] on the red alga *Plocamium cartilagineum*, and several furanones (e.g., **33**) produced by Antarctic *Delisea pulchra* [131] which are known to deter feeding in temperate herbivores [132]. The brown alga *Desmarestia menziesii* elaborates the meroterpenoid menzoquinone which deters sea star herbivores [131].

3.2 Sensory Chemical Ecology

Sensory communication between organisms, and organisms chemically sensing their environments, is a well-developed field of chemical ecology in terrestrial ecosystems. Communication involving marine algae remains relatively unexplored with the exception of a few well studied examples which include sensory communication in algal gametes [37, 63, 133–135] and between macroalgal spores and bacterial biofilms [136–138]. A commonly studied aspect of sensory chemical ecology in terrestrial and freshwater environments is alarm cues. These have rarely been observed in marine algal community processes. Examples come from two species of intertidal brown algae, *Ascophyllum nodosum* and *Fucus vesiculosus*. *A. nodosum* produces defenses in the presence of grazing cues [139, 140], and, in turn, snail grazing attracts its fish and crab predators [141]. In *F. vesiculosus*, waterborne cues increase neighbors' defenses when an individual is being consumed by an isopod herbivore [142].

Pheromones involved in communication between male and female gametes are present in both marine brown algae and freshwater green algae [63, 116, 143, 144]. There are 12 known C_8 or C_{11} hydrocarbon pheromones produced by female gametes in different groups of brown algae [145]. These include ectocarpene [**27**] and pre-ectocarpene [**28**] which is released by female gametes of *Ectocarpus siliculosus* to attract male gametes via a complex chemokinetic behavior and lamoxirene, which is released by female gametes of *Laminaria digitata* to stimulate release of male gametes and attract them via a direct chemotactic behavior [145]. Some of the same brown algal pheromones are produced by mature algae and they or their degradation products can also function as feeding deterrents [146].

Algal spores are also known to sense their chemical and/or physicochemical environments during the process of selecting settlement sites. Kelp spores can exhibit chemotaxis in response to nutrients (swimming either towards beneficial concentrations or away from detrimental levels; [120, 147]), and can also increase settlement rates in the presence of beneficial nutrients [148].

Surface hydrophobicity can be an important cue during settlement of both brown and green algal spores [149–154]. Chemical communication between bacterial biofilms and macroalgae affects spore settlement in *Ulva* spp. [136] as do chemoattractants like fatty acids [155]. Green algal spores can "eavesdrop" on *N*-acyl homoserine lactones (AHLs) released by bacteria as signaling molecules and are much more likely to settle when AHLs are present [137, 138]. The spores reduce swimming speed when close to AHL source if they are in contact with the surface [156]. AHLs produced by bacteria that grow on the surface of *Gracilaria chilensis* can stimulate spore release in the red alga *Acrochaetium* sp. which grows epiphytically on *G. chilensis* [157, 158].

3.3 Defensive Chemical Ecology

The line between sensory and defensive chemical ecology can be muddled [159, 160]. In this section we will discuss defenses against environmental stressors, biofoulers, pathogens, and herbivores. Defense systems arose early in the evolution of eukaryotes, and are conserved in the algae [157]. Innate immune systems are found in eukaryotes from algae to mammals, and include signaling cascades involving oxidative metabolism of polyunsaturated fatty acids in response to external stimuli such as pathogen or microbe associated molecular processes [161], leading to an oxidative burst response [162]. Algae that are known to elicit oxidative bursts, or reactive oxygen species (ROS), in response to pathogen cues include brown algal species in the Laminariales, Desmarestiales, Ectocarpales, and Fucales [163] and six red algal species within the Gracilariaceae [164]. Macroalgae can also produce ROS in response to wounding, including wounding caused by grazers [165]. Eleven of thirteen species of brown and red Antarctic macroalgae released ROS within a minute of wounding, accumulated ROS in tissues at the wound site within 70 min, or both [165]. It is likely that these responses are common in algae, perhaps in wounding signal transduction as in the higher plants [162]. UV exposure, excess light, or temperature change can induce oxidative stress which is associated with the production of volatile halogenated organic compounds (VHOCs) that might act as defense against biofoulers, pathogens, or herbivores [162]. It has also been shown that some species synthesize UV-absorbing compounds after an oxidative burst in response to infection [162].

Light is an essential resource for all algae, but, depending on habitat, photosynthetically active radiation (PAR) is paired with potentially harmful ultraviolet radiation (UVR), including UV-A (400–315 nm) and UV-B (315–280 nm). UVR impacts macroalgal physiology and, therefore, can influence ecological interactions [166]. Compounds that absorb UV are found in algae over 20 μm in diameter [167] because while UV-A may stimulate compound production and repair processes in organisms occasionally initiating harmful secondary photoreactions [168], radiation from UV-B can be damaging. Damage to DNA and proteins directly [169] or through the production of toxic ROS such as superoxide ($O_2^{\bullet -}$), hydroxyl radical ($^{\bullet}OH^-$) or hydrogen peroxide (H_2O_2) ultimately results in oxidative stress of the cell [170–172]. The detrimental effects of UVR induced ROS in cyanobacteria include DNA, protein and lipid damage, photosynthesis inhibition, delay in cell differentiation, motility impairment, lower growth rates, and decreased survivability [173]. High energetic UV-B radiation possesses the greatest potential for the direct damage to DNA and proteins, as well as the production of ROS [174]. MAAs are UV-protectants in red algae, where production is induced by low levels of UV-A and PAR [175]. Phlorotannins are UV-protectants in brown algae, and increase in concentration with increased UVR in some species [139, 176].

Biofouling, or overgrowth by other organisms, is likely to occur for everything immersed in water [176, 177], and may be detrimental to the basal (host) species [178, 179]. Epibiota compete with hosts for basic resources, can decrease growth and reproduction [180, 181], attract grazers [182], increase drag [183] and rigidity, and decrease buoyancy of the thallus [184]. In order to be effective, secondary metabolites or other antifouling compounds need to be present in biologically effective concentrations on the exterior of the alga where fouling organisms would make contact with the basal species, and the compounds must be present in biologically effective concentrations [185]. Correspondingly, algal structures that are either known or hypothesized to produce deterrent compounds can be found on the exterior of thalli. These include corps en cerise in the outer cell layers of *Laurencia obtusa* [186] and gland cells in *Delisea pulchra* and *Asparagopsis armata* [187, 188]. Physodes in brown algae occur most often in the outer cells and may contain compounds with antifouling functions [63]. Antifouling compounds may be polar (constantly exuded into surrounding waters) or lipophilic and may have other functions including defense against pathogens or herbivores. For example *Dictyota menstrualis* produces diterpenoids pachydictyol A [**61**] and dictyol E [**62**] at the thallus surface which inhibit bryozoan settlement at natural concentrations in addition to having anti-herbivore roles [189]. *Ulva reticulata* releases a water soluble antifouling compound which inhibits bryozoan and hydroid larval settlement [190]. Spores of the epi/endophytic filamentous brown alga, *Elachista antarctica* are repelled by and/or have their germination inhibited by compounds produced by sympatric red macrophytes with the exception of its only known host in nature, *Palmaria decipiens* [191].

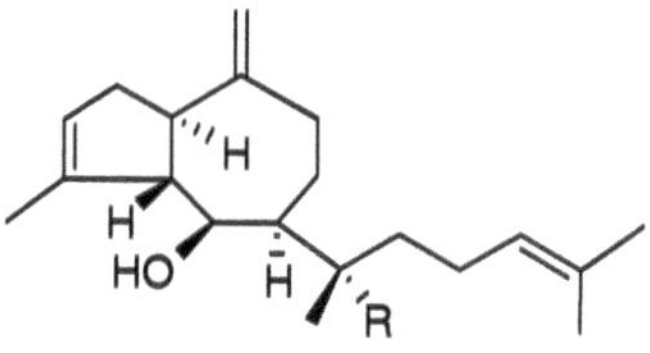

61: R = H
62: R = OH

Pathogens are a threat to macroalgae and resultant physiological responses may be similar to those against biofouling. As with antifouling defenses, defenses against pathogens need to be present on the exterior of thalli or at wound sites. Macroalgal defenses against pathogens have been the subject of several reviews [161, 162, 192–194]. Examples include a tropical brown alga, *Lobophora variegata*, which produces lobophoride (**30**, a polycyclic macrolide) that inhibits fungal settlement and other pathogens [38] and the red alga *Callophycus serratus* which produces diterpene alcohols to inhibit fungal growth [195]. *Delisea pulchra* produces furanones (e.g., **33**) which function as both anti-pathogen and antifouling compounds by interfering with AHL signaling in bacteria and inhibit propagule attachment [159, 187, 196].

In contrast to "inhibitors," some algae produce "inducers" or settlement cues [185]. From a previous example, *Palmaria decipiens* exudes a soluble compound that attracts spores of its epi/endophyte *Elachista antarctica* and induces them to settle on it [191]. Most inducers are primary metabolites: water soluble, or large insoluble macromolecules with soluble components that act as functional cues [197]. A well-studied example of settlement inducers are chemical cues from tropical crustose coralline algae and/or associated bacteria which increase recruitment of scleractinian coral larvae [198–201].

The primary function of algal secondary metabolites is often assumed to be defense against herbivory [160]. Chemical defense can be effective but presumably requires energy or other resources that may be limited. Defenses may be produced constitutively or their production induced only when needed, and they may be produced profusely or localized to specific areas of the thallus. However, location of defenses does not always infer the cause of defense. Different aspects of defense occur on differing time scales and short-term vs. long-term responses may have different adaptive advantages. Short-term responses such as ROS, early signaling cascades, and gene transcription and translation will take place far before secondary metabolite synthesis, and trends in these defenses specifically need more attention [102]. Long-term defensive responses such as secondary metabolite synthesis, transport, bioaccumulation, and secretion can be initiated by quick responses [102, 202], highlighting the significance of classifying short-term responses.

Chemical defense theories allow researchers to test hypotheses in a broader context of the community and explain intra- or interspecies variation in defense allocation over time, space, or within/between populations. Many models exist: the Optimal Defense Theory (ODT), the Growth-Differentiation Balance Hypothesis and the Carbon-Nutrient Balance Hypothesis all focus on intraspecific variation, while the Resource Availability Model and the Plant Apparency Model focus on interspecific variation in more of an evolutionary context [63, 160]. Each of these theories makes assumptions about trade-offs and causality for chemical defense which are not discussed here (*see* [160]), but it should be noted that theories are not mutually exclusive [160]. The ODT is the most thoroughly studied defense model in marine macroalgae, and assumes that chemical defenses come at a cost to other functions (growth, reproduction, etc.) within an organism; therefore defenses should be energetically optimized (in space and time) to avoid impacting fitness. The compounds most often studied under this model are phlorotannins [**43**-**48**] [160]. The Induced Defense Model is commonly considered a corollary of ODT [63, 160, 203]. Induced defenses are only made once an alga has come under attack and consequently are presumed to have lower costs than defenses that are produced constitutively [204]. In addition, overall cost of defense may be lowered because it is not induced throughout a whole thallus [205]. However, induced defenses are likely to be selected for only in response to relatively small herbivores that do not quickly consume the entire algal prey and only when herbivory is variable in space and/or time [206]. If herbivory is common and/or is executed by larger herbivores which would rapidly consume the host, constitutive production of defenses would be advantageous. Induced defenses in algae were first reported in *Fucus distichus* where phlorotannin production was induced in response to natural and artificial grazing [207]. The brown algae remain the group in which induced defenses have most commonly been studied.

Activated defenses are a hybrid of these two defense strategies. In an activated defense, precursor compounds that are less toxic or otherwise bioactive and/or more stable than the final defense compound are produced constitutively and then only activated to the more bioactive form when the alga is under attack. Examples of this strategy are found in tropical siphonous green algae (the calcified *Halimeda* and fleshy rhizoidal, often invasive, *Caulerpa*) where the diterpene halimedatetraacetate [**63**], bearing what is referred to as the masked dialdehyde (e.g., **16**) function, is transformed by hydrolysis of those acetate groups to halimedatrial [**64**] [208–210]. Species that contain DMSP also exhibit activated defenses mechanisms. The green algae *Ulva* spp. produce dimethylsulfide and acrylic acid from DMSP on wounding and these compounds act as feeding deterrents [211].

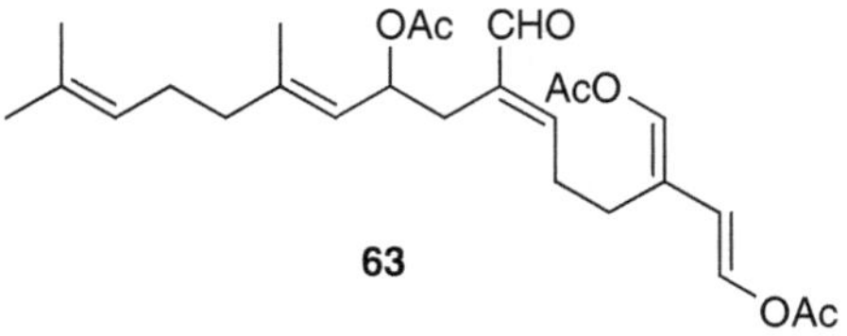

63

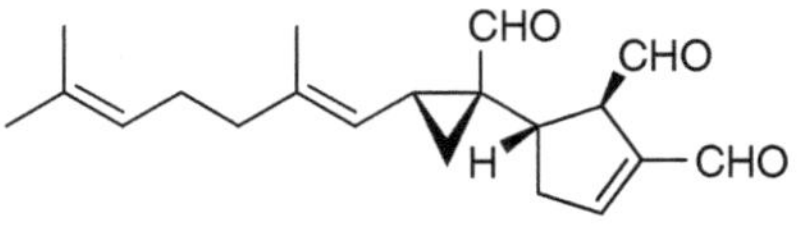

64

3.4 Allelopathy

Allelopathy between algae and sympatric organisms is also called interference competition [212–215]. This form of competition is common in freshwater systems [157, 215], can be extremely intense in all phyla of benthic marine organisms [116], and is present amongst the phytoplankton [132]. Allelochemicals include those that have a negative impact on ecological interactions, as well as antifouling compounds [116]. Examples come from tropical and temperate reefs. On temperate reefs in the North Japan Sea a crustose coralline red algae *Lithophyllum* sp. exudes nonpolar substances that kill zoospores of the brown alga *Saccharina* (*Laminaria*) *religiosa* [214]. Multiple allelopathic chemicals from *Lithophyllum yessoense* inhibit spore settlement in 14 of 17 macroalgal species and spore germination in 13 of the 17 species [216]. This alga also inhibits *Ulva* spp. settlement and may dictate macroalgae distribution on the reefs [217]. On tropical reefs, a monoterpene from *Plocamium hamatum* causes tissue necrosis in soft corals upon physical contact [212], and corals lose symbionts when in contact with thalli or extracts of five sympatric algae [218]. An antibacterial metabolite, the polyketide neurymenolide A [**65**], has been identified as an allelopathic agent in the red alga *Phacelocarpus neurymenioides*, and desorption electrospray ionization mass spectrometry showed highest concentrations of the compound on basal portions of the algal blades where they would be most vulnerable [219]. Also on tropical reefs, a lipid soluble allelochemical from seven common reef algae causes bleaching, a decrease in photosynthetic efficiency, and even death in corals from the genus *Porites* [218]. This chemistry has further been identified as two derivatives of a nor-sesquiterpene loliolide, in the red alga *Galaxaura filamentosa* and two acetylated diterpenes from the green alga *Chlorodesmis fastigiata* [220].

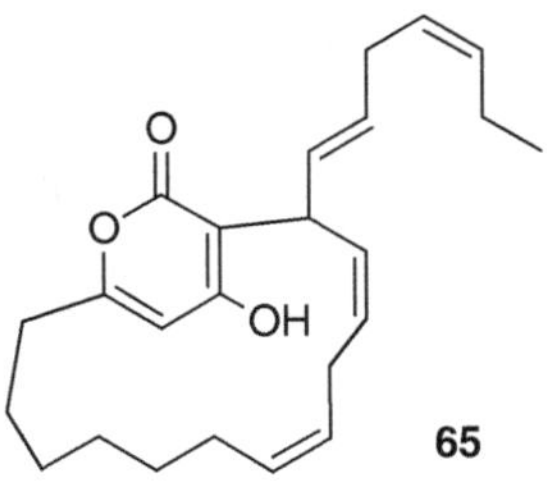

65

3.5 Trophic Interactions

Secondary metabolites may have indirect effects on trophic cascades [221] including carbon and nutrient cycling in the environment. Breakdown of chemically defended macroalgae once they enter detrital foodwebs may be slower than for undefended algae, because compounds selected for as other defenses in living algae may also deter bacterial community formation or metabolism [222], though that carbon does eventually become available to consumers [223]. An important example of secondary metabolite impacts on community dynamics comes from the urchin–sea otter–kelp trophic cascade in the temperate waters of the Northern Pacific Ocean. Algae that grow at depth tend to have more anti-herbivory defenses than shallow-water species, possibly because sea otter foraging is limited by diving time, and therefore they are removing the grazer population in the shallows but not at depth [224]. Kelp-dominated reefs without keystone predators like the sea otter have floras with high phenolic content, like Southwestern Australia [224]. Algae-produced neurotoxins can also function in keystone roles, for example in red tide events which either cause mass die offs of sea urchins which then allow macroalgae to flourish [225], or impact sea otters, releasing benthic invertebrates from predation [226, 227].

Herbivores may benefit from macroalgal chemical defenses if they can sequester them for their own use. Until recently, this had only been observed in Opisthobranches [116, 228, 229]. These molluscs sequester secondary metabolites which may then be modified and elaborated further at higher trophic levels [109, 230]. Their prey include sponges, red, green, and brown algae, and cyanobacteria whose chemistry includes mono-, di-, tri-, sesquiterpenes, nitrogenous compounds, phenolics, halogenated furanones, and steroids [231, 232]. Recently, the Antarctic amphipod *Paradexamine fissicauda* has been shown to consume the heavily defended red alga *Plocamium cartilagineum* and to sequester halogenated defensive metabolites (e.g., **2**, **3**) from the alga for use as defenses against predatory fish [32].

Behavioral sequestration, or associational defense, is another way that other organisms may benefit from algal secondary metabolites. Chemically defended algae can be biodiversity centers or "microsatellites of defense" [222] on tropical reefs where they can increase species richness [109, 114, 233]. Invertebrates can behaviorally exploit algal defensive chemistry by using the algae as a refuge or covering [234–237]. Specific examples of behavioral sequestration include an amphipod, *Pseudamphithoides incurvari*, which encloses itself in the defended brown alga *Dictyota bartayresiana* (*D. bartayresii*) as a protection from predation on Caribbean reefs [234], and sea urchins which cover themselves in red algae as a defense against their anemone predators in McMurdo Sound, Antarctica [236].

3.6 Biotic Interactions

Secondary metabolite chemistry of an alga may also be modified through association with other organisms. For example, the green algae *Ulva* spp. becomes unpalatable to sympatric gastropods when infected with the ascomycete fungus *Turgidosculum* sp. [238]. Similar relationships are often seen between grass and fungal endophytes in terrestrial systems [239–241]. Approximately one-third of described marine fungi are associated with algae [242], with associations ranging from parasitic to facultative to obligate. However, coevolution of these associations is understudied, as is the impact of fungal symbionts on the chemical ecology and secondary metabolites of the host algae.

3.7 Life History Stages

A variety of patterns in alternation between sporophyte and gametophyte stages occurs in the life histories of algae, and may have evolved for a variety of reasons. Selection pressures for maintenance of complex life histories, such as heteromorphic stages, include herbivory and environmental stress [117]. Some isomorphic species have more subtle differences between sexual stages, including secondary metabolite chemistry [243, 244]. In *Asparagopsis armata*, the female gametophytes (and the carposporophytes stage they support) have high levels of anti-herbivore defense in cystocarp walls and carpospores, while the male gametophyte has low levels of defense, but high nutrient content and is readily consumed by its herbivore *Aplysia parvula* [244]. Another example is the rhodophyte *Mazzaella flaccida* where gametophytes are more palatable than sporophytes; reproductive tissues are consumed over non-reproductive tissue due to water soluble chemicals [245].

3.8 Anthropogenic Impacts

Anthropogenic impacts to the marine environment include global warming, decrease in ozone layer (increased UVR light), ocean acidification (increased pCO_2) and changes in water chemistry, change in ocean circulation patterns, sea level rise, and increased runoff [246]. Research to date shows mixed effects of climate change conditions on a variety of macroalgae [247], which is true too for marine invertebrates [248]. Most studies are limited in scope and few encompass impacts on community dynamics that would reveal shifts in chemical ecology of the organisms. One study shows that allelopathy between corals and tropical algae is enhanced in climate change conditions [249]. Other anthropogenic effects include changes in seaweed distribution through human transport and shifts in physical parameters such as nutrification [250]. Invasive species like *Caulerpa* spp. or the red alga *Bonnemaisonia hamifera*, which are chemically defended, may have an advantage over indigenous species because they produce novel weapons against herbivory in their new habitat (Novel Weapons Hypothesis; [251]), and they have the potential to alter structure and function of the invaded ecosystems [252–254].

Caulerpa spp. specifically produces caulerpenyne [**66**] which can be toxic to fish and is unpalatable to urchins [255, 256], and *B. hamifera* produces a 1,1,3,3-tetrabromo-2-heptanone that deters herbivores [251] and provides an association refuge for generalist herbivores in the habitat [253]. These species are less consumed than native species which increases their success in invading new habitats.

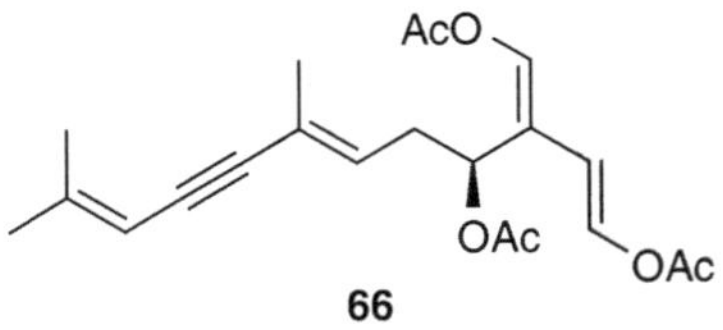

66

4 Summary and Conclusion

Macroalgae produce a wide variety of both simple and complex natural products, many of which have been found with a functional role in ecology of the producing organism. We have tried to highlight here how the phaeophytes, rhodophytes, and chlorophytes display distinct patterns of natural product diversity and use natural products in distinct roles. More importantly, these are exciting times, as advances in technology make the study of both the chemistry and ecology more sensitive and accurate. Taken with a rich accumulation of distribution data, both biological and chemical, from several decades of concerted study, we have a solid basis for data mining to help focus our advanced techniques.

Historically humans have utilized macroalgae both as a food source and medicinally (e.g., [257, 258]), because much like terrestrial plants, they are relatively easy to access and recognize. Since the initial discovery of marine algal products in the 1950s and 1960s, there has been a rapid increase in the description of new algal natural products [2]. On a practical level, these metabolites have value as potential drugs or drug leads [259]. With the emergence of drug resistance [260] to our current arsenal of therapeutics as well as the need for development of new chemotherapies for currently untreatable diseases [261], new and diverse chemical entities are essential for the protection of human health [262]. However, in the context of global climate change, perhaps the strength of studies such as those described herein lies in fostering understanding of our environment. There is little doubt that natural products influence species distribution in a dynamic fashion, and that species distributions in turn influence natural product diversity (e.g., [263]). Thus, it is studies such as these that enable us to know when the environment is changing but also to know why, so that we might take actions to remedy it. All the more reason the further study of structure and function in macroalgal natural products is timely.

Acknowledgements

Preparation of this chapter was supported in part by National Science Foundation awards ANT-0838773 and PLR-1341333 (C.D.A.) and awards ANT-0838776 and PLR-1341339 (B.J.B.) from the Antarctic Organisms and Ecosystems Program.

References

1. Blunt J, Munro MGH (2013) MarinLit. 0813 edn. Christchurch, NZ
2. Leal MC, Munro MH, Blunt JW et al (2013) Biogeography and biodiscovery hotspots of macroalgal marine natural products. Nat Prod Rep 30:1380–1390. doi:10.1039/c3np70057g
3. Blunt JW, Copp BR, Keyzers RA et al (2013) Marine natural products. Nat Prod Rep 30:237–323. doi:10.1039/c2np20112g
4. Stratmann K, Boland W, Müller DG (1992) Pheromones of marine brown algae—a new branch of the eicosanoid metabolism. Angew Chem Int Ed 31:1246–1248
5. Harper MK, Bugni TS, Copp BR et al (2001) Introduction to the chemical ecology of marine natural products. In: McClintock JB, Baker BJ (eds) Marine chemical ecology. CRC, Boca Raton, FL, pp 3–70
6. Maschek JA, Baker BJ (2009) The chemistry of algal secondary metabolism. In: Amsler CD (ed) Algal chemical ecology. Springer-Verlag, Berlin, pp 1–24
7. Fuller RW, Cardellina JH, Kato Y et al (1992) A pentahalogenated monoterpene from the red alga *Portieria hornemannii* produces a novel cytotoxicity profile against a diverse panel of human tumor cell lines. J Med Chem 35:3007–3011
8. Fuller RW, Cardellina JH, Jurek J et al (1994) Isolation and structure/activity features of halomon-related antitumor monoterpenes from the red alga *Portieria hornemannii*. J Med Chem 37:4407–4411
9. Sotokawa T, Noda T, Pi S et al (2000) A three-step synthesis of halomon. Angew Chem Int Ed 39:3430–3432
10. Jung ME, Parker MH (1997) Synthesis of several naturally occurring polyhalogenated monoterpenes of the halomon class. J Org Chem 62:7094–7095. doi:10.1021/Jo971371+
11. Konig GM, Wright AD (1997) *Laurencia rigida*: chemical investigations of its antifouling dichloromethane extract. J Nat Prod 60:967–970. doi:10.1021/Np970181r
12. Juagdan EG, Kalidindi R, Scheuer P (1997) Two new chamigranes from an Hawaiian red alga, *Laurencia cartilaginea*. Tetrahedron 53:521–528. doi:10.1016/S0040-4020(96)01002-2
13. Bansemir A, Just N, Michalik M et al (2004) Extracts and sesquiterpene derivatives from the red alga *Laurencia chondrioides* with antibacterial activity against fish and human pathogenic bacteria. Chem Biodivers 1:463–467. doi:10.1002/cbdv.200490039
14. Sims JJ, Lin GHY, Wing RM (1974) Marine natural products X elatol, a halogenated sesquiterpene alcohol from red alga *Laurencia elata*. Tetrahedron Lett 15:3487–3490
15. Vairappan CS (2003) Potent antibacterial activity of halogenated metabolites from Malaysian red algae, *Laurencia majuscula* (Rhodomelaceae, Ceramiales). Biomol Eng 20:255–259. doi:10.1016/S1389-0344(03)00067-4
16. Lhullier C, Donnangelo A, Caro M et al (2009) Isolation of elatol from *Laurencia microcladia* and its palatability to the sea urchin *Echinometra lucunter*. Biochem System Ecol 37:254–259. doi:10.1016/j.bse.2009.04.004
17. de Nys R, Leya T, Maximilien R et al (1996) The need for standardised broad scale bioassay testing: a case study using the red alga *Laurencia rigida*. Biofouling 10:213–224
18. Steinberg PD, de Nys R, Kjelleberg S (1998) Chemical inhibition of epibiota by Australian seaweeds. Biofouling 12:227–244
19. Viano Y, Bonhomme D, Ortalo-Magné A et al (2011) Dictyotadimer A, a new dissymmetric

bis-diterpene from a brown alga of the genus *Dictyota*. Tetrahedron Lett 52:1031–1035
20. Jiang RW, Lane AL, Mylacraine L et al (2008) Structures and absolute configurations of sulfate-conjugated triterpenoids including an antifungal chemical defense of the green macroalga *Tydemania expeditionis*. J Nat Prod 71:1616–1619. doi:10.1021/Np800307h
21. Kigoshi H, Shizuri Y, Niwa H et al (1981) Laurencenyne, a plausible precursor of various nonterpenoid C_{15}-compounds, and neolaurencenyne from the red alga *Laurencia okamurai*. Tetrahedron Lett 22:4729–4732
22. Butler A, Carter-Franklin JN (2004) The role of vanadium bromoperoxidase in the biosynthesis of halogenated marine natural products. Nat Prod Rep 21:180–188. doi:10.1039/b302337k
23. Fukuzawa A, Aye M, Nakamura M et al (1990) Biosynthetic formation of cyclic bromo-ethers initiated by lactoperoxidase. Chem Lett 19:1287–1290
24. Murai K (1999) Biosynthesis of cyclic bromoethers from red algae. In: Barton DHR, Meth-Cohn O, Nakanishi K (eds) Comprehensive natural product chemistry, vol. 1. Pergamon, Elmsford, NY, pp 303–324
25. Ciavatta ML, Gavagnin M, Puliti R et al (1997) Dactylallene: a novel dietary C_{15} bromoallene from the Atlantic Anaspidean mollusc *Aplysia dactylomela*. Tetrahedron 53:17343–17350
26. Suzuki M, Kurosawa E (1981) Okamurallene, a novel halogenated C_{15} metabolite from the red alga *Laurencia okamurai* Yamada. Tetrahedron Lett 22:3853–3856. doi:10.1016/S0040-4039(01)91327-9
27. Guella G, Chiasera G, Mancini I et al (1997) Twelve-membered O-bridged cyclic ethers of red seaweeds in the genus *Laurencia* exist in solution as slowly interconverting conformers. Chem-Eur J 3:1223–1231. doi:10.1002/chem.19970030809
28. Howard BM, Fenical W, Arnold EV et al (1979) Obtusin, a unique bromine-containing polycyclic ketal from the red marine alga *Laurencia obtusa*. Tetrahedron Lett 20:2841–2844
29. Ayyad SE, Al-Footy KO, Alarif WM et al (2011) Bioactive C_{15} acetogenins from the red alga *Laurencia obtusa*. Chem Pharm Bull 59:1294–1298
30. Braddock DC (2006) A hypothesis concerning the biosynthesis of the obtusallene family of marine natural products via electrophilic bromination. Org Lett 8:6055–6058. doi:10.1021/Ol062520q
31. Ginsburg DW, Paul VJ (2001) Chemical defenses in the sea hare *Aplysia parvula*: importance of diet and sequestration of algal secondary metabolites. Mar Ecol Prog Ser 215:261–274
32. Amsler MO, Amsler CD, von Salm JL et al (2013) Tolerance and sequestration of macroalgal chemical defenses by an Antarctic amphipod: a 'cheater' among mutualists. Mar Ecol Prog Ser 490:79–90
33. Stachowicz JJ (2001) Chemical ecology of mobile benthic invertebrates: predators and prey, allies and competitors. In: McClintock JB, Baker BJ (eds) Marine chemical ecology. CRC, Boca Raton, FL, pp 157–194
34. Müller DG, Jaenicke L, Donike M et al (1971) Sex attractant in a brown alga: chemical structure. Science 171:815–817
35. Boland W, Pohnert G, Maier I (1995) Pericyclic reactions in nature: spontaneous Cope rearrangement inactivates algae pheromones. Angew Chem Int Ed 34:1602–1604
36. Maier I, Pohnert G, Pantke-Böcker S et al (1996) Solid-phase microextraction and determination of the absolute configuration of the *Laminaria digitata* (Laminariales, Phaeophyceae) spermatozoid-releasing pheromone. Naturwissenschaften 83:378–379
37. Pohnert G, Boland W (2002) The oxylipin chemistry of attraction and defense in brown algae and diatoms. Nat Prod Rep 19:108–122
38. Kubanek J, Jensen PR, Keifer PA et al (2003) Seaweed resistance to microbial attack: a targeted chemical defense against marine fungi. Proc Natl Acad Sci U S A 100:6916–6921. doi:10.1073/pnas.1131855100
39. Yotsu-Yamashita M, Haddock RL, Yasumoto T (1993) Polycavernoside A: a novel glycosidic macrolide from the red alga *Polycavernosa tsudai* (*Gracilaria edulis*). J Am Chem Soc 115:1147–1148
40. Yotsu-Yamashita M, Yasumoto T, Yamada S et al (2004) Identification of polycavernoside A as the causative agent of the fatal food poisoning resulting from ingestion of the red alga *Gracilaria edulis* in the Philippines. Chem Res Toxicol 17:1265–1271. doi:10.1021/tx0498556
41. Francavilla M, Franchi M, Monteleone M et al (2013) The red seaweed *Gracilaria gracilis* as a multi products source. Mar Drugs 11:3754–3776. doi:10.3390/md11103754
42. Burreson BJ, Moore RE, Roller PP (1976) Volatile halogen compounds in the alga *Asparagopsis taxiformis* (Rhyodophyta). J Agric Food Chem 24:856–861
43. Paul C, Pohnert G (2011) Production and role of volatile halogenated compounds from marine algae. Nat Prod Rep 28:186–195

44. Giese B, Laturnus F, Adams FC et al (1999) Release of volatile iodinated C_1-C_4 hydrocarbons by marine macroalgae from various climate zones. Environ Sci Technol 33:2432–2439
45. Sugano M, Sato A, Nagaki H et al (1990) Aldose reductase inhibitors from the red alga, *Asparagopsis taxiformis*. Tetrahedron Lett 31:7015–7016
46. Rasmussen TB, Manefield M, Andersen JB et al (2000) How *Delisea pulchra* furanones affect quorum sensing and swarming motility in *Serratia liquefaciens* MG1. Microbiology 146:3237–3244
47. Manefield M, de Nys R, Kumar N et al (1999) Evidence that halogenated furanones from *Delisea pulchra* inhibit acylated homoserine lactone (AHL)-mediated gene expression by displacing the AHL signal from its receptor protein. Microbiology 145:283–291
48. Lorenzo M, Cueto M, San-Martin A et al (2005) Pyranosylmagellanicus a novel structural class of polyhalogenated acetogenins from *Ptilonia magellanica*. Tetrahedron 61:9550–9554
49. Higgs MD, Mulheirn LJ (1981) Hybridalactone, an unusual fatty acid metabolite from the red alga *Laurencia hybrida* (Rhodophyta, Rhodomelaceae). Tetrahedron 37:4259–4262. doi:10.1016/0040-4020(81)85020-X
50. Nagle DG, Gerwick WH (1994) Structure and stereochemistry of constanolactones A-G, lactonized cyclopropyl oxylipins from the red marine alga *Constantinea simplex*. J Org Chem 59:7227–7237
51. Corey EJ, De B, Ponder JW et al (1984) The stereochemistry and biosynthesis of hybridalactone, an eicosanoid from *Laurencia hybrida*. Tetrahedron Lett 25:1015–1018. doi:10.1016/S0040-4039(01)80088-5
52. Gerwick WH (1993) Carbocyclic oxylipins of marine origin. Chem Rev 93:1807–1823. doi:10.1021/Cr00021a008
53. Jang KH, Lee BH, Choi BW et al (2005) Chromenes from the brown alga *Sargassum siliquastrum*. J Nat Prod 68:716–723. doi:10.1021/np058003i
54. Dorta E, Darias J, San MA et al (2002) New prenylated bromoquinols from the green alga *Cymopolia barbata*. J Nat Prod 65:329–333
55. Zhang H, Xiao X, Conte MM et al (2012) Spiralisones A-D: acylphloroglucinol hemiketals from an Australian marine brown alga, *Zonaria spiralis*. Org Biomol Chem 10:9671–9676. doi:10.1039/c2ob26988k
56. Wisespongpand P, Kuniyoshi M (2003) Bioactive phloroglucinols from the brown alga *Zonaria diesingiana*. J Appl Phycol 15:225–228. doi:10.1023/A:1023831131735
57. Francisco C, Banaigs B, Rakba M et al (1986) Cystoseirols: novel rearranged diterpenoids of mixed biogenesis from Cystoseiraceae (brown marine algae). J Org Chem 51:2707–2711
58. Amico V, Cunsolo F, Piattelli M (1984) Balearone, a metabolite of the brown alga *Cystoseira balearica*. Tetrahedron 40: 1721–1725
59. Wessels M, Konig GM, Wright AD (1999) A new tyrosine kinase inhibitor from the marine brown alga *Stypopodium zonale*. J Nat Prod 62:927–930. doi:10.1021/np990010h
60. Sieburth JM, Conover JT (1965) *Sargassum* tannin, an antibiotic which retards fouling. Nature 208:52–53. doi:10.1038/208052a0
61. Lau SCK, Qian PY (1997) Phlorotannins and related compounds as larval settlement inhibitors of the tube-building polychaete *Hydroides elegans*. Mar Ecol Prog Ser 159:219–227. doi:10.3354/Meps159219
62. Targett NM, Arnold TM (1998) Predicting the effects of brown algal phlorotannins on marine herbivores in tropical and temperate oceans. J Phycol 34:195–205. doi:10.1046/j.1529-8817.1998.340195.x
63. Amsler CD, Fairhead VA (2006) Defensive and sensory chemical ecology of brown algae. Adv Bot Res 43:1–91. doi:10.1016/S0065-2296(05)43001-3
64. Pavia H, Cervin G, Lindgren A et al (1997) Effects of UV-B radiation and simulated herbivory on phlorotannins in the brown alga *Ascophyllum nodosum*. Mar Ecol Prog Ser 157:139–146
65. Geiselman JA, McConnell OJ (1981) Polyphenols in brown algae *Fucus vesiculosus* and *Ascophyllum nodosum*: chemical defenses against the marine herbivorous snail, *Littorina littorea*. J Chem Ecol 7:1115–1147
66. Glombitza KW, Zieprath G (1989) Antibiotics from algae. XXXIX. Phlorotannins from the brown alga *Analipus japonicus*. Planta Med 55:171–175
67. Fukuyama Y, Miura I, Kinzyo Z et al (1985) Eckols, novel phlorotannins with a dibenzo-p-dioxin skeleton possessing inhibitory effects on α2-macroglobulin from the brown alga *Ecklonia kurome* Okamura. Chem Lett 14:739–742
68. Glombitza KW, Forstera M, Eckhardt G (1978) Polyhydroxyphenyläther aus der phaeophycee *Sargassum muticum*. Phytochemistry 17:579–580
69. Glombitza KW, Grosse DJ (1985) Antibiotics from algae. XXXIII. Phlorotannins of the brown alga *Himanthalia elongata*. Planta Med 51:42–46

70. Glombitza KW, Schmidt A (1999) Nonhalogenated and halogenated phlorotannins from the brown alga *Carpophyllum angustifolium*. J Nat Prod 62:1238–1240
71. Ragan MA, Glombitza KW (1986) Phlorotannins, brown algal polyphenols. Prog Phycol Res 4:130–241
72. Toth GB, Pavia H (2006) Artificial wounding decreases plant biomass and shoot strength of the brown seaweed *Ascophyllum nodosum* (Fucales, Phaeophyceae). Mar Biol 148:1193–1199. doi:10.1007/s00227-005-0167-2
73. Fairhead VA, Amsler CD, McClintock JB et al (2006) Lack of defense or phlorotannin induction by UV radiation or mesograzers in *Desmarestia anceps* and *D. menziesii* (Phaeophyceae). J Phycol 42:1174–1183. doi:10.1111/j.1529-8817.2006.00283.x
74. Liu M, Hansen PE, Lin X (2011) Bromophenols in marine algae and their bioactivities. Mar Drugs 9:1273–1292. doi:10.3390/md9071273
75. Duan XJ, Li XM, Wang BG (2007) Highly brominated mono- and bis-phenols from the marine red alga *Symphyocladia latiuscula* with radical-scavenging activity. J Nat Prod 70:1210–1213. doi:10.1021/np070061b
76. Xu X, Song F, Wang S et al (2004) Dibenzyl bromophenols with diverse dimerization patterns from the brown alga *Leathesia nana*. J Nat Prod 67:1661–1666. doi:10.1021/np0400609
77. Popplewell WL, Northcote PT (2009) Colensolide A: a new nitrogenous bromophenol from the New Zealand marine red alga *Osmundaria colensoi*. Tetrahedron Lett 50:6814–6817
78. Walsh CT (2008) The chemical versatility of natural product assembly lines. Acc Chem Res 41:4–10. doi:10.1021/ar7000414
79. Shick JM, Dunlap WC (2002) Mycosporine-like amino acids and related gadusols: biosynthesis, accumulation, and UV-protective functions in aquatic organisms. Annu Rev Physiol 64:223–262. doi:10.1146/annurev.physiol.64.081501.155802
80. Bandaranayake WM (1998) Mycosporines: are they nature's sunscreens? Nat Prod Rep 15:159–172
81. Karentz D, McEuen FS, Land MC et al (1991) Survey of mycosporine-like amino acid compounds in Antarctic marine organisms: potential protection from ultraviolet exposure. Mar Biol 108:157–166
82. Hamann MT, Otto CS, Scheuer PJ et al (1996) Kahalalides: bioactive peptide from a marine mollusk *Elysia rufescens* and its algal diet *Bryopsis* sp. J Org Chem 61:6594–6600. doi:10.1021/Jo960877+
83. Hamann MT, Scheuer PJ (1993) Kahalalide F—a bioactive depsipeptide from the sacoglossan mollusk *Elysia rufescens* and the green alga *Bryopsis* sp. J Am Chem Soc 115:5825–5826. doi:10.1021/Ja00066a061
84. Suarez Y, Gonzalez L, Cuadrado A et al (2003) Kahalalide F, a new marine-derived compound, induces oncosis inhuman prostate and breast cancer cells. Mol Cancer Ther 2:863–872
85. Rademaker-Lakhai JM, Horenblas S, Meinhardt W et al (2005) Phase I clinical and pharmacokinetic study of Kahalalide F in patients with advanced androgen refractory prostate cancer. Clin Cancer Res 11:1854–1862. doi:10.1158/1078-0432.Ccr-04-1534
86. Martin-Algarra S, Espinosa E, Rubio J et al (2009) Phase II study of weekly Kahalalide F in patients with advanced malignant melanoma. Eur J Cancer 45:732–735. doi:10.1016/j.ejca.2008.12.005
87. Xu WJ, Liao XJ, Xu SH et al (2008) Isolation, structure determination, and synthesis of galaxamide, a rare cytotoxic cyclic pentapeptide from a marine algae *Galaxaura filamentosa*. Org Lett 10:4569–4672. doi:10.1021/ol801799d
88. Tan LT, Williamson RT, Gerwick WH et al (2000) cis, cis- and trans, trans-Ceratospongamide, new bioactive cyclic heptapeptides from the Indonesian red alga Ceratodictyon spongiosum and symbiotic sponge Sigmadocia symbiotica. J Org Chem 65:419–425
89. Blunt J, Munro MGH, Laatsch H (2013) AntiMarin. 0213 edn. University of Canterbury & University of Göttingen, Christchurch, NZ; Göttingen, Germany
90. Takahashi S, Matsunaga T, Hasegawa C et al (1998) Martefragin A, a novel indole alkaloid isolated from red alga, inhibits lipid peroxidation. Chem Pharm Bull 46:1527–1529
91. Guella G, Mancini I, N'Diaye I et al (1994) 175. Almazole C, a new indole alkaloid bearing an unusually 2,5-disubstituted oxazole moiety, and its putative biogenetic peptidic precursors, from a Senegalese Delesseriacean seaweed. Helv Chim Acta 77:1999–2006
92. Maeda M, Kodama T, Tanaka T et al (1987) Structures of domoilactone A and B, novel amino acids from the red alga, *Chondria armata*. Tetrahedron Lett 28:633–636
93. Sato M, Nakano T, Takeuchi M et al (1996) Distribution of neuroexcitatory amino acids

in marine algae. Phytochemistry 42: 1595–1597

94. Aguilar-Santos G (1970) Caulerpin, a new red pigment from green algae of the genus *Caulerpa*. J Chem Soc C 6:842–843
95. Liu DQ, Mao SC, Zhang HY et al (2013) Racemosins A and B, two novel bisindole alkaloids from the green alga *Caulerpa racemosa*. Fitoterapia 91:15–20. doi:10.1016/j.fitote.2013.08.014
96. Karentz D (2001) Chemical defenses of marine organisms against solar radiation exposure: UV-absorbing mycosporine-like amino acids and scytonemin. In: McClintock JB, Baker BJ (eds) Marine chemical ecology. CRC, Boca Raton, FL, pp 481–520
97. Takano S, Nakanishi A, Uemura D et al (1979) Isolation and structure of a 334 nm UV-absorbing substance Porphyra-334 from the red alga *Porphyra tenera* Kjellman. Chem Lett 26:419–420
98. Tsujino I, Yabe K, Sekikawa I (1980) Isolation and structure of a new amino acid, shinorine, from the red alga *Chondrus yendoi* Yamada Mikami. Bot Mar 23:65–68
99. Garcia-Pichel F, Wingard CE, Castenholz RW (1993) Evidence regarding the UV sunscreen role of a mycosporine-like compound in the cyanobacterium *Gloeocapsa* sp. Appl Environ Microbiol 59:170–176
100. Conde FR, Churio MS, Previtali CM (2000) The photoprotector mechanism of mycosporine-like amino acids. Excited-state properties and photostability of porphyra-334 in aqueous solution. J Photochem Photobiol B-Biol 56:139–144. doi:10.1016/S1011-1344(00)00066-X
101. Whitehead K, Hedges JI (2005) Photodegradation and photo sensitization of mycosporine-like amino acids. J Photochem Photobiol B-Biol 80:115–121. doi:10.1016/j.jphotobiol.2005.03.008
102. Pelletreau KN, Targett NM (2008) New perspectives for addressing patterns of secondary metabolites in marine macroalgae. In: Amsler CD (ed) Algal chemical ecology. Springer-Verlag, Berlin, pp 121–146
103. White SJ, Jacobs RS (1983) Effect of stypoldione on cell cycle progression, DNA and protein synthesis, and cell division in cultured sea urchin embryos. Mol Pharmacol 24: 500–508
104. O'Brien ET, White S, Jacobs RS et al (1984) Pharmacological properties of a marine natural product, stypoldione, obtained from the brown alga *Stypopodium zonale*. In: Bird CJ, Ragan MA (eds) Eleventh international seaweed symposium, vol 22, Developments in hydrobiology. Springer, Netherlands, pp 141–145
105. Pereira RC, Soares AR, Teixeira VL et al (2004) Variation in chemical defenses against herbivory in southwestern Atlantic *Stypopodium zonale* (Phaeophyta). Bot Mar 47:202–208
106. Pavia H, Aberg P, Jacobs RS et al (1996) Spatial variation in polyphenolic content of *Ascophyllum nodosum* (Fucales, Phaeophyta). In: Lindstrom SC, Chapman DJ (eds) Fifth international seaweed symposium, vol 22, Developments in hydrobiology. Springer, Netherlands, pp 199–203. doi:10.1007/978-94-009-6560-7
107. Van Alstyne KL, Puglisi MP (2007) DMSP in marine macroalgae and macroinvertebrates: distribution, function, and ecological impacts. Aquat Sci 69:394–402
108. Paul VJ (1992) Ecological roles of marine natural products. Comstock Publishing Associates, Ithaca, NY
109. Pereira RC, da Gama BAP (2008) Macroalgal chemical defenses and their roles in structuring tropical marine communities. In: Amsler CD (ed) Algal chemical ecology. Springer-Verlag, Berlin, pp 25–55
110. Carpenter RC (1986) Partitioning herbivory and its effects on coral reef algal communities. Ecol Monograph 56:345–363
111. Hay ME (1997) The ecology and evolution of seaweed-herbivore interactions on coral reefs. Coral Reefs 16:S67–S76
112. Morrison D (1988) Comparing fish and urchin grazing in shallow and deeper coral reef algal communities. Ecology 69:1367–1382
113. Hixon MA, Brostoff WN (1996) Succession and herbivory: effects of differential fish grazing on Hawaiian coral-reef algae. Ecol Monograph 66:67–90
114. Littler MM, Taylor PR, Littler DS (1986) Plant defense associations in the marine environment. Coral Reefs 5:63–71
115. Bolser RC, Hay ME (1996) Are tropical plants better defended? Palatability and defenses of temperate vs tropical seaweeds. Ecology 77:2269–2286
116. Jormalainen V, Honkanen T (2008) Macroalgal chemical defenses and their roles in structuring temperate marine communities. In: Amsler CD (ed) Algal chemical ecology. Springer-Verlag, Berlin, pp 57–89
117. Lobban CS, Harrison PJ (1996) Seaweed ecology and physiology. Cambridge University Press, Cambridge

118. Amsler CD, McClintock JB, Baker BJ (2008) Macroalgal chemical defenses in polar marine communities. In: Amsler CD (ed) Algal chemical ecology. Springer-Verlag, Berlin, pp 91–103
119. Wiencke C, Amsler CD (2012) Seaweeds and their communities in polar regions. In: Wiencke C, Bischof K (eds) Seaweed biology novel insights into ecophysiology, ecology and utilization. Springer-Verlag, Berlin, pp 265–291
120. Fukuhara Y, Mizuta H, Yasui H (2002) Swimming activities of zoospores in *Laminaria japonica* (Phaeophyceae). Fish Sci 68:1173–1181
121. Lebar MD, Heimbegner JL, Baker BJ (2007) Cold-water marine natural products. Nat Prod Rep 24:774–797. doi:10.1039/b516240h
122. Wessels H, Hagen W, Molis M et al (2006) Intra- and interspecific differences in palatability of Arctic macroalgae from Kongsfjorden (Spitsbergen) for two benthic sympatric invertebrates. J Exp Mar Biol Ecol 329:20–33
123. Amsler CD, Amsler MO, McClintock JB et al (2009) Filamentous algal endophytes in macrophytic Antarctic algae: prevalence in hosts and palatability to mesoherbivores. Phycologia 48:324–334. doi:10.2216/08-79.1
124. Amsler CD, Iken K, McClintock JB et al (2009) Defenses of polar macroalgae against herbivores and biofoulers. Bot Mar 52:535–545
125. Amsler CD, Iken K, McClintock JB et al (2005) Comprehensive evaluation of the palatability and chemical defenses of subtidal macroalgae from the Antarctic Peninsula. Mar Ecol Prog Ser 294:141–159
126. Nunez-Pons L, Rodriguez-Arias M, Gomez-Garreta A et al (2012) Feeding deterrency in Antarctic marine organisms: bioassays with the omnivore amphipod *Cheirimedon femoratus*. Mar Ecol Prog Ser 462:163–174
127. Amsler CD, McClintock JB, Baker BJ (2014) Chemical mediation of mutualistic interactions between macroalgae and mesograzers structure unique coastal communities along the western Antarctic Peninsula. J Phycol 50:1–10. doi:10.1111/jpy.12137
128. Karsten U (2008) Defense strategies of algae and cyanobacteria against solar ultraviolet radiation. In: Amsler CD (ed) Algal chemical ecology. Springer-Verlag, Berlin, pp 273–296
129. Laturnus F, Wiencke C, Kloser H (1996) Antarctic macroalgae—sources of volatile halogenated organic compounds. Mar Environ Res 41:169–181
130. Laturnus F (2001) Marine macroalgae in polar regions as sources for volatile organohalogens. Environ Sci Pollut Res Int 8:103–108
131. Ankisetty S, Nandiraju S, Win H et al (2004) Chemical investigation of predator-deterred macroalgae from the Antarctic peninsula. J Nat Prod 67:1295–1302. doi:10.1021/Np049965c
132. Wright JT, de Nys R, Poore A et al (2004) Chemical defense in a marine algae: heritability and the potential for selection by herbivores. Ecology 85:2946–2959
133. Amsler CD, Iken KB (2001) Chemokinesis and chemotaxis in marine bacteria and algae. In: McClintock JB, Baker BJ (eds) Marine chemical ecology. CRC, New York, NY, pp 413–430
134. Maier I, Müller DG (1986) Sexual pheromones in algae. Biol Bull 170:145–175
135. Maier I (1995) Brown algal pheromones. Prog Phycol Res 11:51–102
136. Thomas RWSP, Allsopp D (1983) The effects of certain periphytic marine bacteria upon the settlement and growth of *Enteromorpha*, a fouling alga. In: Oxley TA, Barry S (eds) Biodeterioration. Wiley, New York, pp 348–357
137. Tait K, Joint I, Daykin M et al (2005) Disruption of quorum sensing in seawater abolishes attraction of zoospores of the green alga *Ulva* to bacterial biofilms. Environ Microbiol 7:229–240
138. Joint I, Tait K, Callow ME et al (2002) Cell-to-cell communication across the prokaryote-eukaryote boundary. Science 298:1207
139. Toth GB, Pavia H (2000) Water-borne cues induce chemical defense in a marine alga (*Ascophyllum nodosum*). Proc Natl Acad Sci U S A 97:14418–14420
140. Toth GB, Pavia H (2007) Induced herbivore resistance in seaweeds: a meta-analysis. J Ecol 95:425–434
141. Coleman RA, Ramchunder SJ, Moody AJ et al (2006) An enzyme in snail saliva induces herbivore-resistance in a marine alga. Funct Ecol 21:101–106
142. Rohde S, Molis M, Wahl M (2004) Regulation of anti-herbivore defence by *Fucus vesiculosus* in response to various cues. J Ecol 92:1011–1018
143. Maier I (1993) Gamete orientation and induction of gametogenesis by pheromones in algae and plants. Plant Cell Environ 16: 891–907

144. Sekimoto H (2005) Plant sex pheromones. Vit Horm 72:457–478
145. Amsler CD (2008) Algal sensory chemical ecology. In: Amsler CD (ed) Algal chemical ecology. Springer-Verlag, Berlin, pp 297–309
146. Hay ME, Piel J, Boland W et al (1998) Seaweed sex pheromones and their degradation products frequently suppress amphipod feeding but rarely suppress sea urchin feeding. Chemoecology 8:91–98
147. Amsler CD, Neushul M (1989) Chemotactic effects of nutrients on spores of the kelps *Macrocystis pyrifera* and *Pterygophora californica*. Mar Biol 102:557–564
148. Amsler CD, Neushul M (1990) Nutrient stimulation of spore settlement in the kelps *Pterygophora californica* and *Macrocystis pyrifera*. Mar Biol 107:297–304
149. Greer SP, Iken KB, McClintock JB et al (2003) Individual and coupled effects of echinoderm extracts and surface hydrophobicity on spore settlement and germination in the brown alga *Hincksia irregularis*. Biofouling 19:315–326
150. Callow ME, Callow JA, Ista LK et al (2000) Use of self-assembled monolayers of different wettabilities to study surface selection and primary adhesion processes of green algal (*Enteromorpha*) zoospores. Appl Environ Microbiol 66:3249–3254
151. Greer SP, Amsler CD (2002) Light boundaries and the coupled effects of surface hydrophobicity and light on spore settlement in the brown alga *Hincksia irregularis* (Phaeophyceae). J Phycol 38:116–124
152. Greer SP, Amsler CD (2004) Clonal variation in phototaxis and settlement behaviors of *Hincksia irregularis* (Phaeophyceae) spores. J Phycol 40:44–53
153. Finlay JA, Callow ME, Ista LK et al (2002) The influence of surface wettability on the adhesion strength of settled spores of the green alga *Enteromorpha* and the diatom *Amphora*. Integr Comp Biol 42:1116–1122
154. Ista LK, Callow ME, Finlay JA et al (2004) Effect of substratum surface chemistry and surface energy on attachment of marine bacteria and algal spores. Appl Environ Microbiol 70:4151–4157
155. Callow ME, Callow JA (1998) Enhanced adhesion and chemoattraction of zoospores of the fouling alga *Enteromorpha* to some foul-release silicone elastomers. Biofouling 13:157–172
156. Wheeler GL, Tait K, Taylor A et al (2006) Acyl-homoserine lactones modulate the settlement rate of zoospores of the marine alga *Ulva intestinalis* via a novel chemokinetic mechanism. Plant Cell Environ 29:608–618
157. Potin P (2012) Intimate associations between epiphytes, endophytes, and parasites of seaweeds. In: Wiencke C, Bischof K (eds) Seaweed biology: novel insights into ecophysiology, ecology and utilization. Springer, New York, pp 203–234
158. Weinberger F, Beltran J, Correa JA et al (2007) Spore release in *Acrochaetium* sp. (Rhodophyta) is bacterially controlled. J Phycol 43:235–241
159. Kjelleberg S, Steinberg P, Givskov MC et al (1997) Do marine natural products interfere with prokaryotic AHL regulatory systems? Aquat Microbial Ecol 13:85–93
160. Pavia H, Toth GB (2008) Macroalgal models in testing and extending defense theories. In: Amsler CD (ed) Algal chemical ecology. Springer-Verlag, Berlin, pp 147–172
161. Weinberger F (2007) Pathogen-induced defense and innate immunity in macroalgae. Biol Bull 213:290–302
162. Potin P (2008) Oxidative burst and related responses in biotic interactions of algae. In: Amsler CD (ed) Algal chemical ecology. Springer-Verlag, Berlin, pp 245–271
163. Küpper FC, Müller DG, Peters AF et al (2002) Oligoalginate recognition and oxidative burst play a key role in natural and induced resistance of sporophytes of Laminariales. J Chem Ecol 28:2057–2081
164. Weinberger F, Friedlander M, Hoppe H-G (1999) Agar oligosaccharides elicit a physiological response in *Gracilaria conferta* (Rhodophyta). J Phycol 35:747–755
165. McDowell RE, Amsler CD, Dickinson DA et al (2014) Reactive oxygen species and the Antarctic macroalgal wound response. J Phycol 50:71–80
166. Bischof K, Gomez I, Molis M et al (2007) Ultraviolet radiation shapes seaweed communities. In: Amils R, Ellis-Evans C, Hinghofer-Szalkay H (eds) Life in extreme environments. Springer, Netherlands, pp 187–212
167. Garcia-Pichel F (1994) A model for internal self-shading in planktonic organisms and its implications for the usefulness of ultraviolet sunscreens. Limnol Oceanogr 39:1704–1717
168. Hargreaves A, Taiwo FA, Duggan O et al (2007) Near-ultraviolet photolysis of beta-phenylpyruvic acid generates free radicals and results in DNA damage. J Photochem Photobiol B-Biol 89:110–116. doi:10.1016/j.jphotobiol.2007.09.007

169. Häder DP, Kumar HD, Smith RC et al (2007) Effects of solar UV radiation on aquatic ecosystems and interactions with climate change. Photochem Photobiol Sci 6:267–285. doi:10.1039/B700020k
170. He YY, Häder DP (2002) Reactive oxygen species and UV-B: effect on cyanobacteria. Photochem Photobiol Sci 1:729–736. doi:10.1039/B110365m
171. He YY, Häder DP (2002) UV-B-induced formation of reactive oxygen species and oxidative damage of the cyanobacterium *Anabaena* sp.: protective effects of ascorbic acid and N-acetyl-L-cysteine. J Photochem Photobiol B-Biol 66:115–124. doi:10.1016/S1011-1344(02)00231-2
172. He YY, Häder DP (2002) Involvement of reactive oxygen species in the UV-B damage to the cyanobacterium *Anabaena* sp. J Photochem Photobiol B-Biol 66:73–80. doi:10.1016/S1011-1344(01)00278-0
173. Singh SP, Häder DP, Sinha RP (2010) Cyanobacteria and ultraviolet radiation (UVR) stress: mitigation strategies. Ageing Res Rev 9:79–90. doi:10.1016/j.arr.2009.05.004
174. Case RJ, Longford SR, Campbell AH et al (2011) Temperature induced bacterial virulence and bleaching disease in a chemically defended marine macroalga. Environ Microbiol13:529–537.doi:10.1111/j.1462-2920.2010.02356.x
175. Day TA, Neale PJ (2002) Effects of UV-B radiation on terrestrial and aquatic primary producers. Annu Rev Ecol Syst 33:371–396
176. de Nys R, Steinberg PD (1999) Role of secondary metabolites from algae and seagrasses in biofouling control. In: Fingerman M, Nagabhushanam R, Thompson M-F (eds) Recent advances in marine biotechnology, vol. 3. Science Publishers, Enfield, NH, pp 223–244
177. Briand J-F (2009) Marine anti-fouling laboratory bioassays: an overview of their diversity. Biofouling 25:297–311
178. Wahl M, Mark O (1999) The predominantly facultative nature of epibiosis: experimental and observational evidence. Mar Ecol Prog Ser 187:59–66
179. D'Antonio C (1985) Epiphytes on the rocky intertidal alga *Rhodomela larix* (Turner) C. Agardh: negative effects on the host and food for herbivores. J Exp Mar Biol Ecol 86:197–218
180. Cebrian J, Enriquez S, Fortes M et al (1999) Epiphyte accrual on *Posidonia oceanica* (L.) Delile leaves: implications for light absorption. Bot Mar 42:123–128
181. Honkanen T, Jormalainen V (2005) Genotypic variation in tolerance and resistance to fouling in the brown alga *Fucus vesiculosus*. Oecologia 144:196–205
182. Karez R, Engelbert S, Sommer U (2000) 'Co-consumption' and 'protective coating': two new proposed effects of epiphytes on their macroalgal hosts in mesograzer-epiphyte-host interactions. Mar Ecol Prog Ser 205:85–93
183. Anderson LM, Martone P (1989) Biomechanical consequences of epiphytism in intertidal macroalgae. J Exp Biol 217: 1167–1174
184. Wahl M (1989) Marine epibiosis. I. Fouling and antifouling: some basic aspects. Mar Ecol Prog Ser 58:175–189
185. Steinberg PD, de Nys R (2002) Chemical mediation of colonization of seaweed surfaces. J Phycol 38:621–629
186. Sudatti DB, Rodrigues SV, Pereira RC (2006) Quantitative GC-ECD analysis of halogenated metabolites: determination of surface and within-thallus elatol of *Laurencia obtusa*. J Chem Ecol 32:835–843
187. Dworjanyn SA, de Nys R, Steinberg PD (2006) Chemically mediated antifouling in the red alga *Delisea pulchra*. Mar Ecol Prog Ser 318:153–163
188. Nylund GM, Gribben PE, de Nys R et al (2007) Surface chemistry versus whole-cell extracts: antifouling tests with seaweed metabolites. Mar Ecol Prog Ser 329:73–84
189. Schmitt TM, Hay ME, Lindquist N (1995) Constraints on chemically mediated coevolution: multiple functions for seaweed secondary metabolites. Ecology 76:107–123
190. Harder T, Dobretsov S, Qian P-Y (2004) Waterborne polar macromolecules act as algal antifoulants in the seaweed *Ulva reticulata*. Mar Ecol Prog Ser 274:133–141
191. Bucolo P, Amsler CD, McClintock JB et al (2012) Effects of macroalgal chemical extracts on spore behavior of the Antarctic epiphyte *Elachista antarctica* Phaeophyceae. J Phycol 48:1403–1410
192. Lane AL, Kubanek J (2008) Secondary metabolite defenses against pathogens and biofoulers. In: Amsler CD (ed) Algal chemical ecology. Springer-Verlag, Berlin, pp 229–243
193. Goecke FR, Labes A, Wiese J et al (2010) Chemical interactions between marine macroalgae and bacteria. Mar Ecol Prog Ser 409:267–299
194. Bhadury P, Wright PC (2004) Exploitation of marine algae: biogenic compounds for potential antifouling applications. Planta 219:561–578

195. Lane AL, Nyadong L, Galhena AS et al (2009) Desorption electrospray ionization mass spectrometry reveals surface-mediated antifungal chemical defense of a tropical seaweed. Proc Natl Acad Sci U S A 106: 7314–7319
196. Maximilien R, de Nys R, Holmstrom C et al (1998) Chemical mediation of bacterial surface colonisation by secondary metabolites from the red alga *Delisea pulchra*. Aquat Microbial Ecol 15:233–246
197. Steinberg P, de Nys R, Kjellberg S (2002) Chemical cues for surface colonization. J Chem Ecol 28:1935–1951
198. Morse DE, Morse ANC, Raimondi PT et al (1994) Morphogen-based chemical flypaper for *Agaricia humilis* coral larvae. Biol Bull 186:172–181
199. Morse DE, Morse ANC (1991) Enzymatic characterization of the morphogen recognized by *Agaricia humilis* (scleractinian coral) larvae. Biol Bull 181:104–122
200. Heyward AJ, Negri AP (1999) Natural inducers for coral larval metamorphosis. Coral Reefs 18:273–279
201. Negri AP, Webster NS, Hill RT et al (2001) Metamorphosis of broadcast spawning corals in response to bacteria isolated from crustose algae. Mar Ecol Prog Ser 223:121–131
202. Toth GB, Pavia H (2002) Lack of phlorotannin induction in the kelp *Laminaria hyperborea* in response to grazing by two gastropod herbivores. Mar Biol 140:403–409
203. Stamp N (2003) Out of the quagmire of plant defense hypotheses. Q Rev Biol 78:23–55
204. Karban R, Baldwin IT (2007) Induced responses to herbivory. University of Chicago Press, Chicago, IL
205. Potin P, Bouarab K, Salaun JP et al (2002) Biotic interactions of marine algae. Curr Opin Plant Biol 5:308–317
206. Hay ME (1996) Marine chemical ecology: what's known and what's next? J Exp Mar Biol Ecol 200:103–134
207. Van Alstyne KL (1988) Herbivore grazing increases polyphenolic defenses in the intertidal brown alga *Fucus distichus*. Ecology 69:655–663
208. Paul VJ, Van Alstyne KL (1992) Activation of chemical defenses in the tropical green algae *Halimeda* spp. J Exp Mar Biol Ecol 160: 191–203
209. Jung V, Pohnert G (2001) Rapid wound-activated transformation of the green algal defensive metabolite caulerpenyne. Tetrahedron 57:7169–7172
210. Jung V, Thibaut T, Meinesz A et al (2002) Comparison of the wound-activated transformation of caulerpenyne by invasive and non-invasive *Caulerpa* species of the Mediterranean. J Chem Ecol 28:2091–2105
211. Van Alstyne KL (2008) Ecological and physiological roles of dimethylsulfoniopropionate and its products in marine macroalgae. In: Amsler CD (ed) Algal chemical ecology. Springer-Verlag, Berlin, pp 173–194
212. de Nys R, Coll JC, Price IR (1991) Chemically mediated interactions between the red alga *Plocamium hamatum* (Rhodophyta) and the octocoral *Sinularia cruciata* (Alcyonacea). Mar Biol 108:315–320
213. McCook L, Jompa J, Diaz-Pulido G (2001) Competition between corals and algae on coral reefs: a review of evidence and mechanisms. Coral Reefs 19:400–417
214. Suzuki Y, Takabashi T, Kawaguchi T et al (1998) Isolation of an alleolopathic substance from the crustose coralline algae *Lithophyllum* spp., and its effect on the brown alga *Laminaria religiosa* Miyabe (Phaeophyta). J Exp Mar Biol Ecol 225:69–77
215. Gross EM (2003) Allelopathy of aquatic autotrophs. Crit Rev Plant Sci 22:313–339
216. Kim M-J, Choi JS, Kang SE et al (2004) Multiple allelopathic activity of the crustose coralline alga *Lithophyllum yessoense* against settlement and germination of seaweed spores. J Appl Phycol 16:175–179
217. Vermeij MJA, Dailer ML, Smith CM (2011) Crustose coralline algae can suppress macroalgal growth and recruitment on Hawaiian coral reefs. Mar Ecol Prog Ser 422:1–7
218. Rasher DB, Hay ME (2010) Chemically rich seaweeds poison corals when not controlled by herbivores. Proc Natl Acad Sci U S A 107:9683–9688. doi:10.1073/pnas.0912095107
219. Andras T, Alexander TS, Gahlena A et al (2012) Seaweed allelopathy against coral: surface distribution of seaweed secondary metabolites by imaging mass spectrometry. J Chem Ecol 38:1203–1214
220. Rasher DB, Stout EP, Engel S et al (2011) Macroalgal terpenes function as allelopathic agents against reef corals. Proc Natl Acad Sci U S A 108:17726–17731. doi:10.1073/pnas.1108628108
221. Marques LV, Villaca R, Pereira RC (2006) Susceptibility of macroalgae to herbivorous fishes at Rocas Atoll, Brazil. Bot Mar 49:379–385
222. Hay ME (1992) Seaweed chemical defenses: their role in the evolution of feeding

specialization and in mediating complex interactions. In: Paul VJ (ed) Ecological roles for marine secondary metabolites; explorations in chemical ecology series. Cornstock Publishing Associates, Ithaca, pp 93–118

223. Amsler CD, McClintock JB, Baker BJ (2012) Palatability of living and dead detached Antarctic macroalgae to consumers. Antarct Sci 24:589–590
224. Estes JA, Steinberg PD (1988) Predation, herbivory, and kelp evolution. Paleobiology 15:57–60
225. Shears NT, Ross PM (2010) Toxic Cascades: multiple anthropogenic stressors have complex and unanticipated interactive effects on temperate reefs. Ecol Lett 13:1149–1159
226. Kvitek R, Bretz C (2004) Harmful algal bloom toxins protect bivalve populations from sea otter predation. Mar Ecol Prog Ser 271:233–243
227. Kvitek R, DeGange AR, Beitler MK (1991) Paralytic shellfish toxins mediate sea otter food preference. Limnol Oceanogr 36:393–404
228. Cimino G, Ghiselin MT (2009) Chemical defense and the evolution of opisthobranch gastropods. Proc Calif Acad Sci 60:175–422
229. Hay ME (2009) Marine chemical ecology: chemical signals and cues structure marine populations, communities, and ecosystems. Ann Rev Mar Sci 1:193–212
230. Pereira RC, Teixeira VL (1999) Sesquiterpenos das algas marinhas *Laurencia lamouroux* (Ceramiales, Rhodophyta). 1. Significado ecologico. Quim Nova 22:360–374
231. Avila C (1995) Natural products from opisthobranch molluscs: a biological review. Oceanogr Mar Biol 33:487–559
232. Rogers CN, de Nys R, Charlton TS et al (2000) Dynamics of algal secondary metabolites in two species of sea hare. J Chem Ecol 26:721–744
233. Pereira RC, Bianco EM, Bueno LB et al (2010) Associational defense against herbivory between brown seaweeds. Phycologia 49:424–428
234. Hay ME, Duffy JE, Fenical W (1990) Host-plant specialization decreases predation on a marine amphipod: an herbivore in plant's clothing. Ecology 71:733–743
235. Zamzow JP, Amsler CD, McClintock JB et al (2010) Habitat choice and predator avoidance by Antarctic amphipods: the roles of algal chemistry and morphology. Mar Ecol Prog Ser 400:155–163. doi:10.3354/meps08399
236. Amsler CD, McClintock JB, Baker BJ (1999) An antarctic feeding triangle: defensive interactions between macroalgae, sea urchins, and sea anemomes. Mar Ecol Prog Ser 183: 105–114
237. Richardson MG (1979) The distribution of Antarctic marine macroalgae related to depth and substrate. Br Antarct Surv B 49:1–13
238. Cubit JD (1975) Interaction of seasonally changing physical factors and grazing affecting high intertidal communities on a rocky shore. University of Oregon, Eugene, OR
239. Carroll G (1988) Fungal endophytes in stems and leaves: from latent pathogen to mutualistic symbiont. Ecology 69:2–9
240. Clay K (1988) Fungal endophytes of grasses: a defensive mutualism between plants and fungi. Ecology 69:10–16
241. Cheplick GP, Clay K (1988) Acquired chemical defences in grasses: the role of fungal endophytes. Oikos 52:309–318
242. Kohlmeyer J, Kohlmeyer E (1979) Marine mycology. The higher fungi. Academic, New York, NY
243. Payo DA, Colo J, Calumpong H et al (2011) Variability of non-polar secondary metabolites in the red alga *Portieria*. Mar Drugs 9:2438–2468
244. Verges A, Paul NA, Steinberg PD (2008) Sex and life-history stage alter herbivore responses to a chemically defended red alga. Ecology 89:1334–1343
245. Thornber C, Stachowicz JJ, Gaines S (2006) Tissue type matters: selective herbivory on different life history stages of an isomorphic alga. Ecology 87:2255–2263
246. IPCC (2007) Summary for policymakers. In: Solomon S, Qin D, Manning M et al (eds) Climate change 2007: the physical science basis. Contribution of working group I the fourth assessment report of the intergovernmental panel on climate change. Cambridge University Press, New York, NY USA
247. Harley CDG, Anderson KM, Demes KW et al (2012) Effects of climate change on global seaweed communities. J Phycol 48:1064–1078
248. Ries J, Cohen AL, McCorkle DC (2009) Marine calcifiers exhibit mixed responses to CO_2-induced ocean acidification. Geology 37:1131–1134
249. Diaz-Pulido G, Gouezo M, Tilbrook B et al (2011) High CO_2 enhances the competitive strength of seaweeds over corals. Ecol Lett 14:156–162
250. McCook LJ (1999) Macroalgae, nutrients and phase shifts on coral reefs: scientific issues

and management consequences for the Great Barrier Reef. Coral Reefs 18:357–367

251. Enge S, Nylund GM, Harder T et al (2012) An exotic chemical weapon explains low herbivore damage in an invasive alga. Ecology 93:2736–2745
252. Enge S, Nylund GM, Pavia H (2013) Native generalist herbivores promote invasion of a chemically defended seaweed via refuge-mediated apparent competition. Ecol Lett 16:487–492
253. Mack RN, Simberloff D, Lonsdale WM et al (2000) Biotic invasion: causes, epidemiology, global consequences, and control. Ecol Appl 10:689–710
254. Santini-Bellan D, Arnaud PM, Bellan G et al (1996) The Influence of the introduced tropical alga *Caulerpa taxifolia*, on the biodiversity of the Mediterranean marine biota. J Mar Biol Assoc UK 76:235–237
255. Davis AR, Benkendorff K, Ward DW (2005) Responses of common SE Australian herbivores to three suspected invasive *Caulerpa* spp. Mar Biol 146:859–868
256. Boudouresque CF, Lemme R, Mari X et al (1996) The invasive alga *Caulerpa taxifolia* is not a suitable diet for the sea urchin *Paracentrotus lividus*. Aquat Bot 53: 245–250
257. Khalilieh HS, Boulos A (2006) A glimpse on the uses of seaweeds in the Islamic science and daily life during the classical period. Arabic Sci Phil 16:91–101
258. Dillehay TD, Ramirez C, Pino M et al (2008) Monte Verde: seaweed, food, medicine, and the peopling of South America. Science 320:784–786. doi:10.1126/science.1156533
259. Smit AJ (2004) Medicinal and pharmaceutical uses of seaweed natural products: a review. J Appl Phycol 16:245–262
260. Bassetti M, Merelli M, Temperoni C et al (2013) New antibiotics for bad bugs: where are we? Ann Clin Microbiol Antimicrob 12:22–36. doi:10.1186/1476-0711-12-22
261. Pink R, Hudson A, Mouries MA et al (2005) Opportunities and challenges in antiparasitic drug discovery. Nat Rev Drug Discov 4:727–740. doi:10.1038/nrd1824
262. Clardy J, Walsh C (2004) Lessons from natural molecules. Nature 432:829–837. doi:10.1038/nature03194
263. Coley PD, Kursar TA (2014) Ecology. On tropical forests and their pests. Science 343:35–36. doi:10.1126/science.1248110

Chapter 3

Spectrophotometric Assays of Major Compounds Extracted from Algae

Solène Connan

Abstract

This chapter describes spectrophotometric assays of major compounds extracted from microalgae and macroalgae, i.e., proteins, carbohydrates, pigments (chlorophylls, carotenoids, and phycobiliproteins) and phenolic compounds. In contrast to other specific analytical techniques, such as high pressure liquid chromatography (HPLC) or mass spectrometry (MS), commonly applied to purified extracts to reveal more detailed composition and structure of algal compound families, these assays serve as a first assessment of the global contents of extracts.

Key words Carbohydrates, Carotenoids, Chlorophylls, Phenolic compounds, Phycobiliproteins, Proteins, Spectrophotometry

1 Introduction

Many analytical techniques have been developed to analyze the content and composition of algae [1] and to determine their detailed structure (see following chapters for examples of some techniques). Algae live naturally in a very challenging and changing environment which impacts on their morphology, content and composition; many temporal and spatial variations in content and composition of algal compounds have thus been observed (e.g., [2–4]). Algal species produced under controlled conditions also showed a modification in their composition when exposed to changes in their surrounding controlled environment, such as nutrients, temperature, or light (e.g., [5]).

The compounds extracted from macroalgae and microalgae may be analyzed (structure or detailed composition of alginate or pigments for example) using very detailed and specific techniques (as described in the following chapters). However, rapid techniques such as spectrophotometric assays are useful where the global composition of crude algal extracts is needed to characterize them and

Dagmar B. Stengel and Solène Connan (eds.), *Natural Products From Marine Algae: Methods and Protocols*, Methods in Molecular Biology, vol. 1308, DOI 10.1007/978-1-4939-2684-8_3, © Springer Science+Business Media New York 2015

to compare extracts from various algal species or extractions using different solvents or buffers.

This chapter describes the assays of major compounds extracted from algae such as proteins, carbohydrates, phenolic compounds, and pigments (chlorophylls, carotenoids, and phycobiliproteins) using a spectrophotometer. Many different methodologies to assay these compounds have been developed; the purpose of this chapter is not to review all but describe the most commonly applied and/or appropriate assays. These assays, based on either a chromophore (for proteins, carbohydrates, phenolic compounds) or specific molar absorption coefficients (e.g., for pigments), will give an approximate composition of the extract. To obtain a more precise composition of these compounds, purification of the crude extract will be followed by other analytical techniques, e.g., High Pressure Liquid Chromatography (HPLC), Mass Spectrometry (MS), or Nuclear Magnetic Resonance (NMR), as described in the next chapters.

2 Materials

All solvents are ACS reagent grade unless specified otherwise. Use distilled water or ultrapure water (18 MΩ at 25 °C with a Total Organic Carbon level <5 ppb) as specified.

2.1 Proteins

2.1.1 Lowry Assay (See Note 1)

- 0.1 M sodium hydroxide (NaOH): weigh 0.4 g of NaOH and add distilled water up to 100 mL.
- 1 N Folin–Ciocalteu (FC) reagent: dilute 1 volume of commercial FC reagent (2 N) in 1 volume of ultrapure water (1:1, v/v; *see* **Notes 2–4**).
- Reagent A: add 2 g of sodium hydroxide (NaOH) and 10 g of sodium carbonate (Na_2CO_3) to 400 mL of distilled water while stirring until completely dissolved, and adjust the volume to 500 mL.
- Reagent B: add 0.1 g of potassium sodium-tartrate ($KC_4H_4NaO_6{\cdot}4H_2O$) and 50 mg of cupric sulfate ($CuSO_4{\cdot}5H_2O$) to 8 mL of distilled water (*see* **Note 5**). Shake the mixture until completely dissolved, and adjust the volume to 10 mL.
- Complex-forming reagent (Lowry's solution): mix 50 volumes of reagent A + 1 volume of reagent B. Store this reagent at 4 °C (up to 6 months).
- Bovin Serum Albumin (BSA) stock solution (1 mg/mL): dissolve 100 mg of BSA in 100 mL of distilled water.
- Spectrophotometer and semi-micro disposable cuvettes.

2.1.2 Bradford Assay

- 0.1 M sodium hydroxide (NaOH): weigh 0.4 g of NaOH and add distilled water up to 100 mL.
- 96 % ethanol.
- 85 % phosphoric acid (H_3PO_4).
- Whatman no. 1 filter paper.
- Bradford reagent (*see* **Note 6**): dissolve 100 mg of Coomassie Blue G-250 (CBBG) in 50 mL of 96 % ethanol and add 100 mL of 85 % H_3PO_4. Dilute this solution with ultrapure water to 1 L. Filter the reagent through Whatman no. 1 filter paper and store in an amber bottle at room temperature (*see* **Note 7**).
- Bovin γ-globulin stock solution (1 mg/mL; *see* **Note 8**): dissolve 100 mg of bovin γ-globulin in 100 mL of distilled water.
- Spectrophotometer and semi-micro disposable cuvettes.

2.1.3 Bicinchoninic Acid (BCA) Microplate Assay

- Reagent A (*see* **Note 9**): weigh 0.1 g of sodium bicinchoninate ($C_{20}H_{10}N_2O_4Na_2$), 2.0 g of sodium carbonate ($Na_2CO_3 \cdot H_2O$), 0.16 g of sodium tartrate ($C_4H_4Na_2O_6 \cdot 2H_2O$), 0.4 g of sodium hydroxide (NaOH), 0.95 g of sodium bicarbonate ($NaHCO_3$), make up to 100 mL with distilled water. If necessary adjust the pH to 11.25 with $NaHCO_3$ or NaOH. Stable indefinitely at room temperature.
- Reagent B (*see* **Note 9**): dissolve 0.4 g of cupric sulfate ($CuSO_4 \cdot 5H_2O$) in 10 mL of distilled water. Stable indefinitely at room temperature.
- Working reagent: mix 50 volumes of reagent A with 1 volume of reagent B. The solution is apple green in color and is stable at room temperature for 1 week.
- Bovin Serum Albumin (BSA) stock solution (1 mg/mL): dissolve 100 mg of BSA in 100 mL of distilled water.
- Microplate incubator at 30 °C.
- Microplate reader.

2.2 Carbohydrates

2.2.1 Phenol–Sulfuric Acid Method

- 5 % phenol solution (w/v): pour 5 g of phenol in a 100-mL flask and add distilled water up to 100 mL (*see* **Note 10**).
- 15 mL polypropylene tube.
- 95–98 % sulfuric acid (H_2SO_4) (*see* **Note 11**).
- Stock solution of glucose (200 mg/L): dissolve 20 mg of glucose in 100 mL of distilled water.
- Water bath at 35 °C.
- Spectrophotometer and semi-micro disposable cuvettes.

2.2.2 Colorimetric Assay for Simultaneous Quantification of Neutral and Uronic Sugars

- 95–98 % sulfuric acid (H_2SO_4) (*see* **Note 11**).
- 2 % anthrone reagent (w/v) (*see* **Note 12**): dissolve 2 g of anthrone in 100 mL of 95–98 % H_2SO_4.
- 1 M sodium hydroxide (NaOH): weigh 4 g of NaOH and add distilled water up to 100 mL.
- 0.3 M Trizma® hydrochloride (Tris–HCl) pH 8.2 containing 2 % ethylenediaminetetraacetic acid (EDTA): weigh 4.728 g of Tris–HCl and 2 g of EDTA; dissolve in 80 mL of distilled water; adjust the pH to 8.2 with 1 M NaOH; pour into a 100-mL flask and add distilled water to make up to 100 mL.
- Stock solution of glucose (1 g/L): dissolve 0.1 g of glucose in 100 mL of 0.3 M Tris–HCl pH 8.2 with 2 % EDTA.
- Stock solution of D-glucuronic acid (0.5 g/L): dissolve 50 mg of D-glucuronic acid in 100 mL of 0.3 M Tris–HCl pH 8.2 with 2 % EDTA.
- Incubator at 60 °C.
- Microplate reader.

2.3 Phenolic Compounds

2.3.1 Folin–Ciocalteu Assay

- FC reagent [6] (*see* **Note 13**): dissolve 100 g of sodium tungstate ($Na_2WO_4{\cdot}2H_2O$) and 25 g of sodium molybdate ($Na_2MoO_4{\cdot}2H_2O$) in about 700 mL of distilled water; add 100 mL of conc. hydrochloric acid (HCl; 37 %) and 50 mL of 85 % phosphoric acid (H_3PO_4); boil continuously under reflux for 10 h; stop the heating, rinse down the condenser with a small amount of distilled water and then dissolve 150 g of lithium sulfate ($Li_2SO_4{\cdot}4H_2O$) in the slightly cooled solution (*see* **Note 14**); make the final solution to 1 L and filter it through a sintered glass filter to remove insolubles, if necessary (*see* **Note 15**).
- 20 % sodium carbonate solution (Na_2CO_3; w/v): weigh 20 g of Na_2CO_3 in a beaker and add distilled water up to 80 mL; dissolve Na_2CO_3 (*see* **Note 16**); pour in a 100-mL flask and add distilled water up to 100 mL.
- Stock solution of standard (phloroglucinol or gallic acid): dissolve 200 mg of the standard in 100 mL of distilled water.
- Water bath at 70 °C placed in the dark.
- Spectrophotometer and semi-micro disposable cuvettes.

2.3.2 2,4-Dimethoxybenzaldehyde (DMBA) Assay

- Glacial acetic acid >99 %.
- *N*,*N*-dimethylformamide (DMF) (*see* **Note 17**).
- Stock solution of DMBA: dissolve 2 g of DMBA in 100 mL of glacial acetic acid.
- Diluted hydrochloric acid (HCl) solution: pour 16 mL of conc. HCl (37 %) in 84 mL of glacial acetic acid (*see* **Note 18**).

- Prepare the working reagent by mixing equal volumes of the stock solution of DMBA and the diluted HCl solution just before use (*see* **Notes 19** and **20**).
- Stock solution of purified phlorotannin from the species under study: dissolve 5 mg of purified phlorotannin in 10 mL of distilled water.
- Incubator or water bath at 30 °C.
- Spectrophotometer.
- 1-cm glass cuvette + stopper.

2.3.3 Interference Removal Using Polyvinylpolypyrrolidone (PVPP)

- PVPP.
- Pasteur pipettes and teat.

2.4 Pigments (Chlorophylls and Carotenoids)

Various solvents are used for the extraction of pigments from algae and the choice of solvent (and its percentage) depends on the algal species (*see* **Note 21**):

- *N,N* dimethylformamide (DMF; *see* **Notes 17** and **22**).
- Acetone (*see* **Note 23**).
- Methanol (*see* **Note 24**).
- Ethanol (*see* **Note 25**).
- Dimethylsulfoxide (DMSO; *see* **Note 26**).

Other items needed:

- Ball mill or mortar and pestle.
- Ultrasonic bath.
- Pasteur pipettes and teat.
- Spectrophotometer.
- 1-cm glass cuvette + stopper.
- 1 M hydrochloric acid (HCl): pour 80 mL of distilled water into a 100-mL flask, add 0.82 mL of conc. HCl (37 %) and distilled water up to 100 mL.

2.5 Pigments (Phycobiliproteins)

- Spectrophotometer.
- 1-cm quartz glass cuvette.

3 Methods

Many of the assays developed below use a calibration curve of standards: to be in the range of the calibration curve dilute the algal extracts with distilled water if necessary. For the other assays (pigments), the methods detailed below use direct absorbance measurement of extracts and care should be taken to be within the linear range of the spectrophotometer and generally less than 1 absorbance unit; dilute the extract with the extraction solvent if needed.

3.1 Proteins

Different extraction procedures of algal proteins exist using either alkaline lysis at a high temperature followed or not by an acid neutralization [7–9]; a lysis buffer such as Triton X-100 with ethylenediaminetetraacetic acid disodium salt (Na_2-EDTA) and phenyl methyl sulfonyl fluoride [10]; or a phosphate buffer added with sodium dodecyl sulfate (SDS) or β-mercaptoethanol [11]. Ultrasonication of the algal sample-lysis buffer mix will also improve the protein extraction (*see* **Note 27**). Interfering compounds could be removed from the extract by using trichloroacetic acid solution which will precipitate and thus concentrate the protein fraction.

Chapter 4 presents a method optimized for the extraction of proteins from algae.

3.1.1 Lowry Assay

The Lowry assay is based on the Biuret reaction where the copper (Cu^{2+}) reacts with the peptide bonds of the proteins under alkaline conditions to produce Cu^+. These Cu^+ further react with the Folin reagent to reduce the phosphomolybdotungstate into heteropolymolybdenum blue through the oxidation of aromatic amino acids catalyzed by Cu^+. The reactions result in a strong blue color with a maximum absorbance at 750 nm. The intensity of this blue color depends mainly on the tyrosine and tryptophan contents (aromatic amino acids) and to a lesser extent on cystine/cysteine and histidine contents [12]. The method is sensitive to about 0.01 mg of protein/mL; range 0.01–1.0 mg protein/mL. The sensitivity of Lowry assay is moderately constant from protein to protein, thus giving a good approximation of the protein content of a crude extract or a mixture of proteins. However, this assay is subjected to interference from many substances including buffers, detergents, EDTA, nucleic acids and sugars [12]. The assay below is adapted from Barbarino and Lourenço [8] (*see* **Note 28**):

- Suspend precipitated proteins in 2 mL 0.1 M NaOH or use directly the protein extract.
- Prepare calibration curves with Bovin Serum Albumin (BSA) at a max concentration of 1 mg/mL.
- Mix 0.2 mL of sample or standard with 1 mL of the complex-forming reagent. Stir the sample or standard for 2 s with a vortex (*see* **Note 29**).
- Incubate for 20 min at room temperature in the dark.
- Add 0.1 mL of the diluted FC reagent. Stir the sample or standard for 2 s with a vortex.
- Incubate for 35 min at room temperature in the dark.
- Read the absorbance at 750 nm against a blank (0.1 M NaOH or the extraction solvent).

3.1.2 Bradford Assay

The Bradford technique is simpler, faster, and more sensitive than the Lowry method; it is also subject to less interference by common reagents and non protein component of biological samples [13]. The Bradford assay relies on the binding of the dye Coomassie Blue G-250 to proteins. The free dye exists in four different ionic forms, and the more anionic blue dye (which has an absorbance maximum at 595 nm) binds to proteins. The quantity of proteins can thus be estimated by determining the amount of dye in the blue ionic form by measuring the absorbance of the solution at 595 nm. The dye appears to bind more readily to arginine and lysine which can lead to variation in the response of the assay to different proteins [13]. The Bradford assay is relatively free from interference but some compounds still interfere with the reaction, mainly detergents and ampholytes [13]. The method is sensitive down to about 0.01 mg of protein/mL when using the standard assay and 0.001 mg of protein/mL when using the Bradford microassay (*see* **Note 30**); range 0.01–1.0 mg protein/mL. This assay is not suitable for quantifying free amino acids or peptides smaller than 3,000 Da as the dye will not bind to them [13].

The standard method described below is adapted from Barbarino and Lourenço [8] (*see* **Note 28**):

- Suspend precipitated proteins in 0.5 mL 0.1 M NaOH or use directly the protein extract.
- Prepare calibration curves with bovin γ-globulin at a max concentration of 100 μg/0.1 mL.
- Add 5 mL of the Bradford reagent to each 0.1 mL of sample or standard.
- Read the absorbance at 595 nm 5 min after the start of the chemical reaction at room temperature against a blank (0.1 M NaOH or the extraction solvent) (*see* **Notes 31** and **32**).

3.1.3 Bicinchoninic Acid (BCA) Microplate Assay

The bicinchoninic acid (BCA) assay is similar to the Lowry assay: it depends also on the conversion of Cu^{2+} to Cu^{+} under alkaline conditions. The Cu^{+} is detected through the reaction with BCA. Both assays have similar sensitivity but BCA is stable under alkaline conditions. The reaction results in the development of an intense purple color with an absorbance maximum at 562 nm [14]. Since the production of Cu^{+} is a function of the protein concentration and incubation time, the protein content of unknown samples may be determined spectrophotometrically by comparison with known protein standards. This assay is also generally less affected by interfering compounds such as some detergents and denaturing agents (urea, guanidine chloride), which impact the Lowry assay. However, the BCA assay is more sensitive to the presence of reducing sugars and high level of lipids [14]. The following standard method (for samples containing 0.1–1 mg protein/mL; *see* **Note 33**) is adapted from refs. [14, 15]:

- Dilute the samples with distilled water (1:3, v/v).
- Prepare calibration curves with Bovin Serum Albumin (BSA) at a max concentration of 1 mg/mL.
- Pour 10 μL of distilled water with 10 μL of diluted samples/standard in a microplate.
- Add 200 μL of working reagent.
- Cover the microplate and incubate for 60 min at 30 °C in the dark. Cool the microplate at room temperature (*see* **Note 34**).
- Read the absorbance at 562 nm against a blank (distilled water).

3.2 Carbohydrates

Two methods are generally used to assay total carbohydrates in biological samples [16]: the phenol–sulfuric acid or Dubois method [17] and the anthrone method [18]. These assays are based on a first polysaccharide hydrolysis with concentrated sulfuric acid followed by intramolecular dehydration of all the osidic monomers in acidic conditions. The furfural derivatives obtained condense on phenol or anthrone to form colored products which absorb at 490 and 620 nm, respectively. Both assays exhibit different colorimetric response linked to the sugar composition: for example, the intensities of heptose and pentose absorbances are lower than those of hexose and some carbohydrate derivatives do not show any response at all [16]. These procedures are thus highly dependent on the sugar used for the calibration and the carbohydrate contents of samples could thus be over- or underestimated. Also, colorimetric methods for assaying sugars in complex biological samples remain problematic because of the presence of numerous biochemical species that can interfere with sugar quantification, such as nucleic acids mainly when the phenol–sulfuric acid assay is used [16]. The anthrone method is often preferred to assay hexoses; however, the anthrone reagent is expensive and solutions in sulfuric acid are not stable. The phenol–sulfuric acid method gives a more homogeneous response with total sugars and mainly less underestimation of each monosaccharide than with the use of the anthrone method. The phenol–sulfuric acid method is simple, rapid, relatively straightforward, sensitive, and gives reproducible results. The reagent is inexpensive and stable, readily available, and a given solution requires only 1 standard curve for each sugar; moreover, the color produced is permanent and stable [17]. For uronic acid-containing samples, a modified anthrone method has been developed because the phenol–sulfuric acid method involved either an underestimation or an overestimation of the total neutral sugar concentrations [16].

The phenol–sulfuric acid method is directly applied on collected microalgal cells [9]. For macroalgae, several methods of carbohydrate extractions exist such as the use of hot ethanol or water [19], pressurized liquid extraction (PLE; [20]), or enzymes such as the laminarase [21] or α-amylase/amyloglucosidase [22] to assay soluble carbohydrates, laminarin and starch, respectively.

Only the phenol–sulfuric acid method and the modified anthrone method for uronic sugars are described here (*see* **Note 35**).

3.2.1 Phenol–Sulfuric Acid Method

The method described below is adapted from refs. [4, 9, 17] (*see* **Note 28**):

- Prepare a calibration curve using the stock solution of glucose up to 0–100 mg/L.
- Add 0.5 mL of the extract/standard to 0.5 mL of a 5 % phenol solution in a 15 mL polypropylene tube (*see* **Note 36**). Do not mix and be careful to avoid any drops on the walls.
- Add 2.5 mL of 95–98 % H_2SO_4 rapidly directly on the liquid surface in order to achieve good mixing.
- After 10 min of incubation at room temperature, vortex the tube vigorously for 10 s.
- Leave the tube standing for 15 min at room temperature and then place it in a water bath at 35 °C for 30 min (*see* **Note 37**).
- Measure the absorbance of the yellow-orange color at 490 nm against a blank (distilled water or extraction solvent).

3.2.2 Colorimetric Assay for Simultaneous Quantification of Neutral and Uronic Sugars

This assay allows the specific quantification of uronic acids in the presence of neutral sugars and/or proteins. The procedure developed by Rondel et al. [16] is based on the reaction of anthrone with both neutral and uronic acid sugars but a double absorbance reading is proposed, at 560 and 620 nm, in order to quantify the uronic acid and neutral sugars separately. This method takes into account the interference of uronic acids with anthrone reagent and is very sensitive for the quantification of neutral sugars (20–80 mg/L) as well as for simultaneous quantification of uronic sugars (50–400 mg/L).

- Prepare calibration curves of standards: glucose from 0 to 80 mg/L and D-glucuronic acid from 0 to 400 mg/L (*see* **Note 38**).
- Add 200 μL of anthrone reagent to 100 μL of samples dispensed in microplates.
- Incubate these microplates at 60 °C for 30 min.
- Cool the microplate at room temperature for 10 min.
- Measure the absorbance at 560 and 620 nm.

To determine glucose and uronic acid concentrations in samples, resolve the system below:

$$\begin{cases} A_{560} = a_{560} \times [\text{gluc}]_{\text{sample}} + b_{560} \times [\text{GlA}]_{\text{sample}} \\ A_{620} = a_{620} \times [\text{gluc}]_{\text{sample}} + b_{620} \times [\text{GlA}]_{\text{sample}} \end{cases}$$

With A_{560} and A_{620} the absorbances of the sample at 560 and 620 nm, respectively; a_{560} and a_{620} the slopes of the glucose calibration curves at 560 and 620 nm respectively; b_{560} and b_{620} the slopes of the D-glucuronic acid calibration curves at 560 and 620 nm, respectively; $[\text{gluc}]_{\text{sample}}$ and $[\text{GlA}]_{\text{sample}}$ the unknown glucose and D-glucuronic acid concentrations of the sample.

3.3 Phenolic Compounds

Phenolic compounds are secondary metabolites mainly found in brown seaweeds which also have primary function in algae such as strengthening the cell wall in cross-linking with alginates [23]. In brown seaweeds most of these phenolics are polymers of phloroglucinol named phlorotannins which could be halogenated or sulfated, with levels up to 20 % of dry weight of seaweed tissue (% dw; [24]). Quinones such as sargaquinoic acid, have also been found in some brown seaweeds (*Sargassum* spp.; [25, 26]), together with bromophenols [27]. In green and red algae, phenolics, mainly bromophenols, have been observed [27, 28] but in lower contents compared to phlorotannins (for example Flodin et al. [29] found up to 2 μg bromophenol/g fw in *Ulva* sp. and Whitfield et al. [27] up to 2.6 μg/g fw in *Pterocladiella capillacea*). In microalgae, although phenolics have been often extracted and assayed using general spectrophotometric methods as those described below [30], only few studies reported on phenolic composition of microalgae; for example Klejdus et al. [31] found benzoic acid derivatives, hydroxybenzaldehydes and cinnamic acid derivatives in freshwater microalgae and some Cyanobacteria.

Extraction methods of phenolic compounds have been developed mainly for brown seaweeds as these species may contain up to 15 % dw of phenolics ([24]; *see* Chapter 7). Various solvents are used to extract phenolics from seaweeds: acetone, methanol, and ethanol, each at varying proportions in distilled water, with methanol extracting the highest phenolic yield [2, 32–37].

Different assays of algal phenolics have been developed, all based on indirect determination using the formation of a colored chromophore for spectrophotometric measurement: Folin–Denis, Folin–Ciocalteu which is an improvement of Folin–Denis, Prussian Blue, and 2,4-dimethoxybenzaldehyde (DMBA) assay.

The Folin–Denis (FD) assay is one of the chemical assays most commonly used for ecological research [38]. This assay originally developed for protein analyses quantifies concentrations of easily oxidized compounds such as phenolics by color changes accompanying a redox reaction [39]. In the FD reagent, phosphotungstic and phosphomolybdic acids react with hydroxylated aromatic compounds of extracts such as phenolics. In alkaline conditions, the reagent produces a stable blue color which can be quantified spectrophotometrically. Thus, although the reaction does not incorporate the phenols into the colored product, it is stoichiometrically dependent on the number of free hydroxyl groups [24, 40].

The Folin–Ciocalteu (FC) reagent provides an improvement of the FD reagent with the increase of molybdate and mainly the addition of lithium sulfate to the reagent [41] which reduces the amount of precipitate. This precipitate which interferes with the assay is formed when high concentrations of FD reagent are used. Comparison of the FD procedure with that of FC gives greater sensitivity and reproducibility for the FC [6]. Because FD and FC assays were first developed for protein analyses, both assays show interferences from many oxidable compounds, such as tyrosine, tryptophan, ascorbic acid and urea, and even diethylether [40, 42]. However, it is unlikely that amino acid residues from proteins will be a substantial portion of these compounds because proteins are generally insoluble in high concentrations of organic solvents that are usually used in the phenolic extraction procedures [40]. Compared to red and green, marine brown algae contain higher phenolic levels and the interfering substances are thought to account for less than 5 % of the FC reactive compounds [43]. The reaction of phenolics with FC reagent gives different results depending on the composition of the extract: the binding type of phloroglucinol unit (ether, or aryl-aryl) will affect the result due to differences in the number of hydroxyl groups on the phlorotannin molecule [34].

The 2,4-dimethoxybenzaldehyde (DMBA) assay is based on chemistry similar to the vanillin assay [44] in which a colored product is formed in response to the reaction between the aldehyde and the phenolic [45]. The DMBA reacts specifically with 1,3- and 1,3,5-substituted phenolics such as phloroglucinol and phlorotannins forming triphenylmethane pigments after electrophilic substitution [46]. As for the FC assay, the chromophore formed depends on the phlorotannin class: ether or aryl-aryl bonds react differently [34, 36]. DMBA does not react with tannic acid containing ortho- and parahydroxyl-substituted phenolics. The chromophore formed when DMBA reacts with phloroglucinol or phenolics in the algal extract has different UV/Vis maxima. As a consequence, standard and sample solutions are measured at 494 and 515 nm respectively, which could lead to errors in the calculation of phlorotannin content [34]. This assay was developed to avoid interferences from for example ascorbic acid or peptides. Although nonpolar compounds may react in the DMBA assay, color production of such reactions will be slight, and it will not interfere with the estimation of phlorotannin levels [46]. As the reaction is also time- and temperature-dependent, these factors should strictly be controlled [36].

One way of assessing the interference in the different phenolic assays (except the DMBA as the DMBA reagent should bind specifically phlorotannins) is to remove selectively phenolic compounds with polyvinylpolypyrrolidone (PVPP): PVPP is a polymer that can remove tannins and phlorotannins from acidic aqueous solution which can then be removed from solution by filtering or centrifuging [40, 47]. The remaining oxidizable species such as proteins or ascorbic acid are then quantified with FD, FC, Prussian blue assays and subtracted from the results of the assay prior to the addition of PVPP. However, PVPP has a greater affinity for complex phlorotannins such as the polyphenols found in many brown algae, than for low molecular weight compounds such as phloroglucinol which could be the major phenolic compounds of some brown algae [46, 48]. Also, treatment with PVPP failed to completely remove phlorotannins from some fresh algal extracts and from solution of purified phlorotannins extracted from some brown seaweed species [40, 46]. If all phenolics are not precipitated by PVPP, phenolic concentrations are thus underestimated. Extracts treated with PVPP should not yield any color in the DMBA assay since DMBA reacts specifically with phlorotannins [46].

Analyses using the FC reagent are convenient, simple, require only common equipment and have produced a large body of comparable data. Under proper conditions, the assay is inclusive of monophenols and gives predictable (but variable by reactive groups per molecule) reactions with the types of phenols found in algae. However, as phenolics of various structures react to different degrees, expression of the results as a single number such as mg/L phloroglucinol equivalence is necessarily arbitrary [6]. Structural types like fuhalols and hydroxylated phlorethols show various reductive potential due to their different numbers of free hydroxyl groups. The calculated content of phlorotannins may thus vary depending on the structural type of phenolics present in the sample. Hence the polyphenol extracts containing various types of phlorotannins are hardly comparable [34].

DMBA is insensitive to interferences, measuring only phlorotannins, is inexpensive and rapid. One disadvantage is that the variation in reactivity of phlorotannin fractions is somewhat greater than that for the other assays (Folin–Denis, Folin–Ciocalteu, and Prussian Blue). This variation reflects differences in the chemical structures of phlorotannins. Therefore the choice of a reference compound becomes particularly important for the DMBA assay. For most algal species, neither phloroglucinol nor catechin were suitable as references for DMBA because DMBA responds too strongly to these substances relative to other phlorotannins and consequently significantly underestimates the mass of phlorotannin measured [46].

Only FC and DMBA assays and the use of PVPP to measure the interference on the phenolic assay are developed below. Regarding the calibration curves, the choice of standard curve substantially affected the measured estimates of phenolic content for most algal species and mainly brown seaweeds [46]. Phloroglucinol or purified phlorotannin of the species under study should be used for brown macroalgae, as phloroglucinol is the unit of phlorotannin and phlorotannin is the major phenolic compounds observed in those algal species; results are thus expressed as mg/g seaweed eq. phloroglucinol or % eq. phloroglucinol. For red and green macroalgae and microalgae, either phloroglucinol or gallic acid could be used in the FC calibration curve as often the composition of phenolics is unknown.

3.3.1 Folin–Ciocalteu (FC) Assay

The FC assay is based on refs. [2, 32, 35, 49–52] (*see* **Note 28**):

- Mix 100 μL of sample with 50 μL of FC reagent, 200 μL of 20 % of Na_2CO_3 solution (*see* **Note 39**), and 650 μL of distilled water.
- Incubate the mixture at 70 °C in the dark for 10 min.
- After production of a blue color, read the absorbance at 700 nm against a blank with distilled water instead of the seaweed extract (*see* **Note 40**).
- Dilute the stock solution of standard (phloroglucinol or gallic acid): pour 0.0, 1.0, 2.0, 3.0, 4.0 and 5.0 mL of the stock solution into 100-mL flasks and add distilled water. Add reagents as for the seaweed extract.
- Express the total content of phenolic compounds as % of dry weight (% dw) based on a standard curve of phloroglucinol, the monomer of phlorotannin for brown seaweeds, and gallic acid for green and red seaweeds.

3.3.2 2,4-Dimethoxybenzaldehyde (DMBA) Assay

The DMBA assay, specific to phlorotannins, has been developed by Stern et al. [46] and the method below is adapted from refs. [34, 46] (*see* **Note 28**):

- Dilute the stock solution of purified phlorotannins from the species under study with distilled water to prepare the standard curve (0–4 mg/mL; *see* **Note 41**).
- Dissolve algal extract in distilled water (20 mg/mL; *see* **Note 42**).
- Mix 10 μL of extract and standard with 10 μL of DMF (*N,N*-dimethylformamide; *see* **Note 43**) and 2.5 mL of the working reagent.
- Incubate the mixture for 60 min at a temperature between 30 and 33 °C (*see* **Note 44**).
- Read the absorbance at 515 nm against the blank (distilled water; *see* **Note 45**).

3.3.3 Interference Removal Using Polyvinylpolypyrrolidone (PVPP)

The measurement of interference on the phenolic assay is performed using polyvinylpolypyrrolidone (PVPP). The method is adapted from refs. [34, 40, 43]:

- Dissolve algal extract in distilled water (20 mg/mL solution; *see* **Note 46**).
- Add 10 mg/mL of PVPP.

- Stir the mixture for 10 min with a magnetic stirrer.
- Centrifuge the mixture at 5,000×*g* and pipette carefully the supernatant into another tube (*see* **Note 47**). Discard the PVPP.
- Repeat once this PVPP treatment (*see* **Note 48**).
- Assay the remaining supernatant after the second PVPP treatment with FC or DMBA assay; the difference between measurement 1 (without PVPP) and 2 (with PVPP) gives the concentration of phenolic compounds able to interact with PVPP (*see* **Note 49**).

3.4 Pigments (Chlorophylls and Carotenoids)

Instead of using HPLC (*see* Chapter 14), algal pigments can also be determined using a spectrophotometer by measuring the absorbance of the crude extract at wavelengths corresponding to the maximum absorption of the pigment under study and using equations with the molar extinction coefficient of the pigment. Equations for chlorophyll (Chl) determination have been published since 1949 [53] and were further developed [54–60]. For the quantification of carotenoids using a spectrophotometer, equations were developed by Strickland and Parsons [61].

The accuracy of the equations which lead to the simultaneous determination of extracted Chls (*a, b, c, d* and a mixture of these chlorophylls according to the algal used) depends on [62]:

- The accuracy of the chlorophyll extinction coefficient used in the equation.
- The spectral purity of these Chls.
- The solvent used: for example, acetone gives sharper chlorophyll absorption peaks than methanol.
- The presence of water from the extracted cells which will alter the maximum absorbance peak and affect accuracy.
- The relative proportion of the different Chls in the extract.
- The presence of degradation products of Chls.
- The precision of the spectrophotometer used for the absorbance measurements.

When the precise knowledge of algal pigment composition is required or where the sample contains a non-negligible amount of phaeopigments, chromatography (HPLC, TLC, or HPTLC) is the only method which will give accurate level of Chls as well as composition and content of carotenoids [63].

Many solvents are used for the extraction of pigments from algae, mainly *N,N* dimethylformamide (DMF), acetone, methanol, and ethanol (the last three solvents in various proportions).

DMF is a very convenient solvent for Chl extraction since it is effective on intact algal parts (no need for fine grinding) and Chl is quite stable in it (up to 20 days when stored at 4 °C in the dark) [56].

Acetone gives very distinct chlorophyll absorption peaks and is thus the solvent of choice for Chl determination; however acetone may be a poor extractant of Chl from some algae, particularly green algae [60]. Methanol (particularly hot methanol at 60 °C) is a very efficient extractant for Chls, particularly from recalcitrant algae. Like methanol, hot ethanol (60 °C) is an efficient extractant of Chls, even from very resistant materials. There are considerable practical, safety, and economical advantages in using ethanol as the solvent for Chl assay [60].

Commercial acetone and alcohols are often highly acidic leading to the formation of phaeophytin and allomerization products of Chls which are spectrally different and severely interfere with all Chl determinations mainly when using methanol as the solvent. Thus a small amount of magnesium carbonate is added into 100 % methanol to ensure that the solvent is acid-free; the solvent is then filtered and kept at 4 °C [59, 60]. The 90 % solvents (acetone, ethanol, methanol) are also made using a saturated solution of magnesium carbonate hydroxide to neutralize the solvent. Ethanol is not a strong inhibitor of chlorophyllase activity leading to hydrolysis of the phytol chain of phytollated Chls (Chls *a*, *b*, and *d*) to produce the corresponding chlorophyllids when using this solvent. However, chlorophyllase can be easily either removed by filtration or centrifugation because it is not soluble in alcohol; or denaturated by heating above 60 °C. Chlorophyllases are inhibited in acetone if the organic solvent is maintained above 40 % (v/v; [59, 60]).

For brown seaweeds, only few extraction methods and assays have been published (e.g., [64, 65]) and an alternative method has been developed by Seely et al. [64]: dimethylsulfoxide (DMSO) is used as first extractant of fucoxanthin (Fucox), Chl *a*, Chl *c*, and β-carotene as it helps breaking the cell walls, followed by acetone. Acetone extract is further extracted/purified with hexane and equations have been developed for each extract [64].

For carotenoids, which are less polar than Chls, extraction should be performed using 100 % methanol [9].

Phaeopigments could also be assayed using spectrophotometer through acidification techniques with 1 M HCl. These techniques were specifically designed to correct the Chl *a* concentration for interference from pheopigments, but the concentration of pheopigments calculated under low pheopigment/Chl ratios should be viewed with caution [66].

Extraction of pigments is always performed on ground seaweed tissue or pellet using (or not) ultrasound until the pellet is pale in color.

The first trichroic and quadrichroic Chl equations presented below are those developed by Ritchie [60] in a 1-cm path cuvette. The trichroic formulae are simpler than the quadrichroic ones. Regarding the solvents, the acetone and ethanol formulae give very reliable estimates of Chl *a* and Chl *d*. The quadrichroic Chl *b* equations for acetone and ethanol are more reliable than the methanol formulae but both give low Chl *b* values in samples containing no Chl *b*. The methanol algorithms for Chl *b* are very adversely affected by any Ph *a* present. Finally, the quadrichroic Chl *c* equations for acetone and ethanol are more reliable for organisms containing Chl *c1*+*c2* than for those containing only Chl *c2* [60]. These equations could thus be used where sample absorbances are high and degradation products are absent.

The absorbance measurement at 750 nm reflects the absorbance due to the turbidity of the sample and is removed from all other absorbance measurements.

A_{664} = absorption of the sample at 664 nm in a 1-cm path cuvette.

- Trichroic equations for simultaneous determination of Chls *a*, *b*, and *c* in extracts [60].

These equations should be used when a Chl extract has no significant Ph *a* or Chl *d* content.

– 90 % acetone:

$$\text{Chl } a\,(\mu\text{g / mL}) = -0.3002\times(A_{630}-A_{750})-1.7538\times(A_{647}-A_{750})+11.9092\times(A_{664}-A_{750})(\pm 0.0009)$$

$$\text{Chl } b\,(\mu\text{g / mL}) = -1.2942\times(A_{630}-A_{750})+19.8952\times(A_{647}-A_{750})-4.9401\times(A_{664}-A_{750})\ (\pm 0.0094)$$

$$\text{Chl } c\,(\mu\text{g / mL}) = 23.6723\times(A_{630}-A_{750})-7.9057\times(A_{647}-A_{750})-1.5467\times(A_{664}-A_{750})(\pm 0.0079)$$

$$\text{Total Chls}\,(\mu\text{g / mL}) = 22.0780\times(A_{630}-A_{750})+10.2357\times(A_{647}-A_{750})+5.4224\times(A_{664}-A_{750})(\pm 0.0071)$$

– 100 % methanol:

$$\text{Chl } a\,(\mu\text{g / mL}) = -3.2416\times(A_{632}-A_{750})-6.4151\times(A_{652}-A_{750})+16.4351\times(A_{665}-A_{750})\ (\pm 0.0087)$$

$$\text{Chl } b\,(\mu\text{g / mL}) = -3.0228\times(A_{632}-A_{750})+32.1478\times(A_{652}-A_{750})-13.8844\times(A_{665}-A_{750})(\pm 0.0261)$$

$$\text{Chl } c\,(\mu\text{g / mL}) = 34.2247\times(A_{632}-A_{750})-12.8087\times(A_{652}-A_{750})-1.5492\times(A_{665}-A_{750})\ (\pm 0.0139)$$

$$\text{Total Chls}\,(\mu\text{g / mL}) = 27.9603\times(A_{632}-A_{750})+12.9241\times(A_{652}-A_{750})+1.0015\times(A_{665}-A_{750})(\pm 0.0165)$$

– 100 % ethanol:

$$\text{Chl } a\,(\mu\text{g / mL}) = -0.9394 \times (A_{632} - A_{750}) - 4.2774 \times (A_{649} - A_{750}) + 13.3914 \times (A_{665} - A_{750})\ (\pm 0.0090)$$

$$\text{Chl } b\,(\mu\text{g / mL}) = -4.0937 \times (A_{632} - A_{750}) + 25.6865 \times (A_{649} - A_{750}) - 7.3430 \times (A_{665} - A_{750})(\pm 0.0171)$$

$$\text{Chl } c\,(\mu\text{g / mL}) = 28.5073 \times (A_{632} - A_{750}) - 9.9940 \times (A_{649} - A_{750}) - 1.9749 \times (A_{665} - A_{750})\ (\pm 0.0096)$$

$$\text{Total Chls}\,(\mu\text{g / mL}) = 23.4742 \times (A_{632} - A_{750}) + 11.4096 \times (A_{649} - A_{750}) + 4.0735 \times (A_{665} - A_{750})(\pm 0.0175)$$

- Quadrichroic equations for simultaneous determination of Chls *a*, *b*, *c* and *d* in extracts [60].

These equations should be used as the default equations on Chl extracts of unknown composition.

– 90 % acetone:

$$\text{Chl } a\,(\mu\text{g / mL}) = -0.3319 \times (A_{630} - A_{750}) - 1.7485 \times (A_{647} - A_{750}) + 11.9442 \times (A_{664} - A_{750}) - 1.4306 \times (A_{691} - A_{750})\ (\pm 0.0020)$$

$$\text{Chl } b\,(\mu\text{g / mL}) = -1.2825 \times (A_{630} - A_{750}) + 19.8839 \times (A_{647} - A_{750}) - 4.8860 \times (A_{664} - A_{750}) - 2.3416 \times (A_{691} - A_{750})\ (\pm 0.0076)$$

$$\text{Chl } c\,(\mu\text{g / mL}) = 23.5902 \times (A_{630} - A_{750}) - 7.8516 \times (A_{647} - A_{750}) - 1.5214 \times (A_{664} - A_{750}) - 1.7443 \times (A_{691} - A_{750})\ (\pm 0.0075)$$

$$\text{Chl } d\,(\mu\text{g / mL}) = -0.5881 \times (A_{630} - A_{750}) + 0.0902 \times (A_{647} - A_{750}) - 0.1564 \times (A_{664} - A_{750}) + 11.0473 \times (A_{691} - A_{750})\ (\pm 0.0030)$$

$$\text{Total Chls}\,(\mu\text{g/mL}) = 21.3877 \times (A_{630} - A_{750}) + 10.3739 \times (A_{647} - A_{750}) + 5.3805 \times (A_{664} - A_{750}) + 5.5309 \times (A_{691} - A_{750})\ (\pm 0.0056)$$

– 100 % methanol:

$$\text{Chl } a\,(\mu\text{g / mL}) = -2.0780 \times (A_{632} - A_{750}) - 6.5079 \times (A_{652} - A_{750}) + 16.2127 \times (A_{665} - A_{750}) - 2.1372 \times (A_{696} - A_{750})\ (\pm 0.0070)$$

$$\text{Chl } b\,(\mu\text{g / mL}) = -2.9450 \times (A_{632} - A_{750}) + 32.1228 \times (A_{652} - A_{750}) - 13.8255 \times (A_{665} - A_{750}) - 3.0097 \times (A_{696} - A_{750})\ (\pm 0.0212)$$

$$\text{Chl } c\,(\mu\text{g / mL}) = 34.0115 \times (A_{632} - A_{750}) - 12.7873 \times (A_{652} - A_{750}) - 1.4489 \times (A_{665} - A_{750}) - 2.5812 \times (A_{696} - A_{750})\ (\pm 0.0120)$$

$$\text{Chl } d\,(\mu\text{g / mL}) = -0.3411 \times (A_{632} - A_{750}) + 0.1129 \times (A_{652} - A_{750}) - 0.2538 \times (A_{665} - A_{750}) + 12.9508 \times (A_{696} - A_{750})\ (\pm 0.0031)$$

$$\text{Total Chls}\,(\mu g/mL) = 28.6473 \times (A_{632} - A_{750}) + 12.9405 \times (A_{652} - A_{750}) + 0.6845 \times (A_{665} - A_{750}) + 5.2230 \times (A_{696} - A_{750}) \quad (\pm 0.0140)$$

- 100 % ethanol:

$$\text{Chl } a\,(\mu g/mL) = 0.0604 \times (A_{632} - A_{750}) - 4.5224 \times (A_{649} - A_{750}) + 13.2969 \times (A_{665} - A_{750}) - 1.7453 \times (A_{696} - A_{750}) \quad (\pm 0.0053)$$

$$\text{Chl } b\,(\mu g/mL) = -4.1982 \times (A_{632} - A_{750}) + 25.7205 \times (A_{649} - A_{750}) - 7.4096 \times (A_{665} - A_{750}) - 2.7418 \times (A_{696} - A_{750}) \quad (\pm 0.0142)$$

$$\text{Chl } c\,(\mu g/mL) = 28.4593 \times (A_{632} - A_{750}) - 9.9944 \times (A_{649} - A_{750}) - 1.9344 \times (A_{665} - A_{750}) - 1.8093 \times (A_{696} - A_{750}) \quad (\pm 0.0084)$$

$$\text{Chl } d\,(\mu g/mL) = -0.2007 \times (A_{632} - A_{750}) + 0.0848 \times (A_{649} - A_{750}) - 0.1909 \times (A_{665} - A_{750}) + 12.1302 \times (A_{696} - A_{750}) \quad (\pm 0.0023)$$

$$\text{Total Chls}\,(\mu g/mL) = 28.1209 \times (A_{632} - A_{750}) + 11.2884 \times (A_{649} - A_{750}) + 3.7620 \times (A_{665} - A_{750}) + 5.8338 \times (A_{696} - A_{750}) \quad (\pm 0.0135)$$

- Equations for simultaneous determination of Chls *a* and *b* in DMF extracts [56, 67, 68].

These equations have been adapted from Inskeep and Bloom [56] in a 1-cm path cuvette:

$$\text{Chl } a\,(\mu g/mL) = 12.70 \times (A_{664.5} - A_{750}) - 2.79 \times (A_{647} - A_{750})$$

$$\text{Chl } b\,(\mu g/mL) = 20.70 \times (A_{647} - A_{750}) - 4.62 \times (A_{664.5} - A_{750})$$

$$\text{Total Chls}\,(\mu g/mL) = 17.90 \times (A_{647} - A_{750}) + 8.08 \times (A_{664.5} - A_{750})$$

- Equations for determination of Chl *a* and correction for pheopigments [58, 69].

- Extract pigments in 90 % acetone.
- Measure the absorbance of extracts at 665 and 750 nm.
- Acidify the samples with 1 drop of 1 M HCl in the cuvette followed by an invert mixing of the cuvette with the stopper.
- Measure the absorbance again at the same wavelengths as before acidification.
- Resolve the equation below developed by Lorenzen [69] in 90 % acetone extracts:

$$\text{Chl } a\,(\mu g/mL) = 26.73 \times \left[\left(A_{665_o} - A_{750_o} \right) - \left(A_{665_a} - A_{750_a} \right) \right]$$

With A_{665_o} and A_{750_o} the absorbances at 665 and 750 nm before acidification. And A_{665_a} and A_{750_a} the absorbances at 665 and 750 nm after acidification.

- Equation for determination of carotenoids [61, 67, 70].

Strickland and Parsons [61] developed an equation to assay globally carotenoids (mainly β-carotene) using the absorbance of 100 % methanol extracts and an extinction coefficient of 2,500 (100 mL/g/cm):

$$\text{Carotenoid}\left(\mu g / mL\right) = 4 \times \left(A_{480} - A_{750}\right)$$

This equation could be used to obtain global carotenoids level of the sample, but HPLC analysis will be required to determine the carotenoid composition.

- Equations for pigment determination of brown seaweeds [64, 65].

– 4:1 DMSO–water:

$$\text{Chl } a\left(\mu g / mL\right) = 13.74 \times \left(A_{665} - A_{750}\right)$$

$$\text{Chl } c\left(\mu g / mL\right) = 16.18 \times \left[\left(A_{631} - A_{750}\right) + \left(A_{582} - A_{750}\right)\right] - 4.81 \times \left(A_{665} - A_{750}\right)$$

$$\text{Fucox}\left(\mu g / mL\right) = 7.69 \times \left(A_{480} - A_{750}\right) - 5.55 \times \left[\left(A_{631} - A_{750}\right) + \left(A_{582} - A_{750}\right) - 0.297 \times \left(A_{665} - A_{750}\right)\right] - 0.377 \times \left(A_{665} - A_{750}\right)$$

– 10:1 acetone–hexane:

$$\text{Chl } a\left(\mu g / mL\right) = 12.00 \times \left(A_{661} - A_{750}\right)$$

$$\text{Carotenoid}\left(\mu g / mL\right) = 5.18 \times \left(A_{480} - A_{750}\right) - 0.171 \times \left(A_{661} - A_{750}\right)$$

– 3:1:1 acetone–methanol–water:

$$\text{Chl } a\left(\mu g / mL\right) = 13.59 \times \left(A_{664} - A_{750}\right)$$

$$\text{Chl } c\left(\mu g / mL\right) = 16.08 \times \left[\left(A_{631} - A_{750}\right) + \left(A_{581} - A_{750}\right)\right] - 4.82 \times \left(A_{664} - A_{750}\right)$$

$$\text{Fucox}\left(\mu g / mL\right) = 7.09 \times \left(A_{470} - A_{750}\right) - 8.79 \times \left[\left(A_{631} - A_{750}\right) + \left(A_{581} - A_{750}\right) - 0.300 \times \left(A_{664} - A_{750}\right)\right] - 0.195 \times \left(A_{664} - A_{750}\right)$$

3.5 Pigments (Phycobiliproteins)

Phycobiliproteins (or phycobilins) are polar pigments present in some algae, such as Rhodophyta, Cyanobacteria and Cryptophytes where they are the major light-harvesting chromoproteins. They are commonly divided into four classes based on their light absorption properties and types of bilins: phycoerythrins (PEs; λ_{max} = 540–570 nm), phycocyanins (PCs; λ_{max} = 610–620 nm), allophycocyanins (APCs; λ_{max} = 650–655 nm), and phycoerythrocyanins (PECs;

λ_{max} = 560–600 nm) [71]. Within phycoerythrins, three classes are recognized based on their absorption spectrum: R-phycoerythrin (R-PE), B-phycoerythrin (B-PE), and C-phycoerythrin (C-PE) [71, 72].

PEs are commonly extracted in diluted phosphate buffer from algae and then precipitated by salting out with ammonium sulfate at different percentages before further purification using various techniques (expanded bed adsorption chromatography, gel filtration, hydroxyapatite chromatography, a combination of two chromatography modes or preparative electrophoresis) [71]. An extraction and purification procedure of R-PE from red seaweeds is developed in Chapter 5.

Here the equations used for the determination of phycobiliproteins from algae, including PEs but also PC and APC, are presented.

Similar to chlorophyll and carotenoid content determination, phycobiliprotein assays derived from the use of mathematical formulae using the extinction coefficients of the different phycobiliproteins (e.g., [73]). Absorption spectra of samples were measured from 400 to 750 nm in a UV-VIS spectrophotometer (*see* **Note 50**) and the general formula is as follow [74]:

$$\text{Phycobiliprotein}\,(\text{mg / L}) = \frac{(A_{\text{max}} - A_{750})}{\varepsilon d} \times \text{MW} \times 10^3$$

Where, A_{max} is the absorbance maximum of the phycobiliprotein (565 nm for phycoerythrin and 618 nm for phycocyanin); A_{750} the absorbance due to scattering; ε the molar extinction coefficients (PE: 2.41 10^6 L/mol cm; PC: 1.90 10^6 L/mol cm); MW molecular weight of phycobiliproteins (PE: 240,000 g/mol; PC: 264,000 g/mol); d the path length of the cuvette [74]. The R-PE absorption spectra displayed three peaks: two at 495 and 545 nm and one main peak at 565 nm. The spectral profile is commonly used to indicate the non-degradation of R-PE [4].

Beer and Eshel's equations are very commonly used [68, 75, 76] but underestimate the phycobiliprotein contents of some seaweed species such as *Porphyra* spp. and have thus been modified [77]. The equations presented here are adapted from refs. [73, 77, 78]:

$$\text{R-PC}\,(\text{mg / mL}) = 0.154 \times (A_{618} - A_{750})$$

$$\text{R-PE}\,(\text{mg / mL}) = 0.1247 \times [(A_{564} - A_{750}) - 0.4583 \times (A_{618} - A_{750})]$$

$$\text{Purity Index or PI of R-PE} = \frac{A_{564}}{A_{280}}$$

$$\text{APC}\,(\text{mg / mL}) = 0.177 \times (A_{652} - A_{750}) - 0.034 \times (A_{615} - A_{750})$$

4 Notes

1. It is possible to use some assay kits such as the Bio-Rad Dc Protein assay in which the complexing reagent and the Folin reagent are ready to use.
2. The diluted FC reagent could be stored for a long period in a brown bottle (protected from light).
3. The FC reagent contains hydrochloric and phosphoric acids so wear gloves and goggles and work carefully.
4. The FC reagent can also be prepared in the laboratory (*see* Subheading 2.3.1; adapted from Singleton et al. [6]).
5. Use a plastic falcon tube instead of glassware for solutions containing copper.
6. Ready-to-use Bradford reagent is available to purchase from for example Bio-Rad and Sigma.
7. The Bradford reagent is then stable for several weeks.
8. Bovin serum albumin (BSA) is often used as a standard in Bradford assay but it has one big disadvantage: BSA exhibits an unusually large dye response and thus may underestimate the protein content of a sample. On the contrary, bovin γ-globulin exhibits a closer dye binding capacity to the mean of many proteins tested [13].
9. These reagents may be purchased ready-to-use from for example Sigma or Pierce Biotechnology Inc.
10. Phenol is easily absorbed through the skin; fatalities have been documented from skin absorption via a relatively small surface area. Use double gloves, made of nylon, rubber or neoprene. Safety glasses, goggles and face shield should be worn.
11. Use sulfuric acid in a fume hood. Wear nitrile gloves and goggles or face shield.
12. Prepare just before use.
13. You can use commercial FC reagent instead of preparing it.
14. The resultant solution should be clear and intense yellow without a trace of green (blue). Any blue results from traces of reduced reagent and will cause elevated blanks. Refluxing for a short time after adding a couple of drops of bromine followed by removal of the excess bromine by open boiling (in a fume hood) will correct this problem. To avoid excess of bromine, a small amount of 30 % hydrogen peroxide can be substituted to the bromine [6].
15. If protected from reductants, the FC reagent is normally stable indefinitely, even if diluted.
16. Heat the solution if needed as Na_2CO_3 is at its solubilization limit.

17. DMF causes respiratory tract irritation; may cause digestive tract irritation with nausea, vomiting, and diarrhea; causes eye and skin irritation. Prolonged or repeated skin contact may cause dermatitis. This substance has caused adverse reproductive and fetal effects in animals. It may cause liver and kidney damage and has potential cancer hazard. Wear appropriate goggles, gloves and use a fume hood.
18. Use a fume cupboard, cover water bath, and cuvettes with stoppers when using this acid.
19. The working reagent is slightly yellow when mixed, but becomes purple (λ_{max} = 565 nm) upon standing, so it must be prepared fresh each day [46].
20. The working reagent is kept at room temperature before starting the assay [46].
21. For safety reason, work under a fume hood with gloves and goggles when using these solvents.
22. DMF is much more toxic than acetone or ethanol.
23. Acetone is very volatile, highly flammable, causes headache, is narcotic in high concentration, and is a skin irritant (erythema); so care should be taken when using it (gloves, goggles, use of a fume hood). Acetone is also unsuited for fieldwork due to its flammability, propensity for leaking out of containers, volatility, and security concerns makes it problematic to transport particularly by air. Finally, acetone attacks polystyrene and polymethylacrylates; check the resistivity of your laboratory plasticware or use glassware instead [60].
24. Methanol is an insidious and notoriously toxic solvent. Methanol attacks some, but not all, types of plastic commonly used to make laboratory plasticware. Methanol is much easier to transport and easier to handle in the field [60].
25. Ethanol is a much safer solvent than either acetone or methanol: although flammable it is not very toxic. Ethanol does not attack polystyrene so normal plastic container and normal laboratory plasticware can be used [60].
26. DMSO is compatible only with some plasticware such as polypropylene and Teflon but only moderately with polystyrene and not with polycarbonate.
27. When performing a sonic disruption with a sonifier at 70 W, samples should be on ice, and each sample should be sonicated for three cycles of 1 min; between sonication steps, samples should be kept on ice for at least 1 min [11].
28. This assay may be adapted for analysis in a microplate by reducing the volume of reagents and extract accordingly.
29. The vortex-step is critical to get reproducible results and care should be taken to mix well the tubes.

30. In this microassay the volume of the Bradford reagent added is 1 mL instead of 5 mL to each 0.1 mL of sample or standard [13].
31. Do not use quartz cuvette as the dye will bind to it. Traces of dye can be removed from plastic and glassware using methanol or detergent solution.
32. The binding of the dye with the proteins is very quick and the protein-dye complex remains soluble for 1 h [8].
33. A micromethod has also been developed by Walker [14] for samples containing 0.5–10 μg protein/mL with modification in reagents composition and volumes:

    Reagent A: weigh 0.8 g of sodium carbonate ($Na_2CO_3{\cdot}H_2O$), 1.6 g of sodium tartrate ($C_4H_4Na_2O_6{\cdot}2H_2O$), 1.6 g of sodium hydroxide (NaOH), mix and make up to 100 mL with distilled water and adjust pH to 11.25 with 10 M NaOH.

    Reagent B: weigh 4 g of sodium bicinchoninate ($C_{20}H_{10}N_2O_4Na_2$) in 100 mL of distilled water.

    Reagent C: pour 0.4 g of cupric sulfate ($CuSO_4{\cdot}5H_2O$) in 10 mL of distilled water.

    Working reagent: mix 1 vol. of reagent C with 25 vol. of reagent B, then add 26 vol. of reagent A.

    To 100 μL of sample add 100 μL of working reagent. Continue as for the standard method.
34. After heating, the color is stable for at least 1 h.
35. Some kits are also developed by companies to assay sugars such as the glucose determinant assay kit from Megazyme International, Ireland, or Sigma. If you are using these kits, follow manufacturer instructions.
36. Instead of starting from an extract, one method starts from the milled algae [22]: reconstitute 10 mg of freeze-dried and milled biomass (stored at −20 °C before analysis) in 10 mL of water to prepare a known sample concentration for each sample (1 mg/mL). To 1 mL sample add 3 mL of conc. sulfuric acid (72 % w/v) and 1 mL of 5 % phenol (w/v) in a water bath. Incubate the tubes for 5 min at 90 °C. Measure the absorbance at 490 nm.
37. The color is stable for several hours.
38. Both standard solutions and samples have to be diluted in an adequate solvent (called "blank") generally corresponding to the extraction conditions and possibly containing high EDTA concentrations.
39. About pH 10 is desired after combination with the acidic FC reagent and the samples.
40. The absorbance could also be read at 730, 735, 750, 760, or 765 nm instead of 700 nm; all wavelengths give similar phenolic levels as the calibration curve is treated in the same way as the algal extracts.

41. The calibration curve could also be prepared using phloroglucinol. However, the absorbance of standard solutions should be thus measured at a different wavelength from the seaweed extracts: 494 nm instead of 515 nm [34].
42. Small amounts of methanol (4 % methanol) inhibit the reaction between vanillin and phenolics. Also, water or DMF reduce the color yield in the DMBA reaction. When using methanolic or aqueous extracts of algae, controls must be included to account for this variation [46].
43. DMF is necessary for use in protein precipitation assays [46].
44. Reaction temperature and time have to be kept strictly constant.
45. Some color formation could occur in the absence of phlorotannins and should be removed from the assay [46].
46. No effect of solvent type (acetone, methanol, distilled water, or filtered seawater) was observed on the removal of phlorotannins [43].
47. The pH of the supernatant should be in the range between 6 and 7 [34].
48. It is preferable to use low quantities of PVPP repeatedly (10 mg/mL; 2 successive 10 min PVPP incubations) compared to using fewer treatments with a high amount of PVPP [43].
49. Using this method, more than 94 % of phloroglucinol was removed from extracts of two brown seaweeds *Ascophyllum nodosum* and *Fucus vesiculosus* [34]. But this result will depend on the algal species.
50. Use a 1-cm quartz glass cuvette against phosphate buffer as a blank.

References

1. Stengel DB, Connan S, Popper ZA (2011) Algal chemodiversity and bioactivity: sources of natural variability and implications for commercial application. Biotechnol Adv 29: 483–501
2. Connan S, Goulard F, Stiger V et al (2004) Interspecific and temporal variation in phlorotannin levels in an assemblage of brown algae. Bot Mar 47:410–416
3. Le Lann KL, Ferret C, VanMee E et al (2012) Total phenolic, size-fractionated phenolics and fucoxanthin content of tropical Sargassaceae (Fucales, Phaeophyceae) from the South Pacific Ocean: spatial and specific variability. Phycol Res 60:37–50
4. Munier M, Dumay J, Morançais M et al (2013) Variation in the biochemical composition of the edible seaweed *Grateloupia turuturu* Yamada harvested from two sampling sites on the Brittany coast (France): the influence of storage method on the extraction of the seaweed pigment R-phycoerythrin. J Chem. doi:10.1155/2013/568548
5. Guihéneuf F, Stengel DB (2013) LC-PUFA-enriched oil production by microalgae: accumulation of lipid and triacylglycerols containing n-3 LC-PUFA is triggered by nitrogen limitation and inorganic carbon availability in the marine haptophyte *Pavlova lutheri*. Mar Drugs 11:4246–4266
6. Singleton VL, Orthofer R, Lamuela-Raventos RM (1999) Analysis of total phenols and other oxidation substrates and antioxidants by means of Folin-Ciocalteu reagent. Method Enzymol 299:152–178
7. Rausch T (1981) The estimation of microalgal protein content and its meaning to the evaluation of algal biomass I. Comparison of

methods for extracting protein. Hydrobiologia 78:237–251
8. Barbarino E, Lourenço SO (2005) An evaluation of methods for extraction and quantification of protein from marine macro- and microalgae. J Appl Phycol 17:447–460
9. Pruvost J, Van Vooren G, Le Gouic B et al (2011) Systematic investigation of biomass and lipid productivity by microalgae in photobioreactors for biodiesel application. Bioresource Technol 102:150–158
10. González López CV, García MDCC, Fernández FGA et al (2010) Protein measurements of microalgal and cyanobacterial biomass. Bioresource Technol 101:7587–7591
11. Meijer EA, Wijffels RH (1998) Development of a fast, reproductible and effective method for the extraction and quantification of proteins of micro-algae. Biotechnol Tech 12:353–358
12. Waterborg JH (2002) The Lowry method for protein quantification. In: Walker JM (ed) The protein protocols handbook. Humana Press Inc., Totowa, NJ, pp 7–9
13. Kruger NJ (2002) The Bradford method for protein quantification. In: Walker JM (ed) The protein handbook. Humana Press Inc., Totowa, NJ, pp 15–21
14. Walker JM (2002) The bicinchoninic acid (BCA) assay for protein quantitation. In: Walker JM (ed) The protein protocols handbook. Humana Press Inc., Totowa, NJ, pp 11–14
15. Crossman D, Clements K, Cooper G (2000) Determination of protein for studies of marine herbivory: a comparison of methods. J Exp Mar Biol Ecol 244:45–65
16. Rondel C, Marcato-Romain C-E, Girbal-Neuhauser E (2013) Development and validation of a colorimetric assay for simultaneous quantification of neutral and uronic sugars. Water Res 47:2901–2908
17. Dubois M, Gilles KA, Hamilton JK et al (1956) Colorimetric method for determination of sugars and related substances. Anal Chem 28:350–356
18. Dreywood R (1946) Qualitative test for carbohydrate material. Ind Eng Chem 18:499
19. Norikoshi R, Imanishi H, Ichimura K (2008) A simple and rapid extraction method of carbohydrates from petals or sepals of four floricultural plants for determination of their content. J Jpn Soc Hortic Sci 77:289–295
20. Santoyo S, Plaza M, Jaime L et al (2011) Pressurized liquids as an alternative green process to extract antiviral agents from the edible seaweed *Himanthalia elongata*. J Appl Phycol 23:909–917
21. Adams JMM, Ross AB, Anastasakis K et al (2011) Seasonal variation in the chemical composition of the bioenergy feedstock *Laminaria digitata* for thermochemical conversion. Bioresource Technol 102:226–234
22. Laurens LM, Dempster TA, Jones HD et al (2012) Algal biomass constituent analysis: method uncertainties and investigation of the underlying measuring chemistries. Anal Chem 84:1879–1887
23. Deniaud-Bouët E, Kervarec N, Michel G et al (2014) Chemical and enzymatic fractionation of cell walls from Fucales: insights into the structure of the extracellular matrix of brown algae. Ann Bot. doi:10.1093/aob/mcu096
24. Ragan MA, Glombitza KW (1986) Phlorotannins, brown algal polyphenols. In: Round FE, Chapman DJ (eds) Progress in phycological research. Biopress Ltd., Bristol, pp 129–241
25. Hur S, Lee H, Kim Y et al (2008) Sargaquinoic acid and sargachromenol, extracts of *Sargassum sagamianum*, induce apoptosis in HaCaT cells and mice skin: its potentiation of UVB-induced apoptosis. Eur J Pharmacol 582:1–11
26. de la Mare J-A, Lawson JC, Chiwakata MT et al (2012) Quinones and halogenated monoterpenes of algal origin show anti-proliferative effects against breast cancer cells in vitro. Invest New Drugs 30:2187–2200
27. Whitfield FB, Helidoniotis F, Shaw KJ et al (1999) Distribution of bromophenols in species of marine algae from eastern Australia. J Agric Food Chem 47:2367–2373
28. Phillips DW, Towers GHN (1982) Chemical ecology of red algal bromophenols. I. Temporal, interpopulational and within-thallus measurements of lanosol levels in *Rhodomela larix* (Turner) C. Agardh. J Exp Mar Biol Ecol 58:285–293
29. Flodin C, Helidoniotis F, Whitfield FB (1999) Seasonal variation in bromophenol content and bromoperoxidase activity in *Ulva lactuca*. Phytochemistry 51:135–138
30. Li H-B, Cheng K-W, Wong C-C et al (2007) Evaluation of antioxidant capacity and total phenolic content of different fractions of selected microalgae. Food Chem 102: 771–776
31. Klejdus B, Kopecký J, Benešová L et al (2009) Solid-phase/supercritical-fluid extraction for liquid chromatography of phenolic compounds in freshwater microalgae and selected cyanobacterial species. J Chromatogr A 1216: 763–771
32. Koivikko R, Loponen J, Honkanen T et al (2005) Contents of soluble, cell-wall-bound and exuded phlorotannins in the brown alga *Fucus vesiculosus*, with implications on

their ecological functions. J Chem Ecol 31: 195–212
33. Cerantola S, Breton F, Ar Gall E et al (2006) Co-occurence and antioxidant activities of fucol and fucophlorethol classes of polymeric phenols in *Fucus spiralis*. Bot Mar 49:347–351
34. Parys S, Rosenbaum A, Kehraus S et al (2007) Evaluation of quantitative methods for the determination of polyphenols in algal extracts. J Nat Prod 70:1865–1870
35. Connan S, Stengel DB (2011) Impacts of ambient salinity and copper on brown algae: 2. Interactive effects on phenolic pool and assessment of metal binding capacity of phlorotannin. Aquat Toxicol 104:1–13
36. Lopes G, Sousa C, Silva LR et al (2012) Can phlorotannins purified extracts constitute a novel pharmacological alternative for microbial infections with associated inflammatory conditions? PLoS One 7, doi: 10.1371/journal.pone.0031145
37. Trigui M, Gasmi L, Zouari I et al (2013) Seasonal variation in phenolic composition, antibacterial and antioxidant activities of *Ulva rigida* (Chlorophyta) and assessment of anti-acetylcholinesterase potential. J Appl Phycol 25:319–328
38. Folin O, Denis W (1915) A colorimetric method for the determination of phenols (and phenol derivates) in urine. J Biol Chem 22: 305–308
39. Hagerman AE, Butler LG (1989) Choosing appropriate methods and standards for assaying tannins. J Chem Ecol 15:1795–1810
40. Van Alstyne KL (1995) Comparison of three methods for quantifying brown algal polyphenolic compounds. J Chem Ecol 21:45–58
41. Folin O, Ciocalteu V (1927) On tyrosine and tryptophane determinations in proteins. J Biol Chem 73:627–650
42. Tan JBL, Lim YY (2015) Critical analysis of current methods for assessing the in vitro antioxidant and antibacterial activity of plant extracts. Food Chem 172:814–822
43. Toth GB, Pavia H (2001) Removal of dissolved brown algal phlorotannins using insoluble polyvinylpolypyrrolidone (PVPP). J Chem Ecol 27:1899–1910
44. Ribéreau-Gayon P (1972) Plant phenolics, vol 3. Oliver and Boyd, Edinburgh, UK
45. Butler LG, Price ML, Brotherton JE (1982) Vanillin assay for proanthocyanidins (condensed tannins): modification of the solvent for estimation of the degree of polymerization. J Agric Food Chem 30:1087–1089
46. Stern JL, Hagerman AE, Steinberg PD et al (1996) A new assay for quantifying brown algal phlorotannins and comparisons to previous methods. J Chem Ecol 22:1273–1293
47. Yates JL, Peckol P (1993) Effects of nutrient availability and herbivory on polyphenolics in the seaweed *Fucus vesiculosus*. Ecology 74: 1757–1766
48. Jégou C, Culioli G, Kervarec N et al (2010) LC/ESI-MS*n* and ^{1}H HR-MAS NMR analytical methods as useful taxonomical tools within the genus *Cystoseira* C. Agardh (Fucales; Phaeophyceae). Talanta 83:613–622
49. Sanoner P, Guyot S, Marnet N et al (1999) Polyphenol profiles of French cider apple varieties (*Malus domestica* sp.). J Agric Food Chem 47:4847–4853
50. Koivikko R, Eränen J, Loponen J et al (2008) Variation of phlorotannins among three populations of *Fucus vesiculosus* as revealed by HPLC and colorimetric quantification. J Chem Ecol 34:57–64
51. Zubia M, Fabre MS, Kerjean V et al (2009) Antioxidant and cytotoxic activities of some red algae (Rhodophyta) from Brittany coasts (France). Bot Mar 52:268–277
52. Audibert L, Fauchon M, Blanc N et al (2010) Phenolic compounds in the brown seaweed *Ascophyllum nodosum*: distribution and radical-scavenging activities. Phytochem Anal 21: 399–405
53. Arnon DI (1949) Copper enzymes in isolated chloroplasts. Polyphenoloxidase in *Beta vulgaris*. Plant Physiol 24:1–16
54. Jeffrey SW, Humphrey GF (1975) New spectrophotometric equations for determining chlorophylls *a*, *b*, *c1* and *c2* in higher plants, algae and natural phytoplankton. Biochem Physiol Pfl 167:191–194
55. Humphrey GF (1979) Photosynthetic characteristics of algae grown under constant illumination and light-dark regimes. J Exp Mar Biol Ecol 40:63–70
56. Inskeep WP, Bloom PR (1985) Extinction coefficients of chlorophyll a and b in N, N-dimethylformamide and 80 % acetone. Plant Physiol 77:483–485
57. Porra RJ, Thompson WA, Kriedemann PE (1989) Determination of accurate extinction coefficients and simultaneous equations for assaying chlorophylls *a* and *b* extracted with four different solvents: verification of the concentration of chlorophyll standards by atomic absorption spectroscopy. Biochim Biophys Acta 975:384–394
58. Jeffrey SW (1997) Chlorophyll and carotenoid extinction coefficients. In: Jeffrey SW, Mantoura RFC, Wright SW (eds) Phytoplankton pigments in oceanography:

guidelines to modern methods. UNESCO Publishing, Paris, pp 595–596

59. Ritchie RJ (2006) Consistent sets of spectrophotometric chlorophyll equations for acetone, methanol and ethanol solvents. Photosynth Res 89:27–41
60. Ritchie RJ (2008) Universal chlorophyll equations for estimating chlorophylls *a*, *b*, *c* and *d* and total chlorophylls in natural assemblages of photosynthetic organisms using acetone, methanol, or ethanol solvents. Photosynthetica 46:115–126
61. Strickland JDH, Parsons TR (1968) A practical handbook of seawater analysis. Bull Fish Res Bd Can vol. 167
62. Jeffrey SW, Welschmeyer NA (1997) Spectrophotometric and fluorometric equations in common use in oceanography. In: Jeffrey SW, Mantoura RFC, Wright SW (eds) Phytoplankton pigments in oceanography: guidelines to modern methods. UNESCO Publishing, Paris, pp 597–615
63. Mantoura RFC, Jeffrey SW, Llewellyn CA et al (1997) Comparison between spectrophotometric, fluorometric and HPLC methods for chlorophyll analysis. In: Jeffrey SW, Mantoura RFC, Wright SW (eds) Phytoplankton pigments in oceanography: guidelines to modern methods. UNESCO Publishing, Paris, pp 361–380
64. Seely GR, Duncan MJ, Vidaver WE (1972) Preparative and analytical extraction of pigments from brown algae with dimethyl sulfoxide. Mar Biol 12:184–188
65. Stengel DB, Dring MJ (1998) Seasonal variation in the pigment content and photosynthesis of different thallus regions of *Ascophyllum nodosum* (Fucales, Phaeophyta) in relation to position in the canopy. Phycologia 37: 259–268
66. Jeffrey SW (1997) Application of pigment methods to oceanography. In: Jeffrey SW, Mantoura RFC, Wright SW (eds) Phytoplankton pigments in oceanography: guidelines to modern methods. UNESCO Publishing, Paris, pp 127–166
67. Altamirano M, Flores-Moya A, Conde F et al (2000) Growth seasonality, photosynthetic pigments, and carbon and nitrogen content in relation to environmental factors: a field study of *Ulva olivascens* (Ulvales, Chlorophyta). Phycologia 39:50–58
68. Korbee N, Figueroa FL, Aguilera J (2005) Effect of light quality on the accumulation of photosynthetic pigments, proteins and mycosporine-like amino acids in the red alga *Porphyra leucosticta* (Bangiales, Rhodophyta). J Photochem Photobiol B 80:71–78
69. Lorenzen CJ (1967) Determination of chlorophyll and pheopigments: spectrophotometric equations. Limnol Oceanogr 12:343–346
70. Pruvost J, Van Vooren G, Cogne G et al (2009) Investigation of biomass and lipids production with *Neochloris oleoabundans* in photobioreactor. Bioresource Technol 100:5988–5995
71. Munier M, Jubeau S, Wijaya A et al (2014) Physicochemical factors affecting the stability of two pigments: R-phycoerythrin of *Grateloupia turuturu* and B-phycoerythrin of *Porphyridium cruentum*. Food Chem 150: 400–407
72. Mishra SK, Shrivastav A, Mishra S (2011) Preparation of highly purified C-phycoerythrin from marine cyanobacterium *Pseudanabaena* sp. Protein Expres Purif 80:234–238
73. Beer S, Eshel A (1985) Determining phycoerythrin and phycocyanin concentrations in aqueous crude extracts of red algae. Aust J Mar Fresh Res 36:785–792
74. Lawrenz E, Fedewa EJ, Richardson TL (2011) Extraction protocols for the quantification of phycobilins in aqueous phytoplankton extracts. J Appl Phycol 23:865–871
75. Denis C, Morançais M, Li M et al (2010) Study of the chemical composition of edible red macroalgae *Grateloupia turuturu* from Brittany (France). Food Chem 119:913–917
76. Chopin T, Yarish C, Wilkes R et al (1999) Developing *Porphyra*/salmon integrated aquaculture for bioremediation and diversification of the aquaculture industry. J Appl Phycol 11:463–472
77. Sampath-Wiley P, Neefus CD (2007) An improved method for estimating R-phycoerythrin and R-phycocyanin contents from crude aqueous extracts of *Porphyra* (Bangiales, Rhodophyta). J Appl Phycol 19:123–129
78. Senthilkumar N, Suresh V, Thangam R et al (2013) Isolation and characterization of macromolecular protein R-phycoerythrin from *Portieria hornemannii*. Int J Biol Macromol 55:150–160

Chapter 4

Extraction and Enrichment of Protein from Red and Green Macroalgae

Pádraigín A. Harnedy and Richard J. FitzGerald

Abstract

Macroalgae, in particular red and green species, are gaining interest as protein-rich foods for human consumption and sources of proteinaceous biofunctional peptide ingredients. During protein extraction the starting raw material, the cell disruption method utilized and the reagents employed have a major effect on the yield of protein recovered. A method is described herein for extraction and semi-purification of food-grade aqueous and alkaline soluble proteins from red and green macroalgae. Dried milled macroalgae are disrupted by osmotic shock with subsequent removal of aqueous soluble proteins by centrifugation. Alkaline soluble proteins are removed following consecutive treatment of the resultant pellet with an alkaline solution. Aqueous and alkaline soluble proteins are then enriched from the crude extracts by isoelectric precipitation.

Key words Alkaline protein extracts, Aqueous protein extract, Cell disruption, Chlorophyta, Macroalgae, Protein extraction, Rhodophyta

1 Introduction

Macroalgae, a popular food used in many oriental countries, have begun to emerge as an alternative dietary source of protein and protein based biofunctional ingredients [1]. These marine organisms, in particular the red and green species, are reported to contain significant levels of protein (9–47 % (w/w)) [2]. The extraction of macroalgal proteins is hindered, in part, due to inaccessibility of proteins within macromolecular cell wall assemblies, cross-linking via disulfide bonds to polysaccharides within these assemblages and high viscosity and ionic interactions arising from cell wall and intracellular polysaccharides [3, 4]. The latter is particularly relevant for brown macroalgae. The nature of the starting material, the method used for cellular disruption and the use of specific reagents during the protein extraction process can all significantly affect the yield of protein recovered. Furthermore, the intended downstream application of the protein will dictate the types of reagents that can be

Dagmar B. Stengel and Solène Connan (eds.), *Natural Products From Marine Algae: Methods and Protocols*, Methods in Molecular Biology, vol. 1308, DOI 10.1007/978-1-4939-2684-8_4, © Springer Science+Business Media New York 2015

used during extraction (food-grade reagents and processes are required if the macroalgal proteins are intended for human consumption). While the application of food-grade polysaccharidases has been identified as a promising approach for cell disruption and release of proteins, the extraction of intact protein components from macroalgae is often hindered due to the presence of contaminant proteolytic activities within the polysaccharidase preparations [5–8]. Starting with dried milled macroalgae, osmotic shock in aqueous solution and the use of alkaline conditions has been recognized as an effective strategy for the extraction of aqueous soluble protein and solubilization of highly water insoluble hydrophobic macroalgal proteins, respectively [6]. Reducing agents such as *N*-acetyl-L-cysteine could be used to improve cell wall-associated protein extraction. However, inclusion of such agents will depend on the downstream applications of the extracted protein [6].

2 Materials

Prepare all solutions using deionized water and analytical grade reagents. Prepare and store all reagents at room temperature unless otherwise stated.

1. 0.12 M NaOH: dissolve 4.8 g of NaOH in approximately 750 mL of water. When dissolved transfer the solution completely to a 1 L volumetric flask and make up to the mark with water (*see* **Note 1**).
2. 2 M NaOH: dissolve 80 g of NaOH in approximately 750 mL of water. When dissolved transfer as above (*see* **Note 1**).
3. Water at pH 11.0: Place a pH probe in a beaker of stirring dH_2O water. Slowly dropwise add in 2 M NaOH until the pH reaches 11.0 (*see* **Note 2**).

3 Methods

3.1 Extraction of Crude Aqueous Soluble Proteins from Red and Green Macroalgae

1. Weigh exactly 1 g of dried milled seaweed powder and transfer to a 50 mL beaker (*see* **Notes 3** and **4**). Measure out exactly 20 mL of dH_2O. Using the 20 mL of water carefully transfer any remaining seaweed powder from the weigh boat into the beaker. Gently stir for 3 h at 4 °C (*see* **Note 5**).
2. Transfer the macroalgal suspension to a centrifuge tube (*see* **Note 6**) and centrifuge at 4,190 × *g* for 15 min at either 4 °C or room temperature (the temperature will be governed by the temperature at which the aqueous soluble proteins were extracted).

3. Following centrifugation carefully decant the supernatant into a graduated cylinder and take note of the volume (*see* **Notes 7** and **8**). The protein present in the supernatant is termed the crude aqueous protein extract.

3.2 Extraction of Crude Alkaline Soluble Proteins from Red and Green Macroalgae

1. Resuspend the macroalgal cells (pellet) from the previous step at a weight:volume ratio of 1:15 (w/v) (15 mL) with 0.12 M NaOH and stir at room temperature for 1 h (*see* **Notes 9** and **10**).
2. Remove the extract containing alkaline soluble proteins following centrifugation as described above. Subject the pellet from the above to a second alkaline extraction and combine both supernatants and note the volume (*see* **Notes** 7 and **8**). The protein present in the combined supernatant is termed the crude alkaline protein extract.

3.3 Enrichment of the Aqueous and Alkaline Soluble Proteins

1. Transfer the crude aqueous and alkaline soluble protein extracts to separate beakers. Alter the pH of each to pH 3.5 using HCl (*see* **Notes 11** and **12**).
2. After adjustment of both solutions to pH 3.5, allow them stand for 30 min at 4 °C or room temperature for the aqueous protein extract (the temperature used will be governed by the temperature at which the aqueous soluble proteins were extracted) and room temperature for the alkaline protein extract.
3. After 30 min, transfer both extracts to separate centrifuge tubes and centrifuge at 4,190 × *g* for 15 min at the same temperature as used above.
4. Carefully remove the supernatant and resuspend the pellet containing the purified aqueous and alkaline soluble proteins with pH 11.0 water. The majority of the pellet can be resuspended in the centrifuge tube using gentle vortexing (*see* **Note 13**). For aqueous protein extracts that may contain endogenous macroalgal proteolytic activities, the solution should be kept on ice between vortexing. Ensure all the protein has been transferred to a beaker and stir gently until all the protein has gone into solution (*see* **Note 14**). This step may need to be performed at 4 °C for the enriched aqueous protein extract.
5. Once all the protein has gone into solution, measure the volume of each of the purified protein solutions using a graduated cylinder (*see* **Notes** 7 and **8**).
6. Transfer each of the enriched protein solutions to an appropriate container for subsequent storage.

4 Notes

1. Special care is required to prepare a solution of NaOH in water as considerable heat is liberated by the exothermic reaction involved.
2. Water does not have a good buffering capacity; therefore it is difficult to alter the water to exactly pH 11.0. An aqueous solution near to pH 11.0 is sufficient.
3. Drying (freeze- or air-drying) and milling (1 mm screen) the seaweed prior to extraction improves the yield of protein recovered. Not only does it ensure reproducibility, something that is difficult to achieve when using wet blended seaweed, it breaks up the cells and increases the surface area of the seaweed, exposing more surface area for subsequent protein extraction.
4. A ratio of 1:20 (w/v) dried milled seaweed to water was found as the best weight:volume ratio for osmotic shock of macroalgal cells and extraction of aqueous soluble protein [6]. Depending on the scale of the extraction, the weight of seaweed and volume of water can be altered accordingly (e.g., 0.5 g:10 mL or 100 g:2,000 mL).
5. The temperature selected for the extraction of aqueous soluble proteins from macroalgae will depend on the presence or absence of endogenous proteolytic enzymes. If this information is not available the best temperature to perform the extraction of aqueous soluble proteins is 4 °C. If it is known that little or no proteolytic activity is present, the extraction can be performed at room temperature.
6. Centrifuge tubes with a conical base are most suitable for separation of the macroalgal suspension. Their use aids in the formation of a hard pellet. When round bottomed centrifuge tubes were used the pellet was not as hard and when the supernatant was decanted some of the macroalgal cells in the pellet can break away and may end up in the crude protein supernatant.
7. It is critical to record the volume of each of the supernatants in order to determine the quantity of protein present in a given extract following protein quantification analysis.
8. Macroalgal protein extracts can contain components that interfere with colorimetric protein quantification assays. The modified Lowry protein quantification assay can be used to eliminate such interferences [6, 9]. In this assay the protein in the macroalgal extract is precipitated by trichloroacetic acid (TCA) prior to analysis.
9. 0.12 M NaOH, a ratio of 1:15 (w/v) dried milled seaweed to alkaline solution and 1 h extraction duration were found as the

best alkaline concentration, weight:volume ratio and extraction duration for extraction of alkaline soluble protein from macroalgae [6]. Depending on the scale of the extraction, the weight of seaweed and volume of alkaline 0.12 M NaOH can be altered accordingly (e.g., 0.5 g:7.5 mL or 100 g:1,500 mL).

10. In order to ensure the transfer of all the macroalgal cells from the centrifuge tube to the beaker in which the alkaline extraction is to be performed, it is advisable to resuspend the pellet in approximately 2/3 the total volume of 0.12 M NaOH (10 mL if the starting macroalgal weight is 1 g) and transfer the macroalgal suspension to the beaker and wash the centrifuge tube with the remainder of the 0.12 M NaOH (5 mL if the starting macroalgal weight is 1 g).

11. Concentrated (12 M) HCl can be used in the first instance to reduce the pH to approximately 4.0. In order to avoid generation of a solution with too low a pH, 1 M HCl can be used to reduce the pH to pH 3.5.

12. The isoelectric point (pI) for macroalgal aqueous and alkaline proteins is in general between pH 3–4. If the pI value requires checking, the following experiment can be performed. Measure 20 mL of macroalgal extract into beakers (the number depends on the number of pH values chosen; e.g., pH 2.0, 2.5, 3.0, 3.5, 4.0, and 4.5). Alter the pH to the different pH value as described previously. When the desired pH value has been reached allow the solution to stand for 30 min at the required temperature. After 30 min transfer the extracts to separate centrifuge tubes and centrifuge at 4,190 × *g* for 15 min at the required temperature. Remove the supernatant and resuspend the pellet containing the aqueous and alkaline soluble proteins in 15 mL water at pH 11.0 (*see* **Notes 13** and **14**). Measure the volume of each resuspended pellet extract. Determine the concentration of protein in each extract with an appropriate protein quantification assay and express the concentration of protein in each extract as mg of protein. The extract with the highest quantity of protein (i.e., the pH at which the highest quantity of protein precipitated out of solution) is assigned the pI.

13. Using water which has been altered to pH 11.0 is a good way to resolubilize the acidic pellet. Not all the pellet will go into solution at this stage and when decanting the protein suspension into a beaker, it is common that some of the protein particles will stick to the side of centrifuge tube. If particles do stick to the side, use a small volume of water at pH 11.0 to transfer them to the beaker.

14. In some cases water at pH 11.0 is not enough to totally resolubilize the protein pellet; therefore, a small quantity of 2 M

NaOH may need to be added. Use a pH probe to monitor the pH. Depending on the downstream application of the protein extract, the pH can subsequently be adjusted to the desired value.

Acknowledgements

This research (Grant-Aid Agreement No. MFFRI/07/01) was carried out under the *Sea Change* Strategy with the support of the Marine Institute and the Department of Agriculture, Food and the Marine, funded under the National Development Plan 2007–2013.

References

1. Harnedy PA, FitzGerald RJ (2011) Bioactive proteins, peptides and amino acids from macroalgae. J Phycol 47:218–232
2. Fleurence J (2004) Seaweed proteins. In: Yada RY (ed) Proteins in food processing. Woodhead Publishing Limited, Cambridge, pp 197–213
3. Joubert Y, Fleurence J (2008) Simultaneous extraction of protein and DNA by an enzymatic treatment of the cell wall of *Palmaria palmata* (Rhodophyta). J Appl Phycol 20:55–61
4. Deniaud E, Fleurence J, Lahaye M (2003) Preparation and chemical characterization of cell wall fractions enriched in structural proteins from *Palmaria palmata* (Rhodophyta). Bot Mar 46:366–377
5. Fleurence J (2003) R-phycoerythrin from red macroalgae: strategies for extraction and potential application in biotechnology. Appl Biotechnol Food Sci Pol 1:63–68
6. Harnedy PA, FitzGerald RJ (2013) Extraction of protein from the macroalga *Palmaria palmata.* LWT Food Sci Technol 51:375–382
7. Denis C, Le Jeune H, Gaudin P et al (2009) An evaluation of methods for quantifying the enzymatic degradation of red seaweed *Grateloupia turuturu.* J Phycol 21:153–159
8. Fleurence J, Massiani L, Guyader O et al (1995) Use of enzymatic cell wall degradation for improvement of protein extraction from *Chondrus crispus*, *Gracilaria verrucosa* and *Palmaria palmata.* J Appl Phycol 7:393–397
9. Bensadoun A, Weinstein D (1976) Assay of proteins in the presence of interfering materials. Anal Biochem 70:241–250

Chapter 5

Extraction and Purification of R-phycoerythrin from Marine Red Algae

Justine Dumay, Michèle Morançais, Huu Phuo Trang Nguyen, and Joël Fleurence

Abstract

This chapter focuses on the recovery of an R-Phycoerythrin (R-PE)-enriched fraction from marine algae. Since R-PE is a proteinaceous pigment, we have developed a simple and rapid two-step method devoted to the extraction and purification of R-PE from marine red algae. Here we describe a phosphate buffer extraction followed by anion exchange chromatography carried on a DEAE Sepharose Fast Flow column. To ensure the quality and quantity of R-PE recovery, we also indicate different methods to monitor each fraction obtained, such as spectrophotometric indicators, gel filtration, and SDS-PAGE analysis.

Key words Extraction, Pigment, Purification, Red seaweeds, R-PE

1 Introduction

R-Phycoerythrin (R-PE) is a major photosynthetic pigment in red seaweeds. It is an oligomeric water-soluble chromoprotein of 240 kDa, characterized by its absorption spectrum between 400 and 650 nm [1]. This reddish fluorescent pigment is of interest due to its original spectral characteristics [2]. In addition to its color, phycoerythrin emits yellow fluorescence. Applications of R-PE as a bioactive compound, depending on its purity, range from food colorant to moiety in fluorescent energy transfer, fluorescent labels, tags, tracers, and markers [3, 4]. Recently, R-PE biological activities have been found such as antitumoral [2], antioxidant [5, 6] immunosuppressive, or hypertensive [7].

R-phycoerythrin is classically extracted by maceration of algae in buffer or water and then purified by a combination of several techniques, such as ammonium sulfate precipitation or different chromatographic techniques [8–10]. Those techniques have several advantages regardless of algal species chosen over enzymatic

Dagmar B. Stengel and Solène Connan (eds.), *Natural Products From Marine Algae: Methods and Protocols*, Methods in Molecular Biology, vol. 1308, DOI 10.1007/978-1-4939-2684-8_5, © Springer Science+Business Media New York 2015

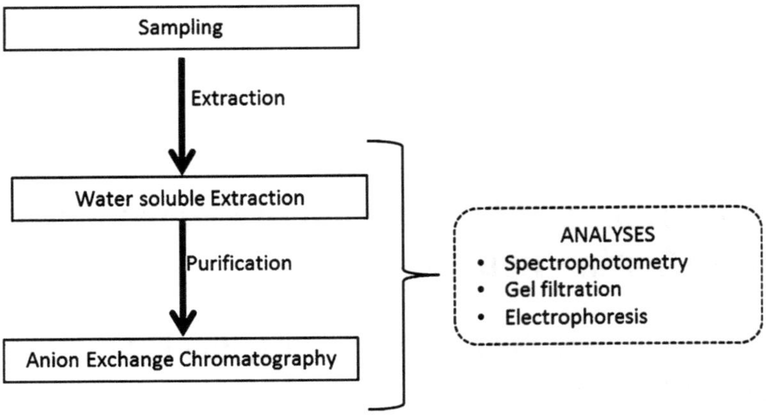

Fig. 1 General scheme of the extraction, purification, and monitoring procedures

procedure even if the latter processes could significantly increase the extraction yield [11].

This chapter deals with a simple method to extract R-Phycoerythrin from red seaweeds in two steps, with high recovery rate and great quality, as commonly used in the laboratory. We also describe the procedure used to monitor and evaluate the quality and quantity of R-PE (Fig. 1).

2 Materials

Prepare all solutions using ultrapure water and analytical-grade reagents. All reagents should be stored in darkness and at 4 °C until use. Diligently follow all waste disposal regulations when disposing waste materials.

2.1 R-phycoerythrin Extraction

1. Extraction buffer: 20 mM phosphate buffer, pH 7. Add 200 mL of ultrapure water into a 1 L glass beaker. Weigh 1.54 g of NaH_2PO_4 and 1.58 g of Na_2HPO_4 and transfer to the glass beaker. Add water to a volume of 700 mL. Mix and adjust pH with HCl or NaOH if required. Make up to 1 L with ultrapure water. Store at 4 °C (*see* **Note 1**).

2.2 Anion Exchange Chromatography

1. DEAE–Sepharose Fast Flow column (Amersham) (26 × 100 mm).
2. Buffer A: 20 mM phosphate buffer, pH 7 (extraction buffer).
3. Buffer B: 20 mM phosphate buffer, 1 M NaCl, pH 7. Add 200 mL of ultrapure water to a 1 L glass beaker. Weigh 1.54 g of NaH_2PO_4, 1.58 g of Na_2HPO_4 and 58.5 g of NaCl and transfer to the glass beaker. Add ultrapure water to a volume of

700 mL. Mix and adjust pH with HCl or NaOH if required. Make up to 1 L with ultrapure water. Store at 4 °C.

4. Spectra/Por Regenerated Cellulose 3.5 kDa.

2.3 Gel Filtration

1. Superdex 200 HR glass column (Amersham) (10 × 300 mm).
2. Gel filtration buffer: 20 mM phosphate buffer, pH 7 (extraction buffer).

2.4 SDS-PolyAcrylamide Gel Electrophoresis (SDS-PAGE)

1. Resolving gel buffer: 1.5 M Tris–HCl, pH 8.8. Add about 100 mL of ultrapure water to a 1 L glass beaker. Weigh 181.7 g of Tris-Base and transfer to the beaker. Add ultrapure water to a volume of 900 mL. Mix and adjust pH with HCl 6 M. Make up to 1 L with ultrapure water. Store at 4 °C.
2. Stacking gel buffer: 0.5 M Tris–HCl, pH 6.8. Weigh 60.6 g of Tris-Base and prepare a 1 L solution as in previous step. Store at 4 °C.
3. 12 % Separating gel: mix 3 mL of ultrapure water, 2.5 mL of resolving gel buffer, 4 mL of 30 % acrylamide/bis acrylamide solution, 100 μL of SDS (*see* **Note 2**), 100 μL of 10 % ammonium persulfate (APS) (w/v in ultrapure water; *see* **Note 3**) and then 10 μL of *N,N,N,N'*-tetramethyl-ethylenediamine (TEMED) (stored at 4 °C) (*see* **Note 4**).
4. 4 % Stacking gel: mix 6.15 mL of ultrapure water, 2.5 mL of stacking gel buffer, 1.33 mL of 30 % acrylamide/bis acrylamide solution, 100 μL of SDS, 100 μL of 10 % APS and then 10 μL of TEMED.
5. Lysis buffer: 0.065 M Tris–HCl, pH 6.8, 25 % glycerol, 2 % SDS, 0.01 % bromophenol blue, 5 % 2-mercaptoethanol. Add the different components in this order: 3.55 mL of ultrapure water, 1.25 mL of stacking gel buffer, 2.5 mL of glycerol, 2 mL of 10 % SDS (v/v in ultrapure water), 0.2 mL of 0.5 % bromophenol blue, and 50 μL of 2-mercaptoethanol. Store at 4 °C (*see* **Note 5**).
6. Running buffer: 0.025 M Tris pH 8.3, 0.192 M glycine, 0.1 % SDS. Prepare 10× native buffer (0.25 M Tris, 1.92 M glycine): weigh 30.3 g of Tris and 144 g of glycine, mix, and make it up to 1 L with ultrapure water. Dilute 100 mL of 10× native buffer to 890 with water and add 10 mL of 10 % SDS (v/v in ultrapure water).
7. Fixing solution: mix 450 mL of methanol, 400 mL of ultrapure water, 100 mL of acetic acid and complete to 1 L with ultrapure water.
8. Dye solution: weigh 2.5 g of brilliant blue and add 450 mL of methanol, 400 mL of ultrapure water, 100 mL of acetic acid, stir and complete to 1 L with ultrapure water.

9. Destaining solution: mix 300 mL of methanol, 550 mL of ultrapure water, 100 mL of acetic acid and complete to 1 L with ultrapure water.

3 Methods

Carry out all laboratory procedures at low light or in darkness as far as possible and at 4 °C to prevent R-PE denaturation.

3.1 Algal Sampling

1. Red algae are commonly found in the low intertidal zone (*see* **Note 6**). Carefully collect samples of the species of interest without epiphytes during low tide.
2. Rinse algae successively with seawater, tap water, and deionized water.
3. Freeze the seaweeds immediately and freeze-dry samples.
4. Grind freeze-dried algae with liquid N_2 to obtain a uniform fine powder.

3.2 R-phycoerythrin Extraction

1. Homogenize the algal powder in extraction buffer following a 1 g/20 mL ratio.
2. Place the system under agitation (150 rpm) during at least 20 min (and no more than 12 h).
3. Centrifuge at 25,000 × g for 30 min to recover the supernatant.
4. If required, perform multiple extractions by using the sludge obtained in the previous step (*see* **Note 7**).
5. Pool all supernatants obtained.

3.3 Purification by Anion Exchange Chromatography

1. Pre-equilibrate the DEAE–Sepharose Fast Flow column with the buffer A at 4 mL/min (around 10 times of the column void volume).
2. Load the supernatant on the top of the column (do not overload the column).
3. Rinse the column with the buffer A at 4 mL/min.
4. Elute the RPE with a three-step increase in the buffer ionic-strength: first at 150 mM NaCl (elution with 15 % buffer A and 85 % buffer B) for 10 min, the second at 200 mM NaCl (elution with 20 % buffer A and 80 % buffer B) for 10 min and the third at 1 M NaCl (elution with 100 % buffer B) for 10 min at 4 mL/min.
5. Monitor the absorption at 280 and 565 nm with a diode array detector.
6. Collect the fraction eluted with 200 mM NaCl.

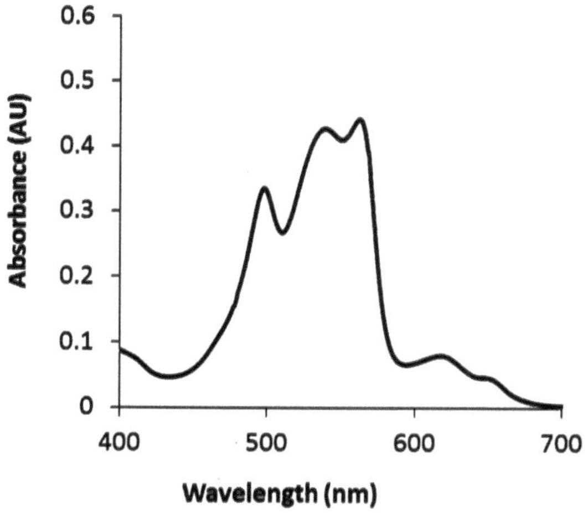

Fig. 2 Absorption spectrum of R-Phycoerythrin in visible light (400–650 nm)

7. Desalt overnight by dialysis using a Spectra/Por Regenerated Cellulose 3.5 kDa against phosphate buffer (*see* **Note 8**).

3.4 Spectrophotometric Characterization

After all steps, read the absorbance of the supernatants to ensure the quality of the fractions.

1. Read absorbance in UV-visible wavelengths from 200 to 800 nm. The absorbance spectrum of native R-PE should typically have peaks at respectively 492, 545, and 565 nm (Fig. 2).
2. Estimate the R-PE concentration (mg/mL) following the Beer and Eshel equation [12]:

$$[\mathrm{R-PE}] = [(A_{565} - A_{592}) - (A_{455} - A_{592}) \times 0.20] \times 0.12$$

3. Estimate the Purity Index (PI) of the fraction using the A_{565}/A_{280} ratio [13–15] (*see* **Note 9**).

3.5 Gel Filtration Analysis

1. Pre-equilibrate the Superdex 200 HR column with the gel filtration buffer at 0.5 mL/min (10 times the void volume of the column).
2. Inject 100 μL of a sample (crude or purified) on the column.
3. Elute with the gel filtration buffer at 0.5 mL/min.
4. Monitor the absorption at 280 and 565 nm with a diode array detector.
5. Demonstrate the purification efficiency comparing chromatograms obtained at each step (Fig. 3) (*see* **Note 10**).

3.6 SDS-PAGE Electrophoresis Analysis

1. Cast rapidly after preparation the 12 % separating gel within a 4 × 7 cm gel cassette. Leave some space for stacking the gel and gently overlay with isobutanol or ultrapure water (*see* **Note 11**).

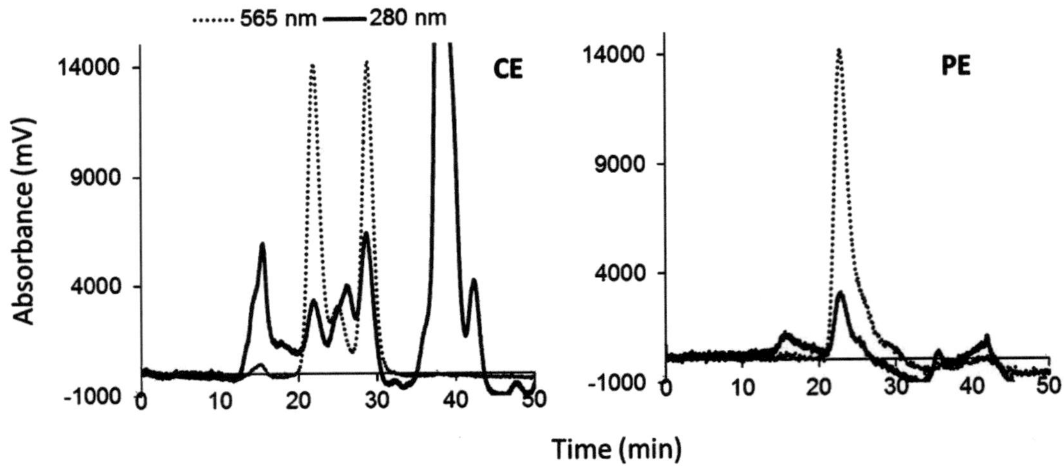

Fig. 3 Superdex HR 200 gel filtration chromatograms at 565 and 280 nm. *CE* crude extract. *PE* purified extract after DEAE chromatography purification

2. When fully polymerized (after 30–45 min), remove the isobutanol and water, and add 5 mL of resolving gel buffer diluted with water (1/4, v/v) to prevent gel dehydration (*see* **Note 12**).
3. Cast rapidly after being prepared the 4 % stacking gel in the gel cassette above the separating gel. Insert a ten-well gel comb immediately without introducing air bubbles.
4. When fully polymerized (after 30–45 min), remove the comb gently and rinse the wells with water or resolving gel buffer.
5. Sample preparation: dilute sample in the lysis buffer in order to approximate 1 mg/mL proteins. Heat between 80 and 100 °C for 5–10 min. Do not treat the pre-stained protein standard. Centrifuge the heated samples at 3,000 × *g* for 30 s. Keep the supernatant.
6. Place the gel carefully on the running electrophoresis module (Mini-PROTEAN, Biorad, or equivalent).
7. Fill the space between gel wells with the running buffer.
8. Gently introduce samples and standard in the appropriate gel wells (*see* **Note 13**).
9. Fill the gel box with the running buffer.
10. Electrophorese at 90 V constant for 30 min and then at 180 V until the dye front (from the bromophenol blue dye in the samples) has reached the bottom of the gel.
11. Following the electrophoresis, open the gel plates with the use of a spatula. The gel remains on one of the glass plates. Carefully transfer the gel to a container with the fixing solution and leave for 5 min. Gently stir the plate to remove the gel from the glass plate.

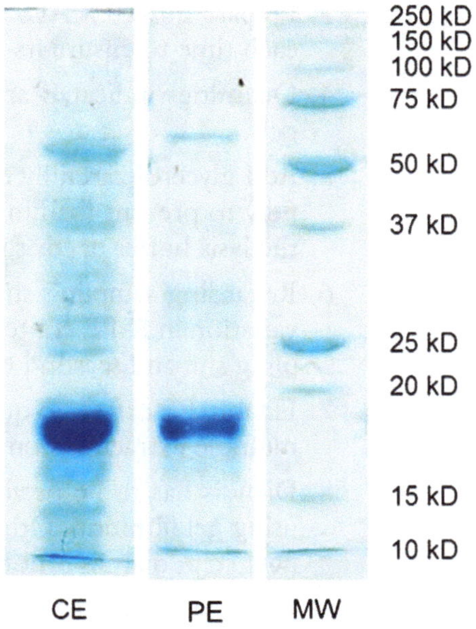

Fig. 4 SDS-PAGE electrophoresis of R-PE extracts following the purification procedure. *CE* crude extract. *PE* purified extract after DEAE chromatographic step. *MW* standard molecular weight protein ladder. CE electrophoresis clearly shows the total protein mix contained in the red algal sample. After purification, PE lane illustrates the efficiency of the method used

12. Place the gel for 2 h in the dye solution previously heated in a water bath at 70 °C and agitate continuously (*see* **Note 14**).
13. Rinse the gel in the fixing solution for 15 min and then in several successive baths with the destaining solution (*see* **Note 15**).
14. Demonstrate the purification efficiency through the comparison of the electrophoresis profile obtained at each step (Fig. 4) (*see* **Note 16**).

4 Notes

1. R-PE is relatively stable across a wide pH range [13, 19], so different extraction buffers could be employed according to the further use of the R-PE crude extract. At industrial scale and for colorant properties, we recommend the use of tap water to prepare the extraction buffer. But for deeper analysis or for high-purity application (flow cytometry, reagent, biological activity assay, purification of the molecule), we recommend to work at a specific pH with extraction buffer prepared with deionized or ultrapure water.
2. Care should be taken when adding SDS solution since bubbles may be created.

3. Prepare this 10 % APS (w/v in ultrapure water) solution fresh each time to ensure its polymerization properties.
4. Quantities indicated are sufficient for the preparation of two gels.
5. Add glycerol carefully due to its high viscosity, and SDS carefully to prevent foaming. SDS precipitates at 4 °C. Therefore, the lysis buffer needs to be warmed prior use.
6. Regarding sampling strategy for sourcing red seaweeds: large variation in R-PE content has been reported according to sampling site and season [16–18].
7. Final R-PE extracted quantity may be improved by performing multiple extractions on the resulting sludge.
8. Dialysis has to be performed in order to analyze the fraction using gel filtration. Otherwise, if salts do not interfere with the next step, dialysis is not necessary.
9. It has been established that a fraction of R-PE could be considered pure when the IP value reaches at least 3.2 [10].
10. Figure 3 illustrates clearly the efficiency of the purification method employed here: several proteinaceous compounds (at 280 nm) have been highlighted on the chromatogram of the crude extract while just one still remains on the chromatogram of the purified extract.
11. This overlay prevents contact with atmospheric oxygen (which inhibits acrylamide polymerization) and help to level the resolving gel solution.
12. At this stage, the resolution gel may be stored overnight at room temperature.
13. When loading samples and standard in the wells, process very carefully to obtain regular deposits at the bottom of each well.
14. Perform all baths (fixing and destaining) in a hermetically closed recipient and agitate carefully.
15. Scan or take a picture of the gel prior to analysis.
16. R-PE possesses characteristic subunits α, β, and γ, at 20, 21, and 30 kDa, respectively.

References

1. Dumay J, Morançais M, Munier M et al (2014) Chapter 11. Phycoerythrins: valuable proteinic pigments in red seaweeds. In: Bourgougnon N (ed) Advances in botanical research – sea plants, vol 71. Elsevier, Amsterdam, pp 321–344
2. Sekar S, Chandramohan M (2008) Phycobiliproteins as a commodity: trends in applied research, patents and commercialization. J Appl Phycol 20:113–136
3. Glazer A (1994) Phycobiliproteins – a family of valuable, widely used fluorophores. J Appl Phycol 6:105–112
4. Isailovic D, Sultana I, Phillips GJ et al (2006) Formation of fluorescent proteins by the attachment of phycoerythrobilin to

R-phycoerythrin alpha and beta apo-subunits. Anal Biochem 358:38–50

5. Fitzgerald C, Gallagher E, Tasdemir D et al (2011) Heart health peptides from macroalgae and their potential use in functional foods. J Agric Food Chem 59:6829–6836
6. Pangestuti R, Kim SK (2011) Biological activities and health benefit effects of natural pigments derived from marine algae. J Funct Foods 3:255–266
7. Cian RE, Martínez-Augustin O, Drago SR (2012) Bioactive properties of peptides obtained by enzymatic hydrolysis from protein byproducts of *Porphyra columbina*. Food Res Int 49:364–372
8. Hilditch CM, Balding P, Jenkins R et al (1991) R-phycoerythrin from the macroalga *Corallina officinalis* (Rhodophyceae) and application of a derived phycofluor probe for detecting sugar-binding sites on cell membranes. J Appl Phycol 3:345–354
9. Wang G (2002) Isolation and purification of phycoerythrin from red alga *Gracilaria verrucosa* by expanded-bed-adsorption and ion-exchange chromatography. Chromatographia 56:509–513
10. Niu JF, Wang GC, Tseng CK (2006) Method for large-scale isolation and purification of R-phycoerythrin from red alga *Polysiphonia urceolata* Grev. Protein Expr Purif 49:23–31
11. Dumay J, Clément N, Morançais M et al (2013) Optimization of hydrolysis conditions of *Palmaria palmata* to enhance R-phycoerythrin extraction. Bioresour Technol 131:21–27
12. Beer S, Eshel A (1985) Determining phycoerythrin and phycocyanin concentrations in aqueous crude extracts of red algae. Aust J Mar Freshw Res 36:785–793
13. Galland-Irmouli AV, Pons L, Luçon M et al (2000) One-step purification of R-phycoerythrin from the red macroalga *Palmaria palmata* using preparative polyacylamide gel electrophoresis. J Chromatogr B 739:117–123
14. Liu L-N, Chen X-L, Zhang X-Y et al (2005) One-step chromatography method for efficient separation and purification of R-phycoerythrin from *Polysiphonia urceolata*. J Biotechnol 116:91–100
15. Rossano R, Ungano N, D'Ambrosio A et al (2003) Extracting and purifying R-phycoreythrin from Mediterranean red algae *Corallina elongata* Ellis & Solander. J Biotechnol 101:289–293
16. Banerjee K, Ghosh R, Homechaudhuri S et al (2009) Seasonal variation in the biochemical composition of red seaweed (*Catenella repens*) from Gangetic delta, northeast coast of India. J Earth Syst Sci 118:497–505
17. Denis C, Morançais M, Li M et al (2010) Study of the chemical composition of edible red macroalgae *Grateloupia turuturu* from Brittany (France). Food Chem 119:913–917
18. Munier M, Dumay J, Morançais M et al (2013) Variation in the biochemical composition of the edible seaweed *Grateloupia turuturu* Yamada harvested from two sampling sites on the Brittany coast (France): the influence of storage method on the extraction of the seaweed pigment R-Phycoerythrin. J Chem. doi:10.1155/2013/568548
19. Munier M, Jubeau S, Wijaya A et al (2014) Physicochemical factors affecting the stability of two pigments: R-Phycoerythrin of *Grateloupia turuturu* and B-Phycoerythrin of *Porphyridium cruentum*. Food Chem 150: 400–407

Chapter 6

Extraction and Analysis of Mycosporine-Like Amino Acids in Marine Algae

Nedeljka N. Rosic, Christoph Braun, and David Kvaskoff

Abstract

Marine organisms use mycosporine-like amino acids (MAAs) as biological sunscreens for the protection from damaging ultraviolet (UV) radiation and the prevention of oxidative stress. MAAs have been discovered in many different marine and freshwater species including cyanobacteria, fungi, and algae, but also in animals like cnidarian and fishes. Here, we describe a general method for the isolation and characterization of MAA compounds from red algae and symbiotic dinoflagellates isolated from coral hosts. This method is also suitable for the extraction and analyses of MAAs from a range of other algal and marine biota.

Key words Coral, LCMS, MAAs, Red algae, *Symbiodinium*, Symbiosis

1 Introduction

UV-absorbing mycosporine-like amino acids (MAAs) have been discovered in many marine and freshwater species [1, 2]. MAAs are found in various animals, cyanobacteria, and algae. These ubiquitous secondary metabolites are transparent, water-soluble compounds with low molecular weight (<400 Da), containing a cyclohexenone or cyclohexenimine chromophore conjugated to an amino acid residue or its imino alcohol (Fig. 1). Maximum absorbance of MAAs is within the UVA and UVB range (310–362 nm) and a molar absorptivity (ε) from 28,100 to 50,000/M/cm. These UV-absorbing compounds are characterized by multiple functions in the organisms. Besides a role in the protection from UV radiation, MAAs are also involved in the anti-oxidative stress response by reducing reactive oxygen species (ROS) and scavenging free radicals, with a potential role as light-harvesting pigments, and as a source of intracellular nitrogen [1, 3–5].

Reef flats are characterized by a very high level of UV radiation reaching an extreme UV index (above 11) during summer. Reef-building corals and their symbiotic dinoflagellates display different MAA profiles depending on species and also symbiotic status. MAAs

Dagmar B. Stengel and Solène Connan (eds.), *Natural Products From Marine Algae: Methods and Protocols*, Methods in Molecular Biology, vol. 1308, DOI 10.1007/978-1-4939-2684-8_6, © Springer Science+Business Media New York 2015

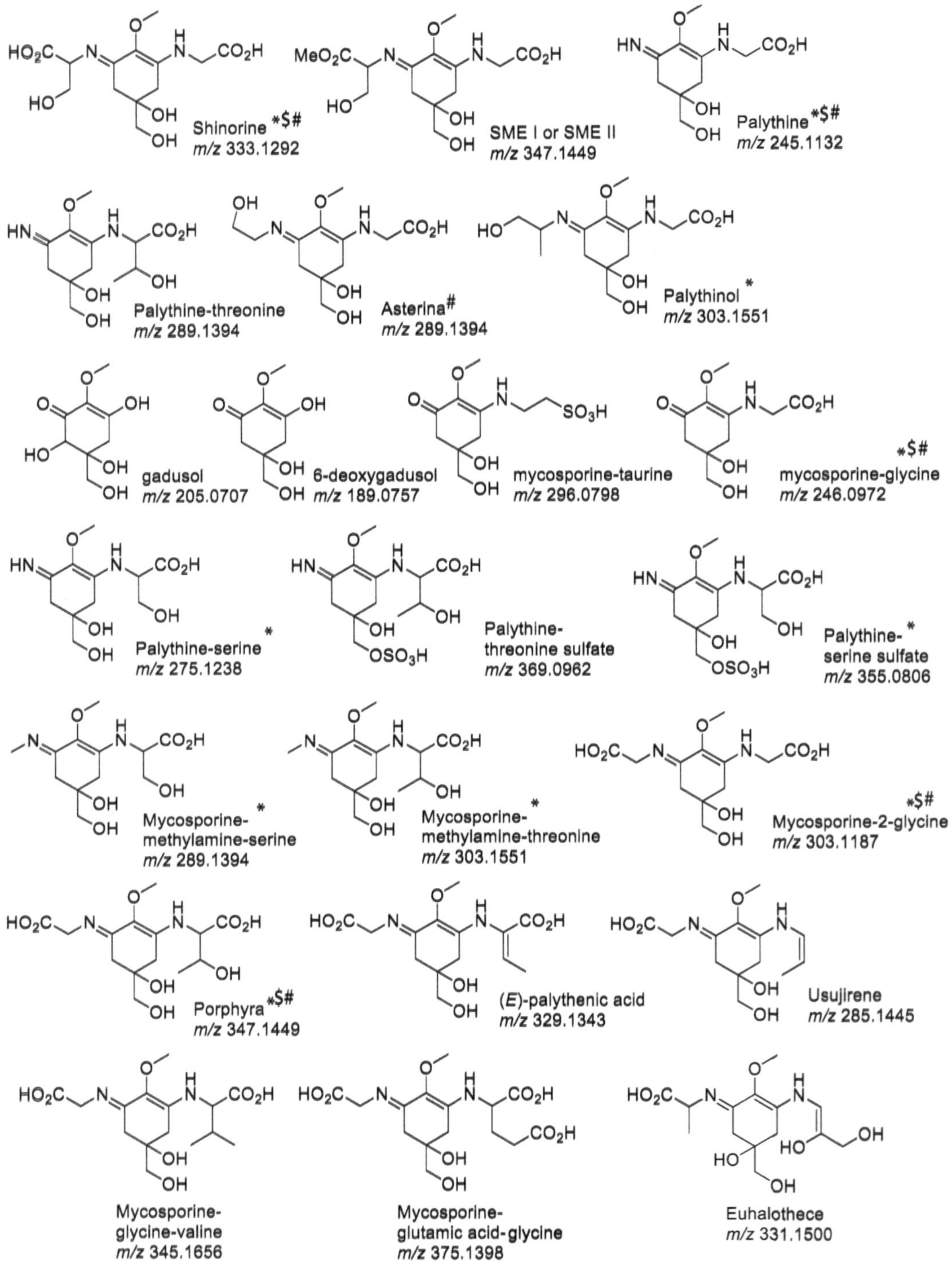

Fig. 1 Chemical structures of known MAAs, adapted from [15], including mass of MAAs commonly found in the red alga *Acanthophora spicifera* (indicated by a hash), *Symbiodinium* spp. (indicated by a dollar), and the hermatypic coral *Stylophora pistillata* (indicated by a star) [7, 11, 14]

from coral-algal symbioses are believed to be produced by microbial endosymbionts such as photosynthetic dinoflagellates from the genus *Symbiodinium*, although the host factor may a play role in MAA synthesis [6–10]. In *Symbiodinium* spp. cultures mainly primary MAAs (mycosporine-glycine, shinorine, porphyra-334, and mycosporine-2-glycine) were detected, which are synthesized during the first part of the MAA biosynthetic pathway, whilst a more diverse MAA profile was observed in symbiosis with coral hosts [6, 7, 11].

A list of known MAAs found in nature including the MAA precursors gadusol and deoxygadusol is shown in Fig. 1. Here, we present a method for the extraction and characterization of MAAs from samples of the red alga *Acanthophora spicifera* and the reef-building coral *Acropora aspera*. Using a modified high-performance liquid chromatography-mass spectrometry (HPLC-MS) method [11], we were able to identify several known MAAs within the tissues of the coral and red algae. Consequently, this method can be efficiently applied for the extraction and characterization of MAAs from different organisms.

2 Materials

Prepare all buffers and solutions using ultrapure water, purified by reverse osmosis, via for example an MQ Ultrapure Water System ("Milli-Q water"). Perform sterilization and degassing of solutions by filtration (0.22 μm) under vacuum (20 mmHg; *see* **Note 1**). Store all reagents at room temperature and dispose chemical waste following the recommended procedures and regulations.

2.1 Biological Materials

1. In the case of the example described in this chapter, material of the red alga *Acanthophora spicifera* was obtained on the Great Barrier Reef (GBR), Australia. The samples were collected (GBRMPA permit G11/33857.1 to LIRS) from a shallow site (water depth <2 m) in the lagoon at Lizard Island (14°40′55.59″S, 145°27′17.81″E) in March 2013.
2. Coral fragments (7 cm long) of *Acropora aspera* harboring *Symbiodinium* spp: ITS-type C3 genotype [12] were collected from the reef flat at Heron Island (GBR), Australia (23°25′S; 152°07′E), in June 2010.

2.2 Chemicals and Reagents for MAA Extraction

1. Artificial seawater with a salinity of 30: Weigh 30 g of NaCl and dissolve in Milli-Q water. Check salinity using salinity meter. Filter-sterilize (using 0.45 μm filter) and autoclave (optional). Store at room temperature.

2. F/2 media for *Symbiodinium* cultures: Use Guillard's (F/2) Marine Water Enrichment Solution that can be obtained from a manufacturer in artificial seawater with a salinity of 30 ([13]; *see* **Note 2**). F/2 media stock is kept at −20 °C.
3. 60 mM phosphate buffer at pH 6.65: Weigh 10.251 g of K_2HPO_4 and 8.165 g of KH_2PO_4 and top up to 1 L with Milli-Q water. Adjust buffer to pH 6.65 by varying the amount of each salt. Filter the buffer using vacuum pump with 0.22 μm filter and store at room temperature.
4. 80 % HPLC-grade methanol in water (v/v).
5. HPLC-grade acetone and methanol.
6. Formic acid.

2.3 Liquid Chromatography–Mass Spectrometry (LC–MS)

1. 10 % methanol in water (v/v).
2. LC–MS instrument: Shimadzu LCMS-2020 single-quadruple mass spectrometer (Kyoto, Japan) (or equivalent), equipped with a UV-photodiode array detector and an electrospray ionization (ESI) source interface. The flow is set at 0.25 mL/min and the autosampler is cooled to 4 °C to preserve samples (*see* **Note 3**).
3. HPLC column: Kinetex XB-C18 column, 100 Å pore size, 50 × 2.1 mm diameter, and 1.7 μm particle size, equipped with a SecurityGuard Ultra C18 pre-column.
4. HPLC conditions: Injection volume of 5 μL (100 μL loop), column temperature at 40 °C to enhance peak shape.
5. Mobile phase A: Aqueous mobile phase of 1 mM ammonium acetate + 0.1 % acetic acid (v/v), pH 3.6: Dissolve 77 mg of LC–MS-grade ammonium acetate in Milli-Q water (1 L) and add 1 mL of LC–MS-grade acetic acid; the pH should be ca. 3.6. Filter on 0.22 μm under vacuum (20 mmHg). Store at room temperature (up to 6 months).
6. Mobile phase B: Acetonitrile/isopropanol (4:1, v/v). Filter on a 0.22 μm filter under vacuum (20 mmHg) for degassing. Use only HPLC-grade chemicals. Store solution after filtering at room temperature.
7. Elution gradient (Table 1): 5 % Mobile phase B for 2 min, increasing to 100 % B over 20 min. From 22.5 to 35 min, the flow rate should be increased to 0.5 mL/min in order to flush the column from late-eluting peaks. The flow rate needs to be adjusted to 0.25 mL/min at 35 min, and the gradient returned to starting conditions (0 % B) at 40 min and allowed to equilibrate for 5 min (total runtime 45 min).
8. UV-photodiode array detector set at 330 nm to follow the elution of MAAs.

Table 1
The LC–MS gradient protocol for the separation of MAAs

Time (min)	Mobile phase A (%)	Mobile phase B (%)	Flow (mL/min)
0	95	5	0.25
2	95	5	0.25
22	0	100	0.50
35.5	0	100	0.25
40	95	5	0.25

Mobile phase A: 1 mM ammonium acetate + 0.1 % acetic acid (v/v), pH 3.6 and *Mobile phase B*: acetonitrile/isopropanol (4:1, v/v). Total flow rate was increased from 0.25 mL/min to 0.5 mL/min from 22.7 to 35.1 min in order to flush the column from late-eluting peaks. Temperature was 40 °C and UV-absorbing peaks were detected at 330 nm

9. Electrospray ionization (ESI) used in positive ion mode switching: Typical ion source parameters: interface temperature 450 °C, desolvation line temperature 250 °C, heating block temperature 200 °C, nebulizing gas flow 1.5 L/min, drying gas flow 12 L/min, interface voltage 4.5 kV, Q-array RF voltage 50 V, desolvation line voltage 0 V, Q-array DC voltage 0 V.
10. Optimization and calibration of the mass spectrometer are achieved with the auto-tuning function using a polypropylene glycol standard solution (obtained from Shimadzu).
11. Data processing may be carried out using the Shimadzu Lab Solutions software (version 1.4), or equivalent, depending on the instrument used.

3 Methods

Carry out all procedures at room temperature whilst keeping samples in dark and cold conditions (−20 °C short-term storage or in −80 °C freezers) before processing.

3.1 Extraction of MAAs

1. Collect red algae on the shore and coral fragments from reef-building corals. After collection, transfer the coral fragments (~7 cm in length) in flow through aquaria for acclimatization in tanks for 48 h. Snap-freeze samples in liquid nitrogen and keep them at −80 °C until further processing.

2. Isolate MAAs from red algae: Use ice-cold 80 % methanol as detailed in [14] (500 μL per 50 μg of dry algal weight) following these steps: (1) freeze-dry red algal samples overnight at −70 °C at 1.5 mbar, (2) grind them, and (3) perform a sequential methanol extraction using 80 % methanol. Repeat the extraction procedure twice by adding 0.5 mL of 80 % methanol to 50 μg of dry algal weight until the pellet appears colorless.
3. Isolation of MAAs from coral nubbins: Perform airbrushing using a stream of air under a pressure mixed with 0.06 M phosphate buffer (pH 6.65), which will remove the tissue from the coral skeleton. Use frozen coral fragments and 12 mL of 0.06 M phosphate buffer (pH 6.65), followed by centrifugation of the homogenate at 4,000 × *g* for 5 min. Resuspend the pellet in filtered artificial seawater (0.45 μm) and separate in aliquots. Use an aliquot for MAA extraction by additional centrifugation (4,000 × *g* for 5 min) and resuspend the pellet in 1 mL of 100 % cold methanol, followed by 10-min sonication on ice-cold water and then centrifugation (4,000 × *g* for 5 min). Transfer supernatant to a new tube and keep it on ice. Apply an additional 1 mL of cold 100 % methanol to the pellet and repeat the previous extraction step until all pigment extracts are removed from the pellet (usually this takes 3–5 subsequent steps) and pale pellet is obtained.
4. In the final stage, filter the extracts using a 0.22 μm filter. Dry the methanol extracts in a vacuum centrifuge at 40 °C and keep them at −20 °C before analyses. Alternatively, methanol extracts can also be kept at −80 °C for a prolonged period.

3.2 MAA Analyses

1. Use filtered methanol extracts in a Shimadzu LCMS 2020 linked to a Shimadzu SPD—UV photodiode array detector (or similar) using the reverse C18 column and modified method described by [11] with mobile phases A and B.
2. Resuspend dried methanol extract in 200 μL of 10 % methanol. Use 5 μL of methanol extracts for a run on an LCMS. Flow rate is in the range of 0.2–0.5 mL/min.
3. The gradient employed is shown in Table 1.
4. Acquire data in the selected ion monitoring (SIM) mode, monitoring the $[M+H]^+$ ions of MAAs (Fig. 2a). Record the MS spectra during 40 min within the mass range of *m/z* 200–450 for positive ions targeting the masses of known MAAs. Analyses are done using nitrogen as a nebulizing (1.1 L/min) and drying gas (10 L/min, 250 °C). Capillary high voltage is set to 4,500 V. Isolation width of 1.0 Da is used for product ion spectra acquisition. Spectra are averaged for each data point (Fig. 3).

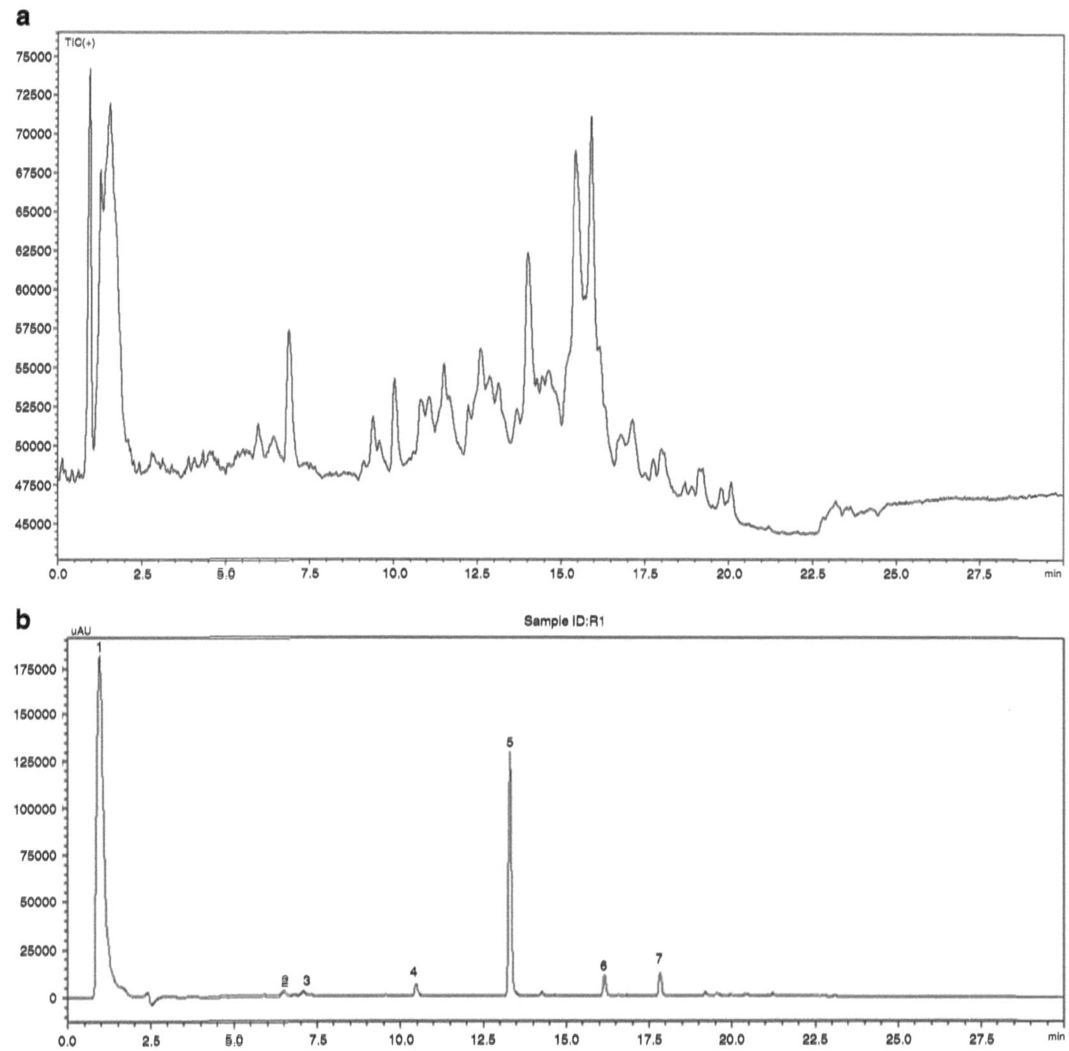

Fig. 2 Total ion chromatogram (TIC) that includes a summary of intensity for the mass range of *m/z* 200–450 of positive ions targeting the masses of known MAAs (**a**), and the HPLC-MS chromatogram at 330 nm of a methanol extract of the red alga *Acanthophora spicifera* (**b**). MAAs' peak identification was based on retention time, absorption maxima (λ_{max}), and *m/z* of positively charged ions $[M+H]^+$. Peak numbers correspond to different MAAs listed in Table 2

A chromatogram at 330 nm of an extract from *Acanthophora spicifera* is shown in Fig. 2b with seven MAAs identified (Table 2). The same LC–MS analysis was applied on a coral extract (*Acropora aspera*) and five MAAs were identified (Fig. 4; Table 3).

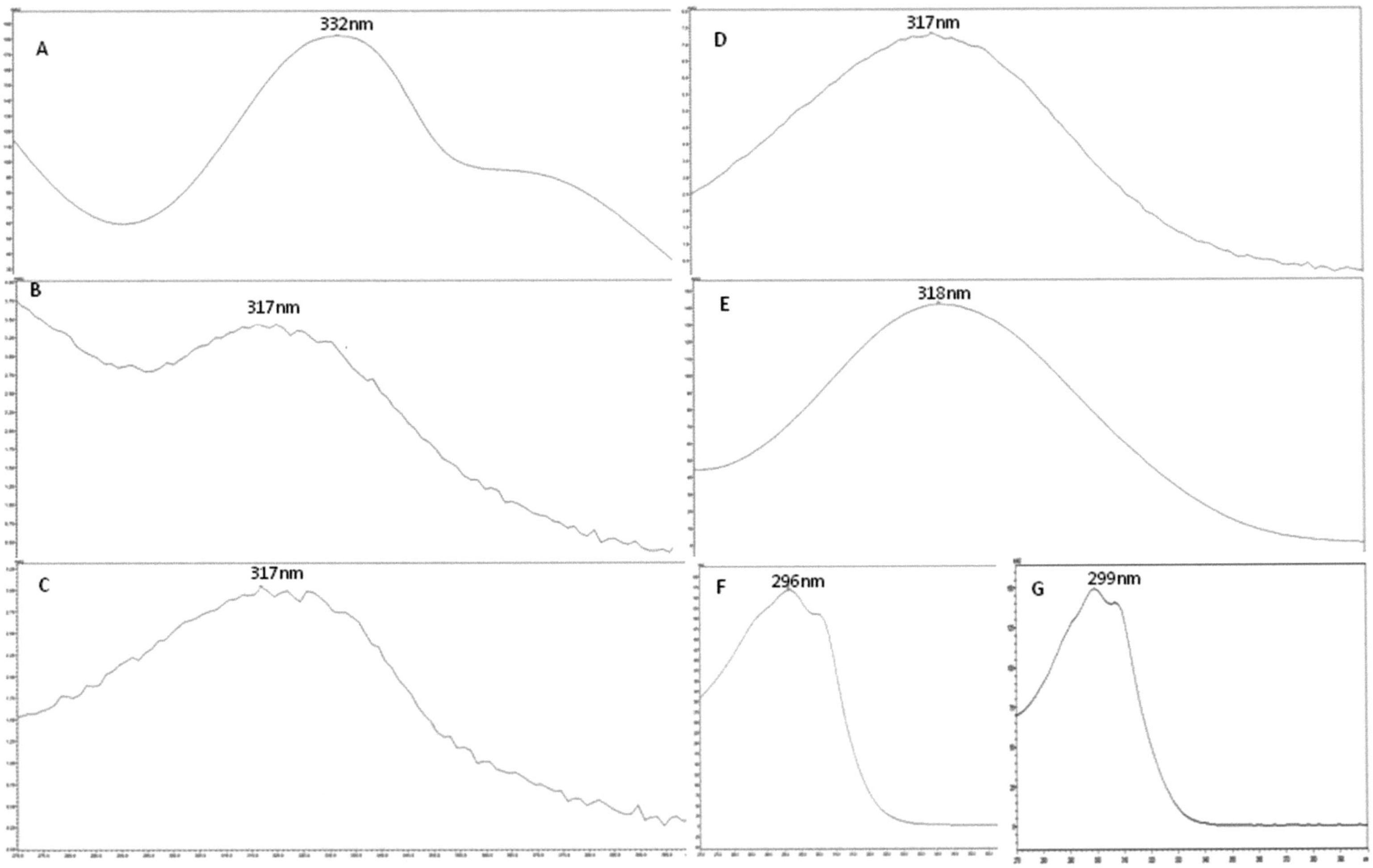

Fig. 3 UV absorption spectra (**a**–**g**) of the different HPLC peaks (1–7) presented in Fig. 2. These peaks were detected in a methanol extract of the red alga *Acanthophora spicifera*

Table 2
LC–MS analysis of MAAs isolated from the red alga *Acanthophora spicifera*

Peak #	MAA name	Retention time (min)	Absorption max (nm)	*m/z* [M + H]$^+$
1	Porphyra-334	0.96	332	347
1	Palythine			245
1	Mycosporine-glycine			246
2	Shinorine	6.50	317	333
3	Palythine-serine-sulfate	7.09	320	355
4	Mycosporine-taurine	10.50	315	296
5	Palythine-threonine	13.30	318	289
6	Gadusol	16.16	296	205
7	Unknown	17.84	299	289

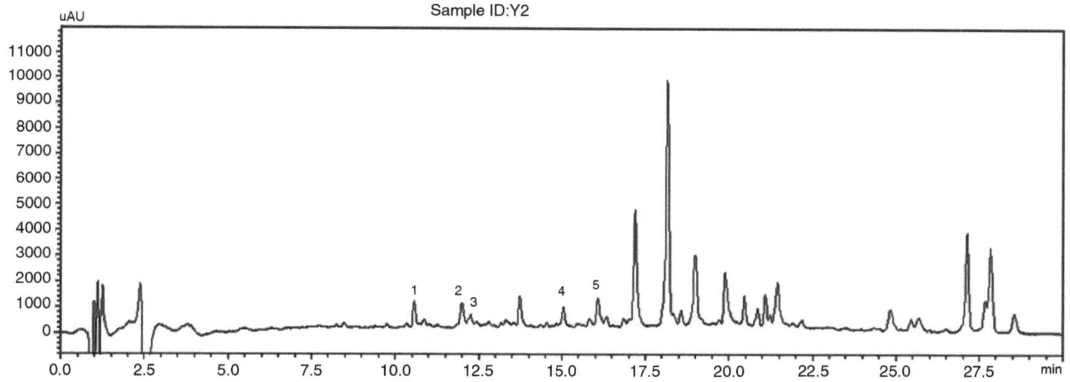

Fig. 4 HPLC-MS chromatogram at 330 nm of a methanol extract of *Acropora aspera*. MAAs' peak identification was based on retention time, absorption maxima (λ_{max}), and *m/z* of positively charged ions [M + H]$^+$ and is listed in Table 3. Other peaks on the chromatogram correspond to other photosynthetic pigments

Table 3
LC–MS analysis of MAAs isolated from the reef-building coral *Acropora aspera*

Peak #	MAA name	Retention time (min)	Absorption max (nm)	*m/z* [M + H]$^+$
1	Palythine-serine	10.58	327	275
2	Palythene	12.01	355	385
3	Porphyra-334	12.26	334	347
4	Palythine	15.05	319	245
5	Shinorine	16.33	325	333

4 Notes

1. If needed, conduct sterilization by autoclaving at 120 °C, with 1.36 atm (20 psi), for 30 min.
2. For F/2 medium use 10 mL of 50× stock (purchased from Sigma) and top up with 490 mL of sterile artificial seawater. This medium can also be prepared using recipes from various websites of algal cultures (e.g., http://www.marine.csiro.au/microalgae/methods/Media%20CMARC%20recipes.htm).
3. In our case the Shimadzu LC–MS system included two LC-20 AD high-pressure pumps, a SIL-20 AC-HT autosampler operated in XL mode, a CTO-20A column oven, a DGU-20A3 degasser unit, and an FCV-11AL solvent selection valve allowing the choice of two solvents per solvent delivery pump. Any other LC–MS system with similar characteristics can be used instead of this system.

Acknowledgments

Discovery Early Career Researcher Award (DE120101412) to N.N.R. from the Australian Research Council supported this research.

References

1. Shick JM, Dunlap WC (2002) Mycosporine-like amino acids and related gadusols: biosynthesis, accumulation, and UV-protective functions in aquatic organisms. Annu Rev Physiol 64:223–262
2. Llewellyn CA, Airs RL (2010) Distribution and abundance of MAAs in 33 species of microalgae across 13 classes. Mar Drugs 8: 1273–1291
3. Singh SP, Kumari S, Rastogi R (2008) Mycosporine-like amino acids (MAAs): chemical structure, biosynthesis and significance as UV-absorbing/screening compounds. Indian J Exp Biol 46:7–17
4. Sinha RP, Häder DP (2002) UV-induced DNA damage and repair: a review. Photochem Photobiol Sci 1:225–236
5. Mayfield AB, Gates RD (2007) Osmoregulation in anthozoan-dinoflagellate symbiosis. Comp Biochem Physiol A 147:1–10
6. Banaszak AT, Santos MGB, LaJeunesse TC et al (2006) The distribution of mycosporine-like amino acids (MAAs) and the phylogenetic identity of symbiotic dinoflagellates in cnidarian hosts from the Mexican Caribbean. J Exp Mar Biol Ecol 337:131–146
7. Banaszak AT, LaJeunesse TC, Trench RK (2000) The synthesis of mycosporine-like amino acids (MAAs) by cultured, symbiotic dinoflagellates. J Exp Mar Biol Ecol 249:219–233
8. Banaszak AT, Trench RK (2001) Ultraviolet sunscreens in dinoflagellates. Protist 152: 93–101
9. Rosic NN (2012) Phylogenetic analysis of genes involved in mycosporine-like amino acid biosynthesis in symbiotic dinoflagellates. Appl Microbiol Biotechnol 94:29–37
10. Rosic NN, Dove S (2011) Mycosporine-like amino acids from coral dinoflagellates. Appl Environ Microbiol 77:8478–8486

11. Carignan MO, Cardozo KHM, Oliviera-Silva D et al (2009) Palythine-threonine, a major novel mycosporine-like amino acid (MAA) isolated from the hermatypic coral *Pocillopora capitata*. J Photochem Photobiol B 94:191–200
12. LaJeunesse TC, Loh WKW, van Woesik R et al (2003) Low symbiont diversity in southern Great Barrier Reef corals, relative to those of the Caribbean. Limnol Oceanogr 48:2046–2054
13. Guillard RR, Ryther JH (1962) Studies of marine planktonic diatoms. I. *Cyclotella nana* Hustedt, and *Detonula confervacea* (cleve) Gran. Can J Microbiol 8:229–239
14. Carefoot TH, Harris M, Taylor BE et al (1998) Mycosporine-like amino acids: possible UV protection in eggs of the sea hare *Aplysia dactylomela*. Mar Biol 130:389–396
15. Carreto JI, Carignan MO (2011) Mycosporine-like amino acids: relevant secondary metabolites. Chemical and ecological aspects. Mar Drugs 9:387–446

Chapter 7

Extraction and Purification of Phlorotannins from Brown Algae

Erwan Ar Gall, Florian Lelchat, Mélanie Hupel, Camille Jégou, and Valérie Stiger-Pouvreau

Abstract

The interest in the physiological roles and bioactivities of plant phenols has increased over the past decades. In seaweeds, many investigations have dealt with phenolic compounds of Phaeophyceae (phlorotannins), even though little is known so far about the ecophysiological variations of their pool or their biosynthetic pathways. We describe here a simple procedure based on the use of water-organic solvent mixtures for the extraction of phlorotannins. Crude extracts are semi-purified and fractionated by separating methods based on both the polarity and the molecular size of compounds. Phenols are then quantified by the Folin-Ciocalteu method and their radical-scavenging activity is characterized using the DPPH test. All along the purification process of phenolic compounds, the efficiency of separation is assessed by ^{1}H-NMR.

Key words Extraction, Folin-Ciocalteu method, Liquid–liquid purification, Molecular size separation, Phlorotannins, Radical-scavenging activity, Seaweed phenols

1 Introduction

Numerous studies have been dedicated to plant phenols, focusing on their structure [1] or their physiological role and biological activities [2, 3]. Increasing attention has also been paid over the past decades to seaweed phenols and particularly to phlorotannins, i.e., oligomers and polymers of phloroglucinol occurring in the Phaeophyceae [4]. Extraction and purification of phlorotannins were mostly undertaken to allow their structural identification [5] or more generally to study both their distribution and their functional characterization [6–13]. To date, extraction of seaweed phenols has usually been conducted on ground fresh or dry seaweeds in either pure or water-mixed organic solvents. Alternative methods to classical solvent/water-consuming extraction have been recently tested, such as subcritical water hydrolysis, centrifugal partition extraction (CPE), supercritical fluid extraction (SFE), or

Dagmar B. Stengel and Solène Connan (eds.), *Natural Products From Marine Algae: Methods and Protocols*, Methods in Molecular Biology, vol. 1308, DOI 10.1007/978-1-4939-2684-8_7, © Springer Science+Business Media New York 2015

again pressurized liquid extraction (PLE) [14, 15], but their applicability to routine isolation of phlorotannins still remains to be seen, due to the low availability of accurate systems and/or high costs. Three main methods have been investigated so far to purify crude extracts and to study the composition of phlorotannin pools: liquid–liquid or solid–liquid (solid-phase extraction, or SPE) separation based on the polarity of molecules [16–18], and molecular size discrimination through dialysis, and/or ultrafiltration steps [19–23]. Chromatographic methods may also be helpful, including low-pressure liquid chromatography [24], thin-layer chromatography [25], high-pressure liquid chromatography (HPLC) [25–27], or again liquid chromatography–high-resolution mass spectrometry [28]. Recent studies on the purification of phlorotannins report on a combination of both polarity- and size-based approaches [20, 29]. The quantification of phlorotannins has been usually performed by the Folin-Ciocalteu method which remains the most suitable and easy-to-perform technique, despite some concerns regarding its specificity for phenols [7, 30]. The evaluation of radical-scavenging activities by the DPPH method has been commonly reported as a functional complement to the quantitative analysis of phlorotannin pools in Phaeophyceae [20, 31], even though other colorimetric and electrochemical methods may be considered more robust [15, 23, 32, 33]. When available, ^{1}H NMR spectroscopy allows a rapid assessment of the efficiency of semi-purification processes, with spectra showing the relative abundance of phenol versus contaminating substances, particularly mannitol [15, 20, 27].

We describe here a simple procedure using mixtures of water and organic solvents to extract phlorotannins from brown algae. Semi-purification and fractionating of crude extracts are performed using techniques based on both the polarity and the molecular size of compounds. Phenol contents are then quantified according to the Folin-Ciocalteu method and characterized by their radical-scavenging activity using the DPPH test. The efficiency of the purification process is assessed step by step using ^{1}H NMR spectra.

2 Materials

2.1 Seaweed Material

1. Collect fresh seaweeds on the shore (*see* **Note 1**). Since geographical and seasonal parameters may both affect phenol contents in Phaeophyceae [7, 9, 21, 22], seaweed material should be collected at a single location and within a period of a few weeks for a given set of experiments.

2.2 Extraction and Purification of Phlorotannins

1. Extraction solvents: Either pure or mixed in various proportions, including deionized water and analytical grade organic solvents (methanol, acetone, ethanol, or ethyl acetate, i.e.,

high to intermediate polarity solvents). For global phenol extractions, deionized water–methanol mixtures (v/v) are preferred (*see* **Note 2**).

2. Liquid–liquid separation solvents: Extra pure-grade hexane, dichloromethane, ethyl acetate, and deionized water.
3. Precipitation solvents: Extra pure-grade acetone and ethanol.
4. Tangential ultrafiltration membranes in cartridges of various sizes and shapes, depending on the system (*see* **Note 3**) and/or dialysis membranes: Due to the large range of molecular sizes within phlorotannins (i.e., between 126 Da for the phloroglucinol unit and up to more than 300 kDa for high-molecular-weight polymers), the use of 50, 10, and 2–3 kDa cutoff size membranes may be recommended as a first approach to study the repartition of the phenol pool between polymers, intermediate compounds, and oligomers.

2.3 Quantification and Qualitative Analysis of Phlorotannins

2.3.1 Folin-Ciocalteu Assay

1. Folin-Ciocalteu reagent (available from national commercial suppliers): Best stored in the dark at 4–5 °C.
2. 200 g/L sodium carbonate: Dissolve 5 g Na_2CO_3 in 25 mL deionized water. The stock solution may be stored at room temperature for several weeks.
3. Standard: Phloroglucinol (1,3,5-trihydroxybenzene).

2.3.2 Radical-Scavenging Activity

1. 90 % Extra pure-grade methanol in deionized water.
2. Commercial solution of 2,2-diphenyl-1-picrylhydrazyl or DPPH·(available from national commercial suppliers). Prepare a dilution at 0.1 mM in 90 % methanol 3 h before conducting the test. Store both pure DPPH and its diluted solution in the dark at 4–5 °C. The solution should not be stored for longer than 2 days.
3. Trolox (6-hydroxy-2,5,7,8-tetramethylchromane-2-carboxylic acid, derivative of vitamin E) and ascorbic acid (vitamin C) (available from national commercial suppliers): Prepare 1 g/L stock solutions in deionized water.

2.3.3 ^1H NMR

1. Deuterium oxide (D_2O, 99.9 %) or deuterated methanol (CD_3OD, 99.9 %) for semipolar and polar fractions and deuterated chloroform ($CDCl_3$, 99.9 %) for relatively nonpolar fractions (Euriso-top, St. Aubin, France): Stored best in the dark at room temperature.
2. Spectrometer, for example a Bruker Avance 400 or 500 MHz equipped with a $^1H/^{13}C/^{15}N$ TCI cryoprobe; in this instance standard pulse sequence available in the Bruker software (Bruker, Wissembourg, France), but also other spectrometers with similar specifications can be used.

3 Methods

3.1 Processing of Algal Material

1. Rapidly clean the thalli with tap water, and remove epiphytes, injured/grazed areas, as well as fertile parts (unless these are of interest). Avoid bases/holdfasts and lower axes/stripes of the thalli (in particular for Fucales and Laminariales) unless these are of particular interest, since phenol contents may vary according to the type, age, and location of the material [10, 34].
2. Cut off 0.5–1 cm long pieces of thalli.
3. Distribute the raw material into freezing bags: 100–250 g fresh weight (FW) per bag from single or pooled individuals, depending on their size and on the foreseen statistical treatment. Do not forget to prepare smaller aliquots to determine the dry weight (DW) of samples and to express contents on a DW basis.
4. Freeze-dry and store in the dark until analyses are conducted (*see* **Note 4**).

3.2 Extraction of Phlorotannins

1. Weigh 15 g of freeze-dried algal material and place them into a 250 mL Erlenmeyer flask.
2. Add 200 mL of solvent and chop the algae thoroughly with a domestic mixer.
3. Agitate at 200 rpm (rounds per minute) in a rotary shaker (*see* **Note 5**) at 30–40 °C (*see* **Note 6**) in the dark (place an aluminum foil around the flask) for 3 h. Keep the flask closed during extraction to avoid excessive solvent evaporation.
4. Filter the resulting mixture using glass wool to discard pieces of algal tissue.
5. Centrifuge the filtrate at ca. 4,000 × *g* for 10–20 min at 4–5 °C and discard the pellet (*see* **Note 7**).
6. Evaporate all alcohol and most of the water down to ca. 40 mL on a rotary evaporator and measure the final volume of extract.
7. Freeze-dry the final aqueous extract. Weigh the resulting dry powder to determine the extracted dry matter and the percentage of phenols after quantification (*see* **Note 8**).

3.3 Liquid–Liquid Semi-purification of Phlorotannins

Liquid–liquid purification is essentially based on the polarity of extracted compounds and therefore on their differential affinity for nonmiscible solvents.

1. Start from ca. 1.5 g of freeze-dried crude extract and dissolve in 40 mL deionized water in a centrifuge tube, or treat directly the aqueous extract (about 40 mL, see above).
2. Add the same volume of hexane, shake the mixture rapidly, and centrifuge at ca. 4,000 × *g* for 10–20 min at 4–5 °C.

3. Keep the aqueous fraction and discard the hexane phase (*see* **Note 9**).
4. Repeat the operation at least two times and pool the aqueous phases.
5. Add the same volume of dichloromethane to the aqueous fraction, shake the mixture rapidly, and centrifuge at ca. 4,000 × *g* for 10–20 min at 4–5 °C (*see* **Note 9**).
6. Repeat the operation at least two times and pool the aqueous phases.
7. Add 2 volumes of ethanol to the aqueous fraction, shake up the mixture, and keep either for 12–24 h at −17 °C or for 1.5 h at −80 °C.
8. Centrifuge the frozen mixture at ca. 4,000 × *g* for 10–20 min at 4–5 °C.
9. Discard the pellet (if present) and remove ethanol from the supernatant. If a pellet occurs, repeat **steps 7–9** twice.
10. Add 3 volumes of acetone to the aqueous fraction, shake up the mixture, and keep either for 12–24 h at −17 °C or for 1.5 h at −80 °C.
11. Centrifuge the frozen mixture at ca. 4,000 × *g* for 10–20 min at 4–5 °C.
12. Discard the pellet (if present) and remove acetone from the supernatant. If a pellet occurs, repeat **steps 10–12** twice.
13. Add the same volume of ethyl acetate to the aqueous fraction, shake the mixture rapidly, and centrifuge at ca. 4,000 × *g* for 10–20 min at 4–5 °C.
14. Separate thoroughly the water and the ethyl acetate phases and repeat the operation at least two times.
15. Pool the aqueous phases and freeze-dry them.
16. Pool ethyl acetate phases and add about 5 mL deionized water. Then evaporate the ethyl acetate on a rotary evaporator and freeze-dry the remaining aqueous solution.

3.4 Size-Based Separation of Phlorotannins

3.4.1 Dialysis

1. Dissolve 20–40 mg of freeze-dried powder in 20 mL deionized water and pour the resulting solution in a dialysis tube (*see* **Note 10**), starting with the highest cutting size membrane.
2. Place the dialysis tube in a 0.5–1 L vessel (Erlenmeyer flask, glass bottle, high beaker), add enough deionized water to cover it, and agitate the water using a magnetic stirrer in the dark at 4–5 °C.
3. Let dialyze for at least 3 days and change the water five to six times during that period. Freeze-dry the concentrate (retentate) and concentrate the dialysate on a rotary evaporator down to a volume of 20 mL. Then, freeze-dry the dialysate.

4. Proceed to a new dialysis step with a lower cutting size membrane from the concentrated dialysate up to the lowest cutting size membrane. After each dialysis, freeze-dry the concentrate (retentate) and concentrate the dialysate on a rotary evaporator down to a volume of 20 mL before freeze-drying.

3.4.2 Tangential Ultrafiltration (UF)

1. Dissolve 50–100 mg of freeze-dried powder in 100 mL deionized water and place the resulting solution in the retentate vessel of a tangential UF system (*see* **Note 4**).
2. Start with the cartridge corresponding to the highest cutting size and ultrafiltrate the sample in the dark at moderate temperature until reaching a retentate volume of 5–10 mL.
3. Add a similar volume of deionized water and repeat the operation at least three times.
4. At the end of the process, remove the retentate and rinse the system with 3–4 cartridge volumes of deionized water, or three cartridge volumes of 20 % ethanol in deionized water and then one cartridge volume of deionized water (*see* **Note 11**).
5. Pool the successive retentates and freeze-dry them.
6. Pool and concentrate the ultrafiltrates by rotary evaporation down to ca. 100 mL and proceed to a new UF step starting at a lower cutting size.
7. Repeat the tangential UF up to the lowest cutting size membrane. After each UF, freeze-dry the concentrate (retentate) and concentrate the dialysate on a rotary evaporator down to a volume of 20 mL before freeze-drying.

3.4.3 Mixed Dialysis and Tangential UF

Dialysis and tangential UF may be alternated, depending on the repartition of oligomers and polymers expected in the semi-purified fractions. In that case, only one dialysis step will be recommended, generally at the beginning of the purification process, i.e., at high cutting sizes.

3.5 Quantification and Qualitative Analysis of Phlorotannins

3.5.1 Folin-Ciocalteu Assay

Assess the phlorotannin content of the crude extract at the successive steps of the semi-purification using the Folin-Ciocalteu method adapted to small volumes in 96-well standard microplates:

1. Prepare a concentrated solution of each sample in deionized water (200 μg/mL, to be diluted if necessary).
2. Put 20 μL of the solution into each well; then add 10 μL of the Folin-Ciocalteu reagent, 40 μL of sodium carbonate (from a 200 g/L solution), and 130 μL of deionized water.
3. Homogenate the reaction mixture, leave for 10 min at 70 °C, and then cool rapidly on ice (5 min).

4. Measure the optical density at 620 nm with a microplate spectrophotometer and plot against a standard curve of phloroglucinol (0–100 μg/mL). Express contents of phlorotannins in mg/g or % of DW.

3.5.2 Radical-Scavenging Activity

Test the radical-scavenging activity of the crude extract and of the fractions following the DPPH·method:

1. Prepare a concentrated solution of each sample in deionized water and dilute it to get at least eight concentrations between 0 and 1 g/L.
2. Pour 100 μL of each dilution per well and measure the absorbance at 540 nm using a microplate photometer; then add 100 μL of a 0.1 mM DPPH solution.
3. Let the mixture develop for 60 min in the dark at a temperature of 15–20 °C and measure the absorbance at 540 nm.
4. Use the solvent of samples as negative control (deionized water here), and Trolox and ascorbic acid (vitamin C) as positive standards. Prepare 5–8 dilutions of positive standards between 0 and 0.1 g/L from the 1 g/L stock solutions.
5. The activity is expressed as the percent inhibition of DPPH based on the difference between the two absorbance measurements at 540 nm. Radical-scavenging activity is expressed as IC_{50} (the concentration of substrate which causes a 50 % loss of the DPPH activity).

3.5.3 ^1H NMR Analysis

If NMR analysis is available, prepare aliquots of each extract/fraction as follows:

1. Weigh 10–20 mg max. of freeze-dried powder; mix it with 700 μL deuterated solvent and shake until complete dissolution.
2. Put the solution in a 5 mm diameter NMR tube and close it with a cap.

^1H NMR is then performed within 1 day, at room temperature on an NMR spectrometer. Resulting spectra will show peaks representing only relative abundances of the various compounds occurring in the fraction, identified by the position (chemical shift or δ) of one to several peaks expressed in parts per million (ppm) (*see* Fig. 1). Figure 1 displays ^1H-NMR spectra illustrating the results of a polarity-based semi-purification procedure, highlighting the evolution of the phenol peaks in the various fractions, in particular with their magnification in the final ethyl acetate fraction versus a global disappearance of mannitol signals. Figure 2 shows spectra obtained along a mixed size- and polarity-based procedure, underlining the interest of both UF and dialysis to discriminate semi-pure size classes of phlorotannins.

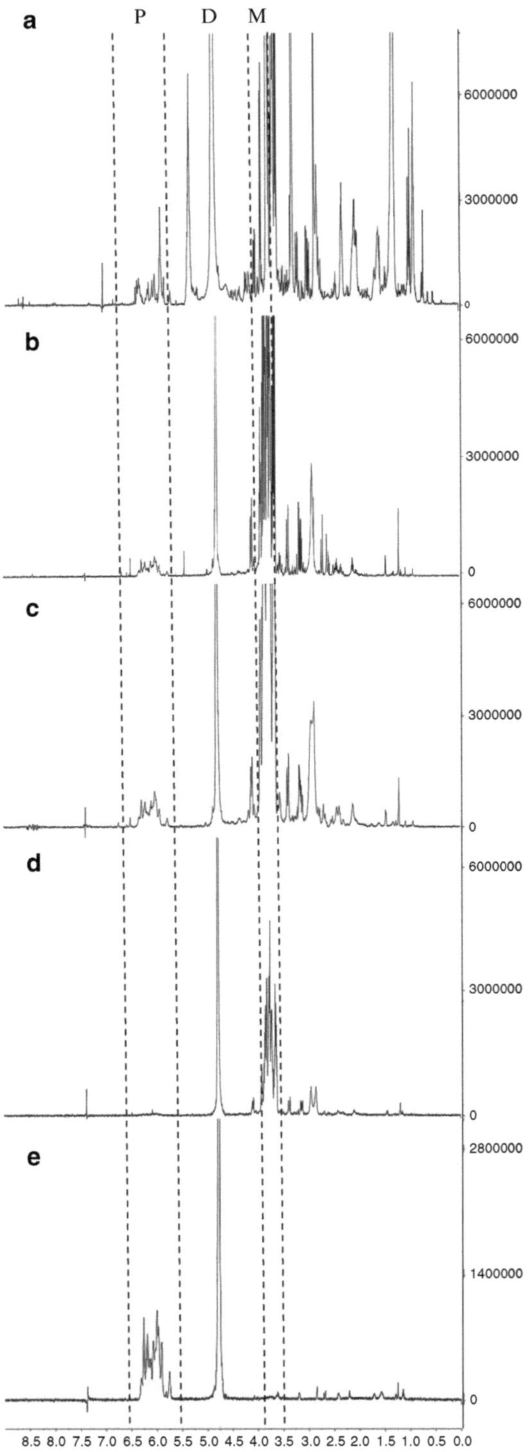

Fig. 1 ^1H NMR spectra of a crude extract and of semi-purified fractions obtained by polarity-based fractionating in *Pelvetia canaliculata* (Fucales, Fucaceae). (**a**) Crude extract. (**b**) Aqueous fraction after washing by both hexane and dichloromethane. (**c**) Aqueous fraction after precipitation by both ethanol and acetone. (**d**) Final aqueous fraction after separation from the ethyl acetate phase. (**e**) Ethyl acetate fraction. P corresponds to the phenol area (5.5–6.5 ppm), D to the D_2O peak, M to the mannitol area

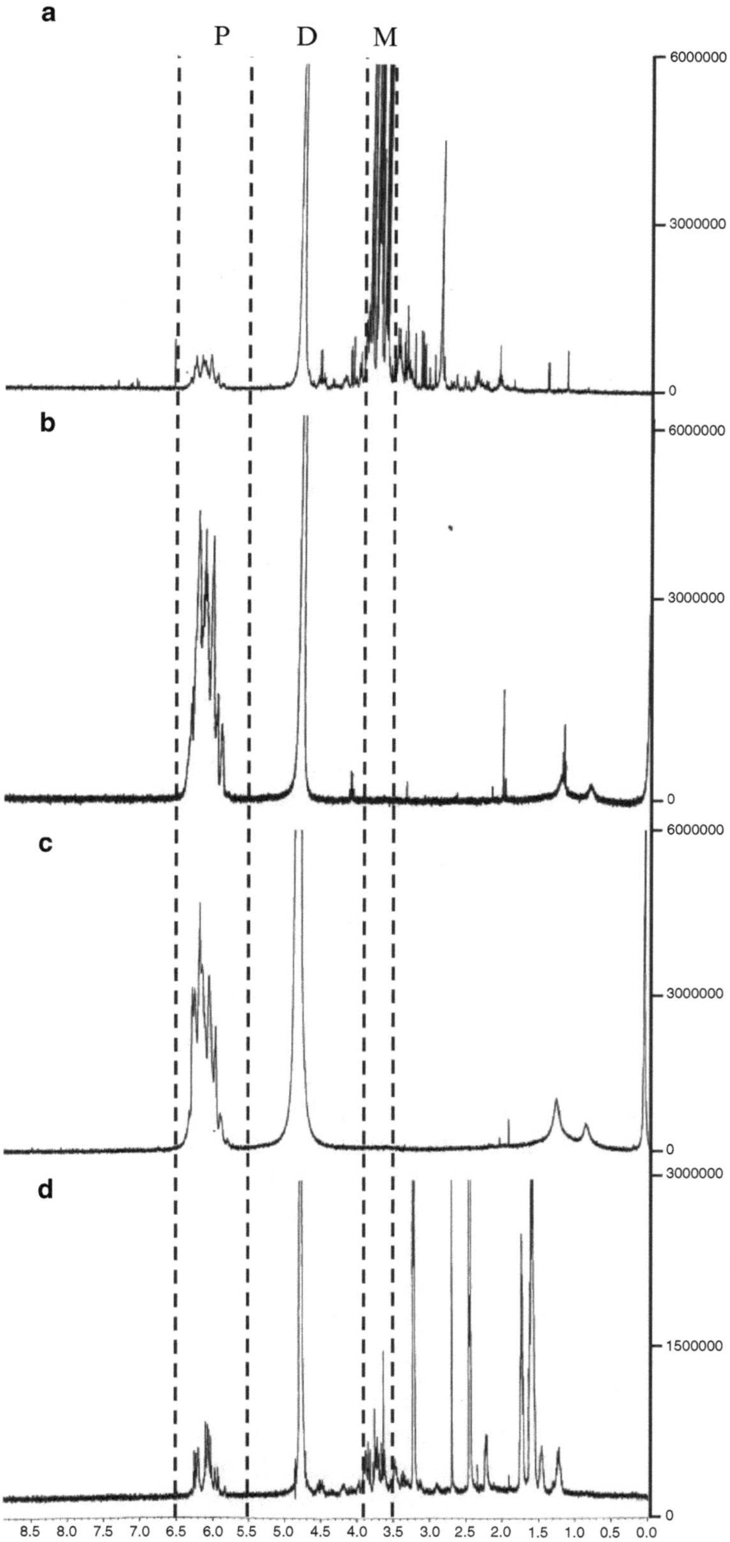

Fig. 2 ^{1}H NMR spectra of a crude extract and of semi-purified fractions obtained by polarity-based and size-based fractionating in *Ascophyllum nodosum* (Fucales, Fucaceae). (**a**) Crude extract. (**b**) Ethyl acetate fraction after hexane and dichloromethane washing. (**c**) Tangential UF concentrate obtained from the ethyl acetate fraction on a 50 kDa cutting size membrane. (**d**) Final 2 kDa dialysate obtained from the 50 kDa ultrafiltrate. Residual mannitol still appears among other molecules smaller than 2 kDa. P corresponds to the phenol area (5.5–6.5 ppm), D to the D_2O peak, M to the mannitol area

4 Notes

1. Test species of Phaeophyceae which are common in your geographical area. In a first approach to phlorotannins, aim for dominating Fucales from the middle to upper intertidal zone which support large populations and biomass. Phenol contents also tend to be higher in species at high bathymetric levels, which, in turn, give better extraction yields and easier purification steps.
2. In view of industrial biomolecule extraction, ethanol is preferred to methanol because of its lower price and toxicity. Similarly, classical solid/liquid extraction may be considered as an exploratory stage prior to the sustainable production of phenols.
3. In tangential ultrafiltration (crossflow ultrafiltration, tangential flow filtration or TFF), a better efficiency will be obtained using systems (modules) able to maintain a constant flow, i.e., fitted with a peristaltic pump. The TFF membranes/cartridges must be adapted to a laboratory scale, i.e., with a capacity ranging between 100 mL and 10 L. Various products are available worldwide, from single cartridges to complete concentrators (for example, Labscale TFF System and Pellicon Holders at Millipore, Kros Flow at Spectrum, Minimate and Centramate at Pall Gelman, Vivaflow cartridges at Sartorius). Most cartridges are both disposable and reusable, and offer a wide range of molecular weight cutoff (MWCO). Polyethersulfone (PES) membranes may be preferable according to their lower binding properties than regenerated cellulose (RC).
4. Seaweeds may be dried gently at 30–40 °C in an aerated hood for 3–5 days, or stored frozen. However, freeze-drying allows better extraction yields as well as more stable biological activities of phlorotannins [11]. When seaweed tissues are kept fresh at 4–5 °C, a rapid degradation of phenolic compounds will occur, whereas freezing at −17 to −25 °C and defrosting are less likely to affect extraction yields.
5. A reciprocal shaker may be used as an alternative to orbital devices, with usually no significant decrease in phenol yields. Make sure to keep a steady motion of about 200 shakes per min.
6. Temperature must be set below 40 °C to preserve the integrity of phenol chains, particularly if the occurrence of small oligophenols is suspected. However, heating over 30 °C will be necessary to maintain extraction time at about 3 h.
7. For a rapid processing, both centrifuge and rotor must be compatible with 250 mL tubes. In that case, maximal speed rarely allows acceleration beyond 4,000 ×*g*. When cooling is not available within the centrifuge, room temperature may be acceptable.

8. The totality of the crude extract may be freeze-dried for long-term storing. However, the semi-purification can be processed from the aqueous extract. In that case, 1–2 mL aliquots will be kept aside for dry mass determination, phenol quantification, DPPH evaluation, and NMR analysis.
9. It may be interesting to analyze the composition and the activity of both hexane and dichloromethane phases. In that case, evaporate the pooled fractions down to 5–10 mL and dilute with methanol to 40–45 mL for Folin-Ciocalteu and DPPH procedures; evaporate aliquots completely under vacuum or by bubbling nitrogen to assess the dry weight and to perform ^{1}H NMR.
10. Dialysis tubes are available commercially with various width and filling volumes per cm. To maximize the exchange surface with deionized water and to reduce the tube length, a 2–4 cm wide tube with a 4 mL/cm capacity will be adequate. Keep 1–2 cm free at each extremity of the tube and close it with 1–2 mm thread (typically for alimentary use). Before cutting off and filling the tube, immerse it for 1–2 h in deionized water to soften it, and then rinse it abundantly with deionized water both inside and outside using a wash bottle.
11. At the end of the procedure, the retentate may become more or less viscous, making necessary several rinsing steps by adding deionized water to achieve a good separation. As for dialysis, the efficiency of the separation will be checked only afterwards, for example by the lack of residual peaks corresponding to mannitol on ^{1}H NMR spectra of retentates (above 650 Da cutting size), or at least their drastic reduction. Mass spectrometry would be decisive in this context [13]; yet it is mainly used to study phlorotannins below 2,000 Da [24]. After use, rinse the system several times with 0.1 M NaOH and store the cartridges closed with 0.01 M NaOH inside. Before reuse, rinse the system several times with deionized water.

Acknowledgements

This work was supported by a grant-in-aid for scientific research (Phlorotan'ING project) from the "Groupement d'Intérêt Scientifique" Europôle Mer to V.S.P. and fellowships from the French Ministry for National Education, Higher Education and Research to M.H. and C.J. The authors are grateful to S. Cérantola, N. Kervarec, and G. Simon for support in NMR analysis and to E. Charpentier.

References

1. Poupard P, Sanoner P, Baron A et al (2011) Characterization of procyanidin B2 oxidation products in an apple juice model solution and confirmation of their presence in apple juice by high-performance liquid chromatography coupled to electrospray ion trap mass spectrometry. J Mass Spectrom 46:1186–1197
2. Lattanzio V, Cardinali A, Linsalata V (2012) Introduction: plant phenolics – secondary metabolites with diverse functions. In: Cheynier V, Sarni-Manchado P, Quideau S (eds) Recent advances in polyphenol research, vol 3. Wiley, New York, pp 1–39
3. Stengel DB, Connan S, Popper ZA (2011) Algal chemodiversity and bioactivity: sources of natural variability and implications for commercial application. Biotechnol Adv 29:483–501
4. Ragan MA, Glombitza K-W (1986) Phlorotannins, brown algal polyphenols. Prog Phycol Res 4:129–241
5. Glombitza KW, Pauli K (2003) Fucols and phlorethols from the brown alga *Scytothamnus australis* Hook. et Harv. (Chnoosporaceae). Bot Mar 46:315–320
6. Pavia H, Toth G (2000) Influence of light and nitrogen on the phlorotannin content of the brown seaweed *Ascophyllum nodosum* and *Fucus vesiculosus*. Hydrobiologia 440:299–305
7. Connan S, Goulard F, Stiger V et al (2004) Phlorotannins in belt forming brown algae of a sheltered shore. Bot Mar 47:410–416
8. Stiger V, Deslandes E, Payri CE (2004) Phenolic contents of two brown algae, *Turbinaria ornata* and *Sargassum mangarevense* on Tahiti (French Polynesia): interspecific, ontogenic and spatio-temporal variations. Bot Mar 47:402–409
9. Connan S, Deslandes E, Ar Gall E (2007) Influence of nycthemeral and tidal cycles on phenol content and antioxidant capacity in three intertidal Phaeophyceae. J Exp Mar Biol Ecol 349:359–369
10. Zubia M, Payri C, Deslandes E (2008) Alginate, mannitol, phenolic compounds and biological activities of two range-extending brown algae, *Sargassum mangarevense* and *Turbinaria ornata* (Phaeophyta: Fucales), from Tahiti (French Polynesia). J Appl Phycol 20:1033–1043
11. Le Lann K, Jégou C, Stiger-Pouvreau V (2008) Effect of different conditioning treatments on total phenolic content and antioxidant activities in two Sargassacean species: comparison of the frondose *Sargassum muticum* (Yendo) Fensholt and the cylindrical *Bifurcaria bifurcata* R. Ross. Phycol Res 56:238–245
12. Plouguerné E, Le Lann K, Connan S et al (2006) Spatial and seasonal variation in density, reproductive status, length and phenolic content of the invasive brown macroalga *Sargassum muticum* (Yendo) Fensholt along the coast of Western Brittany (France). Aquat Bot 85:337–344
13. Ferreres F, Lopes G, Gil-Izquierdo A et al (2012) Phlorotannin extracts from Fucales characterized by HPLC-DAD-ESI-MS*n*: approaches to hyaluronidase inhibitory capacity and antioxidant properties. Mar Drugs 10:2766–2781
14. Meillisa A, Siahaan EA, Jung-Nam Park J-N et al (2013) Effect of subcritical water hydrolysate in the brown seaweed *Saccharina japonica* as a potential antibacterial agent on food-borne pathogens. J Appl Phycol 25:763–769
15. Tanniou A, Serrano Leon E, Vandanjon L et al (2013) Green improved processes to extract bioactive phenolic compounds from brown macroalgae using *Sargassum muticum* as model. Talanta 104:44–52
16. Kubanek J, Lester SE, Fenical W et al (2004) Ambiguous role of phlorotannins as chemical defences in the brown alga *Fucus vesiculosus*. Mar Ecol Prog Ser 277:79–93
17. Cérantola S, Breton F, Ar Gall E et al (2006) Co-occurrence and antioxidant activities of fucol and fucophlorethol classes of polymeric phenols in *Fucus spiralis*. Bot Mar 49:347–351
18. Zubia M, Fabre M-S, Kerjean V et al (2009) Antioxidant and antitumoural activities of some Phaeophyta from Brittany coasts. Food Chem 116:693–701
19. Arnold TM, Targett NM (1998) Quantifying in situ rates of phlorotannin synthesis and polymerisation in marine brown algae. J Chem Ecol 24:577–595
20. Breton F, Cérantola S, Ar Gall E (2011) Distribution and radical scavenging activity of phenols in *Ascophyllum nodosum* (Phaeophyceae). J Exp Mar Biol Ecol 399: 167–172
21. Le Lann K, Connan S, Stiger-Pouvreau V (2012) Phenology, TPC and size-fractioning phenolics variability in temperate Sargassaceae (Phaeophyceae, Fucales) from Western Brittany: native versus introduced species. Mar Environ Res 80:1–11
22. Le Lann K, Ferret C, VanMee E et al (2012) Total phenolic, size-fractionated phenolics and fucoxanthin content of tropical Sargassaceae (Fucales, Phaeophyceae) from the South Pacific Ocean: spatial and specific variability. Phycol Res 60:37–50

23. Tierney MS, Soler-Vila A, Croft AK et al (2013) Antioxidant activity of the brown macroalga *Fucus spiralis* Linnaeus harvested from the West Coast of Ireland. Curr Res J Biol Sci 5:81–90
24. Wang T, Jónsdóttir R, Liu H et al (2012) Antioxidant capacities of phlorotannins extracted from the brown alga *Fucus vesiculosus*. J Agric Food Chem 60:5874–5883
25. Shibata T, Fujimoto K, Nagayama K et al (2002) Inhibitory activity of brown algal phlorotannins against hyaluronidase. Int J Food Sci Technol 37:703–709
26. Koivikko R, Eränen J, Loponen J et al (2008) Variation of phlorotannins among three populations of *Fucus vesiculosus* as revealed by HPLC and colorimetric quantification. J Chem Ecol 34:57–64
27. Audibert L, Fauchon M, Blanc N et al (2010) Phenolic compounds in the brown seaweed *Ascophyllum nodosum*: pool composition and radical-scavenging activities. Phytochem Anal 21:399–405
28. Steevensz AJ, MacKinnon SL, Hankinson R et al (2012) Profiling phlorotannins in brown macroalgae by liquid chromatography–high resolution mass spectrometry. Phytochem Anal 23:547–553
29. Tierney MS, Smyth TJ, Rai DK et al (2013) Enrichment of polyphenol contents and antioxidant activities of Irish brown macroalgae using food-friendly techniques based on polarity and molecular size. Food Chem 139:753–761
30. Parys S, Rosenbaum A, Kehraus S et al (2007) Evaluation of quantitative methods for the determination of polyphenols in algal extracts. J Nat Prod 70:1865–1870
31. Yotsu-Yamashita M, Kondo S, Segawa S et al (2013) Isolation and structural determination of two novel phlorotannins from the brown alga *Ecklonia kurome* Okamura, and their radical scavenging activities. Mar Drugs 11: 165–183
32. Blanc N, Hauchard D, Audibert L et al (2011) Radical-scavenging capacity of phenol fractions in the brown seaweed *Ascophyllum nodosum*: an electrochemical approach. Talanta 84:513–518
33. Keyrouz R, Abasq ML, Le Bourvellec C et al (2011) Total phenolic contents, radical scavenging and cyclic voltammetry of seaweeds from Brittany. Food Chem 126:831–836
34. Connan S, Delisle F, Deslandes E et al (2006) Intra-thallus phlorotannin content and antioxidant activity in Phaeophyceae of temperate waters. Bot Mar 49:39–46

Chapter 8

Enzyme-Enhanced Extraction of Antioxidant Ingredients from Algae

Björn V. Adalbjörnsson and Rósa Jónsdóttir

Abstract

Marine algae are not only a rich source of dietary fibre, proteins, vitamins, and minerals, but also contain a great variety of secondary metabolites with diverse biological activities. Marine macroalgae are a rich source of various natural antioxidants such as polyphenols, especially phlorotannins (made of polyphloroglucinol units) derived from brown algae, which play an important role in preventing lipid peroxidation. In recent years, a number of potent antioxidant compounds have been isolated and identified from different types of edible seaweeds. Extraction methods commonly used for the isolation of antioxidants are based on conventional water or organic solvent extractions. However, recent advances have shown that enzymatic hydrolysis can achieve higher yield of bioactive compounds from algae. Here we describe a method based on enzymatic hydrolysis which both increases yield and decreases cost associated with organic solvents. This method achieves cell wall disruption and breakdown of internal storage components for more effective release of intracellular bioactive compounds. In addition, hydrolysis of proteins produces peptides which may have antioxidant properties, thus enhancing the bioactivity of the algal extract. The method described can be used for production of extracts from red and brown macroalgal species.

Key words Antioxidants, Enzyme hydrolysis, Extraction, Glycosidases, Macroalgae, Proteases

1 Introduction

Extraction via enzymatic hydrolysis is an alternative method to conventional water and organic solvent extraction. The use of hydrolytic enzymes during the extraction process has shown increased extraction yield, improved release of secondary plant metabolites, and a higher bioactivity of extracts [1]. This approach has proven to be successful in extracting bioactive compounds from brown algae [2, 3].

The enzyme-assisted extraction is a simple and efficient method to extract natural antioxidants from seaweeds and is scalable to industrial scale. This method is also environmentally friendly as it does not use any organic solvents or other toxic chemicals and makes thus the extracts safe for food applications. In comparison to water and organic extractions, enzymatic based methods have been

Dagmar B. Stengel and Solène Connan (eds.), *Natural Products From Marine Algae: Methods and Protocols*, Methods in Molecular Biology, vol. 1308, DOI 10.1007/978-1-4939-2684-8_8, © Springer Science+Business Media New York 2015

demonstrated to increase the yield of phenolic compounds in the extracts and increased the antioxidant activity of, e.g., brown macroalgae [2, 4].

Proteases are categorized according to their hydrolyzing mechanism into endopeptidases and exopeptidases [5]. Exopeptidases yield single amino acids as they hydrolyze at the end of polypeptide chains. Endopeptidases hydrolyze proteins around specific amino acids within the polypeptide chain and release peptides of various lengths. The site of hydrolysis depends on the specificity of the peptidase used. These hydrolytic enzymes can help rupturing the algal cell wall and breaking down internal storage materials. This enhances disruption of the algae and release of bioactive compounds. Additionally, peptides, produced from protein hydrolysis, can also increase the antioxidant activities of the extract [3, 6]. Also, bioactive low-molecular-weight polysaccharides produced by hydrolysis of high-molecular-weight polysaccharides, using enzymes such as xylanases, fucosidases, and alginases, may contribute to enhanced antioxidant activities.

It must be kept in mind that the chemical composition of extracts is dependent on several factors: on the seaweed species of interest, the season of seaweed sampling, the environment, and the specificity of the enzyme used [2, 6–8]. As enzymes have evolved to be stable in the environment of their organism, enzymes have different optimal conditions (*see* Table 1). Combined with knowledge of these factors, enzyme-assisted extraction has the potential to significantly enhance the effectiveness of extraction of compounds with antioxidant activity. Commercial enzymes have shown to effectively improve the extraction yield of different enzymatic extracts from *Palmaria palmata* compared to water extract [6]. Additionally, the enzyme-assisted extraction increased the total phenolic content and antioxidant activity evaluated using oxygen radical absorbance capacity (ORAC) of the *P. palmata* extracts as seen in Table 2. *P. palmata* is a species with high protein content and in this example, all the proteases tested significantly enhanced the properties of the extracts compared to carbohydrases and water extraction.

2 Materials

1. 3 % sodium chloride (NaCl) solution: Weigh 3 g of NaCl and add distilled water up to 100 mL.
2. Use freshly collected algae (*see* **Note 1**) and carefully remove contaminants, such as sand and small invertebrates or epiphytes, with seawater or 3 % NaCl. After that, use tap water and distilled water.

Table 1
Examples of the optimum hydrolysis conditions, characteristics, and sources of specific enzymes

Enzyme[a]	Optimum conditions[b]		Characteristics	Source
	pH	Temperature (°C)		
Alacalase 2.4 L FG	8.0		Endopeptidase	*Bacillus licheniformis*
AMG 300 L	4.5	60	Exo-1,4-a-D-glucosidase	*Aspergillus niger*
Celluclast 1.5 L FG	4.5	50	Cellulase	*Trichoderma reesei* ATCC 26921
Flavourzyme 500 MG	7.0	50	Endoprotease and exopeptidase	*Aspergillus oryzae*
Kojizyme 500 MG	6.0	40	Amino- and carboxy-peptidase	*A. oryzae*
Neutrase 0.8 L	6.0	50	Metallo-endoprotease	*Bacillus amyloliquefaciens*
Ultraflo L	7.0	60	Heat-stable multi-active b-glucanase	*Humicola insolens*
Umamizyme	7.0	50	Endo- and exopeptidase complex	*A. oryzae*
Termamyl 120 L	6.0	60	Heat-stable a-amylase	*B. licheniformis*
Viscozyme L	4.5	50	A multienzyme complex (containing arabanase, cellulase, b-glucanase, hemicellulase, and xylanase)	*Aspergillus aculeatus*

Table adapted from ref. 6
[a]Umamizyme is obtained from Amano Enzyme Inc. (Nagoya, Japan). The other enzymes are from Novozymes A/S (Bagsvaerd, Denmark)
[b]Described in refs. 2, 9

3. Blender.
4. 1.0 mm sieve.
5. Water bath.
6. Whatman no. 4 filter paper.
7. 1 M hydrochloric acid (HCl): Pour 8.3 mL of 12 M HCl in 91.7 mL of distilled water.
8. 1 M sodium hydroxide (NaOH): Dissolve 400 mg of NaOH in 90 mL of distilled water and then adjust the volume to 100 mL.
9. Enzymes: Follow the supplier's information regarding the optimal conditions (temperature and pH) for the enzyme.

Table 2
Extraction yield, total phenolic content (TPC), and ORAC values of different enzymatic extracts from *P. palmata* compared to water extract (WE)

		Extraction yield[a] (%)			TPC[b] (% GAE)			ORAC value[c]		
WE	Water extract	37.8	±	1.9	0.52	±	0.00	35.8	±	1.4
Termamyl	Carbohydrase extract	46.5	±	1.3	0.25	±	0.00	25.0	±	1.3
Celluclast	Carbohydrase extract	66.6	±	2.2	0.27	±	0.01	19.3	±	1.9
AMG	Carbohydrase extract	52.3	±	2.0	0.31	±	0.01	31.2	±	1.2
Ultraflo	Carbohydrase extract	68.3	±	1.5	0.34	±	0.01	36.3	±	1.1
Viscozyme	Carbohydrase extract	60.9	±	1.3	0.62	±	0.01	66.0	±	0.8
Flavourzyme	Protease extracts	57.4	±	0.9	0.80	±	0.01	88.1	±	1.3
Neutrase	Protease extracts	55.0	±	2.1	0.86	±	0.01	107.3	±	1.2
Kojizyme	Protease extracts	62.3	±	0.9	0.87	±	0.01	75.8	±	5.0
Protamex	Protease extracts	58.7	±	1.5	0.98	±	0.01	99.8	±	3.1
Alcalase	Protease extracts	58.0	±	1.4	1.00	±	0.01	138.6	±	2.7
Umamizyme	Protease extracts	76.3	±	1.4	1.55	±	0.02	148.6	±	0.9

Values are means ± S.D. ($n = 3$). Data adapted from ref. 6
[a]Extraction yield expressed as g dried extract/100 g dried algal powder
[b]TPC, total phenolic content, % gallic acid equivalents (GAE)
[c]ORAC, oxygen radical absorbance capacity (Trolox equivalents, μmol/g extract)

3 Methods

Methods should be carried out at room temperature unless otherwise specified.

3.1 Preparation of Algae

1. Cut the rinsed algae into small pieces (around 2 × 2 cm) and freeze them at −20 °C.
2. Freeze-dry the small algal pieces.
3. Grind the freeze-dried algae into a powder in a blender.
4. Sieve the powder through a 1.0 mm sieve.
5. Store the powder at −20 °C until extraction for up to 12 months.

3.2 Enzyme Hydrolysis of Algae

1. Add 2 g of algal sample to 50 mL of distilled water and incubate for 10 min in the dark at room temperature.
2. Adjust the pH, using 1 M NaOH and/or HCl, to the optimal pH for the enzyme used (for examples *see* Table 1).

3. Add enzyme (50 mg or 50 μL, depending on in which form the supplier sells the enzyme) to the solution and incubate the enzymatic hydrolysis at the optimal temperature for the enzyme in a water bath for 24 h in the dark (*see* **Note 2**).
4. After hydrolysis the reaction is terminated by boiling the sample in water bath at 100 °C for 10 min (*see* **Note 3**). This is followed by immediate cooling in an ice bath.
5. Centrifuge the hydrolysate at 5,000 × *g* for 10 min at 4 °C.
6. Remove the supernatant and filter through Whatman no. 4 filter paper.
7. Adjust the pH of the filtrate to pH 7.0 with 1 M HCl and/or NaOH.
8. Freeze-dry the extract, weigh it to calculate the extraction yield (*see* **Note 4**), and store at −20 °C (*see* **Note 5**). Extract can be stored at −20 °C for up to 2 years without any significant changes to total polyphenol content or ORAC values.

4 Notes

1. This method can be used with red and brown macroalgal species and enzymes used depend on the algal species and desired product. For hydrolysis of polysaccharide use glycosidase that hydrolyzes the major polysaccharide component of the cell wall, such as use of fucosidases when working with *Fucus vesiculosus.* To include peptides in the extract rather than proteins proteases can be used. This is mostly feasible when working with species with high protein content, such as *P. palmata.*
2. Use a water extract (2 g algal powder extracted with 50 mL of distilled water for 24 h at room temperature) as a control.
3. Heating should be kept to minimum as phlorotannins and phycoerythrin (for example in *F. vesiculosus* and *P. palmata*, respectively) are heat-sensitive molecules and can degrade during the extraction and inactivation of the enzymes.
4. The extraction yield is calculated as below:

$$\text{Yield}\,(\%) = \frac{\text{Weight of freeze dried extract}}{\text{Weight of dried algal sample}} \times 100$$

5. Hydrolysis of polysaccharides and proteins can increase the amount of oligosaccharides and peptides with different bioactivity than desired. The use of alginase, during *F. vesiculosus* extraction, increases the amount of alginate oligosaccharides instead of producing fucoidan oligosaccharides with antioxidant activity.

References

1. Li BB, Smith B, Hossain MM (2006) Extraction of phenolics from citrus peels: II. Enzyme-assisted extraction method. Sep Purif Technol 48:189–196
2. Heo SJ, Park EJ, Lee KW et al (2005) Antioxidant activities of enzymatic extracts from brown seaweeds. Bioresour Technol 96:1613–1623
3. Siriwardhana N, Kim KN, Lee KW et al (2008) Optimisation of hydrophilic antioxidant extraction from *Hizikia fusiformis* by integrating treatments of enzymes, heat and pH control. Int J Food Sci Technol 43:587–596
4. Athukorala Y, Kim KN, Jeon YJ (2006) Antiproliferative and antioxidant properties of an enzymatic hydrolysate from brown alga *Ecklonia cava*. Food Chem Toxicol 44: 1065–1074
5. Shahidi F, Naczk M (2004) Antioxidant properties of food phenolics. In: Shahidi F, Naczk M (eds) Phenolics in food and nutraceuticals. CRC, Boca Raton, FL, pp 403–437
6. Wang T, Jónsdóttir R, Kristinsson HG et al (2010) Enzyme-enhanced extraction of antioxidant ingredients from red algae *Palmaria palmata*. LWT Food Sci Technol 43:1387–1393
7. Koivikko R, Loponen J, Honkanen T (2005) Contents of soluble, cell-wall-bound and exuded phlorotannins in the brown alga *Fucus vesiculosus*, with implications on their ecological functions. J Chem Ecol 31:195–212
8. Wang T, Jónsdóttir R, Ólafsdóttir G (2009) Total phenolic compounds, radical scavenging and metal chelation of extracts from Icelandic seaweeds. Food Chem 116:240–248
9. Sato M, Oba T, Yamaguchi T et al (2002) Antihypertensive effects of hydrolysates of wakame (*Undaria pinnatifida*) and their angiotensin-I-converting enzyme inhibitory activity. Ann Nutr Metab 46:259–267

Chapter 9

Microwave-Assisted Extraction of Fucoidan from Marine Algae

Solange I. Mussatto

Abstract

Microwave-assisted extraction (MAE) is a technique that can be applied to extract compounds from different natural resources. In this chapter, the use of this technique to extract fucoidan from marine algae is described. The method involves a closed MAE system, ultrapure water as extraction solvent, and suitable conditions of time, pressure, and algal biomass/water ratio. By using this procedure under the specified conditions, the penetration of the electromagnetic waves into the material structure occurs in an efficient manner, generating a distributed heat source that promotes the fucoidan extraction from the algal biomass.

Key words Extraction, Degradation, Fucoidan, Fucose, Heat, Irradiation, Marine algae, Microwave, Monosaccharides

1 Introduction

Microwave-assisted extraction (MAE) is a novel extraction technique that has been developed over the past decade and that has attracted significant attention due to its different heating mechanism, moderate capital cost, and good performance (similar or better yields are obtained when compared to conventional extraction processes, using less energy and solvent volume, and shorter extraction times) [1, 2]. For these reasons, the technology of MAE has been applied to extract compounds from a variety of natural resources including algae, plants, and lignocellulosic biomass.

Microwave is an electromagnetic radiation that consists of electric and magnetic fields oscillating perpendicularly to each other in frequencies varying from 300 MHz to 300 GHz [3]. The extraction process using microwaves involves the penetration of the electromagnetic radiation into the material structure, which induces the vibration of water molecules. As a consequence, the temperature of the intracellular liquids increases above boiling point, and the water evaporates and exerts pressure on the material cell wall. The significant pressure developed inside the matrix

Dagmar B. Stengel and Solène Connan (eds.), *Natural Products From Marine Algae: Methods and Protocols*, Methods in Molecular Biology, vol. 1308, DOI 10.1007/978-1-4939-2684-8_9, © Springer Science+Business Media New York 2015

modifies the physical properties of the biological tissue and increases the porosity of the material as a consequence, allowing a better penetration of the extraction solvent through the matrix. As a result of this process, the material cell wall breaks, releasing the intracellular components into the reaction medium [4, 5].

MAE systems are classified as "closed" or "open" systems in order to differentiate between procedures that operate above or under atmospheric pressure, respectively. In the case of the extraction method described in this chapter, a closed system will be used. This system is associated with high pressure and allows a random dispersion of microwave radiation in cavity by a mode stirrer [6, 7]. The extractions are carried out in sealed vessels which, after having been sealed, are placed in a rotating carousel, and this system is introduced into the microwave equipment. The pressure inside each vessel is controlled so that it does not exceed the working pressure, while the temperature can be regulated above the normal boiling point of the extraction solvent [3]. A representation of the closed system used for MAE is illustrated in Fig. 1.

This chapter describes the use of the MAE technique to extract fucoidan from marine algae. Fucoidan is a complex and heterogeneous sulfated water-soluble polysaccharide produced by various marine organisms, mainly by brown macroalgae. In terms of chemical composition, fucoidan is mainly composed of L-fucose and sulfate groups, but it may also contain other hexose and pentose sugars (such as mannose, galactose, glucose, and xylose, among others), uronic acids, acetyl groups, and protein in the composition, which is species related [8]. The size and structure of the fucoidan molecule also differ among algal species and can vary even within the same species. In this chapter, the conditions of pressure, extraction time, and algal biomass/water ratio employed to extract

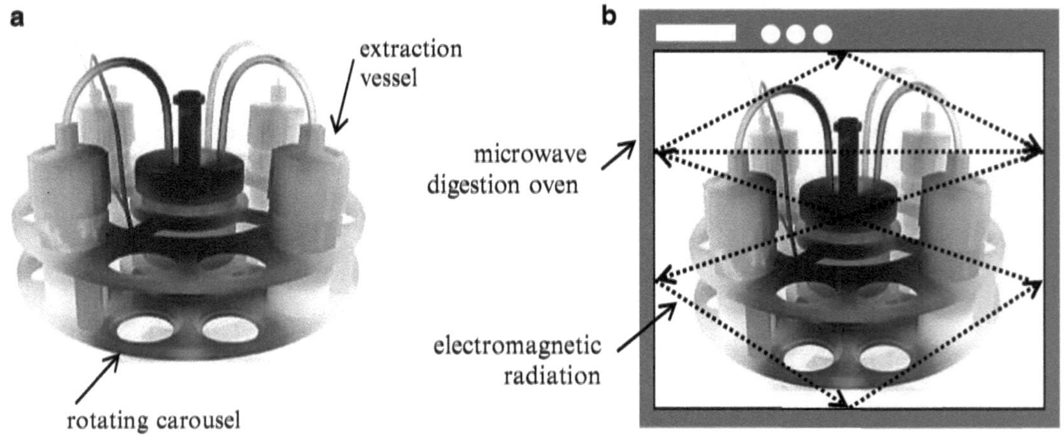

Fig. 1 Representation of the closed system used for MAE. (**a**) Sealed extraction vessels placed in the carousel, and (**b**) microwave digestion oven under operation with the carousel system inside

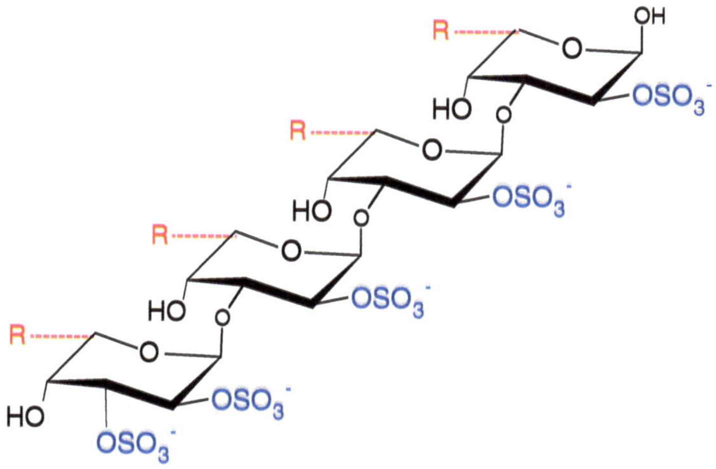

Fig. 2 Structure of the fucoidan molecule. For the brown macroalga *Fucus vesiculosus*, R is usually L-fucose, but it can also include other monosaccharides such as xylose and galactose. For other macroalgae species, R may include a variety of pentose and hexose sugars, such as fucose, mannose, galactose, glucose, and xylose, among others

fucoidan from marine algae by MAE were established by using *Fucus vesiculosus* as macroalgal sample [9]. The structure and chemical composition of fucoidan from *F. vesiculosus* are relatively simple, being mainly composed of L-fucose and sulfate groups (Fig. 2). Fucoidans from other macroalgae have a more complex structure and chemical composition, but the methodology described in this chapter could also be applied to such seaweeds [10].

2 Materials

2.1 Laboratory Supplies

1. Petri dishes.
2. Glass tubes for sample digestion.
3. Qualitative filter paper (example: paper Whatman no. 131).
4. Nylon fiber for filtration (*see* **Note 1**).

2.2 Chemicals

Prepare all solutions using ultrapure water (prepared by purifying deionized water to attain a sensitivity of 18 MΩ cm at 25 °C) and analytical grade reagents. Prepare and store all the solutions at room temperature (unless indicated otherwise). Follow all waste disposal regulations when disposing chemical wastes.

1. 1 % $CaCl_2$ solution (w/v): Dissolve 10 g of anhydrous $CaCl_2$ (or the correspondent amount of the hydrated compound) in 100 mL of water.

2. 2 M trifluoroacetic acid: Add 8.6 mL of trifluoroacetic acid in water in a volumetric flask, and adjust the final volume to 100 mL (*see* **Note 2**).
3. 2 M NaOH solution: Dissolve 8 g of anhydrous NaOH in 100 mL of water (*see* **Note 3**).
4. Ethanol absolute.
5. Acetone.

2.3 Equipments

1. Microwave digestion system equipped with several reaction vessels, valves, and tubing that allow external application of pressure to each vessel from a nitrogen tank. The vessels should be interconnected with tubing and placed in a rotating carousel. At least one of the vessels should be equipped with a pressure sensor that measures and controls the set point within the cell.
2. Milling equipment.
3. Sieves for separation of different particle sizes.
4. Vacuum filtration system.
5. High-performance liquid chromatography (HPLC) system.

3 Methods

Carry out all procedures at room temperature unless otherwise specified.

3.1 Sample Preparation

1. If the marine algae were harvested on coastal areas or beaches, wash the algal material with tap water to remove salt residues, sand, and/or epiphytes. If the algal sample was acquired in a commercial store, proceed to the next step.
2. Dry the algal material at 35 °C to constant weight.
3. Mill the algae to particle sizes of approximately 1 mm. Particles lower than 1 mm should not be used in the analyses (*see* **Note 4**).

3.2 Microwave-Assisted Extraction

1. Mix 1 g of milled algae with 25 mL of distilled water in each microwave extraction vessel (*see* **Notes 5** and **6**).
2. Close the vessels; place them in the rotating carousel, and introduce the system into the microwave equipment (*see* **Note 7**).
3. Turn on the system and irradiate the suspensions at a pressure of 120 psi, with 630 W of microwave energy at a frequency of 2.450 MHz, for 1 min (*see* **Notes 8** and **9**).
4. Wait for the pressure to reach about 5 psi before opening the microwave equipment (*see* **Note 10**).
5. Remove the vessels from the microwave and cool them in an ice bath for approximately 5 min.

6. Filter the content of each vessel through nylon fiber and under vacuum condition to separate the algal residue.
7. Transfer the algal residue to Petri dishes or any other clean recipient and dry it at 35 °C until constant weight is achieved.
8. Weigh the dried material to determine the final mass obtained after extraction, which is needed to calculate the percentage of algae degraded during MAE.
9. Store the dried extracted algae at room temperature and the liquid fraction after extraction in a closed recipient at 4 °C.

3.3 Alginate Precipitation

1. To precipitate the alginate from the liquid fraction after MAE, add the same volume of 1 % $CaCl_2$ solution to the liquid and maintain the mixture at 4 °C for 8 h.
2. Separate the precipitate by centrifugation (8.000 × *g*, 15 min, 4 °C) followed by filtration using qualitative filter paper (*see* **Note 11**).
3. Dry the precipitated alginate in an oven at 35–40 °C until constant weight is achieved and determine the mass recovered by precipitation.

3.4 Fucoidan Precipitation

1. To precipitate the fucoidan from the resulting liquid after alginate removal, add double volume of ethanol absolute to the liquid and maintain the mixture at 4 °C for 8 h.
2. Recover the ethanol-precipitated polysaccharide by centrifugation (8.000 × *g*, 15 min, 4 °C) followed by filtration using qualitative filter paper.
3. Wash the recovered precipitate (fucoidan) in two subsequent steps using ethanol absolute and then acetone. Use the same volume of solvents employed in the precipitation step.
4. Dry the washed precipitate at 35 °C to constant weight and determine the mass of fucoidan recovered.
5. Mill the recovered fucoidan and store in a dry place at room temperature.
6. Discharge the liquid fraction obtained after the precipitation step.

The fucoidan extraction yield (% FEY) and the percentage of algae degraded during MAE (% AD) can be calculated by Eqs. 1 and 2, respectively, where FM is the mass of fucoidan (dry weight basis) obtained after precipitation with ethanol, IAW is the initial weight of dry algae used in the experiment, and FAW is the weight of dry algae recovered after MAE:

$$\%\ \text{FEY} = (\text{FM} / \text{IAW}) \times 100 \quad (1)$$

$$\%\ \text{AD} = ((\text{IAW} - \text{FAW}) / \text{IAW}) \times 100 \quad (2)$$

3.5 Sugar Composition of Fucoidan

1. To determine the sugars present in the fucoidan composition, mix 10–15 mg of the fucoidan sample with 0.5 mL of 2 M trifluoroacetic acid in a glass tube.
2. Seal the glass tube with nitrogen and maintain the reaction at 121 °C for 2 h.
3. Cool the tube in an ice-water bath and centrifuge (3.000 × *g*, 15 min) to eliminate any resulting solid particles.
4. Neutralize the liquid containing the soluble polysaccharide to pH 7 using a 2 M NaOH solution.
5. Inject a neutralized sample in an HPLC system to determine the concentration of monomer sugars present (*see* **Note 12**).

4 Notes

1. A nylon sock can be used for filtration.
2. The trifluoroacetic acid has a strong odor. Inhalation of vapors of trifluoroacetic acid may cause headaches, dizziness, fatigue, and weakness in limbs, coughing, chest pains, nausea, and vomiting. Therefore, the 2 M trifluoroacetic acid solution must be prepared in a chamber with exhaustion of gases. Care must also be taken in order to avoid contact of this acid with the skin and eyes, which may cause irritation and burns.
3. Take care when preparing the 2 M NaOH solution. This reaction releases heat and therefore, this solution should be prepared with the vessel placed in an ice bath for cooling.
4. The algae can be milled using a home blender. Milled materials must be kept in plastic bags or flasks at room temperature. The milling step is important because it ensures uniformly distributed mass and a higher surface-to-volume ratio.
5. The vessel must be dry and free of particulate matter. Drops of liquid or particles will absorb microwave energy, causing localized heating which may char and damage the vessel components.
6. To prevent the co-extraction of other algal components during the fucoidan isolation, the inclusion of a pretreatment step before MAE can be useful, particularly when it is known that the algal material is rich in lipids, terpenes, and/or phenols. Such compounds can be partially extracted from the algal structure together with the fucoidan molecule. For the pretreatment, mix 1 g of milled algae with 20 mL of solvent (chloroform/methanol at 2:1 (v/v)) in a flask and maintain the mixture under magnetic agitation at room temperature (23 °C) for 20 min. Then separate the pretreated algae by centrifugation (2.500 × *g*, 20 min) and dry the material at 35 °C to a constant weight.

7. At least one of the vessels must be equipped with a pressure sensor for measurement and control of the conditions within the cell.
8. Prior to instrument operation: (1) check that the carousel, vessels, and tubing are correctly connected and placed in the system; (2) check the door seal and sealing surfaces of the microwave; and (3) ensure that the door closes properly.
9. The system can take several minutes to attain the desired extraction conditions. The reaction time should be considered only after these conditions are reached.
10. For safety reasons it is necessary to wait for this significant reduction in pressure before opening the microwave equipment.
11. The alginate precipitate adheres to the centrifuge tube, which therefore must be pre-weighed in order to determine the dry weight of alginate after drying the precipitate. However, the liquid fraction must also be filtered after being centrifuged, in order to recover possible mass of precipitated alginate that has not adhered to the centrifuge tube during the centrifugation step.
12. A MetaCarb 87P (300 × 7.8 mm) column operated at 80 °C or any other column with similar characteristics can be used at this stage for the determination of sugars, since it is able to promote an efficient separation of hexose and pentose sugars. Use deionized water as mobile phase at a flow rate of 0.4 mL/min and an RI detector for analyses.

References

1. Bélanger JMR, Paré JRJ (2006) Applications of microwave-assisted processes (MAP™) to environmental analysis. Anal Bioanal Chem 386:1049–1058
2. Eskilsson CS, Björklund E (2000) Analytical-scale microwave-assisted extraction. J Chromatogr A 902:227–250
3. Chan C-H, Yusoff R, Ngoh G-C et al (2011) Microwave-assisted extractions of active ingredients from plants. J Chromatogr A 1218:6213–6225
4. Hahn T, Lang S, Ulber R et al (2012) Novel procedures for the extraction of fucoidan from brown algae. Process Biochem 47: 1691–1698
5. Routray W, Orsat V (2012) Microwave-assisted extraction of flavonoids: a review. Food Bioprocess Technol 5:409–424
6. Dean JR, Xiong G (2000) Extraction of organic pollutants from environmental matrices: selection of extraction technique. TrAC Trends Anal Chem 19:553–564
7. Luque-García JL, Luque de Castro MD (2003) Where is microwave-based analytical equipment for solid sample pre-treatment going? TrAC Trends Anal Chem 22:90–98
8. Morya VK, Kim J, Kim E-K (2012) Algal fucoidan: structural and size-dependent bioactivities and their perspectives. Appl Microbiol Biotechnol 93:71–82
9. Rodriguez-Jasso RM, Mussatto SI, Pastrana L et al (2011) Microwave-assisted extraction of sulfated polysaccharides (fucoidan) from brown seaweed. Carbohydr Polym 86:1137–1144
10. Li B, Lu F, Wei X et al (2008) Fucoidan: structure and bioactivity. Molecules 13:1671–1695

Chapter 10

Extraction and Analysis of Oxylipins from Macroalgae Illustrated on the Example *Gracilaria vermiculophylla*

Dominique Jacquemoud and Georg Pohnert

Abstract

Oxylipins are natural products that are derived by oxidative transformations of unsaturated fatty acids. These metabolites are found in a wide range of organisms from the animal kingdom to plants and algae. They represent an important class of signaling molecules, mediating intra- and intercellular processes such as development, inflammation, and other stress responses. In addition, these metabolites directly function as chemical defense against grazers and pathogens. In the red alga *Gracilaria vermiculophylla*, oxylipin production is initiated by mechanical tissue disruption and can also be induced in intact algae in response to external stress signals. The defense metabolites mostly result from the lipase- and lipoxygenase-mediated conversion of phospho- and galactolipids. Oxylipins can vary greatly in their size, degree of unsaturation, oxidation state, and functional groups. But also isomers with only subtle chemical differences are found. A variety of methods have been developed for separation, detection, and identification of oxylipins. This chapter focuses on the analysis of oxylipins in macroalgae and covers all aspects from sample preparation (including protocols for the investigation of oxylipins in wounded and intact algal tissue), extraction, purification, and subsequent analysis using liquid chromatography coupled to a UV detector or a mass spectrometer. The protocols developed for *G. vermiculophylla* can be readily adapted to the investigation of other macroalgae.

Key words Arachidonic acid, *Gracilaria vermiculophylla*, Hydroxyeicosatetraenoic acids, Leukotrienes, Macroalgae, MS, Oxylipins, Prostaglandins, UPLC, UV

1 Introduction

Oxylipins are found in several marine macroalgae where their function has been mainly attributed to chemical defense [1–5]. But algal oxylipins are also involved in the biosynthesis of pheromones as it is the case for marine brown algae [6]. Especially prolific are red algae that produce a wide variety of mainly C_{20} fatty acid-derived oxylipins that, in part, are also known as regulatory hormones in mammals [7]. Detailed investigations on the oxylipins of *Gracilaria* spp. have been reported over the last decades. A summary figure describing the main pathways involved in the biosynthesis of *Gracilaria* oxylipins is presented in Fig. 1: the cyclooxygenase (COX) pathway leads to the prostaglandins E_2

Dagmar B. Stengel and Solène Connan (eds.), *Natural Products From Marine Algae: Methods and Protocols*, Methods in Molecular Biology, vol. 1308, DOI 10.1007/978-1-4939-2684-8_10, © Springer Science+Business Media New York 2015

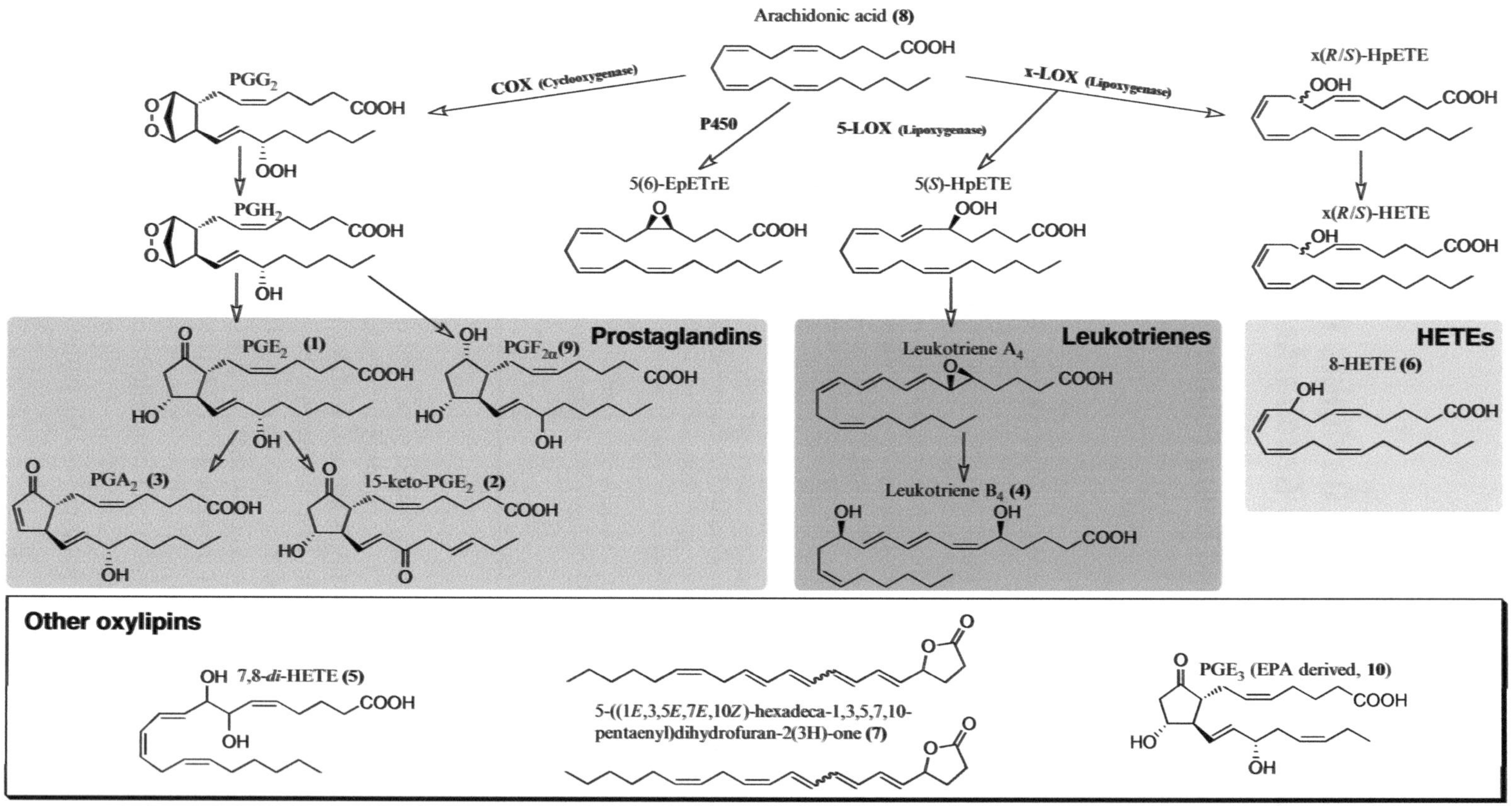

Fig. 1 Main pathways of arachidonic acid metabolism relevant to oxylipin biosynthesis in *G. vermiculophylla*. *PG* prostaglandin, *H(p)ETE* hydro(pero)xy eicosatetraenoic acid, *EpETrE* epoxyeicosatrienoic acid

(PGE_2, **1**), 15-keto E_2 (15 k-PGE_2, **2**), PGA_2 (**3**), PGF_{a2} (**9**), and PGE_3 (**10**) (numbers correspond to structures in Fig. 1) [2, 8, 9]. Lipoxygenases (x-LOX) can insert a hydroperoxide at the position x=5, 8, 9, 11, 12, or 15 and the resulting hydroperoxy-fatty acids can be further transformed to alcohols, ketones, or epoxides. Starting from arachidonic acid (**8** in Fig. 1) these transformations typically lead to hydroxyeicosatetraenoic acids (HETE), keto-eicosatetraenoic (KETE) acids, and epoxy eicosatrienoic acids. Starting from HPETEs, hydroperoxide isomerases can lead to the formation of dihydroxyeicosatetraenoic acids (like 7,8-*di*-HETE, **5** in Fig. 1) [10] and the formation of leukotrienes like leukotriene B_4 (LTB_4, **4** in Fig. 1) can be rationalized from epoxy intermediates. Other arachidonic acid-derived metabolites have been identified, but their biosynthesis remains uncertain. Another important pathway regarding oxylipin synthesis that is based on cytochrome P450-mediated transformations has not been described in *Gracilaria* spp.

Here, we introduce a protocol for the comprehensive analysis of oxylipins that was developed for *Gracilaria vermiculophylla* (Ohmi) Papenfuss (formerly *G. verrucosa*), which can also be applied to the investigation of oxylipins from other algae (with minor modifications). The red alga *G. vermiculophylla* is native to the north-west Pacific but has become invasive over the last 20 years and is now also distributed along the coasts of the Atlantic and Pacific oceans, as well as in the North and Baltic Seas [11]. It is thus readily available for method testing. Its invasive success has been attributed to its robustness and apt defenses. Attention has mainly focused on defense metabolites produced in response to wounding and induction. But since *Gracilaria* spp. are used as food in human nutrition and several of the detected oxylipins can cause severe food poisoning, the method described here can also directly be applied for monitoring of toxicity in this context.

The protocol presented here introduces methods for the identification of wound-activated as well as constitutively produced oxylipins. We also describe a minor variation that can be used to identify precursor fatty acids of algal oxylipins by using isotope-labeled probes. The method generally allows following the enzymatic conversion of lipids and arachidonic acid (AA, **8** in Fig. 1), leading to an array of oxylipins [1, 4]. We introduce extraction and purification protocols as well as chromatographic and mass spectrometric procedures for an in-depth analysis of the oxylipin profiles.

2 Materials

Prepare all solutions and medium using ultrapure water (0.55 μS) and/or analytical grade solvents. Use HPLC/MS-grade solvents for liquid chromatography and analytical grade solvents for extraction.

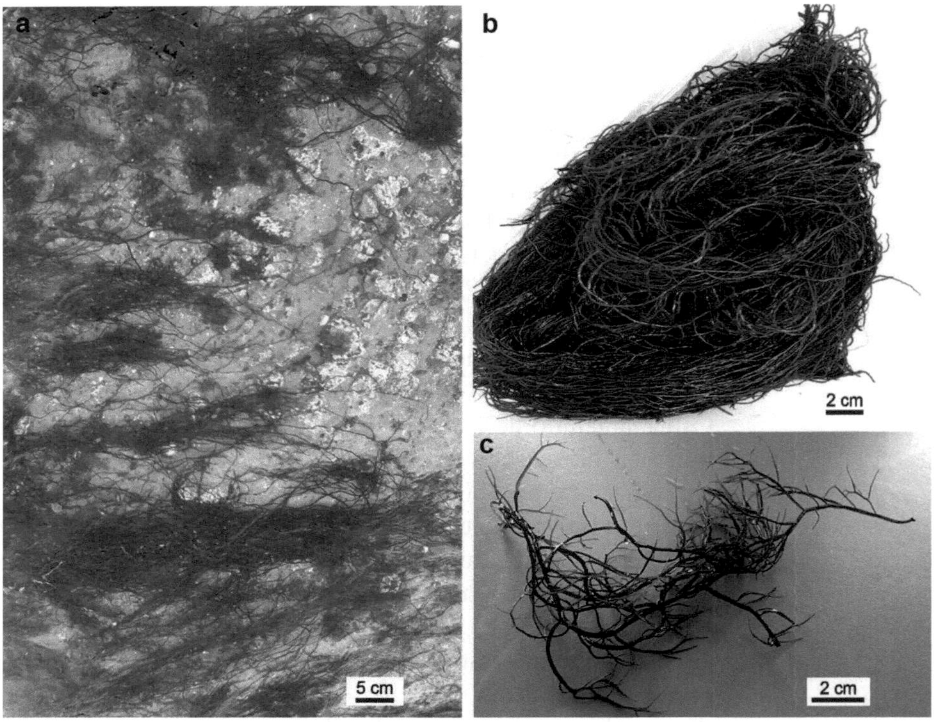

Fig. 2 *Gracilaria vermiculophylla*. (**a**) *G. vermiculophylla* invasive along the German North Sea coasts. (**b**) Whole thallus. (**c**) Detail of a thallus

2.1 Algal Material

1. *Gracilaria vermiculophylla* is a perennial red macroalga (Fig. 2) that can be collected throughout the year. It is usually found in intertidal regions, which makes collection easy. The alga is often encountered in muddy locations where water forms small ponds. It can be found with different degree of epiphytism by microorganisms, barnacles, and other seaweeds.
2. Remove the major fouling mass by brushing with a soft toothbrush in order to minimize contamination of the extracts. During cleaning utmost care has to be taken that no tissue damage occurs.
3. Since *G. vermiculophylla* is very robust it can also be kept in culture before further investigation (*see* **Note 1**). However, it has to be noted that some chromatographic peaks found in algae freshly collected from the sea were not encountered in individuals kept longer than a month in culture. Analysis of samples should thus ideally be conducted as soon as possible after collection if properties of fresh material are of concern.

2.2 Grinding and Freeze- Drying

1. A standard porcelain mortar and pestle.
2. Liquid nitrogen.

3. Freeze-dryer, Alpha 1-2 LDplus (Christ, Osterode am Harz, Germany) (*see* **Note 2**).

2.3 Labeling the Products of Arachidonic Acid Metabolism

1. [5,6,8,9,11,12,14,15-$^2H_8$]-arachidonic acid (Cayman Chemical, Tallinn, Estonia).
2. 2-[$^2H_1$] or 2-[$^2H_2$]-arachidonic acid synthesized according to reference [12].

2.4 Extraction

1. Methanol.
2. 1.5 mL microtubes with clamps (*see* **Note 3**).
3. Centrifuge for microtubes.
4. Ultrasonic bath.

2.5 Enrichment

1. OASIS HBL (*N*-vinylpyrrolidone-divinylbenzene-copolymer) 1 cc/10 mg (Waters, Eschborn, Germany).
2. Heptane.
3. Diethyl ether.
4. Ethyl acetate.
5. Methanol.
6. Ultrapure water.
7. Vacuum manifold system.
8. Vacuum pump and controller.
9. 4 mL glass vials.

2.6 LC/MS

1. Prepare mobile phases for UPLC analyses by mixing the solvents for mobile phases A and B, respectively, that are aliquoted in glass measuring cylinders using UPLC/MS-grade solvents: Mobile phase A: 0.1 % formic acid (w/v) in 2 % acetonitrile in ultrapure water (v/v); mobile phase B: 0.1 % formic acid (w/v) in acetonitrile.
2. An Acquity ultraperformance liquid chromatography system (UPLC, e.g., Waters, Manchester, UK, or equivalent) equipped with a 5 cm BEH C18 UPLC column (2.1 mm, 1.7 μm; Waters or a similar C18 column) at 30 °C: In our case, the LC was coupled to a Waters 2996 PDA UV-Vis detector and a time-of-flight Q-TOF micro mass spectrometer (Waters) equipped with electrospray ionization (ESI) (*see* **Note 4**). MassLynx V4.1 software (Waters) was used to pilot the instrument and process the data.
3. ESI-MS parameters: Set the capillary voltage to 2,700 V in negative mode. In positive mode, set the capillary to 3,000 V (*see* **Note 5**). Set the sample cone voltage to 25 V, the cone gas flow to 20 L N_2/h, and the desolvation gas flow to 700 L N_2/h. Set the source temperature to 120 °C and the desolvation

temperature to 300 °C. Set the collision energy to 10 V. Values are instrument dependent and should be adjusted by tuning before measurements.

4. Record UV-Vis spectra between 190 and 300 nm with a resolution of 1.2 nm at a rate of 1 scan/s, interpolating at 656 nm.

3 Methods

The wound-activated formation of oxylipins occurs rapidly in *G. vermiculophylla*. If the oxylipin content in intact tissue is concerned, shock-freezing in liquid nitrogen and handling in the cold until the sample contains no remaining water after lyophilization are essential (**Procedure A**). For investigation of the sum of oxylipins in intact tissue and those formed after wounding an alternative, **Procedure B** is recommended. In **Procedure C** a guideline for labeling experiments with biosynthetic precursors is described.

This protocol will be presented for thalli of approximately 5 cm in length and a wet weight of ca. 200 mg. Under the conditions described, enough extract is obtained to perform various analytical experiments and repeated LC/MS measurements. The protocol can however be scaled to accommodate different amounts of sample. Especially for the preparative purification of oxylipins a scale-up is required.

3.1 Collection and Processing of Algal Material

1. Collect intact *G. vermiculophylla* from field sites or laboratory cultures. In case of epiphytized material clean algae with a soft toothbrush without damaging the tissue. Samples can be incubated in seawater for several days to study the impact of various stress conditions (epiphytism, grazing, etc.). It is advised to collect samples which match the desired size and mass without having to cut them, as this would cause local wounding. Select pieces of ca. 5 cm in length (*see* **Note 6**). Rinse briefly under tap water.
2. Dry the sample between layers of tissue.
3. Precool the mortar with liquid nitrogen, put the algal thallus into the mortar, and pour in liquid nitrogen. Grind until a very fine powder is obtained. Add liquid nitrogen if required to keep the sample cold at all times (*see* **Note 7**).
4. **Procedure A:** To monitor oxylipins in intact tissue, cool a preweighed microtube with liquid nitrogen. Cool the spatula by dipping in liquid nitrogen, and transfer the cold material in the microtube. Proceed immediately with point 5.

   Procedure B: If the wound-activated production of oxylipins is to be investigated, let the sample thaw in the mortar and incubate for 10 min at room temperature (*see* **Note 8**). During thawing the grayish powder will turn brown, and then take on a red/brown color. The consistence changes to a slimy texture. Transfer to a pre-weighed microtube and proceed with 5.

Procedure C: If biosynthetic investigations are to be conducted, suspend 2 mg of labeled precursor molecules, such as deuterated arachidonic acid, in 15 μL of water assisted by sonication. Add to the sample. Mix by additional grinding, let the sample thaw in the mortar, and incubate for 10 min at room temperature (*see* **Note 8**). During thawing the grayish powder turns brown and then takes on a red/brown color. The consistence changes to a slimy texture. Transfer to a pre-weighed microtube and proceed with 5.

5. Close the microtube, seal with a clamp, and store it in liquid nitrogen until freeze-drying (*see* **Note 3**).
6. Open the lid and transfer the cold sample in a freeze-dryer. Proceed to full dryness of the sample (at least 6 h).

3.2 Oxylipin Extraction

1. Weigh the tube and determine dry weight (ca. 20 mg) by subtracting weight of the microtube.
2. Add 400 μL of methanol to the powder.
3. Sonicate in ultrasonic bath for 5 min (*see* **Note 9**).
4. Centrifuge for 90 s at 16,000 × *g*.
5. Transfer the supernatant to a microtube.
6. Repeat from **step 2** three times. Pool the supernatants. The powder should turn from brownish to a pinkish shade of red. The solvent should be almost colorless after the last extraction step (*see* **Note 10**).

3.3 Enrichment and Partial Purification of the Oxylipin Fraction

Enrichment of the extract and partial purification can be realized on a Waters OASIS® HLB SPE cartridge (*see* **Note 11**).

1. Dilute the samples to a final 10 % methanol content (v/v), using deionized water.
2. Place the SPE cartridge on the vacuum manifold. A vacuum of 600 Torr should be applied.
3. Condition the cartridge with 4 mL of methanol.
4. Equilibrate with 4 mL of 10 % methanol in water (v/v).
5. Load the sample on the cartridge (*see* **Note 12**).
6. Wash the cartridge with 4 mL of water, to remove salts.
7. Wash with 4 mL of 10 % methanol.
8. Perform a first elution using 4 mL of heptane. Discard this unpolar fraction not containing oxylipins (*see* **Note 13**).
9. Elute the oxylipin fraction using 4 mL of diethyl ether and collect the eluate in a 4 mL glass vial.

Table 1
Elution program

Time (min)	Mobile phase A (%)	Mobile phase B (%)
0.00	100.0	0.0
0.50	100.0	0.0
1.00	75.0	25.0
5.50	40.0	60.0
6.50	0.0	100.0
8.00	0.0	100.0
8.50	100.0	0.0
9.00	100.0	0.0

Mobile phase A: 0.1 % formic acid and 2 % acetonitrile in water
Mobile phase B: 0.1 % formic acid in acetonitrile
Flow 0.6 mL/min

10. Elute more polar metabolites by flushing the column with 4 mL ethyl acetate, and then 4 mL methanol (*see* **Note 14**).
11. The cartridge can be reconditioned and used again (*see* **Note 15**).
12. Remove solvent under a gentle flow of nitrogen.
13. Redissolve the fractions in methanol to give equal concentrations normalized to the determined dry weight (add 1 mL of methanol per 20 mg dry weight).

3.4 Sample Analysis: Liquid Chromatography/Mass Spectrometry

1. Set the flow rate used at 0.6 mL/min and elute samples using a gradient of the two mobile phases A and B. The elution program is presented in Table 1.
2. Inject 2 μL of the sample; this volume should provide strong signals (*see* **Note 16**).
3. UV-Vis detection: Record spectra with a DAD detector monitoring UV wavelength range between 190 nm and 300 nm (Fig. 3 presents UV spectra of the identified oxylipins).
4. Mass spectrometry: Record mass spectra in negative ionization mode (*see* **Note 5**) throughout the chromatographic separation at a scan rate of 0.32 scans/s with an inter scan delay of 0.1 s and a scan range from 60 to 1,000 *m/z*.

Figure 4 presents a typical base peak intensity chromatogram of the extract from wounded *G. vermiculophylla*. Oxylipins elute mainly between 2 and 5.5 min. The mass spectra of the identified oxylipins and arachidonic acid are presented in Fig. 5. If labeled arachidonic acid is applied (**Procedure C**) oxylipins with >20 % labeling can be clearly detected in the mass spectra. Although 7,8-*di*-HETE is labeled in *G. vermiculophylla* [4], it is not found as

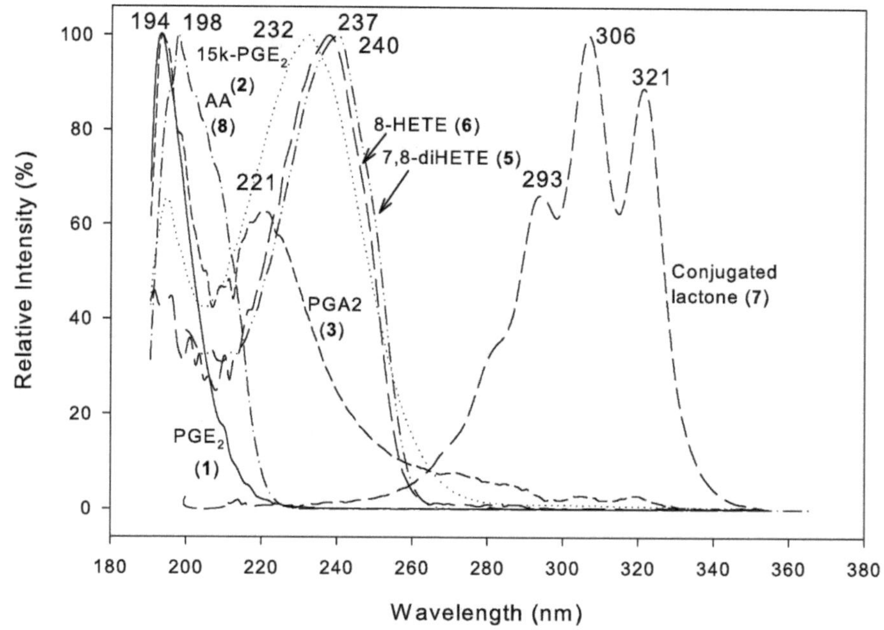

Fig. 3 UV spectra of main *G. vermiculophylla* oxylipins. *Numbers in brackets* refer to compounds detailed in Fig. 1

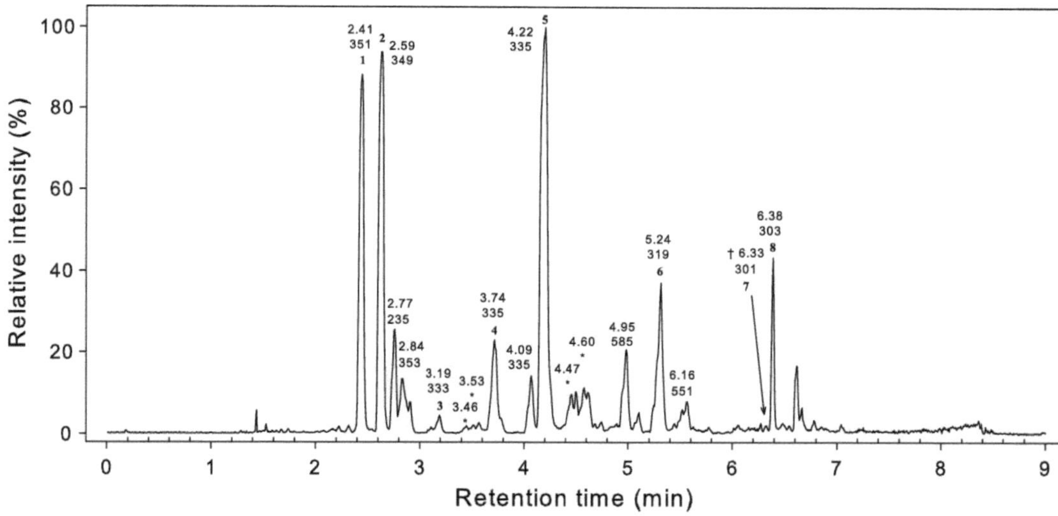

Fig. 4 Typical UPLC-MS in base peak monitoring mode of wounded *G. vermiculophylla*. † Peak only detectable in ESI-positive mode. For each peak, retention time, *m/z*, and numbers referring to Fig. 1 are given

labeled compound in its close relative *G. chilensis*, suggesting an origin other than metabolic conversion of free arachidonic acid. It has been hypothesized that 7,8-*di*-HETE is directly released from galactolipids [1].

Table 2 summarizes the known oxylipins identified in *G. vermiculophylla*, their retention time, maxima of UV absorption, and significant fragment ions in ESI-MS.

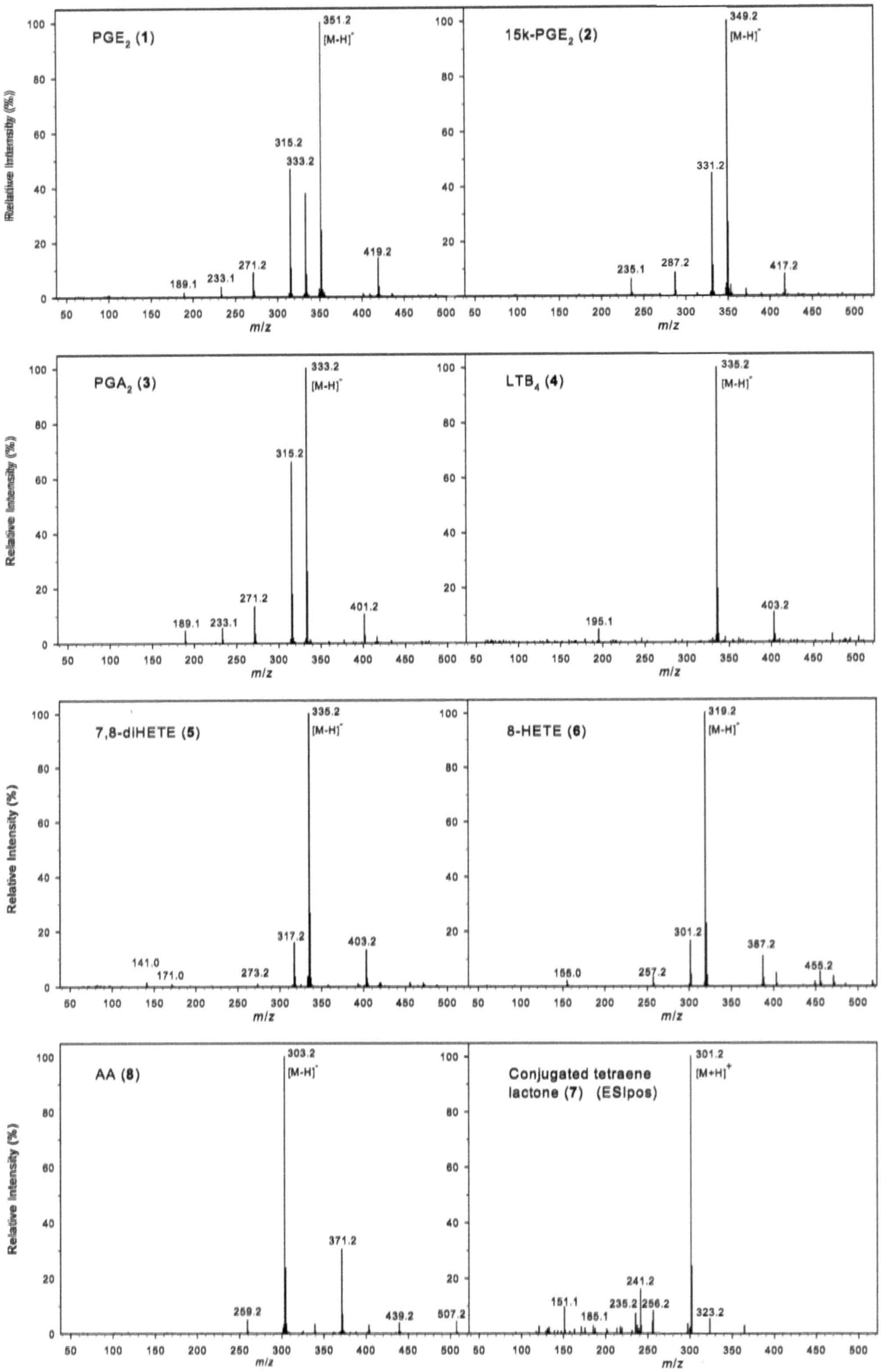

Fig. 5 Mass spectra (ESI-negative ionization) of *G. vermiculophylla* oxylipins unless mentioned otherwise. *Numbers in brackets* refer to compounds detailed in Fig. 1

Table 2
Summary table of oxylipins identified from *G. vermiculophylla*. *m/z* are arranged according to descending intensity

Metabolite	Retention time (min)	UV (nm)	ESI-negative (*m/z*)	ESI-positive (*m/z*)
PGE_2 (**1**)	2.41	194	351.2 $[M–H]^-$, 315.2 $[M–H–2H_2O]^-$, 333.2 $[M–H–H_2O]^-$, 419.2 $[M–H+HCOONa]^-$, 271.2 $[M–H–H_2O–CO_2]^-$, 233.1	375.2 $[M+Na]^+$, 299.2 $[M+H–3H_2O]^+$, 317.2 $[M+H–2H_2O]^+$, 335.2 $[M+H–H_2O]^+$
15 k-PGE_2 (**2**)	2.59	194, 232	349.2 $[M–H]^-$, 331.2 $[M–H–H_2O]^-$, 287.2 $[M–H–H_2O–CO_2]^-$, 235.1, 417.2 $[M–H+HCOONa]^-$	373.2 $[M+Na]^+$, 315.2 $[M–H–2H_2O]^+$, 333.2 $[M+H–H_2O]^+$
PGA_2 (**3**)	3.16	193, 221	333.2 $[M–H]^-$, 315.2 $[M–H–H_2O]^-$, 271.2 $[M–H–H_2O–CO_2]^-$, 401.2 $[M–H+HCOONa]^-$, 233.1, 189.1	317.2 $[M+H–2H_2O]^+$, 299.2 $[M+H–H_2O]^+$, 357.2 $[M+Na]^+$
LTB_4 (**4**)	3.74	Not measured	335.2 $[M–H]^-$, 317.2 $[M–H–H_2O]^-$, 403.2 $[M–H+HCOONa]^-$, 273.2 $[M–H–H_2O–CO_2]^-$, 171.0	Not measured
7,8-diHETE (**5**)	4.16	240	335.2 $[M–H]^-$, 317.2 $[M–H–H_2O]^-$, 403.2 $[M–H+HCOONa]^-$, 273.2 $[M–H–H_2O–CO_2]^-$, 171.0	359.2 $[M+Na]^+$, 375.2 $[M+K]^+$, 301.1 $[M+H–2H_2O]^+$, 319.2 $[M+H–H_2O]^+$
8-HETE (**6**)	5.24	237	319.2 $[M–H]^-$, 301.2 $[M–H–H_2O]^-$, 387.2 $[M–H+HCOONa]^-$, 257.2 $[M–H–H_2O–CO_2]^-$, 455.2 $[M–H+HCOONa]^-$, 155.0	303.2 $[M+H–H_2O]^+$, 359.2 $[M+K]^+$, 343.2 $[M+Na]^+$
Conjugated tetraene lactone (**7**)	6.33	293, 306, 321	*n.d.*	301.2 $[M+Na]^+$, 241.2, 151.1, 323.2 $[M+Na]^+$
AA (**8**)	6.38	198	301.2 $[M–H]^-$, 371.2 $[M–H+HCOONa]^-$, 259.2 $[M–H–CO_2]^-$	305.2 $[M+H]^+$, 287.2 $[M–H_2O]^+$

4 Notes

1. The alga can withstand desiccation and is tolerant against temperature variation between 5 and 25 °C. It can be cultivated in waters with salinity ranging from 15 to 35. Instant Ocean medium (Aquarium Systems, France) can be used to keep the

alga over several weeks. The medium has to be exchanged regularly every week and the concentration of the alga should be kept below 50 g/L of (artificial) seawater, ideally 10 g/L.

2. Lyophilization allows easy handling of the samples. If a freeze-dryer is not available, **Procedure B** and **Procedure C** can still be followed without bringing any change to the protocol described here. **Procedure A**, however, needs to be carried out under constant cooling, taking care never to let the sample temperature rise above –20 °C to prevent enzyme activity. In this case samples would have to be kept in liquid nitrogen or on dry ice. The extraction solvent should be precooled to –20 °C.
3. Even though safe-lock microtubes are designed to withstand liquid nitrogen, it is advisable to use clamps to ensure that the overpressure caused by the evaporation of liquid nitrogen and that of solvents during sonication will not cause the opening of the tube lid during manipulation or centrifugation.
4. Metabolic profiling can be performed on any ESI LC/MS systems with adaptation of tuning parameters, solvent, and gas flows.
5. Electrospray ionization in negative ion mode is highly recommended, as the sensitivity for nearly all oxylipins is much higher than in the positive mode. However, metabolite 7 is only detected in positive mode.
6. It is advised to select one 5 cm fragment for each analysis if appropriate material is available. If the length of the thalli is smaller pool several thalli. Determining the fresh weight of such comparably small pieces of algal tissue is problematic. Normalization by exact weighing after freeze-drying gives more accurate results.
7. It is of utmost importance to work in liquid nitrogen. Working on ice does not prevent enzymatic activity. We do not recommend heat treatment to inactivate enzymes since this causes damage to the thermolabile oxylipins and does not lead to reproducible results. Working with liquid nitrogen can cause severe freeze burns. Always wear proper protective gloves, lab coat, long pants, and heavy shoes. Safety goggles are required.
8. For larger samples it is recommended to help the thawing process by stirring the powder with a spatula. The mortar can be partially immersed in a warm water bath to accelerate the thawing process.
9. If no ultrasonic bath is available, use a vortex instead.
10. Samples stored in methanol at –20 °C are stable over several months.
11. Other types of cartridges can be used. Chromabond EASY (modified polystyrene-divinylbenzene copolymer) 1 cc/10 mg

cartridges (Macherey-Nagel, Düren, Germany) and Bakerbond Octadecyl (C18) 1 cc/10 mg cartridges (J.T. Baker, Phillipsburg, NJ, USA) were also tested. They are suitable for oxylipin extraction; however the EASY cartridges required more solvent to completely elute the oxylipin fraction. The C18 cartridges yielded more non-oxylipin peaks in the oxylipin fraction.

12. Even if the flow-through is green, it was verified using ESI-MS that it contains no detectable oxylipins. This procedure was tested with extract from up to 1 g of algal material extracted with a volume of 10 mL of methanol, diluted to 100 mL without any losses. SPE cartridge size can be adjusted for larger sample sizes.
13. Arachidonic acid is partially eluted at this step. If arachidonic acid is to be quantified, reserving an aliquot of the extract is advised.
14. Indirect evidence indicates that unknown polar oxylipins might be contained in this fraction as well. Two metabolites of *m/z* 458.2 and 456.2 with retention times of 1.63 and 1.70 min, respectively, elute in the final methanol fraction, independently of the SPE cartridge type used. Given their even mass, it is probable that these metabolites carry a peptide group.
15. OASIS HLB cartridges tolerate desiccation and can be dried after rinsing with methanol, to be used another time. C18 SPE cartridges tend not to tolerate desiccation. While they can be reused directly to process several extracts in a row, it is not recommended to dry them and store them for later usage.
16. Adapt injection volume to the LC/MS system if the peaks' intensities are not at their optima.

Acknowledgements

This work was supported by the Friedrich Schiller University Jena and the Volkswagen foundation. We thank Florian Weinberger and Ester Rickert for help in collecting the algae.

References

1. Lion U, Wiesemeier T, Weinberger F et al (2006) Phospholipases and galactolipases trigger oxylipin-mediated wound-activated defence in the red alga *Gracilaria chilensis* against epiphytes. ChemBioChem 7:457–462
2. Rempt M, Weinberger F, Grosser K et al (2012) Conserved and species-specific oxylipin pathways in the wound-activated chemical defense of the noninvasive red alga *Gracilaria chilensis* and the invasive *Gracilaria vermiculophylla*. Beilstein J Org Chem 8:283–289
3. Weinberger F, Lion U, Delage L et al (2011) Up-regulation of lipoxygenase, phospholipase, and oxylipin-production in the induced

chemical defense of the red alga *Gracilaria chilensis* against epiphytes. J Chem Ecol 37:677–686
4. Nylund G, Weinberger F, Rempt M et al (2012) Metabolomic assessment of induced and activated chemical defence in the invasive reg alga *Graciliaria vermiculophylla*. PLoS One 6:e29359
5. Andreou A, Brodhun F, Feussner I (2009) Biosynthesis of oxylipins in non-mammals. Prog Lipid Res 48:148–170
6. Pohnert G, Boland W (2002) The oxylipin chemistry of attraction and defense in brown algae and diatoms. Nat Prod Rep 19: 108–122
7. Gerwick WH (1994) Structure and biosynthesis of marine algal oxylipins. Biochim Biophys Acta 1211:243–255
8. Rybin VG, Svetashev VI, Imbs AB et al (2013) Detection of prostaglandin E-3 in the red alga *Gracilaria vermiculophylla*. Chem Nat Compd 49:535–536
9. Varvas K, Kasvandik S, Hansen K et al (2013) Structural and catalytic insights into the algal prostaglandin H synthase reveal atypical features of the first non-animal cyclooxygenase. Biochim Biophys Acta 1831:863–871
10. Gerwick WH, Moghaddam M, Hamberg M (1991) Oxylipin metabolism in the red alga *Gracilariopsis lemaneiformis*: mechanism of formation of vicinal dihydroxy fatty-acids. Arch Biochem Biophys 290:436–444
11. Nyberg CD, Thomsen MS, Wallentinus I (2009) Flora and fauna associated with the introduced red alga *Gracilaria vermiculophylla*. Eur J Phycol 44:395–403
12. Rempt M, Pohnert G (2010) Novel acetylenic oxylipins from the moss *Dicranum scoparium* with antifeeding activity against herbivorous slugs. Angw Chem Int Ed Engl 49:4755–4758

Chapter 11

Lipids and Fatty Acids in Algae: Extraction, Fractionation into Lipid Classes, and Analysis by Gas Chromatography Coupled with Flame Ionization Detector (GC-FID)

Freddy Guihéneuf, Matthias Schmid, and Dagmar B. Stengel

Abstract

Despite the number of biochemical studies exploring algal lipids and fatty acid biosynthesis pathways and profiles, analytical methods used by phycologists for this purpose are often diverse and incompletely described. Potential confusion and potential variability of the results between studies can therefore occur due to change of protocols for lipid extraction and fractionation, as well as fatty acid methyl esters (FAME) preparation before gas chromatography (GC) analyses. Here, we describe a step-by-step procedure for the profiling of neutral and polar lipids using techniques such as solid–liquid extraction (SLE), thin-layer chromatography (TLC), and gas chromatography coupled with flame ionization detector (GC-FID). As an example, in this protocol chapter, analyses of neutral and polar lipids from the marine microalga *Pavlova lutheri* (an EPA/DHA-rich haptophyte) will be outlined to describe the distribution of fatty acid residues within its major lipid classes. This method has been proven to be a reliable technique to assess changes in lipid and fatty acid profiles in several other microalgal species and seaweeds.

Key words FAME analysis, GC-FID, Lipid class separation, Microalgae, Seaweeds, Thin-layer chromatography

1 Introduction

Seaweeds and microalgae are known to synthesize large amounts of long-chain polyunsaturated fatty acids (LC-PUFA), such as eicosapentaenoic (EPA, 20:5 n-3) and docosahexaenoic (DHA, 22:6 n-3) acids, with important neutraceutical and pharmaceutical applications [1, 2]. Microalgae also have the ability to accumulate oil (triacylglycerols, TAG) as potential feedstock for biodiesel production [3]. However, due to the metabolic plasticity of algae, their lipid and fatty acid profiles are highly variable between species, and strongly affected by abiotic factors and other habitat changes [4]. Indeed, some microalgal species have the ability to produce substantial amounts (i.e., 20–50 % dry cell weight) of TAG under stress conditions, such as high light, alkaline pH, and

Dagmar B. Stengel and Solène Connan (eds.), *Natural Products From Marine Algae: Methods and Protocols*, Methods in Molecular Biology, vol. 1308, DOI 10.1007/978-1-4939-2684-8_11, © Springer Science+Business Media New York 2015

nutrient depletion, reaching up to ~80 % in some species [4, 5]. Recent studies have also revealed that the fatty acid content and composition in seaweeds can vary spatially and temporally [1]. Prior to considering the production of n-3 LC-PUFA for human health applications, or biodiesel from autotrophic algae, it is crucial to identify, and better understand, the mechanisms by which biochemical and environmental factors trigger oil and LC-PUFA accumulation in algae.

As in higher plants, polar lipids (i.e., glycolipids and phospholipids) are the major constituents of cellular membranes in algae [6, 7]. Glycolipids are located predominantly in photosynthetic membranes, and plastid lipids in algae are mainly represented by monogalactosyldiacylglycerol (MGDG), digalactosyldiacylglycerol (DGDG), and sulfoquinovosyldiacylglycerol (SQDG). The major phospholipids in most algal species are phosphatidylcholine (PC), phosphatidylethanolamine (PE), and phosphatidylglycerol (PG). In addition, phosphatidylserine (PS), phosphatidylinositol (PI), and diphosphatidylglycerol (DPG) may also be found in substantial quantities. In addition to these common glycolipids and phospholipids, other lipids have been reported from some algal species (i.e., sulfoquinovosylmonogalactosylglycerol, SQMG; diacylglyceryl hydroxymethyl-*N*,*N*,*N*-trimethyl-β-alanine, DGTA; diacylglyceryl carboxyhydroxymethylcholine, DGCC; diacylglyceryl glucuronide, DGGA). The presence of betain lipids (BL i.e., DGTA, DGCC, DGGA) in some algal species even suggested that these compounds might replace PC in terms of metabolic role [8, 9]. However, in most photoautotrophic algae, LC-PUFA are mainly accumulated into complex polar lipids constituting membranes, while triacylglycerols (TAG) are predominantly constructed of saturated (SFA) and monounsaturated (MUFA) fatty acids [10–12]. Only few species (e.g., *Pavlova lutheri*) have the ability to accumulate lipids and triacylglycerols containing n-3 LC-PUFA under specific growth conditions such as high inorganic carbon supply and nitrogen limitation [11, 13]. These species are therefore prominent candidates for n-3 LC-PUFA-oil production.

In this context, research on algal lipids has gathered significant interest over the last few years, aiming to better understand the mechanisms of oil or valuable fatty acids accumulation in algae, and methods for effective lipid and fatty acid analysis are essential. Today, expensive techniques and instruments such as thin-layer chromatography coupled to flame ionization detector (TLC-FID), thin-layer chromatography matrix-assisted laser desorption and ionization-mass spectrometry (TLC-MALDI-MS), or high performance liquid chromatography coupled to atmospheric pressure chemical ionization-mass spectrometry (HPLC-ACPI-MS) are available and these are used to separate and quantify lipids. However, we describe in this chapter a more conventional and

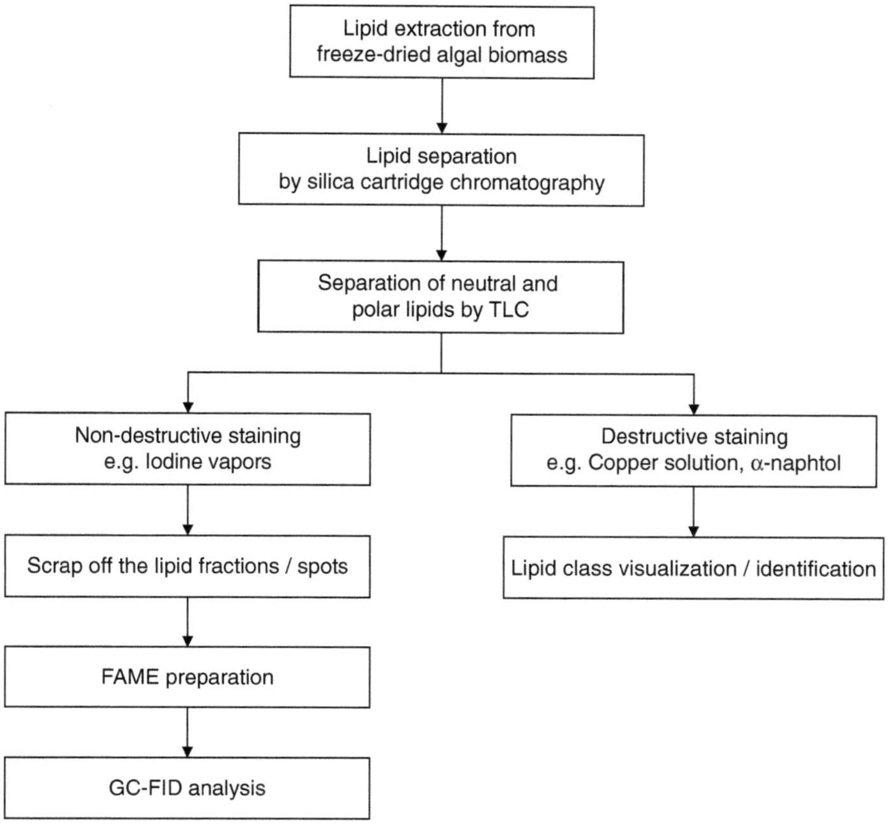

Fig. 1 Schematic representation of neutral and polar lipid analysis from freeze-dried algal biomass (seaweed or microalgae)

less-expensive step-by-step procedure (Fig. 1) for neutral and polar lipid profiling using techniques such as thin-layer chromatography (TLC) and gas chromatography coupled with flame ionization detector (GC-FID). Briefly, the total lipids are extracted from algal biomass by solid–liquid extraction (SLE), the major lipid classes are fractionated using silica cartridges and TLC, and the individual lipid classes are scraped from the TLC plates and converted to fatty acyl methyl esters (FAME) by direct or nondirect transmethylation, before analysis by GC-FID. The marine microalga *Pavlova lutheri* (Haptophyta) is used as a model to illustrate the analytical procedure described thereby.

2 Materials

All chemicals should be analytical or HPLC grades. Use borosilicate glassware and plastic products that are resistant to chemicals and compatible with the solvents used.

2.1 Main Equipment Required

1. Freeze-dryer system.
2. Food blender.
3. Laboratory test sieve (250 or 500 μm).
4. Ultrasonic bath.
5. Centrifuge.
6. Rotavapor (equipped with a heating bath and a vacuum pump).
7. Nitrogen evaporator system.
8. Silica cartridges 500 mg (e.g., Bond Elut™ JR-SI).
9. Glass syringes 20 mL.
10. TLC chambers.
11. TLC plates (e.g., Silica Gel 60, 20×20 cm, 0.25 mm thickness, Merck, Darmstadt, Germany).
12. TLC reagent sprayer.
13. 15 mL Pyrex® borosilicate glass tubes equipped with Teflon (PTFE)-lined screw cap.
14. Electrical sand-bath (e.g., magnetic hotplate stirrer + Pyrex® borosilicate glass beaker containing sand).
15. Gas chromatography system equipped with a flame ionization detector and a fused silica capillary column (e.g., DB-WAXETR 0.25 mm × 30 m × 0.25 μm, Agilent Technologies, USA).
16. Gas chromatography vials, and specific inserts (~100–200 μL content).

2.2 Gas chromatography Standards

1. GC internal standard (e.g., pentadecanoic acid 15:0, 99 %). Add 50 mg of fatty acid standard in 10 mL of methanol. Store if needed at −20 °C for a maximum of 1 month in a 15 mL borosilicate glass tubes equipped with Teflon (PTFE)-lined screw cap, to avoid solvent evaporation and change of concentration.
2. GC FAME standards (e.g., Supelco™ 37 Component FAME Mix, Supelco, USA; Menhaden Oil FAME, Supelco, USA; FAME Mix 189-1 and 189-15, Sigma-Aldrich, France). Dilute as required in hexane and store at −20 °C in a 15 mL borosilicate glass tubes equipped with Teflon (PTFE)-lined screw cap. These standards could be stored at −20 °C up to several months if they are used only for peak identification.

2.3 Solvents

All solutions should be prepared just before each experiment using deionized water (prepared by purifying distilled water to obtain a resistivity of 18 ΩM/cm at 25 °C). Solutions containing highly volatile solvents (e.g., chloroform) must be stored in borosilicate glass bottles with Teflon insert (for safety, *see* **Note 1**).

1. Chloroform.
2. Methanol.
3. Chloroform:methanol (1:1, v/v).
4. Hexane.
5. *Lipid extraction*: chloroform:methanol (1:2, v/v). Add 100 mL of chloroform to 200 mL of methanol, mix and store at room temperature in a 500 mL borosilicate glass storage bottle.
6. *Neutral lipid fractionation* (*solvent system I*): petroleum ether:diethyl ether:acetic acid (80:20:1, v/v/v). Mix 320 mL of petroleum ether with 80 mL of diethyl ether and 4 mL of acetic acid, and store at room temperature in a 500 mL borosilicate glass storage bottle.
7. *Polar lipid fractionation* (*solvent system II*): chloroform:acetone: methanol:acetic acid:water (50:20:10:10:5, v/v/v/v/v). Mix 200 mL of chloroform with 80 mL of acetone, 40 mL of methanol, 40 mL of acetic acid and 20 mL of Milli-Q water, and store at room temperature in a 500 mL borosilicate glass storage bottle.
8. *Transmethylation solution*: 2 % (v/v) sulphuric acid in methanol. Fill half of a 100 mL volumetric flask with methanol, add 2 mL of concentrated H_2SO_4, and fill up with methanol until etched ring graduation mark. Store at room temperature in a 100 mL borosilicate glass storage bottle (*see* **Note 2**).

2.4 TLC Visualization Reagents

1. *Iodine vapor*: place iodine crystals in a TLC chamber and wait until vapor saturation.
2. *Sulfuric acid solution*: 10 % (v/v) sulphuric acid in methanol. Fill half of a 100 mL volumetric flask with methanol, add 10 mL of concentrated H_2SO_4, and fill up with methanol until etched ring graduation mark. Store at room temperature in a 100 mL borosilicate glass storage bottle.
3. *Copper solution*: 10 % (w/v) copper sulfate in 8 % (v/v) aqueous phosphoric acid. Fill half of a 100 mL volumetric flask with Milli-Q water, add 10 g of copper sulfate ($CuSO_4$) and 8 mL of concentrated phosphoric acid (H_3PO_4), and fill up with Milli-Q water until etched ring graduation mark. Store at room temperature in a 100 mL borosilicate glass storage bottle.
4. *α-naphtol solution*: 2.4 % α-naphtol (w/v), 10 % sulphuric acid (v/v) and 80 % ethanol (v/v) solution. Using a 100 mL volumetric flask, mix 2.4 g α-naphtol with 80 mL ethanol, 10 mL of concentrated H_2SO_4, and fill up with Milli-Q water until etched ring graduation mark. Store at room temperature in a 100 mL borosilicate glass storage bottle.

5. *Molybdenum-blue solution*: the reagent is prepared as two solutions:
   (a) Add 40.11 g of molybdenum trioxide (MoO_3) to 1 L of 12.5 M sulphuric acid and boil gently on a hot plate until dissolved; let the light yellow solution cool to ambient temperature overnight, the solution will turn light blue.
   (b) Add 1.78 g of powdered molybdenum to 500 mL of the solution I and boil gently on a hot plate for 15 min, and cool down at room temperature.

   For preparation of the spray reagent (dark green solution), add equal volumes of solutions I and II with two volumes of water. The solution is stable for several months. Store at room temperature in a 1 L borosilicate glass storage bottle.

3 Methods

Every step of this procedure has to been done under a fume hood following strict safety regulations (*see* **Notes 1** and **2**).

3.1 Preparation of Algal Samples

3.1.1 Seaweed Samples

1. Select macroalgal samples carefully from the shore, choosing individuals without epiphytes or obvious grazing marks.
2. Store the macroalgal biomass collected in an ice box containing ice packs during transportation from the sampling site to the laboratory.
3. Clean algal materials and rinse quickly under tap water to remove the salt and possible sand residues.
4. Freeze seaweed samples at −20 °C before freeze-drying.
5. Store freeze-dried samples at −20 °C until processing (up to 2 years; or at −80 °C for longer).
6. Just before analysis, grind freeze-dried samples using a traditional food blender, and filter the resulting powder through a sieve (250 or 500 μm) to achieve even particle size.

3.1.2 Microalgal Samples

1. Harvest cells from a known volume of microalgal culture (*see* **Note 3**) by gentle centrifugation (1,200 × *g* for 10 min) using conical plastic tubes.
2. Freeze the microalgal pellet obtained at −20 °C before freeze-drying.
3. Store tubes containing freeze-dried samples at −20 °C prior to analysis.

3.2 Total Lipid Extraction

Total lipid extraction by solid–liquid extraction (SLE) procedure is based on a modified version of Bligh and Dyer's method [14].

1. Weigh 1 g of freeze-dried ground macroalgal biomass, or use a freeze-dried microalgal pellet corresponding to approximately 100–200 mL of culture according to the cell density (~100–200 mg freeze-dried biomass). Record the exact weight or volume.
2. Add 6 mL of chloroform:methanol (1:2 v/v) to the powder. Add a magnetic bar and close the tube under nitrogen gas (N_2) to limit oxidation.
3. Disrupt the cells using ultrasounds water bath during 30 min at room temperature, and mix with a magnetic stirrer the mixture overnight in a cold room set up at 4 °C.
4. Add 2 mL chloroform and vortex well.
5. Add 3.6 mL of Milli-Q water to give a final chloroform: methanol:water ratio of 1:1:0.9 (v/v/v).
6. Vortex well and centrifuge at 4 °C (1,000 ×*g* for 3 min).
7. Collect the organic/chloroform phase containing lipids (lower phase) in a round-bottom flask.
8. Repeat the extraction 2–3 times by adding 4 mL of chloroform to the remaining methanol:water phase, and collect the organic phase as previously explained.
9. Evaporate the combined organic phase obtained (~12–16 mL) to dryness using a rotavapor under reduced pressure at 40 °C and dissolve the lipid extract in 2–3 mL of chloroform.
10. Transfer the lipid extract to a small pre-weighed vial, dry under a flow of nitrogen gas (N_2) using a nitrogen evaporator, then freeze-dry and measure the total crude lipid content gravimetrically by weighing the freeze-dried vial and subtracting the weight of the vial (*see* **Note 4**).
11. The dry lipid extract can be stored at –20 °C under nitrogen gas (N_2) for approximately 1 month without lipid degradation.

3.3 Lipid Class Separation

The lipid class separation procedure combines the method described in [15] using silica cartridges to separate the neutral and polar lipids, and TLC procedures adapted from [16–18] to separate individual lipid classes.

3.3.1 Lipid Separation by Silica Cartridge Chromatography

1. Dissolve the lipid extract in a small volume of chloroform.
2. Insert a silica cartridge in a holder, connect a 20 mL glass syringe and wash/equilibrate the cartridge with 5 mL of methanol and 5 mL of chloroform by pressing the plunger down.
3. Remove the syringe and load the full lipid extract sample dissolved in chloroform (100–200 μL) using a micropipette onto the cartridge until fully absorbed. Wash the vial with small volume of chloroform and load onto the cartridge.

4. Connect the syringe and fill it with 20 mL of chloroform. Insert the plunger and press slowly and carefully to washout the neutral lipids (nonpolar fraction). Use a round-bottom flask to collect the eluted fraction.
5. Remove the syringe carefully, remove the plunger, reconnect the syringe and fill it with 20 mL of methanol. Elute the polar lipids in a separate round-bottom flask (*see* **Note 5**).
6. Evaporate the neutral and polar fractions obtained to dryness using a rotary evaporator and dissolve the lipid extracts in 2–3 mL of chloroform.
7. Transfer the lipid extracts to small vials; dry under a flow of nitrogen gas (N_2).
8. The dry lipid fractions can be stored at −20 °C under nitrogen gas (N_2) for approximately 1 month without lipid degradation.

3.3.2 Neutral and Polar Lipid Fractionation by Thin-Layer Chromatography (TLC)

Both neutral and polar lipids are subjected to one-dimensional TLC for lipid class separation, using TLC plates coated with silica gel. Neutral lipids are separated using the solvent system I containing petroleum ether:diethyl ether:acetic acid (80:20:1 v/v/v). Polar lipids are separated into individual lipids using solvent system II containing a five-component mixture of chloroform:acetone: methanol:acetic acid:water (50:20:10:10:5 v/v/v/v/v).

1. Prepare 500 mL of each solvent system.
2. Using a sharp carbon pencil, trace two lines at 2 cm from each side of the TLC plates.
3. There are two general preparation steps for TLC plates: pre-washing and activation (*see* **Note 6**).

   Sometimes pre-washing of the silica gel coated TLC plates is necessary to remove impurities usually originating from the binder: pre-wash a plate by blank development in a TLC chamber. Use either the mobile phase for the separation, or methanol, or mixtures of chloroform:methanol (1:1 v/v). Then let the TLC plates dry in air under a fume hood. Mark the plate to show the direction of blank development: when the plate is then later used for chromatography of the sample, the same direction of development should be employed.

   On the day of the analysis, activate the TLC plates with silica gel coated on glass by heating in a hot-air oven for 30 min at 110 °C, or TLC plates with silica gel coated on aluminum or plastic at 90 °C. It should be performed on a solid metal or glass surface to ensure uniform heat distribution. Then allow cooling before loading your samples.
4. Dry two clean TLC chambers, cut chromatography papers (20×20 cm), and place one sheet along the side of each chamber. This paper sheet will help the TLC chamber to be saturated in solvent vapors. Pour ~100 mL of freshly prepared solvent

system I or II up to 1 cm into two different chambers. Wait ~30–45 min for vapor saturation while loading the plates.

5. Dissolve the neutral and polar lipid fractions in ~50–100 μL of chloroform or methanol (neutral and polar lipids, respectively).
6. For neutral lipid separation, load 50–100 μL of the total lipid extract or neutral lipid fraction as a line with a 100 μL syringe on a first TLC plate. Allow samples to dry. Place the plate with the samples facing down in the chamber to develop using solvent system I (*see* **Note 7**).

   For polar lipid separation, similarly load 50–100 μL of the total lipid extract or polar lipid fraction on a second TLC plate. Allow samples to dry. Place the plate in the chamber to develop using solvent system II (*see* **Note 7**).
7. Close immediately the chamber with the cover and let run for approximately 1 h, until the solvent front has reached the upper line. Remove the plate and leave to dry under the fume hood.
8. Upon complete dryness, visualize the lipids by brief exposure to iodine vapor under the fume hood: place the TLC plates in the TLC chamber containing iodine. Proceed until lipid spots turn yellow-brown in color (*see* **Note 8**). Mark directly the edges of the spots with a carbon pencil.
9. Scrape the silica containing lipids from each spot off the plate using a sharp razor blade, and place the silica powder in a 15 mL borosilicate glass tube properly labeled and close using a Teflon (PTFE)-lined screw cap (*see* **Note 9**).
10. Identify individual lipid classes by comparison of their R_f values to R_f values provided in literature [19], by using specific lipid charring reagents, and/or by running commercial lipid standards along with the samples (*see* **Notes 10** and **11**).
11. Two unspecific lipid charring:

    Spray the plate with the sulphuric acid solution under the fume hood; allow to dry for a few minutes, and bake at 120 °C for 15 min or until spots appear.

    Spray the plate with the copper solution under the fume hood, allow to dry, and bake at 180 °C for 15 min.
12. Two specific lipid charring:

    Glycolipids staining using the α-naphtol solution: spray the plate with the α-naphtol solution under the fume hood, allow to dry, and bake at 120 °C for 3–5 min until glycolipid bands appear in blue or purple.

    Phospholipids staining using the molybdenum-blue reagent: spray the plate with molybdenum-blue reagent under the fume hood and allow to dry. Compounds containing phosphate ester will appear immediately as blue spots.

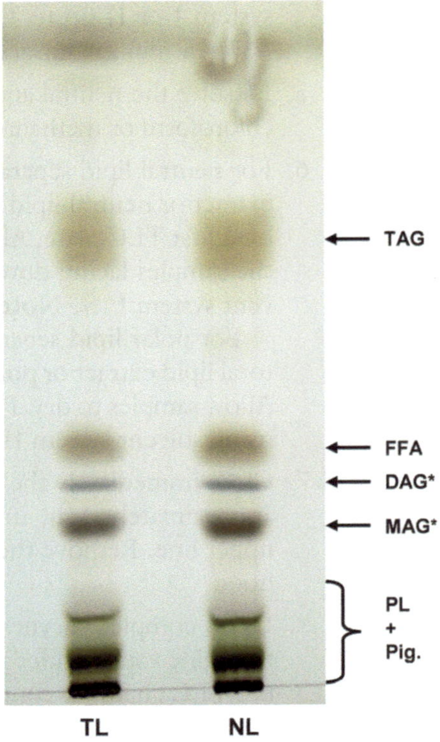

Fig. 2 Example of separation of neutral lipids from *P. lutheri* using one-dimensional TLC. Lipid-spots were stained using the copper solution. Spots marked with an *asterisk* symbol represent lipid classes with uncertain identification: spots were not identified by co-chromatography with pure standards, but based on R_f values from the literature where possible. *DAG* diacylglycerols, *FFA* free fatty acids, *MAG* monoacylglycerols, *NL* neutral lipids, *Pig.* pigments, *PL* polar lipids, *TAG* triacylglycerols, *TL* total lipids

Examples of separation of neutral and polar lipids from *P. lutheri* using one-dimensional TLC are shown on Figs. 2 and 3a (*see* **Note 12**).

3.4 Fatty Acid Analyses by Gas Chromatography Coupled with Flame Ionization Detection (GC-FID)

Fatty acid methyl esters (FAME) are obtained by transmethylation of an aliquot of total lipid extract, or by direct-transmethylation of the silica powder containing the different lipid fractions and/or the freeze-dried algal powder (*see* **Note 13**).

3.4.1 FAME Preparation

1. Add 2 mL of dry methanol containing 2 % (v/v) H_2SO_4 to the tubes containing an aliquot of total lipid extract, silica powder with lipids, or known weight of freeze-dried algal powder (~50 mg), and 10–50 μL of 5 mg/mL pentadecanoic acid 15:0 with a 50 μL syringe as internal standard (*see* **Note 14**).

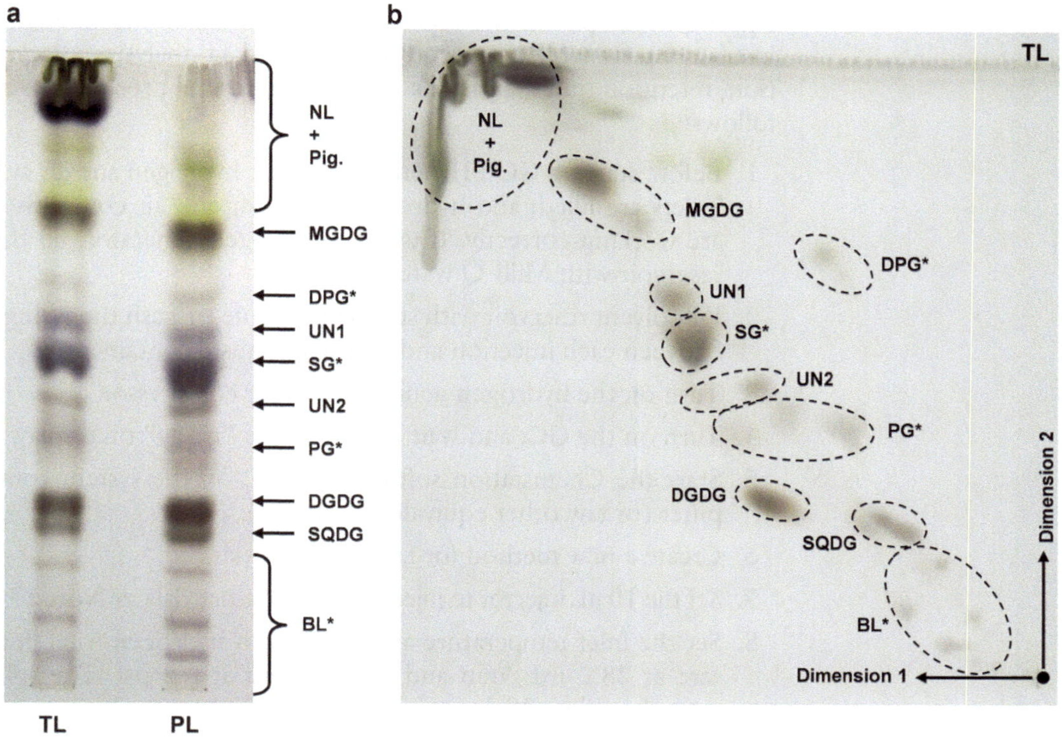

Fig. 3 Example of separation of polar lipids from *P. lutheri* using one- (**a**) and two- (**b**) dimensional TLC. Lipid-spots were stained using the copper solution. Spots marked with an *asterisk* symbol represent lipid classes with uncertain identification. *BL* betain lipids, *DGDG* digalactosyldiacylglycerols, *DPG* diphosphatidylacylglycerols, *MGDG* monogalactosyldiacylglycerols, *NL* neutral lipids, *Pig.* pigments, *PG* phosphatidylglycerols, *SG* acylated sterol glycosides, *SQDG* sulfoquinovosyldiacylglycerols, *TL* total lipids, *UN1-2* unidentified lipid classes

2. Add a small magnetic bar, and close glass tubes tightly with Teflon-lined caps under nitrogen gas (N_2) (*see* **Note 15**).
3. Incubate glass tubes at 80–90 °C for 1.5 h with continuous stirring in a sand bath.
4. After the tubes have cooled down, add 1 mL of Milli-Q water followed by 1–2 mL hexane and vortex vigorously. Centrifuge samples at 1,000×*g* for 5 min.
5. Under a fume hood, transfer the hexane upper-layer, containing FAME, to a GC glass vial and cap tightly (*see* **Note 16**).
6. Samples can be stored at 4 °C for short-term (~24 h) and at −20 °C for few days.

3.4.2 FAME Analysis by GC-FID

In the case of example results outlined in this chapter, gas chromatographic analysis of FAME was performed on an Agilent 7890A GC/5975C MSD Series equipped with a flame ionization detector and a fused silica capillary column DB-WAXETR. Samples were injected using an Agilent auto-injector 7683B series.

Agilent MSD productivity ChemStation software was used for instrument control, data acquisition, and data analysis (integration, retention times, and peak areas). The general procedure is as follows:

1. Before starting GC-FID, ensure that the hydrogen and air cylinders are filled, and hydrogen generator and air compressor are working correctly. If using a hydrogen generator, fill the reservoir with Milli-Q water.
2. Fill solvent reservoir with sufficient hexane to wash the syringe between each injection and empty the waste container.
3. Turn on the hydrogen generator and air compressor.
4. Turn on the GC, and wait until it shows "ready" on display.
5. Start the Chemstation software for GC on the system computer (or any other equivalent software).
6. Create a new method for fatty acid analysis.
7. Set the 10 μL injector to inject 2 μL sample per vial (*see* **Note 17**).
8. Set the inlet temperature at 250 °C with hydrogen total flow rate at 28.2 mL/min and the pressure at 5.5 psi. The split ratio should be 20:1.
9. Set initially the oven temperature to 140 °C. Program this oven temperature at 140 °C for 1 min, and raise it from 140 to 200 °C by a rate of 15 °C/min and then from 200 to 250 °C at a rate of 2 °C/min. Hydrogen should be used as a carrier gas at a flow rate of 1.2 mL/min. One run takes approximately 30 min.
10. Set the temperature of the flame ionization detector to 300 °C with a hydrogen flow rate of 40 mL/min and air flow rate at 400 mL/min.
11. Save your FAME analysis method.
12. Load your method for FAME analysis, and wait until it shows "ready".
13. Create a run sequence table with the number of vials and sample names. Choose a directory file to save your data. Give a name and save your sequence.
14. Place the vials into the auto-sampler tray in the same order as in the run sequence table.
15. When the instrument is ready, start the run sequence.

3.4.3 GC-FID Data Analysis

1. Identification of FAME is obtained by co-chromatography with authentic commercially available FAME standards, and retention time relationships.
2. Fatty acid contents are quantified by comparison of peak areas between a known amount of added pentadecanoic acid 15:0 as internal standard (*see* **Note 18**).

For example, to calculate the EPA content:

$$\mu g_{(EPA)} = \left(\text{peak area}_{EPA} \times \mu g_{15:0\ added}\right) / \text{peak area}_{15:0\ added}$$

For example, to calculate the total fatty acid (TFA) content:

$$\mu g_{(TFA)} = \sum\left[\left(\text{peak area}_{each\ fatty\ acid} \times \mu g_{15:0\ added}\right) / \text{peak area}_{15:0\ added}\right]$$

3. Results of fatty acid composition are expressed as % of TFA.
   Figure 4 shows an example of GC-FID chromatogram of the FAME obtained by direct-transmethylation of *P. lutheri* freeze-dried biomass (*see* **Note 19**), and Table 1 for the fatty acid composition (expressed as % of TFA) of total lipids and major individual lipid classes in *P. lutheri* (*see* **Note 20**).
4. The percentage of each lipid class is calculated by dividing the total FAME for one lipid class with the total FAME of all lipid classes (TFA residues).
   For example, to calculate the ratio of TAG:

$$\text{TAG } (\%\ \text{TFA}) = \left[\sum\left(\text{FAMEs}_{into\ TAG}\right) / \sum\left(\text{FAMEs}_{into\ total\ lipids}\right)\right] \times 100$$

Figure 5 shows an example of lipid class composition, expressed as % of TFA residues per class, of the total crude lipid extract from *P. lutheri* (*see* **Note 21**).

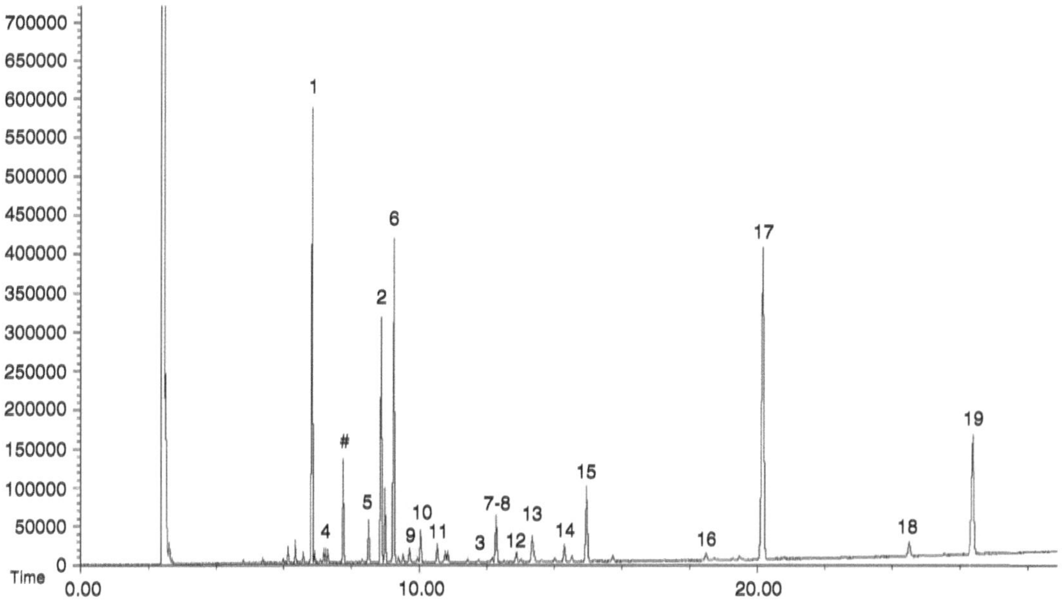

Fig. 4 GC-FID chromatogram of the FAME obtained by direct-transmethylation of freeze-dried *P. lutheri* biomass (*see* Table 1 for peak identification). # Pentadecanoic acid (15:0) used as an internal standard

Table 1
Example of fatty acid composition (% TFA) of total lipids and major individual lipid classes (neutral and polar) in *P. lutheri* batch-cultivated using a relatively high sodium bicarbonate supply (1 g/L) and harvested at the beginning of the stationary phase

			Neutral	Polar		
n°	Fatty acids (% TFA)	TL	TAG	MGDG	DGDG	SQDG
1	14:0	12.23 ± 0.29	12.24 ± 0.02	4.12 ± 0.05	19.16 ± 0.31	29.03 ± 0.12
2	16:0	14.34 ± 0.37	10.76 ± 0.17	7.96 ± 0.02	11.39 ± 0.13	23.33 ± 0.10
3	18:0	0.25 ± 0.03	0.22 ± 0.01	1.59 ± 0.05	1.28 ± 0.09	4.37 ± 0.04
	Sum of SFA	26.82 ± 0.15	23.55 ± 0.13	14.55 ± 0.16	32.86 ± 0.20	57.83 ± 0.19
4	14:1	0.51 ± 0.05	0.74 ± 0.03	0.60 ± 0.03	0.37 ± 0.02	0.25 ± 0.01
5	15:1	1.80 ± 0.11	0.14 ± 0.01	0.21 ± 0.01	0.12 ± 0.01	0.61 ± 0.01
6	16:1 n-7	16.52 ± 0.08	23.05 ± 0.46	11.71 ± 0.34	14.70 ± 0.57	6.91 ± 0.48
7	18:1 n-9	2.90 ± 0.01	2.93 ± 0.09	1.66 ± 0.02	0.62 ± 0.02	2.61 ± 0.01
8	18:1 n-7	0.30 ± 0.01	0.52 ± 0.06	1.31 ± 0.02	2.72 ± 0.14	2.46 ± 0.05
	Sum of MUFA	22.03 ± 0.02	27.38 ± 0.34	15.49 ± 0.34	18.53 ± 0.76	12.83 ± 0.11
9	16:2 n-6	1.21 ± 0.03	0.60 ± 0.01	1.08 ± 0.11	0.44 ± 0.06	0.21 ± 0.02
10	16:2 n-4	1.44 ± 0.01	1.49 ± 0.03	1.98 ± 0.18	1.28 ± 0.05	0.63 ± 0.01
11	16:4 n-3	1.10 ± 0.01	1.23 ± 0.03	1.09 ± 0.10	0.41 ± 0.02	0.30 ± 0.01
12	18:2 n-6	0.66 ± 0.01	0.85 ± 0.02	1.75 ± 0.10	0.35 ± 0.02	1.40 ± 0.02
13	18:3 n-6	2.01 ± 0.02	2.67 ± 0.05	1.85 ± 0.03	0.78 ± 0.05	0.30 ± 0.04
14	18:3 n-3	0.92 ± 0.02	0.94 ± 0.02	1.67 ± 0.04	0.31 ± 0.04	0.56 ± 0.01
15	18:4 n-3	4.99 ± 0.03	7.04 ± 0.19	10.86 ± 0.08	2.19 ± 0.12	0.54 ± 0.02
16	20:4 n-3	0.46 ± 0.02	0.61 ± 0.03	0.56 ± 0.04	0.36 ± 0.03	0.69 ± 0.01
17	20:5 n-3	26.78 ± 0.10	20.76 ± 0.08	38.77 ± 0.77	32.60 ± 0.29	11.70 ± 0.13
18	22:5 n-3	0.71 ± 0.01	1.76 ± 0.07	0.19 ± 0.02	0.08 ± 0.04	0.43 ± 0.02
19	22:6 n-3	10.23 ± 0.08	10.82 ± 0.07	9.39 ± 0.28	9.42 ± 0.25	9.67 ± 0.12
	Sum of PUFA	50.49 ± 0.10	48.78 ± 0.46	69.18 ± 0.17	48.23 ± 0.44	26.45 ± 0.31
	Others	0.67 ± 0.07	0.62 ± 0.04	1.66 ± 0.06	1.42 ± 0.12	3.99 ± 0.14
	Sum of n-3	44.72 ± 0.05	42.56 ± 0.40	61.97 ± 0.36	45.02 ± 0.66	23.21 ± 0.28
	Sum of n-6	4.33 ± 0.06	4.73 ± 0.09	5.34 ± 0.01	1.93 ± 0.16	2.61 ± 0.04
	Ratio n-6/n-3	0.10 ± 0.01	0.11 ± 0.01	0.08 ± 0.01	0.04 ± 0.01	0.11 ± 0.02

Results are expressed as ±SD ($n = 2$) obtained from two analytical repetitions from the same crude lipid extract
DGDG digalactosyldiacylglycerols, *MGDG* monogalactosyldiacylglycerols, *SQDG* sulfoquinovosyldiacylglycerols, *TAG* triacylglycerols, *TL* total lipids. Others correspond to unidentified peaks or unknown fatty acids

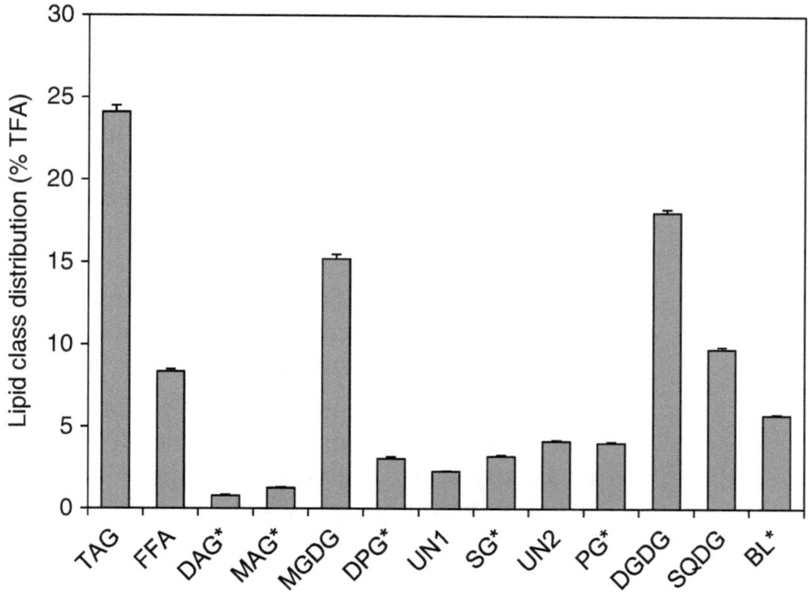

Fig. 5 Lipid class composition (% TFA) of *P. lutheri* batch-cultivated using a relatively high sodium bicarbonate supply (1 g/L) and harvested at the beginning of the stationary phase (*see* Figs. 2 and 3 for lipid class abbreviations). Spots marked with an *asterisk* symbol represent lipid classes with uncertain identification. Results are expressed as ±SD ($n = 2$) obtained from two analytical repetitions from the same crude lipid extract

4 Notes

1. Due to health hazards associated with solvent exposure (e.g., chloroform, methanol), avoid being exposed to vapors by working under a fume hood.
2. Regarding health and safety requirements for solvents, always wear protective goggles, gloves, and a lab coat when handling acids and bases, in particular concentrated H_2SO_4 which can cause serious damage to skin and clothing, and respiratory system if inhaled.
3. It might be important to know the cell density, dry weight (DW), and ash-free dry weight (AFDW) of the culture if you want to express the TFA content or specific fatty acids per cell, per DW, or per AFDW.
4. Gravimetric determination of total lipid content is an not fully quantitative technique to measure the total crude lipid extracted from freeze-dried biomass. Using this method usually leads to an overestimation of the total lipid content expressed per dry weight. Indeed, the crude lipid extract usually contains pigments and other residual impurities.
5. At this stage, the cartridge should be almost colorless.

6. Handle the plates carefully: no fingerprints—these will be visible after exposure to iodine vapor.
7. Make sure the loading lines remain smaller than 4 mm in thickness and are above the solvent mixture used. A hair-dryer at low temperature can be use to dry the spots during loading. In our case, 2–3 samples were loaded on each 20 × 20 cm TLC plates with a minimum ~2 cm space between each. After full development, remove the plate and leave to dry under the fume hood. Overloading your TLC plates can induce a poor separation of the different lipid classes, it is crucial to estimate the right amount of sample to load.
8. A short exposure to iodine vapor is required to avoid any oxidation of lipids as iodine may covalently modify polyunsaturated fatty acids. A glass wool plugged Pasteur pipette with iodine crystals through which N_2 is blown over individual lipid-spots can also be used in order to better control the time of iodine exposure.
9. If required for further analysis, such as HPLC-MS, lipid fractions can be extracted from silica gel: insert a small amount of glass wool into a labelled Pasteur pipette used as a column, carefully transfer the silica gel containing lipids into appropriate column, and sequentially extract the lipids with 1 mL of chloroform:methanol (2:1, v/v) followed by 1 mL of chloroform:methanol (1:1, v/v). The elute lipid fraction is recovered into a small vial, and solvents evaporate off under a fume hood using a stream of nitrogen gas (N_2). The dry lipid extracts closed under N_2 and store at −20 °C can be kept for approximately 1 month without lipid degradation.
10. Lipids separated by TLC can be either reversibly and nondestructively stained with iodine for quantitative analysis, or irreversibly and destructively stained by spraying with specific reagents.
11. Differences in the solvent composition can lead to altered mobility of the different components and possible R_f value errors. A common problem occurs when the TLC chamber is not completely sealed, and the solvents can evaporate slowly. Since evaporation rate depends on the boiling point, solvents like chloroform will evaporate faster. As a result, the mobile phase becomes more polar. If possible, running standards on the same plate will always help a positive identification of your spots.
12. Due to the number of lipid classes in *P. lutheri*, two-dimensional TLC separation may be performed to increase the resolution/separation between spots and to better identify the different polar lipid classes (Fig. 3b), using chloroform:methanol:water (65:25:4, v/v/v) for the first direction (mix 260 mL of chloroform with 100 mL of methanol and 16 mL of Milli-Q water, and store at room temperature in a 500 mL borosilicate glass storage bottle), and solvent system II for the second direction.

13. To avoid possible loss of material throughout the lipid extraction steps, the direct-transmethylation protocol using freeze-dried ground biomass can be used to prepare the FAME and determine the total fatty acid (TFA) composition and content of the algal biomass.
14. The volume of internal standard added depends of the amount of lipid extract or algal biomass used.
15. Tubes need to be sealed under nitrogen gas (N_2) to limit oxidation, and evaporation of the solvent.
16. At this stage, the sample can be concentrated by evaporation to dryness under a flow of nitrogen gas (N_2), the residue re-dissolved in a small volume of hexane (~100 μL) and transferred to a GC vial equipped with a specific insert (~100–200 μL content) and the cap is closed tightly.
17. The injection volume can be changed according to your sample concentrations. Be aware of the maximum vapor volume capacity of your inlet.
18. Peak areas on the GC traces should be within limits linearly proportional to the amount (by weight) of material eluting from the column (*21*).
19. Using the DB-WAXETR column, FAME with short carbon chains and fewer double bonds (i.e., SFA and MUFA) have shorter retention times and appear before LC-PUFA (i.e., EPA and DHA).
20. All examples presented in this chapter were obtained from *P. lutheri* batch-cultivated using a relatively high sodium bicarbonate supply (1 g/L) and harvested at the beginning of the stationary phase. Under these conditions, the major LC-PUFA were stearidonic acid (18:4 n-3), EPA (20:5 n-3) and DHA (22:6 n-3) accounting respectively for ~5, ~27 and ~10 % of TFA (Table 1). SFA were mainly represented by myristic (14:0) and palmitic (16:0) acids (~12 and ~14 % of TFA, respectively), while the major MUFA was palmitoleic acid (16:1 n-7) with ~17 % of TFA. EPA was mainly partitioned into MGDG and DGDG constituting the chloroplastic membranes; DHA was equally found in all lipid classes analyzed. Interestingly, n-3 LC-PUFA (i.e., 18:4 n-3, EPA and DHA) were also incorporated into TAG (~7, ~20 and ~11 % of TFA, respectively).
21. In *P. lutheri*, grown under the conditions previously described (*see* **Note 20**), the neutral lipids were mainly composed of TAG and free fatty acids (FFA) accounting, respectively, for 24 and 8 % of the TFA residues (Fig. 5). The polar lipids were mostly composed of the three main glycolipids: MGDG, DGDG, and SQDG, accounting for 15, 18, and 10 % of TFA residues, respectively (Fig. 5).

Acknowledgements

This work was supported by NutraMara, the Irish Marine Functional Foods Research Initiative (Grant-Aid Agreement No. MFFRI/07/01) carried out under the Sea Change Strategy with the support of the Marine Institute and the Department of Agriculture, Food and the Marine (DAFM), funded under the National Development Plan 2007–2013 for Ireland.

References

1. Schmid M, Guihéneuf F, Stengel DB (2013) Fatty acid contents and profiles of 16 macroalgae collected from the Irish coast at two seasons. J Appl Phycol 26:451–463
2. Mimouni V, Ulmann L, Pasquet V et al (2013) The potential of microalgae for the production of bioactive molecules of pharmaceutical interest. Curr Pharm Biotechnol 13:2733–2750
3. Hu Q, Sommerfeld M, Jarvis E et al (2008) Microalgal triacylglycerols as feedstocks for biofuel production: perspectives and advances. Plant J 54:621–639
4. Stengel DB, Connan S, Popper ZA (2011) Algal chemodiversity and bioactivity: sources of natural variability and implications for commercial application. Biotechnol Adv 29:483–501
5. Breuer G, Lamers PP, Martens DE et al (2012) The impact of nitrogen starvation on the dynamics of triacylglycerol accumulation in nine microalgae strains. Bioresour Technol 124:217–226
6. Guschina IA, Harwood JL (2009) Algal lipids and effect of the environment on their biochemistry. In: Kainz M, Brett MT, Arts MT (eds) Lipids in aquatic ecosystems. Springer, New York. doi:10.1007/978-0-387-89366-2_1
7. Harwood JL, Guschina IA (2009) The versatility of algae and their lipid metabolism. Biochimie 91:679–684
8. Kato M, Sakai M, Adachi K et al (1996) Distribution of betaine lipids in marine algae. Phytochemistry 42:1341–1345
9. Liu B, Benning C (2013) Lipid metabolism in microalgae distinguishes itself. Curr Opin Biotechnol 24:300–309
10. Alonso DL, Belarbi EH, Rodriguez-Ruiz J et al (1998) Acyl lipids of three microalgae. Phytochemistry 47:1473–1483
11. Tonon T, Harvey D, Larson TR et al (2002) Long chain polyunsaturated fatty acid production and partitioning to triacylglycerols in four microalgae. Phytochemistry 61:15–24
12. Guihéneuf F, Fouqueray M, Mimouni V et al (2010) Effect of UV stress on the fatty acid and lipid class composition in two marine microalgae *Pavlova lutheri* (Pavlovophyceae) and *Odontella aurita* (Bacillariophyceae). J Appl Phycol 22:629–638
13. Guihéneuf F, Stengel BD (2013) LC-PUFA-enriched oil production by microalgae: accumulation of lipid and triacylglycerols containing n-3 LC-PUFA is triggered by nitrogen limitation and inorganic carbon availability in the marine haptophyte *Pavlova lutheri*. Mar Drugs 11:4246–4266
14. Bligh EG, Dyer WJ (1959) A rapid method of lipid extraction and purification. Can J Biochem Physiol 37:911–917
15. Sukenik A, Carmeli Y, Berner T (1989) Regulation of the fatty acid composition by irradiance level in the eustigmatophyte *Nannochloropsis* sp. J Phycol 25:689–692
16. Williams JP (1978) Glycerolipids and fatty acids on algae. In: Craigie JS, Hellebust JA (eds) Handbook of phycological methods. Cambridge University Press, Cambridge, UK, pp 99–107
17. Christie WW (1982) Lipid analysis, 2nd edn. Pergamon, Oxford, UK
18. Henderson RJ, Tocher DR (1992) Thin-layer chromatography. In: Hamilton RJ, Hamilton S (eds) Lipid analysis: a practical approach. IRL, Oxford, UK, pp 65–111
19. Christie WW (ed) (1989) Gas chromatography and lipids: a practical guide. The Oily Press Ltd., Ayr., Scotland, p 307

Chapter 12

HRMAS NMR Analysis of Algae and Identification of Molecules of Interest via Conventional 1D and 2D NMR: Sample Preparation and Optimization of Experimental Conditions

Gaëlle Simon, Nelly Kervarec, and Stéphane Cérantola

Abstract

Nuclear magnetic resonance (NMR) has become an astounding tool for molecular characterization. Thanks to the development of probes and the increase of magnetic field, NMR has entered the field of biology and facilitated the identification of natural compounds. Indeed, this nondestructive NMR tool makes possible the complete characterization of less and less quantities of material via 1D and 2D sequences on many nuclei (e.g., ^{1}H, ^{13}C, ^{31}P, ^{15}N). More recently, the development of high-resolution magic-angle spinning (HRMAS) probes have permitted direct analysis of living tissue (e.g., a piece of algae) without prior extraction providing information on both the total content and the ratio of different molecules within the sample; thus HRMAS facilitates a wide range of analyses, such as species differentiation or studies of metabolomics according to various environmental or experimental conditions. This chapter describes the specific sample preparation, based on an algal sample or extract, required for all NMR analyses in order to optimize the NMR response and obtain the most valuable information.

Key words Algae, HRMAS, NMR, Quantification, Structural elucidation

1 Introduction

Over the past 50 years, nuclear magnetic resonance (NMR) has become the predominant technique for determining the structure of organic compounds. NMR is a versatile powerful analytical tool, used to determine concentrations, dynamics, folding, interactions, and structures of a wide variety of molecules. Of all spectroscopic methods, NMR allows complete characterization of compound(s). Although larger amounts of sample are needed than, e.g., for mass

Dagmar B. Stengel and Solène Connan (eds.), *Natural Products From Marine Algae: Methods and Protocols*, Methods in Molecular Biology, vol. 1308, DOI 10.1007/978-1-4939-2684-8_12, © Springer Science+Business Media New York 2015

spectroscopy, NMR is nondestructive, and with modern instruments structural analyses may be obtained from samples weighing less than a milligram. Indeed, to retrieve even better data, sensitivity can be boosted by increasing the magnetic field but also by using new types of probes such as cryogenic probes which reduce the thermal noise when detecting the NMR signals. This recent progress has allowed NMR to contribute to the field of biological research and become a precious tool for metabolomics, as well as drug screening as low concentrations of biological and drug-like molecules can quickly be analyzed.

NMR-based analyses are typically performed on liquid extracts and 1D and 2D sequences on many nuclei (e.g., ^{1}H, ^{13}C, ^{31}P, ^{15}N). NMR is now commonly used in studies on algae such as structure elucidation of new natural compounds [1–3], isolation of compounds of interest [4, 5], metabolomic studies according to environmental or experimental conditions [6, 7], as well as biofuel research [8, 9]. Moreover, the development of high-resolution magic-angle spinning (HRMAS) probes, and thus direct analyses on living tissue or whole cells, allows biologists to access, faster and without the need for prior extraction, useful information about the content of, e.g., a plant fragment, mammalian tissue (e.g., muscle, liver, skin), or a bacterial pellet.

HRMAS NMR is at the interface of the liquid phase and the solid-state NMR. An HRMAS probe and a MAS pneumatic unit are added to a standard high-resolution spectrometer (Fig. 1) to allow analyses of a larger range of samples, either solid or liquid, of biological, chemical, or pharmaceutical interest. HRMAS technique is based on sample orientation at magic angle (54.7°), as well as the fast rotation of this sample (from 1,000 to 15,000 Hz) after levitation. Thus, magic-angle spinning facilitates liquid homogeneity by the reduction of some phenomena such as chemical shift anisotropy and dipolar couplings leading to broad and poorly resolved peaks. For instance, spinning at the magic angle averages the anisotropic interactions of free motion molecules, and results lead to narrower NMR signals (as those obtained in solution); on the other hand cellular membranes, constituted by molecules with low mobility, cause very broad signals on the spectra which often mix with the noise. In Fig. 2, the ^{1}H NMR spectrum of a brown macroalga, *Pelvetia canaliculata*, obtained by classical NMR exhibits broad signals (Fig. 2b) whereas the ^{1}H NMR spectrum obtained by HRMAS NMR displayed clear peaks (Fig. 2a), making interpretation of the spectrum easier. HRMAS is of great interest to biology allowing direct observation of the total content and the ratio of specific molecules within a sample, and thus the differentiation between algal species [10], and the follow-up of the

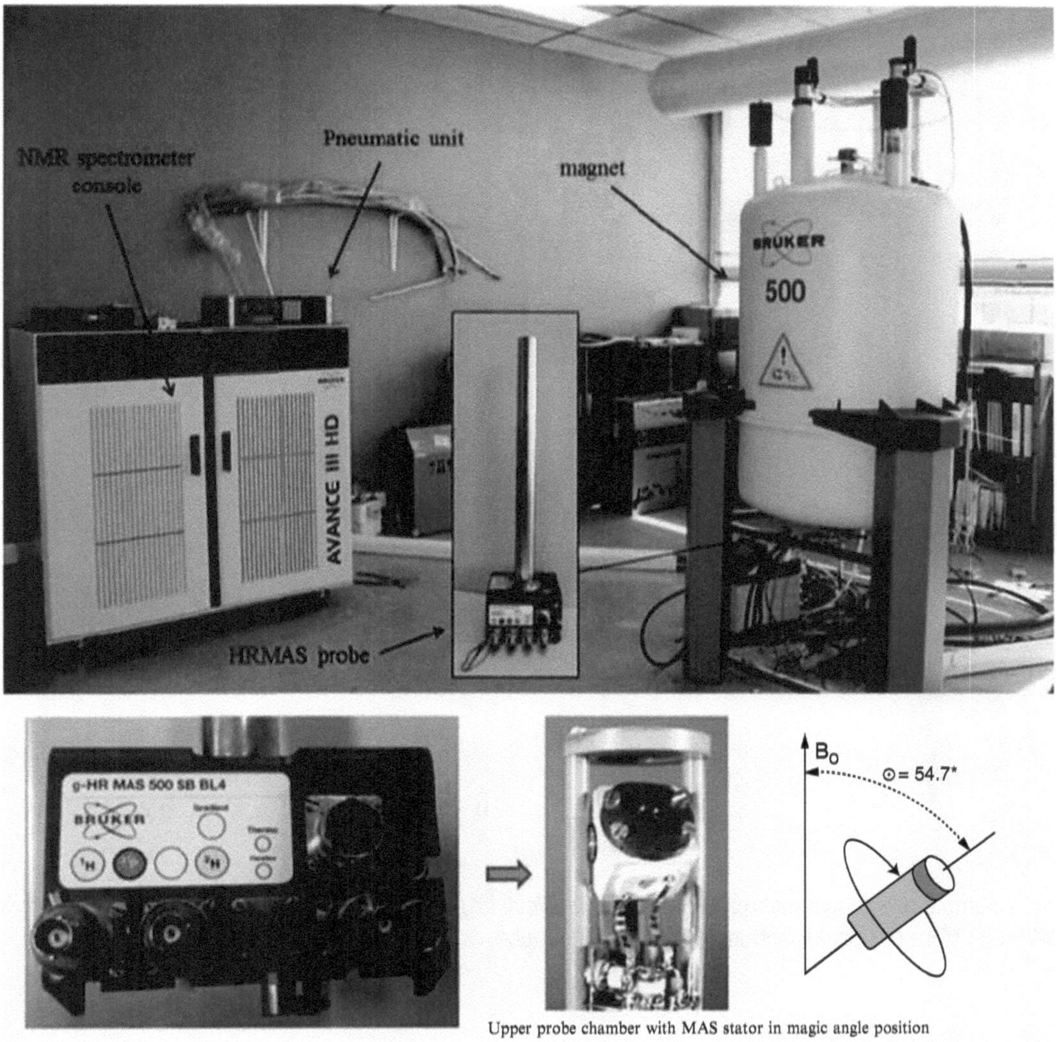

Fig. 1 500 MHz NMR spectrometer (11.7 T) equipped with HRMAS probe

synthesis of compounds of interest according to various environmental or experimental conditions (pH, salinity, season, species, pollutant, etc.) [11]. However, when the identification of specific metabolites is required, an extraction procedure according to an established protocol is generally followed.

This chapter describes the specific preparation of an algal sample and extract which is required prior to all NMR analyses in order to optimize NMR response and obtain the best information.

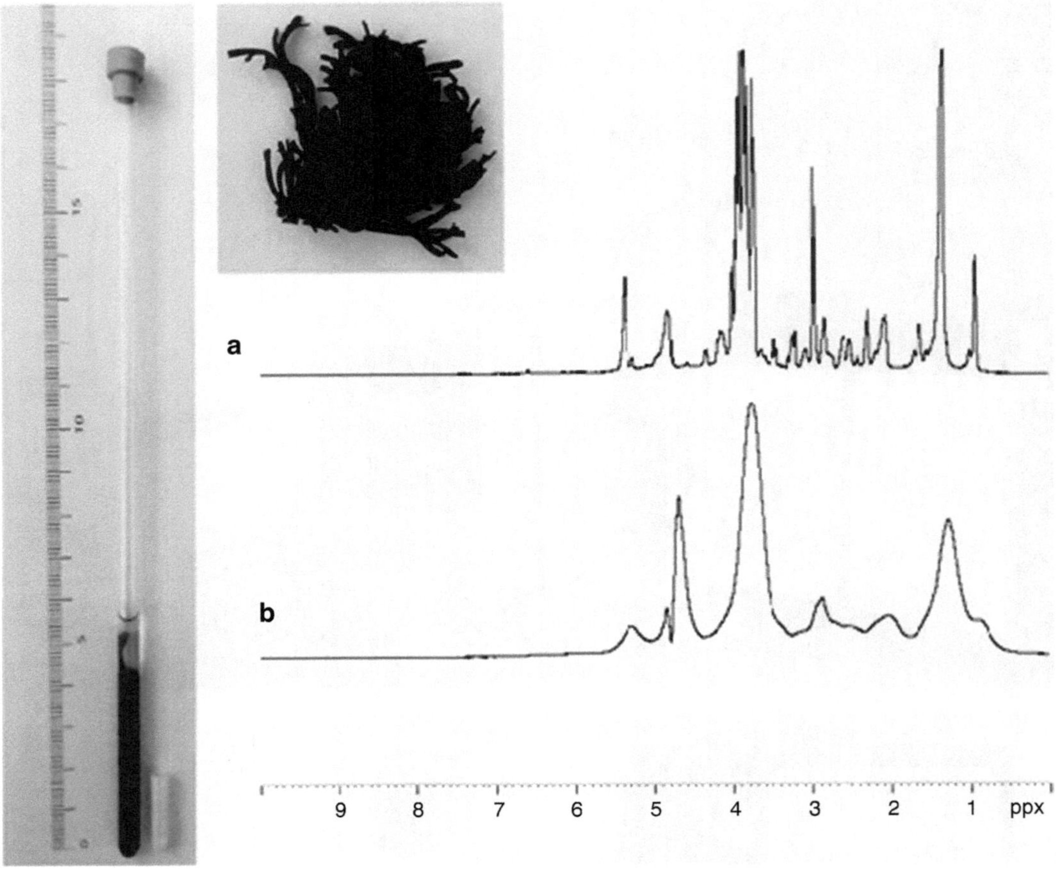

Fig. 2 Sample of a brown macroalga, *Pelvetia canaliculata*. (**a**) ^1H HRMAS NMR (500 MHz) spectrum; (**b**) ^1H NMR (500 MHz) classical spectrum in D_2O in a 5 mm tube

2 Materials

2.1 HRMAS NMR

1. NMR spectrometer equipped with an HRMAS probe (*see* **Notes 1** and **2**).
2. MAS tools (Fig. 3): Rotor, insert, cap, MAS positioning tool for insert, MAS filling tool, and cap remover.
3. Commercial deuterated water (D_2O) (*see* **Note 3**).

2.2 Classical NMR

1. Spectrometer equipped with a classical probe (e.g., QNP, TBI, TXO, BBO) or a cryoprobe.
2. NMR tubes (*see* **Note 4**).
3. Commercial NMR solvents (*see* **Note 5**). In Table 1 are presented properties of some deuterated NMR solvents.
4. Cotton filter (pipette and cotton).

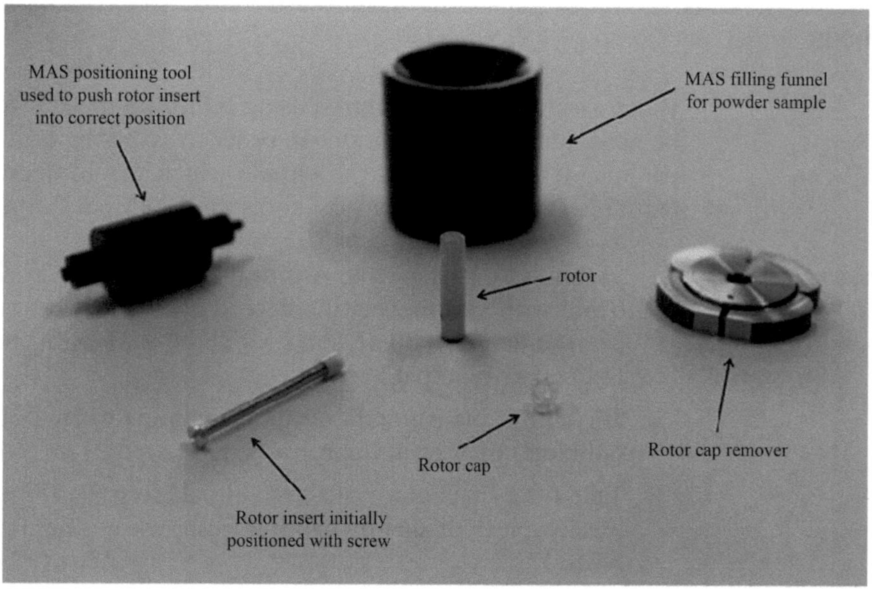

Fig. 3 MAS tools

Table 1
Properties of some deuterated NMR solvents

Solvent	B. p. (°C)	F. p. (°C)	Residual ^{1}H signal (δ in ppm)	Residual ^{13}C signal (δ in ppm)
Acetone-d_6	56	−95	2.04	206.0, 29.8
Acetonitrile-d_3	82	−44	1.93	118.2, 1.3
Benzene-d_6	80	6	7.15	128.0
Chloroform-d	61	−64	7.26	77.0
Cyclohexane-d_{12}	81	7	1.38	26.4
Dichloromethane-d_2	40	−95	5.32	53.8
Dimethylsulfoxide-d_6	189	19	2.49	39.5
Dimethylformamide-d_7	153	−60	2.74, 2.91, 8.01	30.1, 35.2, 162.7
Methanol-d_4	65	−98	3.30, 4.78	49.0
Nitromethane-d_3	101	−29	4.33	62.8
Pyridine-d_5	115	−42	7.19, 7.55, 8.71	123.5, 135.5, 149.9
Tetrahydrofuran-d_8	66	−66	1.73, 3.58	25.3, 67.4
Toluene-d_8	111	−95	2.09, 6.98, 7.00, 7.09	20.4, 125.2, 128.0, 128.9, 137.6
Water	100	0	4.78	

B.p. boiling point, *F.p.* freezing point

3 Methods

When a sample will be analyzed using NMR, no metallic tools must be used during its preparation. In order to avoid the occurrence of paramagnetic particles in the sample which would degrade spectrum resolution, plastic or glass tools must be used throughout.

3.1 Analysis of an Algal Piece

1. Samples are prepared in 4 mm diameter rotors in zirconium oxide with volume varying according to their internal shape from some μL to 80 μL (max.). Fill the rotor with commercial D_2O (Fig. 4, **step 1**).
2. Stir with a diamagnetic needle to eliminate any potential air bubble(s) present in the rotor (Fig. 4, **step 2**; *see* **Note 6**).
3. Take off a small piece of algae (Fig. 4, **step 3**). If lyophilized, break a piece of alga by hand; if fresh, use a ceramic knife (*see* **Note 7**).
4. Insert the algal fragment in the rotor (Fig. 4, **step 4**).
5. Complete the volume of the rotor with commercial D_2O (Fig. 4, **step 5**).

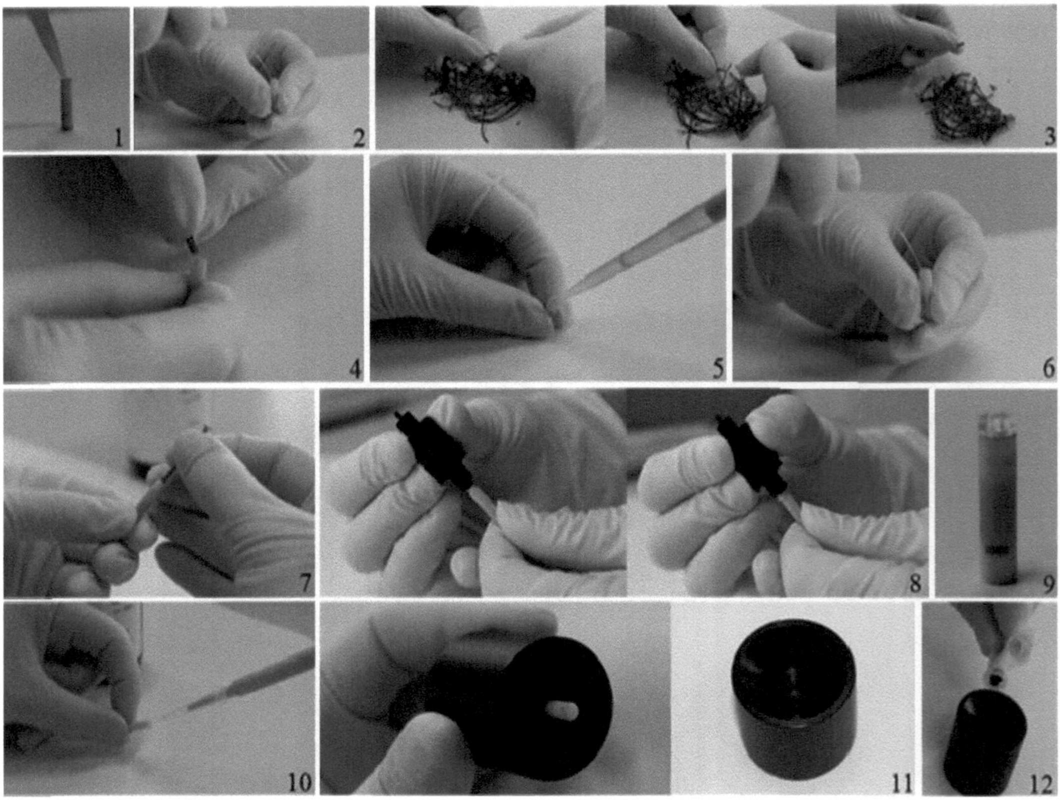

Fig. 4 HRMAS rotor preparation steps

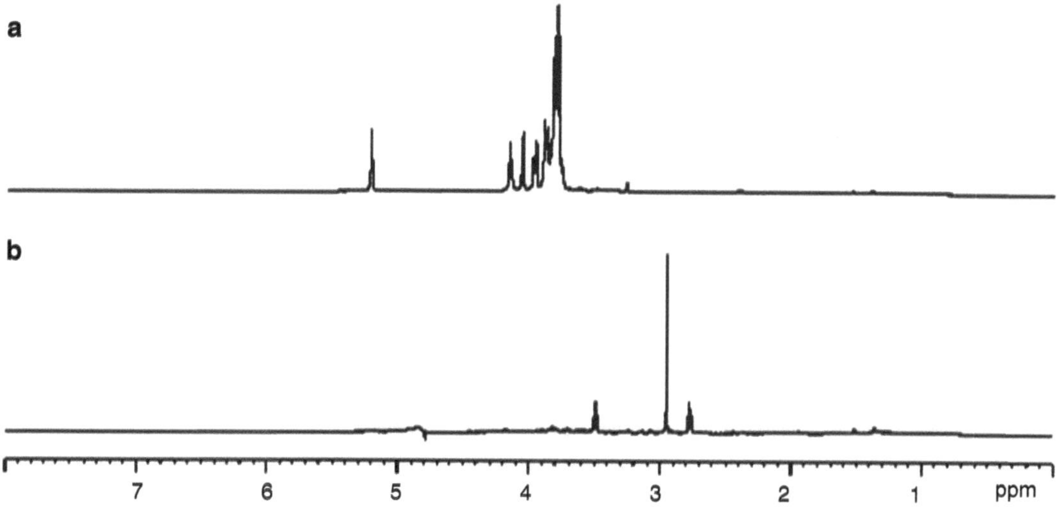

Fig. 5 500 MHz ^{1}H HRMAS NMR spectra with presaturation of the water signal. (**a**): A red macroalga *Palmaria palmata*; (**b**) a green macroalga *Ulva* sp

6. Stir with a diamagnetic needle to eliminate potential air bubble present in the rotor (Fig. 4, **step 6**).
7. Place the insert at the top of the rotor with the help of the screw (Fig. 4, **step 7**).
8. Push the insert in the rotor until a drop of water comes out (Fig. 4, **step 8**).
9. Remove all excess water. The rotor must be completely dry (*see* **Note 8**).
10. Push in the rotor cap (Fig. 4, **step 9**).
11. Put the rotor in the magnet for NMR analysis.

An example of HRMAS NMR spectra of two macroalgal species is presented in Fig. 5. Note that ^{1}H NMR profiles can differ greatly between species.

3.2 Analysis of an Algal Powder

1. Place 10 μL of commercial D_2O in the rotor (Fig. 4, **step 10**).
2. Insert the rotor in the filling tool (Fig. 4, **step 11**).
3. Place a few milligrams of algal powder into the rotor (Fig. 4, **step 12**). Tap the vial to move the power down. If you need to use a spatula, it must be glass or plastic.
4. Repeat all steps from **step 5** in Subheading 3.1.

An example of HRMAS NMR analysis of algal powder, i.e., of the brown macroalga *Cystoseira tamariscifolia*, is presented in Fig. 6. The use of the HRMAS tool has allowed following the production of phenolic compounds directly on powder of *C. tamariscifolia* collected monthly from January to December [12].

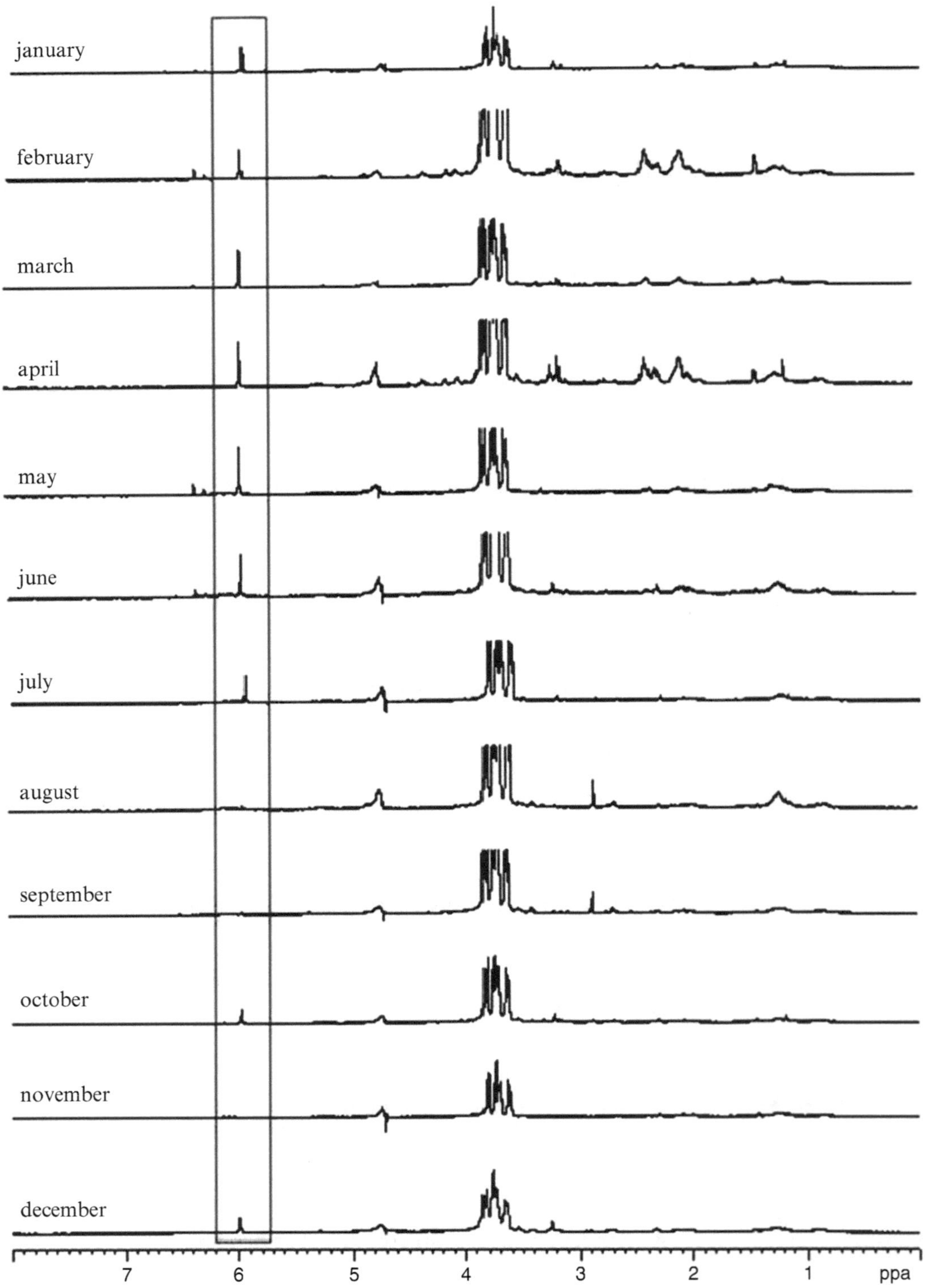

Fig. 6 500 MHz ^{1}H HRMAS NMR spectra with presaturation of the water signal. Seasonal variation in the quantity of phloroglucinol within the brown macroalga *Cystoseira tamariscifolia* [12]

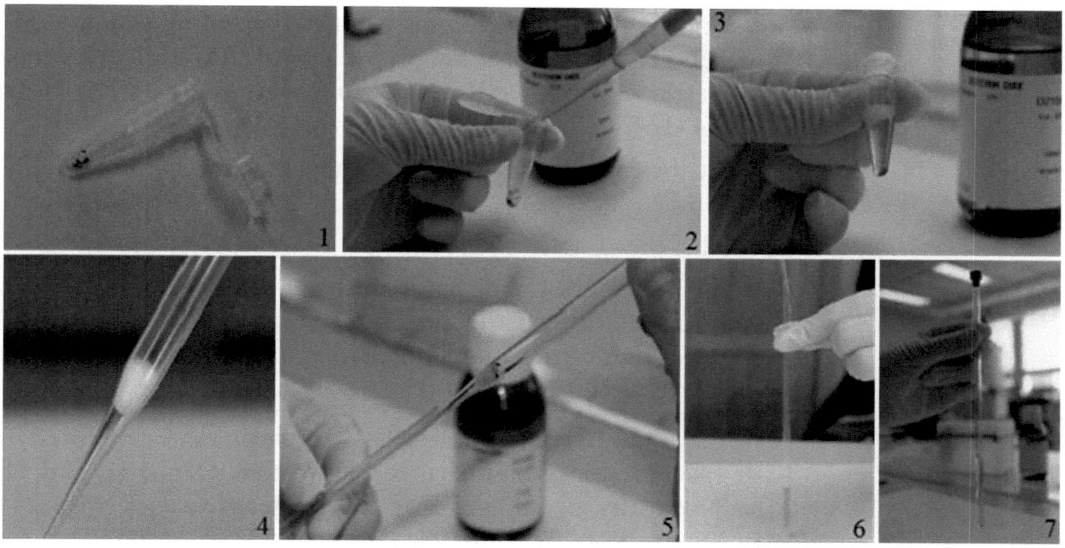

Fig. 7 NMR tube preparation steps

Phloroglucinol (1,3,5-trihydroxybenzene), the base unit of phlorotannins (phloroglucinol polymers) which are intensively investigated in brown algae, was identified as the major phenolic compound. ^{1}H HRMAS NMR spectra have been recorded for each sample. The intensity of the phloroglucinol signal throughout the months is highlighted in Fig. 6.

3.3 Analysis of an Algal Extract for Characterization of a Molecule of Interest

1. Place the dry extract in a small vial (Fig. 7, **step 1**).
2. Dissolve the extract in 700 μL of an appropriate NMR solvent (Fig. 7, **step 2**; *see* **Note 9**). The sample can be agitated, sonicated, or vortexed and even heated in order to achieve complete dissolution (*see* **Notes 10** and **11**).
3. If the sample contains any solid material (e.g., precipitates, dust) (Fig. 7, **step 3**), it must be filtered. Indeed, suspensions present in solution will prevent a correct shimming and thus reduce spectral resolution, i.e., increase the line width of the spectrum. A small amount of cotton wool inside a Pasteur pipette (Fig. 7, **step 4**) can be used to filter the extract (*see* **Note 12**).
4. Make sure that the NMR tube is clean inside and outside and is in good condition with no cracks or chips (*see* **Note 13**). The NMR tube must not be shorter than 16 cm.
5. After the dissolution of the extract, it can be transferred directly to a 5 mm NMR tube by passing the solution through the cotton filter (Fig. 7, **step 5**).

6. Finally, the sample volume can be adjusted by adding NMR solvent up to a final sample volume of 700 μL (or a sample height of 55 mm in the NMR tube) (Fig. 7, **step 6**; *see* **Notes 14–16**).
7. Put the cap on and shake to homogenize the solution (Fig. 7, **step 7**).
8. Label the sample (*see* **Note 17**).
9. The NMR tube can be put in the magnet for NMR analysis. 1D and 2D NMR experiments on many nuclei (e.g., ^{1}H, ^{13}C, ^{31}P, ^{15}N) can be carried out. Temperature of analysis can be adjusted between +100 and −100 °C.

Care must be taken with samples that contain water (*see* **Note 18**), salt (*see* **Note 19**), or exchangeable protons OH, NH, and SH (*see* **Note 20**), or samples that are pH sensitive (*see* **Note 21**), viscous (*see* **Note 22**), unstable (*see* **Note 23**), or analyzed quantitatively (*see* **Note 24**).

1D NMR analyses (^{1}H and ^{13}C) rarely allow detailed structural determination. This can be accomplished by means of two-dimensional NMR experiments which can be divided into two types, homonuclear and heteronuclear. Each type can provide information, either through bond (correlation spectroscopy [COSY], heteronuclear single-quantum correlation [HSQC], heteronuclear multiple-quantum correlation [HMQC], heteronuclear multiple-bond correlation [HMBC], total correlated spectroscopy [TOCSY]) or through space (nuclear overhauser effect spectroscopy [NOESY], rotating frame Overhauser effect spectroscopy [ROESY]) coupling information. COSY, HMQC/HSQC, and HMBC represent the leading 2D methods for molecular identification. If necessary, other 2D NMR analyses can be conducted to achieve complete characterization of molecules.

An example of conventional NMR analyses of an extract of the red macroalga *Grateloupia turuturu* is depicted in Fig. 8.

4 Notes

1. Most common NMR magnets are 400 and 500 MHz but proton resonance frequency in NMR ranged between 200 MHz and 1 GHz.
2. Common probes are $^{1}H/^{13}C/^{15}N$ and $^{1}H/^{31}P$.
3. In order to avoid occurrence of additional proton signals which may overlap with signals of interest, deuterated solvents must be used as deuterium is not observed in a spectrometer tuned to, e.g., protons or carbon. Moreover, deuterated solvents allow locking the magnetic field and so prevent its drift during the analysis which avoids possible signal degradation.

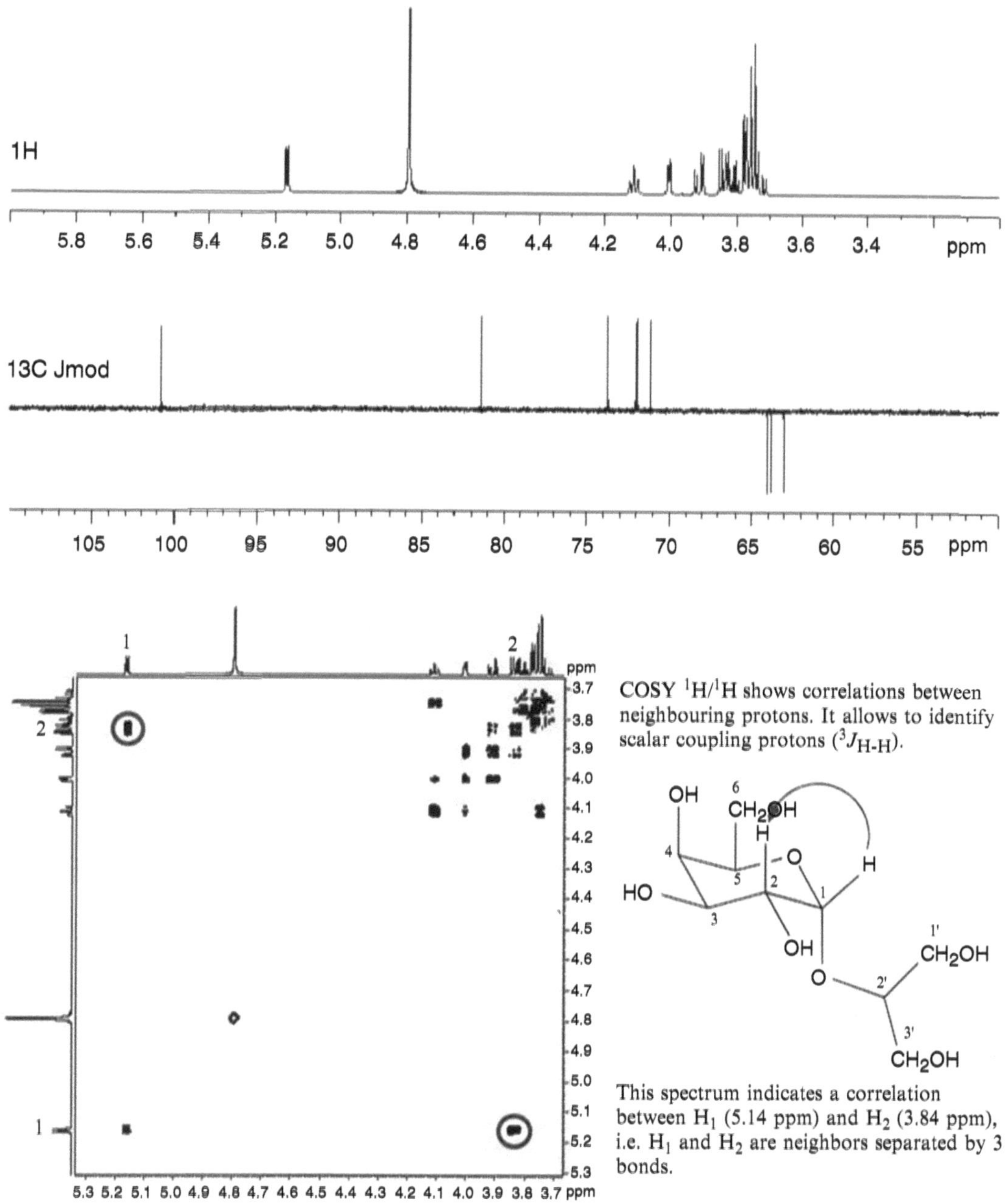

Fig. 8 500 MHz ^1H, ^{13}C Jmod, COSY $^1H/^1H$, HMQC $^1H/^{13}C$, and HMBC $^1H/^{13}C$ of floridoside isolated from the red macroalga *Grateloupia turuturu*

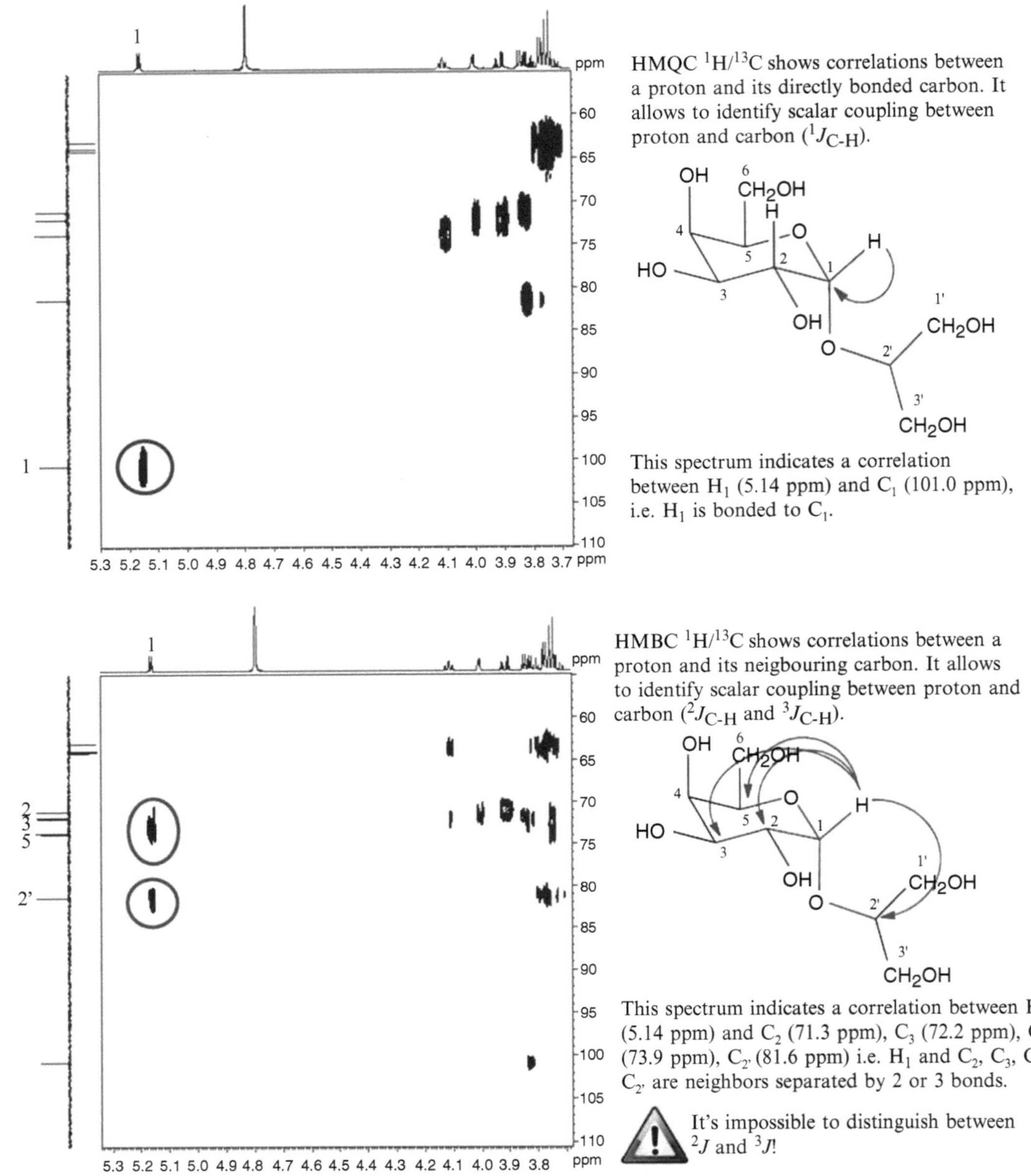

Fig. 8 (continued)

4. High-resolution NMR samples are routinely prepared and run in 5 mm NMR tubes.
5. The most common NMR solvent used is $CDCl_3$.
6. The presence of air bubbles may prevent the rotor rotation and thus lead to very poor resolution of the NMR spectrum.
7. A new and clean scalpel can be used to take a piece of algae but take care not to introduce metal into the sample.

8. Humidity can prevent the rotor rotation.
9. Some NMR solvents readily absorb moisture from the atmosphere and thus give water signal on the spectrum. Bottles of these solvents should thus be tightly capped when not in use (molecular sieve can be added in bottles and/or solvents can be stored under nitrogen or argon).
10. The concentration of the sample should be high enough to give a good signal-to-noise ratio on the spectrum, thus minimizing exchange effects found at low concentrations. The exact concentration of your sample in the lock solvent will depend on the sensitivity of the spectrometer used. If you have no guidelines for a specific spectrometer, use one drop of extract for liquids and one or two crystals for solid extracts.
11. Extremely viscous samples will have broad lines which is one of the reasons for not adding too much sample. Try to avoid preparing very viscous samples if possible.
12. Medical cotton is reported to be free of compounds that can be extracted by organic solvents [13].
13. NMR tubes that are dirty on the outside will result in poor spectra and must not be placed into the NMR instrument in order to avoid probe degradation.
14. Trying to concentrate your sample by reducing the sample volume is not recommended. It will cause difficulties for the NMR spectrometer to shim the sample. Furthermore, spectra recorded from samples which have been made up in a too small volume of deuterated solvent will almost always exhibit line broadening relative to a correctly prepared sample.
15. If only little material is available, specific NMR tubes can be used, such as 3 mm or Shigemi®, to limit the quantity of solvent; the use of cryoprobe will also improve the signal-to-noise ratio to overcome little quantity of sample.
16. Do not fill the NMR tube with solvent. This will dilute your sample and waste solvent, and make shimming more difficult.
17. Permanent marker can be used to write a reference on the top of the NMR tube or stick a label on the top of the NMR tube with the reference of the sample and the used solvent.
18. "Water suppression" allows the elimination of unwanted solvent resonances and the conservation of all others. It is commonly employed in the study of biological molecules where the water signal often dominates all others.
19. Extracts of marine tissues contain salt in variable quantities which causes difficulties in NMR observation even to the point of preventing any signal detection. There are nevertheless several possibilities to obtain spectra of saline samples, such as a preliminary desalination of the sample, the reduction of sample

concentration, or the replacement of the 5 mm NMR tube by, e.g., a 3 mm NMR tube in order to reduce the volume of the analyzed sample and thus the quantity of salt.

20. When compounds containing OH, NH, or SH protons are placed in protonated solvents (D_2O, MeOD), these hydrogens often exchange with deuterium spontaneously. Because of this, the signal of these protons "disappears" from the spectrum. By contrast, solvents like DMSO and acetone usually lead to discrete OH, NH, or SH signals. In many samples, OH, NH, and SH protons can be recognized from their characteristic chemical shifts or broadened appearance. Their chemical shifts vary over a wide range depending on experimental conditions.
21. Samples, especially tissue extracts, contain pH-sensitive metabolites (amino acids and other classical compounds of metabolism). Such molecules are particularly sensitive to cellular pH variation which tends to complicate NMR spectra interpretation, because chemical shifts and signal shape can be greatly modified according to pH conditions.
22. NMR spectra of viscous samples (e.g., polymers, polysaccharides) show often large signals at room temperature due to poor resolution. Sample can be diluted in order to improve resolution. If the resolution is still not improved or if the detection limit is reached, a spectrum can be recorded at a higher temperature to observe better resolved signals. Of course, the probe temperature must not come close to the boiling point of the solvent used.
23. There are two possibilities to treat an unstable sample: (a) record the spectra at low temperature especially if the analysis is likely to take long (for example 2D characterization); and (b) quickly analyze the sample at room temperature using a cryoprobe. Moreover, some biological samples undergo microbial degradation in aqueous solutions. To overcome this problem, sodium azide can be added as biocide, except for ^{15}N NMR analyses.
24. NMR quantification requires the addition of an internal or external reference to the sample. When acquisition has ended, the spectrum is carefully integrated and integral(s) of the molecule to quantify and integral(s) of the reference compound of known concentration are compared to calculate the concentration of the molecule studied. This method cannot be used when signals overlap: integrated signals must be isolated. Results are obtained quickly with a precision close to that obtained by HPLC post-column peak integration.

References

1. Urban S, Daniel A (2013) NMR spectroscopy: structure elucidation of cycloelatanene A: a natural product case study. Methods Mol Biol 1055:99–116
2. Abou-El-Wafa GSE, Shaaban M, Shaaban KA et al (2013) Pachydictyols B and C: new diterpenes from *Dictyota dichotoma* Hudson. Mar Drugs 11:3109–3123
3. Bilan MI, Grachev AA, Shashkov AS et al (2013) Preliminary investigation of a highly sulfated galactofucan fraction isolated from the brown alga *Sargassum polycystum*. Carbohydr Res 377:48–57
4. Ioannou E, Quesada A, Rahman MM et al (2012) Structures and antibacterial activities of minor Dolabellanes from the brown alga *Dilophus spiralis*. Eur J Org Chem 27: 5177–5186
5. Kwon T-H, Kim T-W, Kim C-G et al (2013) Antioxidant activity of various solvent fractions from edible brown alga, *Eisenia bicyclis* and its active compounds. J Food Sci 78: C679–C684
6. Simon-Colin C, Kervarec N, Pichon R et al (2004) NMR ^{13}C-isotopic enrichment experiments to study carbon-partitioning into organic solutes in the red alga *Grateloupia doryphora*. Plant Physiol Biochem 42:21–26
7. Bondu S, Cerantola S, Kervarec N et al (2009) Impact of the salt stress on the photosynthetic carbon flux and ^{13}C-label distribution within floridoside and digeneaside in *Solieria chordalis*. Phytochemistry 7052:173–184
8. Patel B, Hellgardt K (2013) Hydrothermal upgrading of algae paste: application of ^{31}P NMR. Environ Prog Sustain Energ 32: 1002–1012
9. Beal CM, Webber ME, Ruoff RS et al (2010) Lipid analysis of *Neochloris oleoabundans* by liquid state NMR. Biotechnol Bioeng 106:573–583
10. Le Lann K, Kervarec N, Payri CE et al (2008) Discrimination of allied species within genus *Turbinaria* (Fucales, Phaeophyceae) using HRMAS NMR spectroscopy. Talanta 74: 1079–1083
11. Bondu S, Kervarec N, Deslandes E et al (2007) The use of HRMAS NMR spectroscopy to study the in vivo intra-cellular carbon/nitrogen ration of *Solieria chordalis*. J Appl Phycol 20:223–229
12. Jégou C (2011) The genus *Cystoseira* from the coasts of Brittany: taxonomy, ecology and natural products. PhD thesis, University of Western Brittany, Brest
13. Derome AE (1987) Modern NMR techniques for chemistry research. Pergamon, Oxford, UK

Chapter 13

Extraction, Purification, and NMR Analysis of Terpenes from Brown Algae

Marc Gaysinski, Annick Ortalo-Magné, Olivier P. Thomas, and Gérald Culioli

Abstract

Algal terpenes constitute a wide and well-documented group of marine natural products with structures differing from their terrestrial plant biosynthetic analogues. Amongst macroalgae, brown seaweeds are considered as one of the richest source of biologically and ecologically relevant terpenoids. These metabolites, mostly encountered in algae of the class Phaeophyceae, are mainly diterpenes and meroditerpenes (metabolites of mixed biogenesis characterized by a toluquinol or a toluquinone nucleus linked to a diterpene moiety).

In this chapter, we describe analytical processes commonly employed for the isolation and structural characterization of the main terpenoid constituents obtained from organic extracts of brown algae. The successive steps include (1) extraction of lipidic content from algal samples; (2) purification of terpenes by column chromatography and semi-preparative high-performance liquid chromatography; and (3) structure elucidation of the isolated terpenes by means of 1D and 2D nuclear magnetic resonance (NMR). More precisely, we propose a representative methodology which allows the isolation and structural determination of the monocyclic meroditerpene methoxybifurcarenone (MBFC) from the Mediterranean brown alga *Cystoseira amentacea* var. *stricta*. This methodology has a large field of applications and can then be extended to terpenes isolated from other species of the family Sargassaceae.

Key words 1D and 2D NMR, Chromatographic techniques, *Cystoseira amentacea* var. *stricta*, Meroditerpenes, Phaeophyceae, Terpenes

1 Introduction

Phaeophyceae (brown algae) constitute one of the classes of marine organisms first studied by natural product chemists for the determination of their phytochemical composition [1]. These algae are widely known for their propensity to produce hydrocolloids and common lipids, but also have the ability to biosynthesize a large array of secondary metabolites, mainly terpenoids and phenolic derivatives. As shown in Fig. 1, terpenoids isolated from species of the class Phaeophyceae account for almost 40 % of all algal

Dagmar B. Stengel and Solène Connan (eds.), *Natural Products From Marine Algae: Methods and Protocols*, Methods in Molecular Biology, vol. 1308, DOI 10.1007/978-1-4939-2684-8_13, © Springer Science+Business Media New York 2015

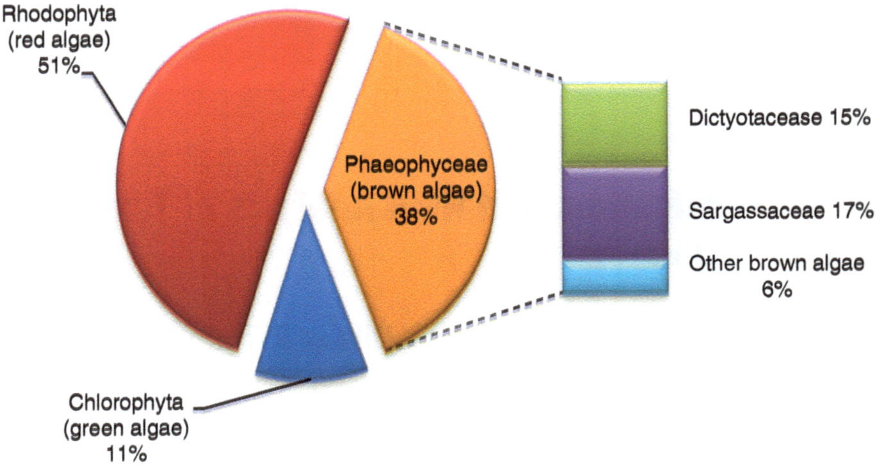

Fig. 1 Taxonomic distribution of secondary metabolites produced by macroalgae

metabolites described to date [2]. The main producers belong to the families Sargassaceae—a taxonomic group composed of the recent merger of the previous Sargassaceae and Cystoseiraceae families [3]—and Dictyotaceae, from which more than 80 % of all the compounds described from brown algae have been isolated.

From a chemical point of view, sargassacean species biosynthesize mainly merediterpenes and linear diterpenes, whereas most compounds described from Dictyotacean species are cyclic diterpenoids, sesquiterpenes, and various types of meroterpenes. In Dictyotaceae, a majority of the terpenes are described from species of the genus *Dictyota* (Fig. 2a), these compounds being exclusively cyclic diterpenes which could be classified into three main chemical groups: dolabellanes, prenylated guaianes, and xenicanes [4, 5] (Fig. 3). In the same family, *Dictyopteris* and *Taonia* spp. are known to produce sesquiterpenes and meroterpenes, whereas species of the genus *Stypopodium* yield meroditerpenes. The family Sargassaceae is composed by several genera being the most prolific in terms of terpene yield. In species belonging to the genera *Cystoseira*, *Sargassum*, and *Halidrys*, meroditerpenoids constitute the most common metabolites [6, 7], while linear diterpenoids are predominant in the genus *Bifurcaria* [8], and various irregular isoprenoid derivatives are found in *Cystophora* spp. (Figs. 2b and 3).

More precisely, in the genus *Cystoseira*, meroditerpenoids could be classified into four main groups depending on the structure of their diterpene side chain: linear, monocyclic, bicyclic, or rearranged. All these compounds are derived from a common linear biosynthetic precursor, geranylgeranyltoluquinol, which could lead to a series of oxidized linear analogues by specific oxidations occurring predominantly at the C-5, C-12, and C-15 positions. Some of these activated linear meroditerpenes may be subjected to

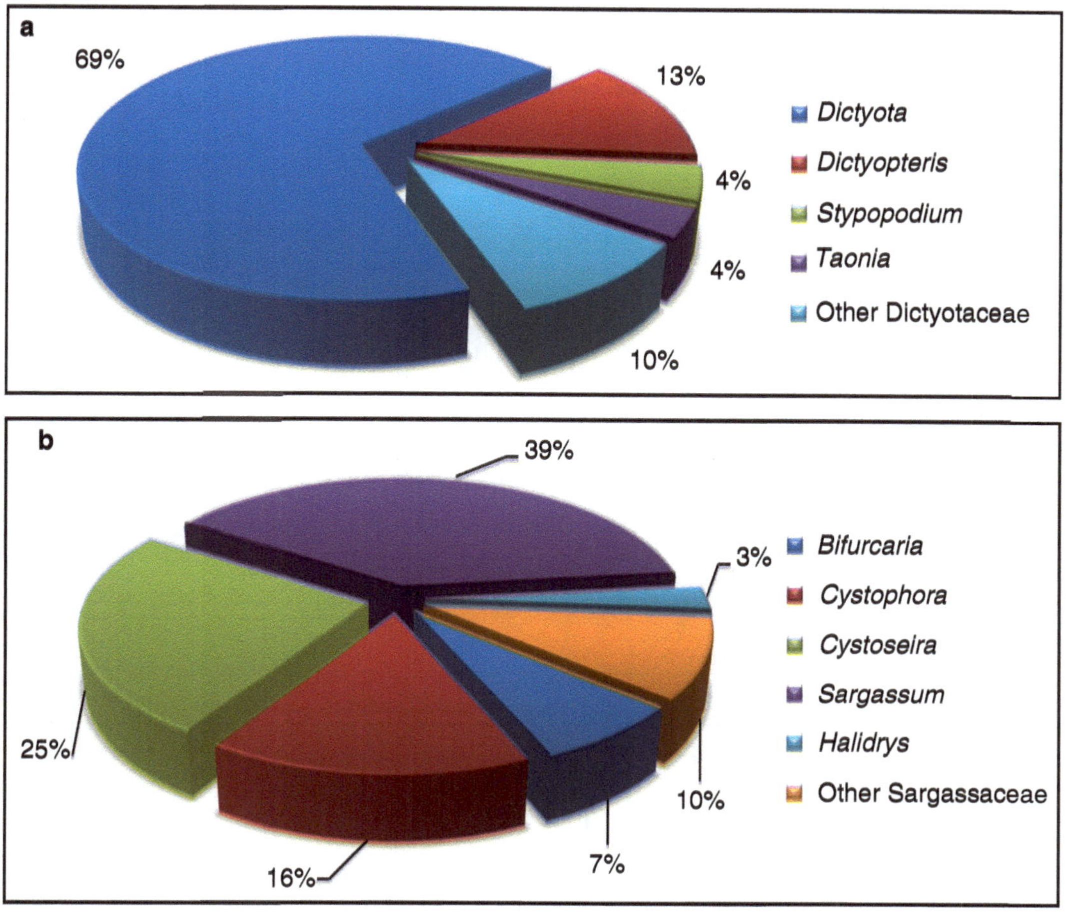

Fig. 2 Taxonomic distribution of natural products in Dictyotaceae (**a**) and Sargassaceae (**b**)

cyclization, almost exclusively between C-7 and C-11, and their resulting monocyclic congeners then play a significant role in the biosynthesis of more complex *Cystoseira* metabolites [6, 7, 20] (Fig. 4).

Brown algae have adapted to a wide variety of marine ecological niches and their secondary metabolites may have ecological roles that are still poorly documented, especially for terpenoids. However, some of these compounds produced by several Dictyotaceae and Sargassaceae species have proved to exhibit antifouling effects against a great variety of colonizing organisms [9, 21–24]. In other ecological studies, the anti-herbivore activity of some terpenes has been demonstrated [25–28].

This chapter focuses on the Mediterranean brown alga *Cystoseira amentacea* var. *stricta* which has been widely studied for its ability to biosynthesize not only monocyclic but also a wide range of bicyclic and rearranged meroditerpenes. In particular, its organic extracts contain high amounts of methoxybifurcarenone

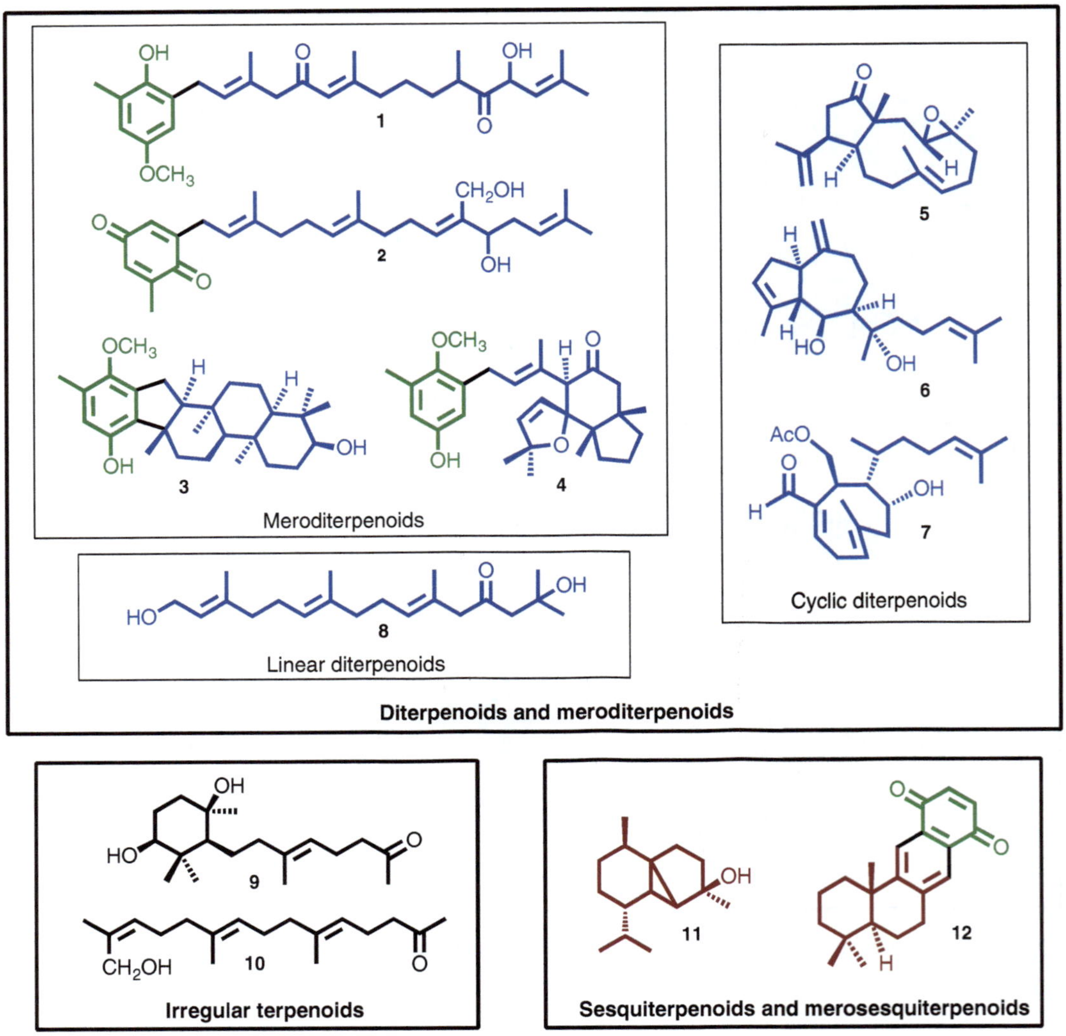

Fig. 3 Selected examples of common terpenes isolated from brown algae: *1* was isolated from *Halidrys siliquosa* [9], *2* from *Sargassum fallax* [10], *3* from *Stypopodium flabelliforme* [11], *4* from *Cystoseira* sp. [12], *5* from *Dictyota dichotoma* [13] and *Dictyota spiralis* [14], *6* from *D. dichotoma* and several other Dictyotaceae [13], *7* from *D. dichotoma* [15], *8* from *Bifurcaria bifurcata* [16], *9* and *10* from *Cystophora moniliformis* [17], *11* from *Taonia atomaria* [18], and *12* from *Dictyopteris undulata* [19]

(MBFC, Fig. 5), a meroterpene bearing a monocyclic side chain, which is considered as a key intermediate in metabolic pathways leading to structurally more complex meroditerpenoids [29]. This compound has been previously described for its potent antifungal, antibacterial, and enzymatic inhibitory activities, and for its toxicity against larvae of *Artemia salina* [30, 31]. It should be mentioned that its demethyl analogue, bifurcarenone isolated from the brown alga *Bifurcaria galapagensis*, was found to exhibit antibacterial activity and cytotoxicity against fertilized sea urchin eggs [32].

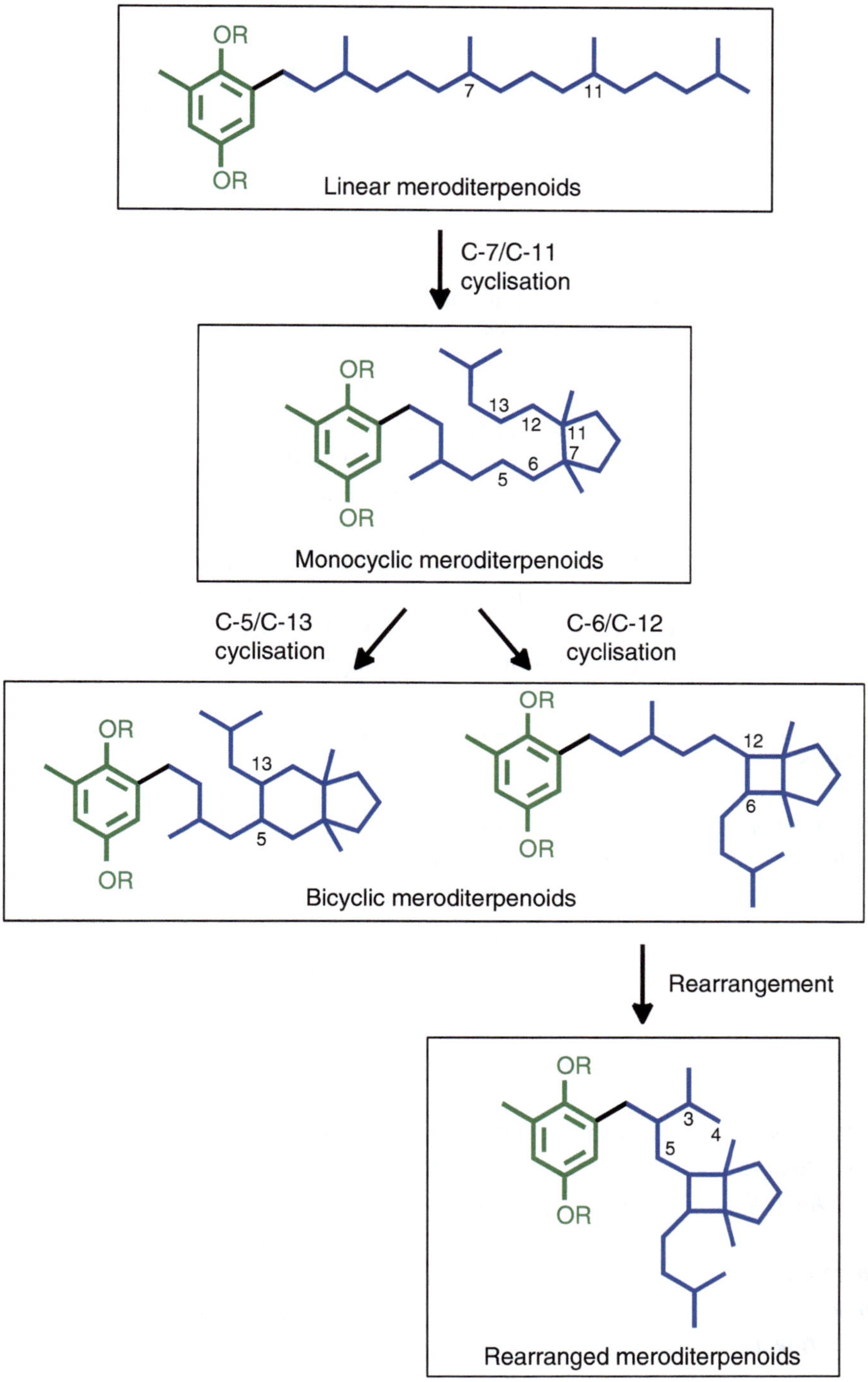

Fig. 4 Chemical classification of meroditerpenes isolated from *Cystoseira* spp. (adapted from ref. 6)

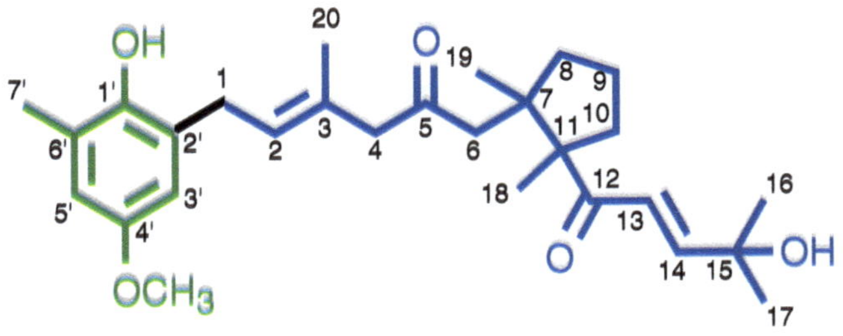

Fig. 5 Chemical structure of methoxybifurcarenone (MBFC) isolated from the brown alga *Cystoseira amentacea* var. *stricta*

In this chapter, we provide detailed protocols for the efficient extraction and isolation of MBFC from the lipidic extracts of *C. amentacea* var. *stricta* collected off the French Mediterranean coast. The NMR study implemented to elucidate the chemical structure of MBFC is also fully developed. This methodology could be easily generalized for all members of this chemical family which can be isolated from other species of Sargassaceae, and more largely from other brown algae.

2 Materials

2.1 Algal Treatment and Extraction

1. Brown alga *Cystoseira amentacea* var. *stricta* collected from the Mediterranean Sea (*see* **Note 1**), air-dried (oven), or stored at −20 °C before being freeze-dried.
2. Ceramic mortar with a pestle or commercial Waring blender.
3. Dichloromethane and methanol of analytical grade.
4. Ultrasonic bath.
5. Blank filter paper and glass funnel.

2.2 Fractionation by Column Chromatography (CC), Analysis by Thin-Layer Chromatography, and Purification by High-Performance Liquid Chromatography

1. Ethyl acetate, isooctane, methanol, and sulfuric acid of analytical grade, acetonitrile of high-performance liquid chromatography (HPLC) grade, and MilliQ water.
2. For CC fractionation, Si gel 60 (particle size 63–200 μm), glass column with a coarse frit (*see* **Note 2**).
3. Several isooctane/ethyl acetate and then ethyl acetate/methanol solutions (500 mL of each) of increasing polarity should be prepared for CC elution, starting from 20 % ethyl acetate in isooctane up to 100 % ethyl acetate (by increasing 20 % ethyl acetate at each step), and then from 100 % ethyl acetate up to 100 % methanol (by increasing 20 % methanol at each step). Almost 50 mL of elute should be collected in each round flask (100 mL).

4. For thin-layer chromatography (TLC), pre-coated TLC aluminum sheets Si60 F254 (20 × 10 cm), using isooctane/ethyl acetate (3:2, v/v) as eluting solvent, and an adapted TLC development chamber are used. TLC revelation is performed by treatment with a solution of sulfuric acid (1 M) followed by gradual heating on a heating plate.
5. Final purification is performed using HPLC, e.g., using a Varian HPLC system, with refractometric detection, and Rheodyne syringe loading injector equipped with a 20 µL sample loop, semi-preparative VWR column, PurospherStar C18 5 µm (250 × 10 mm), flow rate 3.0 mL/min using a mixture of acetonitrile in water as eluent (4:1, v/v).

2.3 Structural Analysis by NMR

1. 5 mm NMR sample tube for 500 MHz magnetic field.
2. Deuterated chloroform ($CDCl_3$, 99.96 atom % D) in sealed glass ampoule.
3. NMR spectrometer, Bruker Avance DRX 500 MHz: All experiments are run at 298 K with a direct probe head (5 mm $^{13}C/^{1}H$ Z-Grd) (*see* **Note 3**).
4. All NMR experiments are carried out using pulse sequences supplied by the spectrometer manufacturer (Bruker, TopSpin 2.1). Spectrum calibration is performed by using deuterated solvent as internal reference ($CDCl_3$: 7.26 ppm for ^{1}H NMR and 77.16 ppm for ^{13}C-NMR) (*see* **Note 4**).

3 Methods

3.1 Algal Treatment and Extraction

1. Just after collection, clean the algal material from epiphytes and immediately (1) squeeze and dry the seaweed in air on paper filters (or in a ventilated oven at 30 °C), or (2) freeze it at −20 °C and freeze-dry.
2. Grind coarsely the dried samples.
3. Extract the algal powder three times with dichloromethane/methanol (1:1, v/v) at room temperature (3 × 500 mL) in an ultrasonic bath (20 min for each extraction). Pool the extracts, filter them on a paper filter (through the funnel), and evaporate them under vacuum (*see* **Note 5**).

3.2 Fractionation by Column Chromatography (CC), Analysis by Thin-Layer Chromatography, and Purification by High-Performance Liquid Chromatography

1. Fractionate the resulting crude extract through CC (Si60 gel) using isooctane/ethyl acetate/methanol gradient elution of increasing polarity to afford several fractions, some of them containing terpenes (*see* **Note 6**).
2. Elute fractions with mixtures from 40 to 80 % of ethyl acetate in isooctane, which are those containing the main terpenoid constituents. Analyze those fractions by TLC and select fractions on the basis of the retardation factor (R_f) of their chemical constituents (*see* **Note 7**).

3. Further purification of the selected fractions is achieved by semi-preparative RP-18 HPLC: Dissolve the selected fractions in methanol and inject them in the HPLC system. Repetitive injections allow the purification of methoxybifurcarenone (MBFC) in sufficient amount (*see* **Note 8**).

3.3 NMR Analysis

3.3.1 Sample Preparation

1. Filter the HPLC subfraction containing MBFC and remove solvents under vacuum (*see* **Note 9**).
2. Dissolve 15 mg of MBFC in deuterated chloroform and transfer in the NMR tube (*see* **Note 10**).

3.3.2 NMR Experiments (See **Note 11**)

1. Acquire ^{1}H NMR spectrum (*see* **Note 12**) using the following parameters: Spectral width (SW): 6.3 kHz, center of SW (O1): 2,785 Hz, number of complex data points: 64 k, acquisition time (aq): 5.3 s, relaxation delay (D1): 2 s, number of scan (ns): 120, number of dummy scan (ds): 4, and a 30° flip angle pulse width.
2. Acquire $^{13}C\{^{1}H\}$ NMR spectrum (*see* **Note 13**) with the following settings: SW: 30 kHz, O1: 12,625 Hz, number of complex data points: 64 k, aq: 0.87 s, D1: 2 s, ns: 10,000, and a 30° flip angle pulse width. ^{1}H decoupling is achieved using Waltz 16 pulse sequence. Prior to Fourier transformation, the fids are multiplied by an exponential line broadening function of 1 Hz.
3. Acquire DEPT 135 and DEPT 90 NMR spectra (*see* **Note 14**) using the following: SW: 21.9 kHz, O1: 10,359 Hz, number of complex data points: 64 k, aq: 1.09 s, D1: 2 s, ns: 5,000 and a 90° flip angle pulse width. ^{1}H decoupling is achieved using Waltz 16 pulse sequence. Prior to Fourier transformation, the fids are multiplied by an exponential line broadening function of 1 Hz.
4. Acquire ^{1}H-^{1}H gs-COSY spectrum (*see* **Note 15**) with the following parameters: SW: 6.3 kHz in both dimensions, O1: 2,785 Hz in both dimensions, number of complex data points in F2: 2 k, number of t1 increments: 256 (ten scans by increment), aq: 0.16 s, D1: 1.5 s. Prior to Fourier transformation, data are zero filled in F1.
5. Obtain phase-sensitive gs-HSQC (echo-antiecho mode) spectrum (*see* **Note 16**) with SW in F1: 6.3 kHz, O1: 2,785 Hz, number of complex data points in F2: 1 k, SW in F2: 21.3 kHz, O2: 10,060 Hz, and number of t1 increments: 256 (48 scans by increment). Other main parameters are aq: 0.08 s, D1: 1.4 s, and J_1 (H–C): 145 Hz. Prior to Fourier transformation, a QSINE window function (SSB = 2) is applied in both dimensions and the data are zero filled and predicted linearly (NC = 32) to 1 k data points in F1.

6. Acquire gs-HMBC spectrum (*see* **Note 17**) with the following settings: SW in F1: 6.3 kHz, O1: 2,785 Hz, number of complex data points in F2: 1 k, SW in F2: 28.17 kHz, and number of t1 increments: 256 (80 scans by increment). Other main parameters are the following: aq: 0.32 s D1: 1.5 s, and mixing time: 62.5 ms (optimized delay for a long-range coupling of $J=8$ Hz). Prior to Fourier transformation, a SINE window function (SSB = 0) is applied in both dimensions and the data are zero filled and predicted linearly (NC = 32) to 1 k data points in F1.
7. Acquire phase-sensitive ^{1}H-^{1}H gs-NOESY (States-TPPI mode) experiment (*see* **Note 18**) using SW in F1: 6.3 kHz, O1: 2,785 Hz in both dimensions, number of complex data points in F2: 2 k, number of t1 increments in F1: 256 (32 scans by increment), aq: 0.16 s, D1: 2 s, and mixing time (D8): 0.6 s. Prior to Fourier transformation, a QSINE window function (SSB = 2) is applied in both dimensions and the data are zero filled and predicted linearly (NC = 32) to 1 k data points in F1.

3.3.3 NMR Data Analysis

1D and 2D NMR spectra are used to determine the chemical structure of MBFC and, more generally, those of similar algal meroditerpenes.

1. Usually, the terpene nature of the isolated compounds is easily defined thanks to characteristic ^{1}H NMR isoprene signals such as singlet methyls observed from 0.7 to 2.0 ppm and/or broad triplet olefinic methines between 5.0 and 6.0 ppm. In the case of MBFC, five methyl groups are revealed by singlet signals at δ_H 1.16, 1.17, 1.35, 1.36, and 1.72, whereas its spectrum shows only one signal for a trisubstituted double bond at δ_H 5.31 (t, $J=6.5$ Hz, H-2). These data are in agreement with the occurrence of a diterpene chain which is modified on three out of its four isoprene units. If the isolated compound is a meroterpene, typical signals for protons of the toluquinol (or toluquinone) moiety are detected as (1) doublets between 6.0 and 7.0 ppm for methines (H-3′ and H-5′), (2) a singlet from 2.0 to 2.5 ppm for the methyl group (H_3-7′), and, in some case, (3) singlet(s) between 3.4 and 4.0 for methoxyl group(s). Such signals are observed on the ^{1}H NMR spectrum of MBFC at δ_H 6.58 (d, $J=3.0$ Hz, H-5′), 6.53 (d, $J=3.0$ Hz, H-3′), 2.26 (s, H_3-7′), and 3.74 (s, CH_3O). These data allow ascertaining that MBFC is a meroditerpene and suggest that a methoxyl and a free phenol groups are present in the toluquinol. Other ^{1}H NMR signals on the spectrum of MBFC are identified as two scalar-coupled olefinic protons at δ_H 6.85 (d, $J=15.0$ Hz, H-14) and 6.63 (d, $J=15.0$ Hz, H-13), two deshielded sp^3 methylenes at δ_H 3.34 (t, $J=6.5$ Hz, H_2-1) and 3.03 (s, H_2-4), and non-equivalent methylene and/or methine protons with complex coupling patterns (Table 1).

Table 1
NMR data ($CDCl_3$, 500 MHz) for MBFC

No.	δ_C	Mult.	δ_H (*J* in Hz)	No.	δ_C	Mult.	δ_H (*J* in Hz)
1	31.0	CH_2	3.34 *t* (6.5)	15	71.3	C	–
2	127.8	CH	5.31 *t* (6.5)	16	29.6	CH_3	1.36 *s*
3	131.1	C	–	17	29.6	CH_3	1.35 *s*
4	55.9	CH_2	3.03 *s*	18	20.4	CH_3	1.17 *s*
5	209.2	C	–	19	21.4	CH_3	1.16 *s*
6	47.6	CH_2	a: 2.42 *d* (15.5) b: 2.25 *d* (15.5)	20	16.9	CH_3	1.72 *s*
7	47.0	C	–	1′	146.7	C	–
8	37.1	CH_2	a: 1.91 *m* b: 1.71 *m*	2′	128.1	C	–
9	20.2	CH_2	1.72 *m*	3′	113.2	CH	6.53 *d* (3.0)
10	34.5	CH_2	a: 2.32 *m* b: 1.53 *m*	4′	153.3	C	–
11	60.1	C	–	5′	114.2	CH	6.58 *d* (3.0)
12	204.8	C	–	6′	126.4	C	–
13	122.8	CH	6.63 *d* (15.0)	7′	16.7	CH_3	2.26 *s*
14	152.8	CH	6.85 *d* (15.0)	OCH_3	55.8	CH_3	3.74 *s*

δC ^{13}C NMR chemical shifts, *Mult* multiplicity of carbon atoms, *δH* ^{1}H NMR chemical shifts (multiplicity of protons: *s* for singlet, *d* for doublet, *t* for triplet, and *m* for multiplet), *J* coupling constants

2. ^{13}C and DEPT spectra allow assessment of the number and multiplicity of the carbon atoms of the studied compound. Sesquiterpenoids and diterpenoids typically include 15 and 20 carbon atoms, respectively. This number could be supplemented by seven carbon atoms if a toluquinol part is also present, and by one or two additional carbon atoms for each subsequent methoxyl or acetate group, respectively (*see* **Note 19**). HSQC spectrum allows the unambiguous assignment of carbon resonances to their directly attached protons (Table 1). For MBFC, 28 signals are observed in the ^{13}C NMR spectrum and further characterized by DEPT experiments as seven methyls, six sp^3 methylenes, five olefinic methines, and ten quaternary carbons (which include five olefinic, two carbonyl, and three sp^3 carbon atoms). These data confirm the presence of the toluquinol moiety, and therefore the meroditerpenoid nature of MBFC, but also demonstrate the occurrence of two ketone carbonyls at δ_C 204.8 (C-12) and 209.2 (C-5).

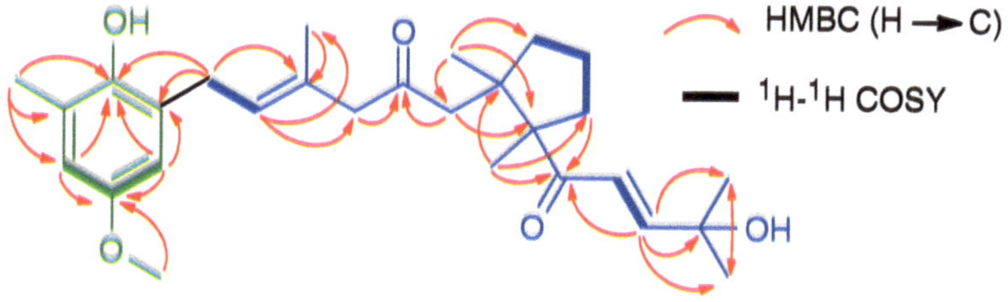

Fig. 6 Key ^{1}H–^{1}H COSY and HMBC correlations observed for MBFC

3. With a full analysis of ^{1}H-^{1}H COSY and HMBC spectra, the spin-coupled systems are sequentially connected (Fig. 6), thus enabling complete assignments of NMR signals and the elucidation of the chemical structure of MBFC (*see* **Note 20**).
4. ^{1}H-^{1}H NOESY spectrum evidences spatial coupling and is then used to complete the structural characterization of terpenoids defining the *E/Z* configuration of double bonds and the relative configuration of potential stereocenters. These data also contribute to attribute ambiguous ^{1}H signals and to assert the localization of some chemical groups. The configuration of the Δ^1 double bond is assigned as *E* on the basis of correlations between H_3-20/H_2-1 and H-2/H_2-4 on the ^{1}H-^{1}H NOESY spectrum of MBFC but also due to the upfield ^{13}C NMR signal of the methyl carbon of the isoprene unit at δ_C 16.4 (CH_3-20) (*see* **Note 21**) [33, 34]. The *E* configuration of the Δ^{13} double bond is deduced from the value of the $^3J_{HH}$ coupling constant (15.0 Hz). The relative configuration of the two CH_3-18 and CH_3-19 bridgehead methyls is assigned as *cis* by comparison with other structurally similar compounds [29, 32, 35–37].
5. These meroditerpenoids are subject to transformation into the C-3 epimeric mixtures of their corresponding chromenes and chromanes when subjected to acidic conditions (Fig. 7). Such degradation is observed for MBFC within a few hours in the NMR tube filled with $CDCl_3$ (see **Note 22**).

4 Notes

1. Due to their great ecological interest and their scarcity, all species of the genus *Cystoseira* (except *C. compressa*) are now protected as amended by the Convention for the Protection of the Marine Environment and the Coastal Region of the Mediterranean (Barcelona Convention, 1992).
2. A minimum ratio of 40:1 (w/w) stationary phase/crude extract should be used. An increase of this ratio (e.g., 100:1) would allow a better separation of the phytochemical constituents. The size of the glass chromatographic column should be

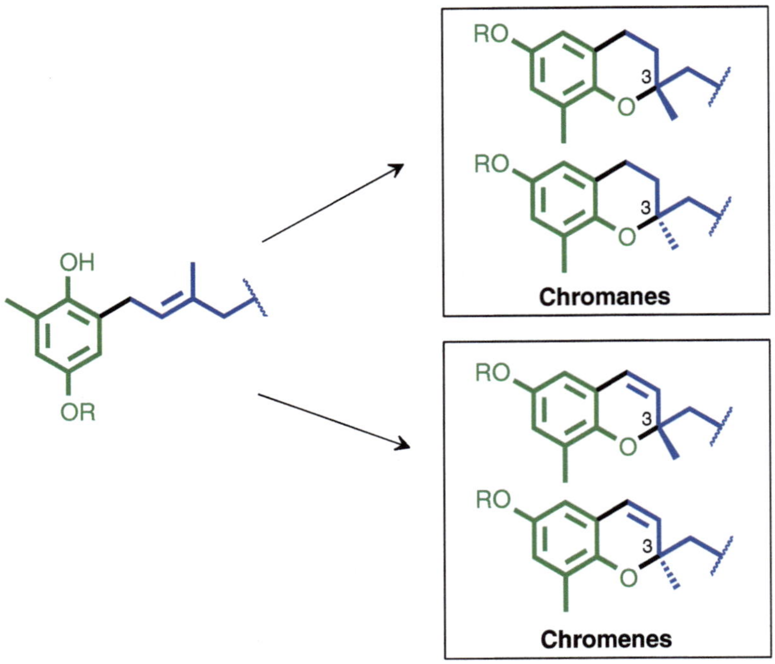

Fig. 7 Degradation of meroterpenoids into the C-3 epimeric mixtures of their chromane and chromene derivatives

selected depending on the quantities of stationary phase involved in the separation process.

3. In order to increase sensitivity, different probes could be used: direct probe head for ^{13}C NMR spectra and inverse probe head for 1D and 2D NMR experiments, such as ^{1}H, ^{1}H-^{1}H COSY, HSQC, HMBC, or ^{1}H-^{1}H NOESY.

4. Usually zero calibration is done with tetramethylsilane (TMS). But, in the case of small amounts of analyte, it is recommended to conduct the procedure without TMS in order to obtain an optimized gain value. At the end of the analysis, TMS could be added to set the zero of the chemical shift scale.

5. For these algae and using this method, extraction yield could fluctuate depending on collection site, season, and species, but common values range from 4 to 8 % (w/w).

6. Dry loading, a common technique in chromatography is used when most of the components of the crude extract are sparingly soluble in the first eluting mixture (in this case, isooctane/ethyl acetate, 4:1, v/v). This step is performed by total solubilisation of the crude extract in the solvent extraction mixture (dichloromethane/methanol, 1:1, v/v), addition of stationary phase (Si60, in the ratio stationary phase/crude extract 1:1, w/w), and removal of the solvents under reduced pressure. The obtained dry sample is loaded on the top of the glass chromatographic column.

7. With such chromatographic conditions, carotenes, chlorophylls, sterols, and xanthophylls are eluted with R_f of 0.9-0.8, 0.7-0.6, 0.5, and 0.3, respectively. Sterols are unambiguously identified by their pinkish coloring after the TLC plate being treated with sulfuric acid and heated. The main terpenes, including MBFC, are eluted with R_f from 0.5 to 0.3.
8. The concentration of the sample should be cautiously adjusted in order to be sufficient to limit the number of injections but not too high to avoid column saturation. Commonly, concentration values between 1 and 20 mg/mL might be acceptable. As in the case of the crude extracts, extraction yield of MBFC from samples of *C. amentacea* var. *stricta* could drastically fluctuate depending on the specific characteristics of the sample (site, season, preservation mode, etc.) but also, due to the degradability of MBFC in acidic conditions, extraction time and length of the separation processes. Typical extraction yields of MBFC could range from 0.1 to 1 % (w/w) of the algal sample.
9. A sample with suspended solid particles will cause broadening of the NMR signals and indistinct spectra. One of the reasons that may explain this phenomenon is the poor magnetic field homogeneity of the sample cause by difference of magnetic susceptibilities between solid and solution.
10. The appropriate quantity of compound required for the acquisition of good NMR spectra is about 5–20 mg. At lower concentrations, spectra will be dominated by signals of common contaminants.
11. In order to obtain the best results, sample preparation (quality of NMR tube and deuterated solvent, solution homogeneity, etc.), field homogeneity (shimming could be done manually or automatically), sample temperature, adjustment of the probe (tuning and matching), and determination of the correct radiofrequency pulse (P90, 90° pulse length for ^1H) require special attention.
12. A first ^1H NMR spectrum allows the characterization of the different sets of protons of the molecule and also the optimization of three crucial parameters for the following 2D NMR experiments (SW, O1, and P90). The best way to proceed is firstly to determine SW and O1, then to measure P90, and finally to acquire a ^1H spectrum with optimized values for these parameters.
13. This spectrum allows the identification of the carbon atoms within the molecule. A larger SW value should be chosen to be sure to see all signals. By the way, with this experiment, two important parameters, SW and O1, can be optimized for further 2D NMR (HMBC). If only small quantities of analyte are available or in the case of freely soluble samples, the use of Shigemi or capillary NMR tubes might increase the sensitivity.

Information given by this experiment can also be indirectly obtained via F1 projection of some 2D NMR data (HSQC and HMBC).

14. When an unknown sample is analyzed, both DEPT 135 and DEPT 90 experiments may be performed. Two useful parameters (SW and O1) for 2D NMR (HSQC) can be optimized. For both analytical procedures, the scan number used to acquire the ^{13}C spectrum can be divided by 2.

15. For this analysis, P90 ^{1}H pulse length should be used. If informative COSY cross peaks are likely to be found close to the diagonal, the alternative ^{1}H-^{1}H gs-COSY 45 experiment can be used.

16. P90 ^{1}H pulse length should be used for the acquisition of this spectrum. The advantage of this experiment is to have a DEPT 135 in the 2D map correlation.

17. This analysis needs the use of P90 ^{1}H pulse length. Various mixing times (coupling constant) should be used in order to detect all the correlations. The initial value of this parameter can be set at 62.5 ms (8 Hz). If some correlations are missing, higher (100 ms, 5 Hz) or lower (50 ms, 10 Hz) values of mixing time might be tried. If only small quantities of analyte are available, it can be useful to reduce the number of t1 increments in F1 dimension (TD1) to 128 and increase the scan number (ns).

18. For this analysis, P90 ^{1}H pulse length should be used. For this kind of molecule (MW < 1,000 Da), a first value of mixing time (D8) can be fixed at 600 ms (8.3 Hz). If none or low correlations are observed, it can be useful to (1) lower the temperature, (2) degas the sample, and (3) optimize the value of D8 (measure of ^{1}H spin-lattice relaxation time, T1). If correlation peaks appear with the same sign as the diagonal peaks, the 2D ^{1}H-^{1}H ROESY NMR experiment should be preferred. The 1D NMR version of this experiment (sel-NOESY) is also interesting in case of overlapped spin systems. Then, it is possible to identify more precisely specific spin systems, measure coupling constants, and quantify NOE effects (a way to determine interproton distances).

19. This number can be reduced in case of symmetry (but it is quite rare in such compounds) or isochronous nuclei.

20. It can be pointed out that, as previously argued for diterpenes and meroditerpenes isolated from *Cystoseira* spp. [6, 7], MBFC shows oxidized functions at the preferred positions C-5, C-12, and C-15.

21. Before the current use of NOESY data, assignments of the geometry of the double bonds were based upon the ^{13}C

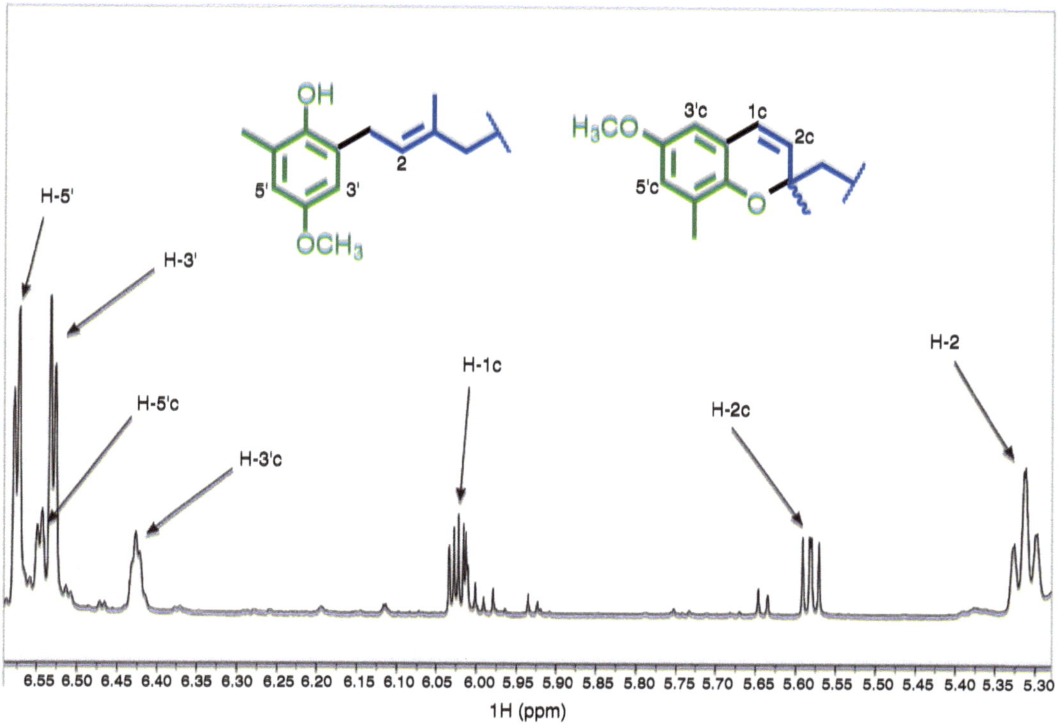

Fig. 8 Expansion (olefinic protons region) of the ^{1}H NMR spectrum ($CDCl_3$, 500 MHz) of degraded MBFC

chemical shifts of the methyl groups of the terpene unit: the configurations were assigned as *E* when the ^{13}C NMR chemical shift of the methyl signal was lower than 20 ppm and *Z* when this value was upper than 20 ppm.

22. In order to avoid such a phenomenon (Fig. 8), NMR spectra might be recorded in C_6D_6. Moreover, in this solvent ^1H NMR signals are shifted and some of them which are overlapped in $CDCl_3$ become clearly separated. Nevertheless, for comparison with literature data, it should be very informative to use NMR data obtained in $CDCl_3$.

Acknowledgments

For this work, Gérald Culioli was partially supported by CNRS (Côte d'Azur delegation—DR 20). Two of the authors (Annick Ortalo-Magné and Gérald Culioli) are indebted to Pr. Louis Piovetti (Université de Toulon, MAPIEM) and Pr. Robert Valls (Aix-Marseille Université, LISA) for the many helpful discussions and the knowledge they have shared with us.

References

1. Blunt JW, Copp BR, Keyzers RA et al (2013) Marine natural products. Nat Prod Rep 30:237–323, And previous reports in the series
2. Department of Chemistry, University of Canterbury (2013) MarinLit database, version 17.3. Department of Chemistry, University of Canterbury, Christchurch, New Zealand, http://www.chem.canterbury.ac.nz/marinlit/marinlit.shtml
3. Rousseau F, de Reviers B (1999) Phylogenetic relationships within Fucales (Phaeophyceae) based on combined partial SSU+LSU rDNA sequence data. Eur J Phycol 34:53–64
4. Vallim MA, De Paula JC, Pereira RC et al (2005) The diterpenes from Dictyotacean marine brown algae in the Tropical Atlantic American region. Biochem Syst Ecol 33:1–16
5. Teixeira VL, Kelecom A (1988) A chemotaxonomic study of diterpenes from marine brown algae of the genus *Dictyota*. Sci Total Environ 75:271–283
6. Valls R, Piovetti L (1995) The chemistry of the Cystoseiraceae (Fucales: Pheophyceae): chemotaxonomic relationships. Biochem Syst Ecol 23:723–745
7. Amico V (1995) Marine brown algae of family Cystoseiraceae: chemistry and chemotaxonomy. Phytochemistry 39:1257–1279
8. Muñoz J, Culioli G, Köck M (2013) Linear diterpenes from the marine brown alga *Bifurcaria bifurcata*: a chemical perspective. Phytochem Rev 12:407–424
9. Culioli G, Ortalo-Magné A, Valls R et al (2008) Antifouling activity of meroditerpenoids from the marine brown alga *Halidrys siliquosa*. J Nat Prod 71:1121–1126
10. Reddy P, Urban S (2009) Meroditerpenoids from the southern Australian marine brown alga *Sargassum fallax*. Phytochemistry 70:250–255
11. Sabry OMM, Andrews S, McPhail KL et al (2005) Neurotoxic meroditerpenoids from the tropical marine brown alga *Stypopodium flabelliforme*. J Nat Prod 68:1022–1030
12. Norte M, Sanchez A, Gonzalez AG (1993) Claraenone, a new meroditerpene from brown alga. Tetrahedron Lett 34:3485–3486
13. Amico V, Oriente G, Piattelli M et al (1980) Diterpenes based on the dolabellane skeleton from *Dictyota dichotoma*. Tetrahedron 36: 1409–1414
14. Ioannou E, Quesada A, Rahman MM et al (2011) Dolabellanes with antibacterial activity from the brown alga *Dilophus spiralis*. J Nat Prod 74:213–222
15. Enoki N, Ishida R, Matsumoto T (1982) Structures and conformations of new nine-membered ring diterpenoids from the marine alga *Dictyota dichotoma*. Chem Lett 11: 1749–1752
16. Ortalo-Magné A, Culioli G, Valls R et al (2005) Polar acyclic diterpenoids from *Bifurcaria bifurcata* (Fucales, Phaeophyta). Phytochemistry 66:2316–2323
17. Reddy P, Urban S (2008) Linear and cyclic C_{18} terpenoids from the southern Australian marine brown alga *Cystophora moniliformis*. J Nat Prod 71:1441–1446
18. De Rosa S, De Giulio A, Iodice C et al (1994) Sesquiterpenes from the brown alga *Taonia atomaria*. Phytochemistry 37:1327–1330
19. Kurata K, Taniguchi K, Suzuki M (1996) Cyclozonarone, a sesquiterpene-substituted benzoquinone derivative from the brown alga *Dictyopteris undulata*. Phytochemistry 41: 749–752
20. Gouveia V, Seca AML, Barreto MC et al (2013) Di- and sesquiterpenoids from *Cystoseira* genus: structure, intra-molecular transformations and biological activity. Mini Rev Med Chem 13:1150–1159
21. Barbosa JP, Fleury BG, da Gama BAP et al (2007) Natural products as antifoulants in the Brazilian brown alga *Dictyota pfaffii* (Phaeophyta, Dictyotales). Biochem Syst Ecol 35:549–553
22. Viano Y, Bonhomme D, Camps M et al (2009) Diterpenoids from the Mediterranean brown alga *Dictyota* sp. evaluated as antifouling substances against a marine bacterial biofilm. J Nat Prod 72:1299–1304
23. Mokrini R, Mesaoud MB, Daoudi M et al (2008) Meroditerpenoids and derivatives from the brown alga *Cystoseira baccata* and their antifouling properties. J Nat Prod 71:1806–1811
24. Schmitt TM, Lindquist N, Hay ME (1998) Seaweed secondary metabolites as antifoulants: effects of *Dictyota* spp. diterpenes on survivorship, settlement, and development of marine invertebrate larvae. Chemoecology 8:125–131
25. Hay ME, Duffy JE, Fenical W et al (1988) Chemical defense in the seaweed *Dictyopteris delicatula*: differential effects against reef fishes and amphipods. Mar Ecol Prog Ser 48:185–192
26. Hay ME, Duffy JE, Pfister CA et al (1987) Chemical defense against different marine herbivores: are amphipods insect equivalents? Ecology 68:1567–1580

27. Barbosa JP, Teixeira VL, Pereira RC (2004) A dolabellane diterpene from the brown alga *Dictyota pfaffii* as chemical defense against herbivores. Bot Mar 47:147–151

28. Pereira RC, Soares AR, Teixeira VL et al (2004) Variation in chemical defenses against herbivory in southwestern Atlantic *Stypopodium zonale* (Phaeophyta). Bot Mar 47:202–208

29. Mesguiche V, Valls R, Piovetti L et al (1997) Meroditerpenes from *Cystoseira amentacea* var. *stricta* collected off the Mediterranean coasts. Phytochemistry 45:1489–1494

30. Bennamara A, Abourriche A, Berrada M et al (1999) Methoxybifurcarenone: an antifungal and antibacterial meroditerpenoid from the brown alga *Cystoseira tamariscifolia*. Phytochemistry 52:37–40

31. Abourriche A, Charrouf M, Chaib N et al (2001) Bioactivities of methoxybifurcarenone: a new meroditerpenoid from the brown alga *Cystoseira tamariscifolia*. Recent Res Dev Phytochem 5:185–189

32. Sun HH, Ferrara NM, McConnell OJ et al (1980) Bifurcarenone, an inhibitor of mitotic cell division from the brown alga *Bifurcaria galapagensis*. Tetrahedron Lett 21: 3123–3126

33. Coates RM, Ley DA, Cavender PL (1978) Synthesis and carbon-13 nuclear magnetic resonance spectra of *all-trans*-geranylgeraniol and its nor analogues. J Org Chem 43: 4915–4922

34. Couperus PA, Clague ADH, Van Dongen JPCM (1976) ^{13}C chemical shifts of some model olefins. Org Magn Res 8:426–431

35. Amico V, Oriente G, Neri P et al (1987) Tetraprenyltoluquinols from the brown alga *Cystoseira stricta*. Phytochemistry 26: 1715–1718

36. Mori K, Uno T (1989) Synthesis and structure revision of bifurcarenone a unique monocyclic diterpene in combination with a hydroquinone C_7 unit as an inhibitor of mitotic cell division. Tetrahedron 45:1945–1958

37. Mori K, Uno T, Kido M (1990) Determination of the absolute configuration of bifurcarenone by the synthesis of its (1′*R*,2′*R*)-isomer. Tetrahedron 46:4193–4204

Chapter 14

Extraction, Isolation, and Identification of Sesquiterpenes from *Laurencia* Species

Angélica Ribeiro Soares

Abstract

The knowledge about the chemical structure of the secondary metabolites and their relative abundances in algae is very important to several fields of basic and applied research in biology, chemistry, and many other disciplines. The attainment of such knowledge requires special attention to the origin of the organism in question and the methodology applied. Here, we present a protocol to obtain and identify some sesquiterpenes from *Laurencia* species based on traditional methodologies, such as flash and thin-layer chromatographies, nuclear magnetic resonance spectroscopy, and gas chromatography mass spectrometry. Red algae of the genus *Laurencia* are known to produce structurally diverse terpenes; most of them are halogenated compounds with important ecological functions and significant potential for the discovery of new biotechnological applications.

Key words Chromatography fingerprints, Flash chromatography, GC–MS, Halogenated sesquiterpenes, NMR, Sesquiterpenes, Thin-layer chromatography

1 Introduction

Red algae of the genus *Laurencia* (family Rhodomelaceae, order Ceramiales) are known to produce structurally diverse natural products, most of them halogenated compounds of distinct chemical classes such as sesquiterpenes, diterpenes, triterpenes, and acetogenins [1–3]. Numerous reports have highlighted the important ecological function and the potential biotechnological application of these metabolites [3–5]. Although the genus *Laurencia* is considered the most heavily studied algal group, it is still a prolific producer of new compounds from the marine environment [2, 3]. This remarkable chemical diversity is observed not only between species but also for different populations, geographically close or distant [6–8]. Information on chemical diversity can be used to guide future efforts to select populations aiming to obtain specific compounds from *Laurencia* species, including the identification of secondary metabolites with biotechnological potential.

Dagmar B. Stengel and Solène Connan (eds.), *Natural Products From Marine Algae: Methods and Protocols*, Methods in Molecular Biology, vol. 1308, DOI 10.1007/978-1-4939-2684-8_14, © Springer Science+Business Media New York 2015

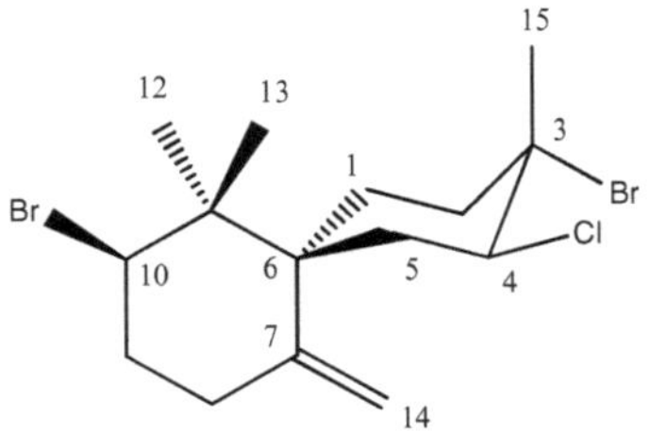

(1)

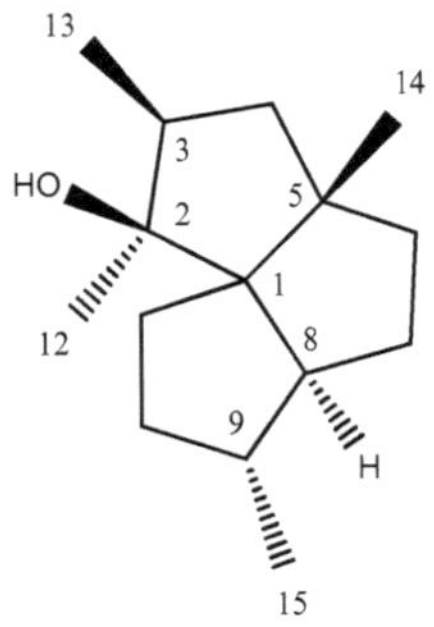

(2)

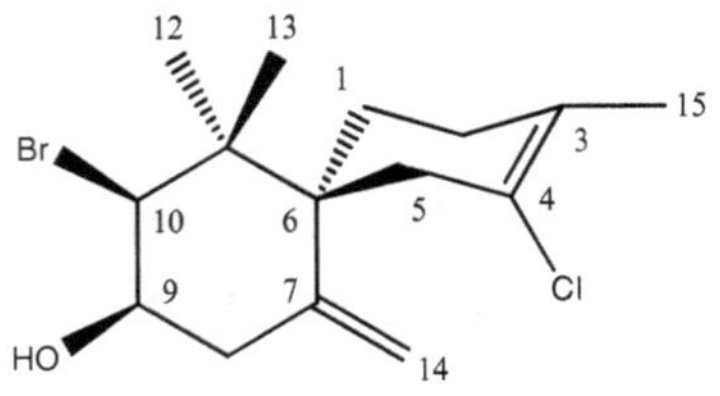

(3)

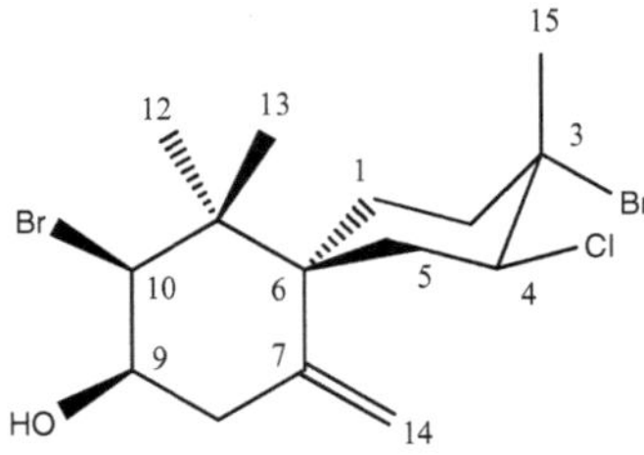

(4)

Fig. 1 Structures of obtusane (**1**), triquinane derivative (**2**), elatol (**3**), and obtusol (**4**)

However, the processes employed to isolate and identify algae secondary metabolites strongly determine the final outcome of the analysis. In general, samples obtained from algae are complex mixtures and the qualitative as well as the quantitative content of compounds may be very variable. The number of compounds obtained and their relative amounts depend on (a) how the biological material was treated, (b) what extraction and isolation techniques were employed, (c) how the chemical structures of the compounds were determined, and (d) what technique was used to assess the chemical variability of the sample. Here we describe an simple and commonly applied strategy to obtain sesquiterpenes from *Laurencia* spp. These compounds are widely distributed and especially abundant within this genus, and represent around 50 % of all secondary metabolites already isolated [2, 3]. Then we exemplify the isolation and identification of four sesquiterpenes: obtusane (**1**), triquinane derivative (**2**), elatol (**3**), and obtusol (**4**), one non-halogenated molecule, and three halochamigranes distributed in a wide range of species of the genus *Laurencia* collected from the Brazilian coast. Besides their ecological importance, these compounds have been shown a broad spectrum of pharmacological activity [3, 4] (Fig. 1).

2 Materials

Use only pure organic solvents obtained commercially or by distillation. Diligently follow all waste disposal regulations when disposing waste materials. Use a rotary evaporator to remove the solvent from the samples.

2.1 Extraction

1. *n*-Hexane:ethyl acetate 50:50 (v/v).
2. Blender.
3. Ultrasonic bath.
4. Sintered glass funnel.

2.2 Thin-Layer Chromatography

1. Thin-layer chromatography (TLC): Aluminum plates pre-coated with silica gel 60 GF_{254} (10×10 cm, thickness: 0.25 mm).
2. TLC mobile phase: *n*-Hexane:dichloromethane (v/v).
3. Dichloromethane (or ethyl acetate).
4. Pasteur pipette or a capillary tube.
5. 5 % sulfuric acid in ethanol: Add w/v of concentrated sulfuric acid (H_2SO_4 % ≥ 98.0) to ethanol (99.5 %, ACS) (*see* **Note 1**).
6. UV-lamp with two different wavelengths (254 and 365 nm).

2.3 Gas Chromatography–Mass Spectrometry

1. Ethyl acetate.
2. 0.45 μM PTFE syringe filter.
3. Gas chromatography coupled to mass spectrometry (GC–MS) in the electron impact mode (70 eV) using an Rtx-1MS capillary column (60 m×0.25 mm; film thickness: 0.1 μm, Restek). Helium should be used as carrier gas.

2.4 Proton and Carbon Nuclear Magnetic Resonance Spectroscopy

1. 5 mm NMR tubes.
2. Pasteur pipettes.
3. Nuclear magnetic resonance spectrometer operating at 300.13 MHz for ^1H and 75.47 MHz for ^{13}C. Chemical shifts are reported in parts per million (δ) downfield from the internal standard tetramethylsilane (TMS).
4. Deuterated chloroform ($CDCl_3$).

2.5 Flash Chromatography

1. Silica gel (SiliaFlash® F60 40–63 μm (230–400 mesh) 60 Å, Silicycle, Quebec City, QC, Canada).
2. *n*-Hexane.
3. Glass columns of different diameters (ø = 1.2, 1.8 and 5.0 cm) containing a sintered glass disc (grade 3) at the bottom.

4. A vacuum-filtering flask (Erlenmeyer shape).
5. Dichloromethane.
6. *n*-Hexane:dichloromethane solutions: 100:0, 75:25, 50:50, 0:100 (v/v).
7. Dichloromethane:methanol solutions: 50:50, 0:100 (v/v).
8. Fractionation of fraction F2: *n*-Hexane:dichloromethane solutions: 60:40, 55:45, 50:50, 48:52, 45:55, 0:100 (v/v).

2.6 Equipment for Chemical Identification

1. Polarimeter: Optical rotations should be measured in methanol using a polarimeter with a sodium lamp at 598 nm and 25 °C.
2. Fourier transform infrared spectrophotometer, using bromide potassium (KBr) pellets.
3. High-resolution mass spectrometer with electrospray ionization (ESI) or atmospheric-pressure chemical ionization (APCI).
4. NMR spectrometer.

3 Methods

Carry out all procedures at room temperature.

3.1 Extraction

1. Clean the freshly collected biological material using seawater to eliminate any associated organisms, sand, etc.
2. To air-dry the algae at room temperature spread them out on paper towel and keep them protected from sunlight or strong light sources. If a freeze-dryer is available it could be used, although many sesquiterpenes can be extracted by simple air-drying.
3. Grind the dry material using a blender.
4. In a screw cap Erlenmeyer flask, add 100.0 g of the milled material with 0.8 L of the extraction solvent (8 mL of solvent to each 1 g of dry algae).
5. Sonicate the Erlenmeyer for 20 min in an ultrasonic bath at a constant temperature of 25 °C. Thereafter close the flask and keep it at room temperature for solvent extraction overnight on the bench, protected from light.
6. Separate supernatants using a sintered glass funnel (vacuum can be used if necessary).
7. Remove the solvent of the supernatant liquid fraction at reduced pressure with the assistance of a rotary evaporator (*see* **Note 2**).
8. Transfer the concentrated material with the solvent assistance to a tared small vial. The solvent removed by this equipment

could be added again to the Erlenmeyer containing the biomass to continue the process of extraction. To increase the yield, the same biomass should be extracted at least three times. The material obtained from all extractions should be combined within the same tared small vial.

9. Evaporate the residual solvent from the sample in the small vial by using a rotary evaporator, or a centrifugal vacuum concentrators or a nitrogen flow. After evaporation of solvent determine the weight and calculate the crude extract yield.

3.2 Chemical Profile of the Crude Extract

Since every biological sample is different, it is important to determine the chemical profile of the crude extract before starting any effort to obtain a natural compound. The information on the complexity and the kinds of compounds present within a sample can be very useful to guide the isolation of the target compound. There are many different methods to obtain this information; chromatographic techniques may also be coupled with spectroscopy methods such as mass spectrometry and nuclear magnetic resonance which are most commonly used. Since all crude extracts are complex mixtures, chromatography is the most effective way to separate them into their components for further identification. However, spectroscopy by nuclear magnetic resonance has emerged as an important technique for compound identification. Despite its importance for structural elucidation of a pure compound, information about the chemical traits of the major compounds can be very useful to identify quickly the class of secondary metabolites in a sample. Here we outline the use of three useful techniques to guide chemical characterization work of secondary metabolites in *Laurencia* spp.: TLC (the simplest chromatographic method), GC–MS, and ^{1}H nuclear magnetic resonance spectroscopy (^{1}H NMR).

3.2.1 Chemical Profile by TLC

1. Cut a TLC plate to 5×2 cm (height×width) with the assistance of one scissor (if it is a plastic- or aluminum-backed plate) (*see* **Note 3**).
2. Prepare a TLC tank (*see* **Note 4**) by lining with a filter paper and adding the mobile phase (*see* **Note 5**) to a depth of about 0.3 cm.
3. Prepare a sample of your crude extract for TLC (ca. 1–2 % solution) by dissolving a small quantity (1.0–1.5 mg) in dichloromethane (or ethyl acetate).
4. Using a soft pencil, gently mark a dot about 0.5 cm from the bottom. Apply the sample solution on the TLC at the dot using a Pasteur pipette or a capillary tube (*see* **Note 6**). Wait for solvent evaporation.
5. Under a fume hood, place the TLC plate in the TLC tank. Allow the solvent to creep up the TLC plate until it is about

0.3 cm from the top and then remove it from the tank. Keep the plate under the fume hood until complete solvent evaporation. Before evaporation, mark the initial level of the solvent front on the plate with the aid of a pencil. Wait for the complete solvent evaporation from the plate.

6. To visualize the spots after the run, first place the TLC plate under the UV lamp (*see* **Note 7**) to show any UV-active spots. Plates that have been impregnated with a fluorescent indicator will show spots for the compounds under an ultraviolet light due to quenching of the fluorescence by the substance on the plate. Use a pencil to surround the spots. Then spray the plates with 5 % sulfuric acid in ethanol (*see* **Note 8**) and heat the plate at 100 °C until the spots appear.
7. After the reaction with sulfuric acid obtusane, triquinane derivative, elatol, and obtusol will appear as yellow, pink, navy, and purple spots, respectively. The retardation factor values (R_f; *see* **Note 9**) of obtusane, triquinane derivative, obtusol, and elatol using SiOH TLC plate in *n*-Hexane:dichloromethane solution 50:50 (v/v) are around 0.85, 0.55, 0.50, and 0.45, respectively (*see* **Note 10**).

3.2.2 Chemical Profile by GC–MS

1. Dissolve 2.0 mg of crude extract in 1.0 mL of ethyl acetate and filter it through a 0.45 μM PTFE syringe filter prior to injection.
2. Use the following conditions for GC–MS analysis: Temperature of injector, interface, and ion source should be set at 260, 260, and 240 °C, respectively. The temperature of the oven should be programmed to keep at 100 °C for 2 min and then increase the oven temperature, at a rate of 8 °C/min, to 300 °C. The flow rate of the carrier gas should be set to 1.8 mL/min. Inject the sample in split mode (1 μL; ratio 1:20).
3. Obtain the chromatogram at the "full scan" mode monitoring the molecular weight between *m/z* 100 and 600. The compounds will be identified based on mass spectral and retention time comparison with the isolated compounds. The expected retention times for obtusane (**1**), triquinane derivative (**2**), elatol (**3**), and obtusol (**4**) are 16.4, 6.7, 14.8, and 18.0 min, respectively. Examples of chromatograms from three different populations (**A–C**) of *Laurencia dendroidea* collected at the Brazilian coast are illustrated in Fig. 2, showing qualitative and quantitative differences. The sesquiterpenes obtusane (**1**), triquinane derivative (**2**), elatol (**3**), and obtusol (**4**) were observed in parts of the samples and, when present, at different concentrations. Life cycle, ecological factors, as well as genetic variation can influence the secondary metabolite composition of an organism and contribute to the chemical diversity

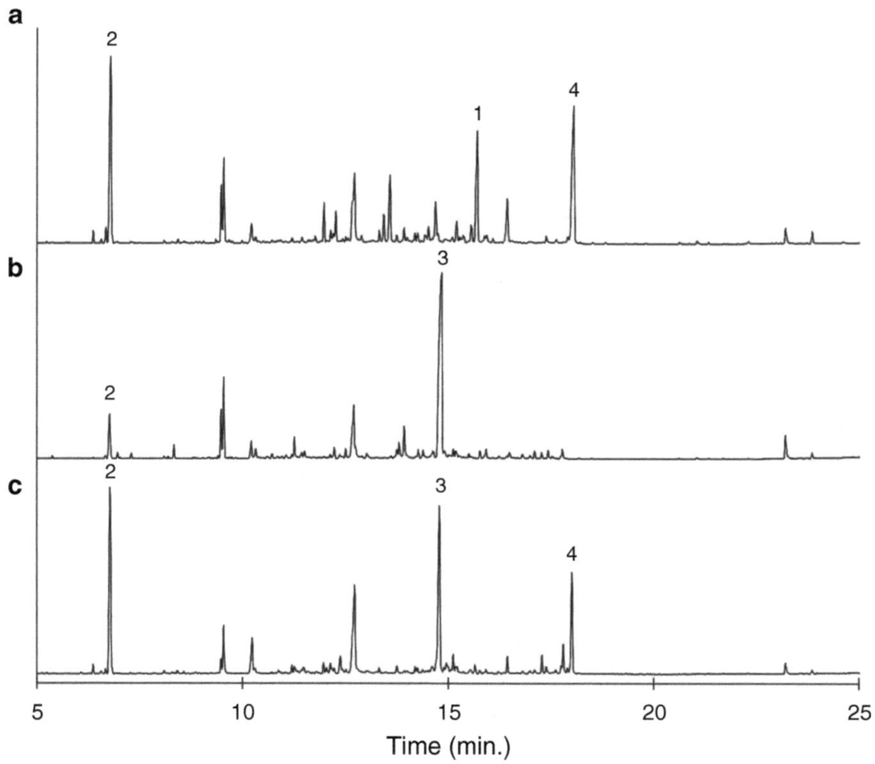

Fig. 2 Chromatograms of crude extracts of *Laurencia dendroidea* from different populations (A–C) collected from the Brazilian coast. Obtusane (**1**), triquinane derivative (**2**), elatol (**3**), and obtusol (**4**)

observed within populations. Such chemical variation between samples should be considered before starting any phytochemical work.

3.2.3 Chemical Profile by ¹H NMR

1. Weigh 5.0 mg of the crude extract. Dry the sample for 2 h in a freeze-dryer to eliminate any water or solvent residues.
2. Dry a clean NMR tube and the Pasteur pipette for 15 min in an oven at 70 °C.
3. Dissolve the sample in 500 μL of $CDCl_3$ containing 0.05 % TMS (v/v) as an internal standard.
4. Transfer the solution with the assistance of a dry Pasteur pipette into the NMR tube (*see* **Notes 11** and **12**).
5. The NMR analysis should be operating at a proton NMR frequency of 300.13 MHz at either 20 or 25 °C. Obtain each spectrum using the 256 scans and 300 k data points, with a pulse width of 8.0 μs (30°) and relaxation delay of 2.0 s.

In general, a rapid inspection of the ^{1}H NMR spectrum of *Laurencia* crude extract will allow the detection of the halogenated chamigrane sesquiterpenes. The group of signals (in the most cases singlet or doublet) are between 5.5–3.2 ppm (characteristic of halogen and oxygen substitution as well as dehydrogenations at different positions in the skeleton) and 2.0–0.8 ppm (methyl groups bonded to tertiary carbons) and can be used as diagnostics signals to detect halogenated chamigrane sesquiterpenes in the sample. The complete ^{1}H and ^{13}C NMR signals for compounds obtusane (**1**), triquinane derivative (**2**), elatol (**3**), and obtusol (**4**) are shown in Table 1.

3.3 Fractionation and Purification of Crude Extracts from *Laurencia* Species Containing Sesquiterpenes as Major Compounds

1. Homogenize 60.0 g of silica gel in 100 mL of *n*-Hexane (*see* **Note 13**).
2. Pour the mixture gently into the glass column (ø = 5.0 cm) or glass Buchner funnel (*see* **Note 14**) and connect it to a vacuum-filtering flask with the assistance of a rubber adapter. Pressurize the column gently to pack the silica (*see* **Notes 15** and **16**).
3. Solubilize 3.0 g of crude extract (*see* **Note 17**) in a minimum amount of dichloromethane (around 5.0 mL) and mix it with a small amount of silica (enough to form a 3–5 mm layer at the top of the column). Dry the mixture using a gentle airflow or a rotary evaporator (*see* **Note 18**).
4. Very carefully load the dry sample-impregnated silica (*see* **Note 19**) evenly onto the top of the column with the assistance of a spatula. During this process it is essential to create a flat surface of the sample-impregnated silica (*see* **Note 20**).
5. Elute the column in *n*-hexane:dichloromethane (100:0, 75:25, 50:50, 0:100; v/v) and dichloromethane:methanol (50:50, 0:100; v/v) using a volume of 200 mL for each solution (*see* **Notes 21** and **22**). The elution results in six fractions (F1–F6).
6. Remove the solvents from each fraction with the assistance of a rotatory evaporator.
7. Analyze the column fractions obtained by running TLCs (*see* Subheading 3.3) to know the chemical complexity of each sample and identify all target compounds. However, as each biological sample is different (e.g., Fig. 2) it is likely that different compounds can be found in each fraction and some modifications should be applied during the purification steps to improve results. Before starting a new flash column it is first necessary to determine the amount of silica gel required and the solvent system to be used (*see* **Notes 5** and **23**). The fraction F1 (obtained in 100 % *n*-hexane) can be re-chromatographed using an isocratic elution system.
8. Mount the silica gel column (ø = 1.2 cm), containing a sintered glass disc at the bottom, vertically using a clamp stand and connected it to a vacuum-filtering flask.

Table 1
^1H (300 MHz) and ^{13}C (75 MHz) NMR spectroscopic data of the compounds obtusane (1), triquinane derivative (2), elatol (3), and obtusol (4), in $CDCl_3$ as solvent and TMS used as internal standard

C/H	1		2		3		4	
	δ_C, Mult	δ_H (Mult, J in Hz)	δ_C, Mult	δ_H (Mult, J in Hz)	δ_C, Mult	δ_H (Mult, J in Hz)	δ_C, Mult	δ_H (Mult, J in Hz)
1	25.5, CH_2	1.87 (m) 1.71 (m)	69.9, C	–	25.6, CH_2	1.62 (ddd,13.4, 12.5, 7.6) 1.82 (ddd,13.4, 4.1, 2.1)	25.6, CH_2	1.74 (m)
2	440.4, CH_2	2.26 (m) 2.17 (m)	85.3,C	–	29.3, CH	1.81 (ddd,14.8, 12.5, 2.1) 1.96 (ddd,14.8, 7.6, 4.1)	40.5, CH_2	2.30 (m)
3	68.4, C	–	41.0, CH	1.62 (m)	128.0, C	–	67.6, C	–
4	67.9, CH	4.71 (dd. 12.4, 4.7)	47.3, CH_2	1.61 (m) 1.34 (m)	124.1, CH	–	68.1, CH	4.70 (dd, 10.8, 2.9)
5	37.2, CH_2	2.27 (m) 1.96 (dd, 14.4, 12.6)	49.7, C	–	38.6, CH_2	2.59 (brd, 16.1) 2.37 (brd, 16.1)	37.1, CH_2	1.94 (dd, 12.7) 2.36 (brs)
6	50.4, C	–	41.8, CH_2	1.46 (m) 1.46 (m)	49.1, CH_2	–	50.3, C	–
7	145.7, C	–	29.0, CH_2	1.51 (m) 1.29 (m)	140.8, C	–	141.2, C	–
8	33.5, CH_2	2.32 (m) 2.16 (m)	57.7, CH	1.50 (m)	38.0, CH_2	2.63 (dd, 14.4, 3.0) 2.51 (dd, 14.4, 3.0)	38.5, CH_2	2.62 (dd, 14.1, 3.1) 2.49 (dd, 14.1, 3.1)
9	35.8, CH_2	2.27 (m) 2.05 (m)	42.5, CH	1.38 (m)	72.2, CH	4.14 (brq, 3.0)	71.9, CH	4.10 (brs)
10	63.5, CH	4.44 (dd.12.7, 4.5)	36.3, CH_2	1.72 (ddd, 2.2, 5.7, 11.3) 1.15 (dt, 6.9, 11.4)	70.8, CH	4.60 (d, 3.0)	70.1, CH	4.47 (d, 3.0)

(continued)

Table 1
(continued)

C/H	1		2		3		4	
	δ_C, Mult	δ_H (Mult, J in Hz)	δ_C, Mult	δ_H (Mult, J in Hz)	δ_C,Mult	δ_H (Mult, J in Hz)	δ_C, Mult	δ_H (Mult, J in Hz)
11	43.9, C	–	27.9, CH_2	1.86 (ddd, 2.3, 6.8, 13.5) 1.42 (m)	43.1, C	–	44.2, C	–
12	23.8, CH_3	1.14 (s)	22.3, CH_3	1.10 (s)	24.2, CH_3	1.08 (s)	24.2, CH_3	1.08 (s)
13	17.4, CH_3	0.96 (s)	12.5, CH_3	0.87 (d, 6.1)	20.7, CH_3	1.07 (s)	20.6, CH_3	1.08 (s)
14	114.8, CH_2	5.26 (s) 4.87 (s)	26.7, CH_3	1.07 (s)	115.9, CH_2	5.12 (brs) 4.79 (brs)	117.8, CH_2	5.39 (s) 5.05 (s)
15	29.6, CH_3	1.83 (s)	19.5, CH_3	0.97 (d, 6.4)	19.4, CH_3	1.70 (brs)	23.9, CH_3	1.83 (s)
OH	–	–	–	–	–	2.19 (brs)	-	1.56 (brs)

Chemical shifts (δ, ppm) and coupling constants (J in Hz)
Resonance multiplicities (Mult) abbreviated as follows: *s* singlet, ***d*** doublet, *t* triplet, *q* quartet, ***dd*** double doublet, ***ddd*** double double doublet, ***dq*** double quartet, ***dt*** double triplet, ***m*** multiplet, *br* broad

9. Pack the silica gel (20.0 g) as described in **step 1**.
10. Solubilize the sample (200 mg) in a minimum amount of *n*-Hexane and load the solution as described in **step 3**.
11. Elute the column in 100 % *n*-Hexane (150 mL) and collect fractions with 10–15 mL each. Obtusane should be identified in the subfraction 7.
12. To continue the fractionation process to obtain other sesquiterpenes, the fraction F2 (1.2 g), obtained in **step 5**, should be re-chromatographed using flash chromatography.
13. Use a gradient elution system on silica gel in a column (ø = 1.8 cm) containing a sintered glass disc at the bottom.
14. Homogenize the silica (60.0 g) in *n*-hexane:dichloromethane (60:40; v/v) and prepare the column as described in **step 1**.
15. Prepare the sample as described in **step 2** and apply the dry sample-impregnated silica according to **step 3**.
16. Elute the column with a gradient of elution hexane:dichloromethane (60:40; 55:45; 50:50; 48:52; 45:55; 0:100; v/v) with 100 mL of each mixture. Collect fractions with 20–25 mL each.
17. Analyze all the fractions obtained by TLC (as described in Subheading 3.2.1) and combine the fractions with the same chemical profile. Triquinane derivative is usually obtained in hexane:dichloromethane (55:45; v/v). Elatol and obtusol are usually obtained in hexane:dichloromethane (50:50; v/v) and in hexane:dichloromethane (48:52; v/v), respectively.

3.4 Structural Identification of Some Sesquiterpenes

1. Common methods, such as infrared spectroscopy (IR), high-resolution mass spectra (HRMS), and nuclear magnetic resonance (NMR) are used for determining the chemical structure of a compound. To identify the compounds obtained, analyze the purified fractions by optical rotation, IR, HRMS, ^{1}H NMR, and ^{13}C NMR and compare the data with the spectroscopic data of each compound (Fig. 1) listed below and in Table 1:

   Obtusane (**1**):

   White gum.

$$[\alpha]_D^{25} + 18.8 (c\, 0.05, CHCl_3).$$

   IR (KBr) ν_{max} 2926; 1716; 1456; 911; 870; 804; 738 cm^{-1}.

   HRMS: *m/z* 398.6037 (calcd. for $C_{15}H_{23}Br_2Cl$).

   GC–MS *m/z* (rel int. %): 402 (1), 400 (4), 398 (5), 396 (2), 385 (3), 383 (5), 381 (2), 319 (11), 318 (15), 316 (10), 283 (21), 281 (22), 239 (10), 237 (28), 202 (12), 201 (54), 109 (100), 107 (39).

   ^{1}H and ^{13}C NMR spectroscopic data, *see* Table 1.

Triquinane derivative (**2**):

Colorless oil.

$$[\alpha]_D^{25} -11.31 (c\,0.67, CHCl_3).$$

IR (thin film) ν_{max} 3035, 2932, 2865, 1720, 1457, 1375, 1239, 1165, 1082, 1006, 900 cm^{-1}.

HRMS: *m/z* 222.3669 (calcd. for $C_{15}H_{26}O$).

GC–MS *m/z* (rel int. %): 222 (3), 207 (1), 189 (1), 135 (37), 86 (100), 85 (14), 81 (32), 79 (10).

^1H and ^{13}C NMR spectroscopic data, *see* Table 1.

Elatol (**3**):

Colorless oil.

$$[\alpha]_D^{25} -66.2 (c\,0.13, CHCl_3).$$

IR (thin film) ν_{max} 3458, 2970, 2947, 1718, 1676, 1437, 1346, 1215, 1082, 1029, 898, 817; 736 cm^{-1}.

HRMS: *m/z* 334.6995 (calcd. for $C_{15}H_{23}BrClO$).

GC–MS *m/z* (rel int. %): 319 (2), 317 (1), 299 (3), 297 (3), 281 (2), 253 (8), 238 (7), 237 (40), 236 (18), 235 (100), 217 (7), 209 (15), 207 (29), 200 (9), 199 (36), 193 (8), 181 (6), 179 (13), 173 (6), 172 (5), 171 (21), 169 (13), 167 (9), 165 (17), 161 (5), 159 (10), 158 (6), 157 (27), 156 (5), 155 (17), 153 (29), 145 (11), 144 (8), 143 (25), 142 (8), 141 (26), 139 (19), 135 (12), 133 (25), 155 (12), 130 (6), 129 (19), 128 (12), 127 (23), 125 (6), 121 (18), 119 (23), 117 (19), 116 (7), 115 (28), 109 (25), 108 (6), 107 (36), 105 (36), 95 (12), 93 (29), 91 (58), 85 (76).

^1H and ^{13}C NMR spectroscopic data, *see* Table 1.

Obtusol (**4**):

White powder.

$$[\alpha]_D^{25} +9.61 (c\,0.05, CHCl_3).$$

IR (KBr) ν_{max} 3465, 2969, 1640, 1441, 1384, 1350, 1314, 1200, 1089, 1024, 907, 813, 792 cm^{-1}.

HRMS: *m/z* 414.6030 (calcd. for $C_{15}H_{23}Br_2ClO$).

GC–MS *m/z* (rel int. %): 319 (25), 318 (17), 317 (100), 316 (13), 315 (76), 299 (17), 297 (18), 235 (23), 217 (12), 200 (18), 199 (47), 173 (13), 171 (13), 169 (12), 163 (11), 159 (10), 157 (34), 145 (20), 143 (30), 141 (13), 135 (15), 134 (10), 133 (43), 131 (17), 129 (21), 127 (11), 121 (13), 119 (52), 117 (21), 115 (19), 109 (12), 107 (70), 106 (11), 105 (67), 93 (55), 91 (71), 85 (97).

^1H and ^{13}C NMR spectroscopic data, *see* Table 1.

2. After obtaining and identifying the compounds analyze each one by co-injection with standard compounds under the same GC–MS conditions. This will provide the correct identification of the chromatographic signals in chromatogram.

4 Notes

1. The solution is stable at room temperature.
2. Avoid bath temperatures higher than 40 °C because excessive heat can lead to the decomposition of the compounds.
3. If more than one sample is analyzed on a TLC plate, the appropriate width of plate should be enough to leave a 0.5 cm gap between individual spots.
4. Commercial TLC tanks can be used, but it is usually much cheaper and convenient to use a 100 mL beaker with a watch glass or a Petri plate as lid.
5. The polarity of the TLC solvent system (mobile phase) can be changed depending on the polarity of the compounds in the sample and the distance between their spots on the TLC plate. The polarity of the mixture can be adjusted easily by changing the proportions of the two solvents used. For example, if the compounds analyzed are polar (they will bind more strongly to the silica on the TLC plate) and do not travel much with the solvent system used, then a larger amount of the polar solvent (or a more polar solvent) should be used, e.g., 100 % dichloromethane.
6. Load a small amount of the solution in the pipette. Usually the absorbed capillary content is enough. This will prevent the spreading of the sample on the TLC forming a large spot. A good spot should have an internal diameter of 1.0–1.5 mm.
7. Wear UV-protective glasses and do not look directly at the light.
8. To see the spots on the TLC plate other universal reagents can be used, such as vanillin dissolved in ethanol (6 g in 250 mL) with sulfuric acid (2.5 mL), or ceric sulfate (15 % aqueous sulfuric acid saturated with ceric sulfate). *Caution*: All solutions must be prepared in an ice bath since the reactions are exothermic and can cause overheating.
9. The distance that a compound travels up a TLC plate is called the retardation factor (R_f) and can be calculated dividing the distance of the center of the spot from the baseline by the distance of solvent front from baseline.
10. Since the R_f value of a compound is not very accurate, the best way to compare compounds by TLC is by running them on the same TLC plate. The inclusion of a pure compound on the

TLC plate applied over the investigated sample (*co-spot*) helps to identify the compound in the sample.

11. If any solid is observed in the solution, it should be removed before by filtering it using a Pasteur pipette plugged with cotton and placed into the NMR tube.
12. Avoid storing the sample with solvent in the NMR tube for long periods, even in the refrigerator. Some compounds are unstable in some organic solvents. Try to prepare and analyze the sample in the same day.
13. Swirl the mixture to ensure that all trapped air is removed.
14. Wash any silica residues into the column using more solvent.
15. It is important to ensure that no air is trapped in the column. This will lead to the reduction of the chromatographic resolution. If you see any air bubbles in the silica body placed in the column, let a column volume of solvent percolate through the silica under gravity. Then, pressurize and flush the solvent until no air remains in the column.
16. Be careful not to allow the solvent to drop below the level of the silica at any time during the elution as it will affect the chromatographic resolution. To prevent this from happening, always leave a minimum volume of solvent above the silica (at least 2 cm from silica level in the column). If necessary, close the valve of the column during the elution and add more solvent.
17. The amount of silica and solvents used should be proportionally adjusted to the mass of crude extract used. This is also important during all other steps of the chromatography processes where the initial amount of the sample is variable.
18. Remove the solvent completely and mix the mixture with the assistance of a glass rod until obtaining a homogenous dry powder. The presence of dichloromethane in the powder can cause elution problems because it can drag more polar components through the column.
19. Before applying the dry mixture, reduce the solvent level to 3–5 cm above the silica level in the column. After applying the mixture, elute the solvent slowly to ensure the complete incorporation of the sample in the silica (*see* also **Note 16**).
20. Use the minimum amount of solvent to rinse off any remaining sample from the flask, adding it gently to avoid disturbing the even distribution of the sample on the silica body.
21. Add the solvent *very carefully* to the top of the column using a Pasteur pipette to drip the solvent onto the walls of the column.
22. Add a sufficiently large amount of each solution to provide a safe volume of solvent over the silica before applying the pressure necessary to give a fast solvent flow rate. Collect each fraction continuously.

23. To set up the chromatographic conditions, first run a series of TLCs to find the solvent system that will give a good separation of the components of the mixture under study. If two components of interest are running close together, a difference of R_f values around 0.2–0.3 between them will indicate a satisfactory solvent system. There are often irrelevant impurities that could be either very polar or very nonpolar; these can be largely ignored. The next step is the amount of silica to be used. Usually, if the component required is well separated from other components, a ratio of 20:1 (silica:sample) should be used. If the R_f difference between spots is less than 0.2, the separation may be improved by increasing the amount of silica according to the ratios shown in Fig. 3.

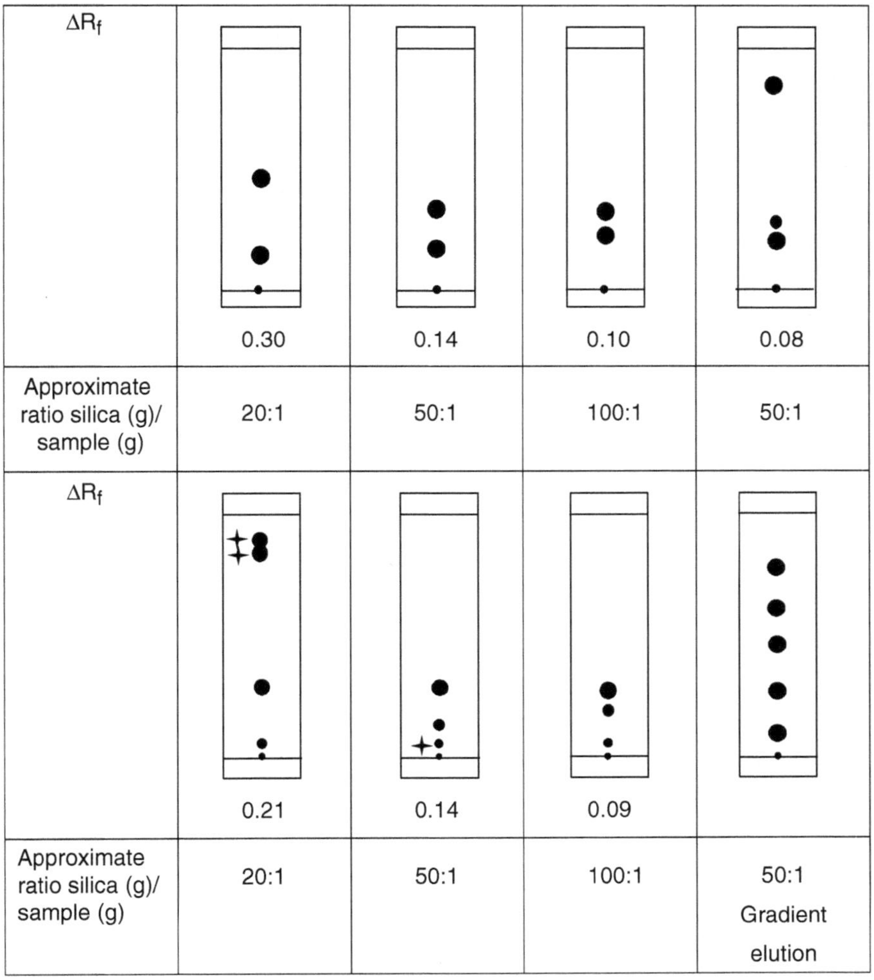

Fig. 3 Illustrative guide of silica:sample ratios for flash chromatography. *Spots marked with a star* are not required products

Acknowledgments

This work was supported by Conselho Nacional de Desenvolvimento Científico e Tecnológico (CNPq) and FINEP (3175/06). A.R. Soares is also grateful to CNPq for a research fellowship. The author thanks Dr. Heitor M. Duarte and Dr. Alessandra L. Valverde for valuable comments and corrections. The author also thanks Nathalia N.P. Carneiro and Fernanda L.S. Machado for some analyzes.

References

1. Blunt JW, Copp BR, Munro HG et al (2011) Marine natural products. Nat Prod Rep 28:196–268
2. Ji N-Y, Wang B-G (2013) Nonhalogenated organic molecules from *Laurencia* algae. Phytochem Rev. doi:10.1007/s11101-013-9326-0
3. Wang B-G, Gloer JB, Ji N-Y (2013) Halogenated organic molecules of Rhodomelaceae origin: chemistry and biology. Chem Rev 113: 3632–3685
4. Machado FLS, Kaiser CR, Costa SS et al (2010) Biological activity of the secondary metabolite from marine algae of the genus *Laurencia*. Braz J Pharmacogn 20:441–452
5. Fujii MT, Cassano V, Stein EM et al (2011) Overview of the taxonomy and of the major secondary metabolites and their biological activities related to human health of the *Laurencia* complex (Ceramiales, Rhodophyta) from Brazil. Braz J Pharmacogn 21:268–282
6. Abe T, Masuda M, Suzuki T (1999) Chemical races in the red alga *Laurencia nipponica* (Rhodomelaceae, Ceramiales). Phycol Res 47:87–95
7. Masuda M, Abe T, Sato S et al (1997) Diversity of halogenated secondary metabolites in the red alga *Laurenica nipponica* (Rhodomelaceae, Ceramiales). J Phycol 33:196–208
8. Caccamese S, Azzolina R, Toscano RM (1981) Variations in the halogenated metabolites of *Laurencia obtusa* from Eastern Sicily. Biochem Syst Ecol 9:241–246

Chapter 15

The Use of HPLC for the Characterization of Phytoplankton Pigments

José L. Garrido and Suzanne Roy

Abstract

HPLC is still the technique of choice for the analysis and characterization of phytoplankton pigments. In this chapter we describe procedures for sample preparation and pigment extraction, and the use of octyl silica columns and pyridine-containing mobile phases to separate chlorophylls and carotenoids. The identification of pigments on the basis of their retention times and visible spectra, the preparation of pigment standards, and the quantitative analysis by either external or internal standard procedures are also described.

Key words Carotenoids, Chlorophylls, HPLC, Octyl silica, Photodiode array detector, Phytoplankton

1 Introduction

The analysis of photosynthetic pigments has become a source of essential information for studies of the physiology and ecology of marine microalgae. Examples of applications of detailed pigment information include the study of changes in natural plankton populations (associated for example with climate change), the ground-truthing of satellite-derived algal biomass estimations, the photosynthetic responses to the changing aquatic light environment, or the trophic transfer from primary producers.

The photosynthetic pigments of algae belong to three chemical families: phycobiliproteins, carotenoids, and chlorophylls (Chls) [1]. Whereas phycobiliproteins are water soluble, carotenoids and Chls are lipid soluble. Most of the time, the term "phytoplankton pigment analysis" refers to the joint analysis of chlorophylls and carotenoids because both types of compounds occur in all photosynthetic algae, they are easily extracted into organic solvents, and they can be detected with high sensitivity in the visible range.

Chls and carotenoids are present in different algal taxa with variable degrees of specificity: some of them occur in several algal classes whereas others are restricted to one or only a few algal groups, making them unambiguous chemotaxonomic markers.

Dagmar B. Stengel and Solène Connan (eds.), *Natural Products From Marine Algae: Methods and Protocols*, Methods in Molecular Biology, vol. 1308, DOI 10.1007/978-1-4939-2684-8_15, © Springer Science+Business Media New York 2015

Liquid chromatography is the method of choice for the analysis of algal Chls and carotenoids. The analysis of complex algal pigment extracts, especially those derived from phytoplankton samples from natural waters, still constitutes a challenge for chromatographic techniques. Algal Chls and carotenoids span a wide range of polarities in which some of them only differ by small structural features (in some cases, only the position of a double bond).

An excellent short course on phytoplankton pigments analysis is available on the Internet [2], but for those readers interested either in broad knowledge on pigment biology or in detailed information on their analysis, a recent and comprehensive monograph is strongly recommended [3]. This book, together with the preceding one [4], covers most aspects in pigment characterization, chemotaxonomy, and applications in oceanography. The latest advances in chromatographic analysis of algal pigments have been recently reviewed [5]. In this chapter we describe a specific protocol for the analysis of algal pigments based on that proposed by [6]. Other protocols exist (e.g., the one developed by [7]) but here we describe the one we are most familiar with. The advantages and disadvantages of these and other alternative methods have been discussed elsewhere [1, 8].

2 Materials

2.1 Sample Preparation

1. Sampling: Seawater is filtered on glass fiber filters (GF/F type, nominal pore size 0.7 μm). A regulated vacuum pump is needed (*see* **Note 1**).
2. Pigment extraction solvent: Prepare extraction solvent by either mixing 9 vol of acetone with 1 vol of water or by mixing 9.5 vol of methanol (MeOH) with 0.5 vol of water (*see* **Note 2**). Use graduate cylinders and HPLC-grade solvents.
3. Extraction: Use Pyrex® screw cap glass tubes with polytetrafluoroethylene (PTFE)-lined caps. A centrifuge and a probe sonicator or an ultrasonic bath can be necessary, depending on the extraction procedure (*see* **step 2** of Subheading 3.1 and **Note 2**). Use PTFE syringe filters with 0.22 μm pore size to clarify extracts before injection in the HPLC system.

2.2 Chromatographic Equipment

1. Pumps: A pumping system able to deliver at least binary gradients is needed. This can be achieved either with a high-pressure mixing system or with a four-solvent, low-pressure mixing gradient pump.
2. Detection: A diode-array detector is essential for pigment characterization, especially in natural water samples. Fluorescence detection (very sensitive and selective towards Chls) can be used in addition to diode array.

3. Columns: Octylsilica (C_8) stationary phases with particle size 3.5 μm or less should be used, packed in columns of 150×4.6 mm.

2.3 Solvents

1. Use HPLC-grade MeOH, acetonitrile (AcN), acetone, and water. Pyridine and acetic acid must be of reagent grade or better (*see* **Note 2**).
2. 0.25 M pyridinium acetate (pH=5.0) stock solution: Add 10 mL of acetic acid and 20 mL of pyridine to 900 mL of HPLC-grade water in a 1 L beaker, mixing with a magnetic stirrer. Continuously monitor pH with a pH meter. Add acetic acid dropwise until pH value is 5.0. Transfer the mixture to a 1 L volumetric flask and adjust volume (the final pyridine concentration is 0.248 M).
3. 0.025 M pyridinium acetate working solution: Dilute ten times the stock solution with HPLC-grade water. Filter this solution through a 0.45 μm filter before use (*see* **Note 3**).

2.4 Pigment Standards

1. Commercial pigment standards: Pigments can be obtained from several companies, including Carotenature, Chromadex, DHI Lab Products, Frontier Scientific Inc., Sigma-Aldrich, and VWR. DHI Lab Products (Hørsholm, Denmark) offers solutions of typical marine algae pigments that can be used directly for HPLC calibration. Standards can also be isolated from reference algal cultures.
2. Reference algal cultures: If you wish to isolate pigment standards from algal cultures, recommendations have been made for phytoplankton unialgal cultures of known pigment composition [9]. Extracts from these cultures can also serve to fix retention times and online spectra of their characteristic pigments.
3. Isolation of pigment standards: Longer columns with larger size C_8 particles (5 μm) can be used in the preparative work for the isolation and purification of pigment standards. Octadecylsilica (C_{18}) solid-phase extraction cartridges (for example Waters SeP-Pak, Waters, Milford, MA, USA) are used for concentration. A source of dry N_2 gas is needed.
4. Internal standard (if calibration with an internal standard is used see Subheading 3.5 below): *Trans*-ß-apo-8′-carotenal (Sigma Chemical Company, St. Louis, MO, USA). Other internal standards (such as vitamin E acetate) are available.
5. Quantification: A UV-Vis spectrophotometer is needed to determine the concentration of pigment solutions. Optical glass cuvettes with tight PTFE stoppers should be used to avoid solvent evaporation.

3 Methods

3.1 Sample Preparation (See Note 4)

1. Sampling: Filter seawater or algal culture samples onto glass fiber filters making sure that vacuum is not higher than 200 mm Hg (*see* **Note 5**). Deep freeze the filters immediately and keep them frozen until extraction (*see* **Note 6**).
2. Pigment extraction: Place frozen filters from algal cultures or natural samples in PTFE-lined screw-capped tubes, and then add the extraction solvent (acetone:water 9:1, v/v, or MeOH:water 9.5:0.5; *see* **Note 2**) to each tube: at least 5 mL if 47 mm diameter filters were used to collect the plankton sample, a minimum of 3 mL for 25 mm filters, or not less than 1.5 mL for 13 mm filters. If using an internal standard for calibration (see Subheading 3.5 below), add it at this point. Grind the filter using a stainless steel spatula. Place the tube in an ultrasonic bath with ice and water for 5 min. Centrifuge for 5 min at 3,500×*g* (this step can be omitted, *see* **Note 2**). Alternatively, use an ultrasonic probe to disrupt the filter and break the retained cells: put the probe (set at 50 W) inside the solvent and move it up and down for 60 s while keeping the tube in a beaker with ice to prevent heating. Centrifuge for 5 min at 3,500×*g* (*see* **Note 2**). Whichever procedure is used, filter extracts through a 0.22 μm PTFE syringe filter before injecting them into the HPLC system.

3.2 Chromatography

1. Injection: Mix aliquots of pigment extracts with water (one volume of 90 % aqueous acetone extract with 0.4 vol of water or 1 vol of 95 % aqueous MeOH extract with 0.2 vol of water (*see* **Note 7**)) to avoid shape distortion of earlier eluting peaks [10] and inject the sample immediately (*see* **Note 7**). If a programmable autosampler is available, program it to perform water mixing, preferably in the sample loop. In this case, water addition can be divided into 2 vol, to be loaded in the loop before and after the sample extract (*see* **Note 7**).
2. Elution: If a high-pressure mixing system is employed, mix solvents in a graduate cylinder to prepare two mobile phases, as follows: for eluent A mix MeOH:AcN:0.025 M aqueous pyridinium acetate (pH 5.0) (50:25:25; v/v/v). Eluent B is composed of MeOH:AcN:acetone (20:60:20; v/v/v). Filter both eluents through a 0.45 μm filter before use. Set flow rate at 1 mL/min. Program the gradient profile according to Table 1.
3. If a quaternary low-pressure mixing system is available there is no need to prepare mixed eluents. Place each solvent in an eluent line. Set flow rate at 1 mL/min. Program the gradient profile according to Table 2 (*see* **Note 8**).

Table 1
Gradient profile for binary chromatographic systems

Time (min)	% A MeOH: AcN: 25 mM aq. pyridinium acetate (pH 5) (50:25:25 v/v/v)	% B MeOH:AcN:acetone (20:60:20 v/v/v)
0	100	0
22	60	40
28	5	95
38	5	95
40	100	0

Table 2
Gradient profile for quaternary chromatographic systems

Time (min)	% MeOH	% AcN	% 25 mM aq. pyridinium acetate (pH 5)	% acetone
0	50.0	25.0	25.0	0.0
22	38.0	39.0	15.0	8.0
28	21.4	58.3	1.3	19.0
38	21.4	58.3	1.3	19.0
40	50.0	25.0	25.0	0.0

4. Detection: Program the diode array detector to get a full visible spectrum (at least from 400 to 700 nm) at each chromatographic point. Monitor the chromatogram at a wavelength that allows general detection of Chls and carotenoids (435–440 nm). Chromatograms at different wavelengths can also be obtained, allowing either a general or a selective detection (*see* Subheading 3.4). Many systems can provide a chromatographic trace summing the response at several wavelengths, which increases the sensitivity but reduces the selectivity of the detection. If fluorescent detection is employed, use broad excitation and emission bandwidths to increase sensitivity and detection of all Chl derivatives.

3.3 Pigment Standards

Preparation of standards: Wet SPE cartridge with 5 mL acetone and then 5 mL of a water:acetone mixture (7:3, v/v). Collect pigment fractions from preparative separation after passing HPLC detector. Immediately before processing, dilute fractions by adding half their volume of water, and then pass through the cartridge. Observe the formation of a colored zone in the cartridge. If the pigment passes through, add more water to the eluate and pass it again through the cartridge (*see* **Note 9**). Eliminate excess solvent by blowing the cartridge with N_2 until completely dry. Then quickly elute the pigment with acetone (or any other adequate solvent in which extinction coefficients are available [11]) (*see* **Note 10**).

1. Characterization of standards: Check the purity of each standard by injecting an aliquot in the HPLC system. If more than one peak appears or if the spectrum is not homogeneous at different times of peak elution (*see* **Note 11**), a further purification step with a different chromatographic system is needed (*see* **Note 12**).

3.4 Pigment Identification

1. Identify pigments by:
   (a) Their retention behavior: Compare pigment retention times with those of standards (or pigment profiles of standard cultures)—*see* Fig. 1. Considering that the same retention time can be shared by several pigments if they co-elute in the same peak (*see* **Note 11**) further identification criteria are needed.
   (b) Their visible spectrum: Compare the spectrum of each peak in the sample with that of the corresponding pigment standard eluting at the same retention time. Examine the full visible spectrum in different sections of each peak, for its characterization in terms of pigment identity and peak purity (*see* **Note 13**).

For comparison purposes, reference retention times and spectral data are available [1–4, 6, 7, 11–13]. Selective detection can be achieved by extracting chromatograms at selected wavelengths (e.g., 435, 470, and 665 nm [1]) (*see* **Note 14**).

3.5 Quantitative Analysis

1. Prepare a primary standard solution of each pigment in a solvent in which extinction coefficients are available (usually acetone or ethanol [11]). Determine the concentration spectrophotometrically (absorbance reading at the coefficient wavelength should be between 0.1 and 0.8 absorbance units), taking care of keeping tubes or volumetric flasks and spectrophotometer cuvettes tightly stopped between transfers to minimize solvent evaporation.
2. System calibration, external standard method [14]: Prepare a "concentrated pigment mixture" by carefully mixing known

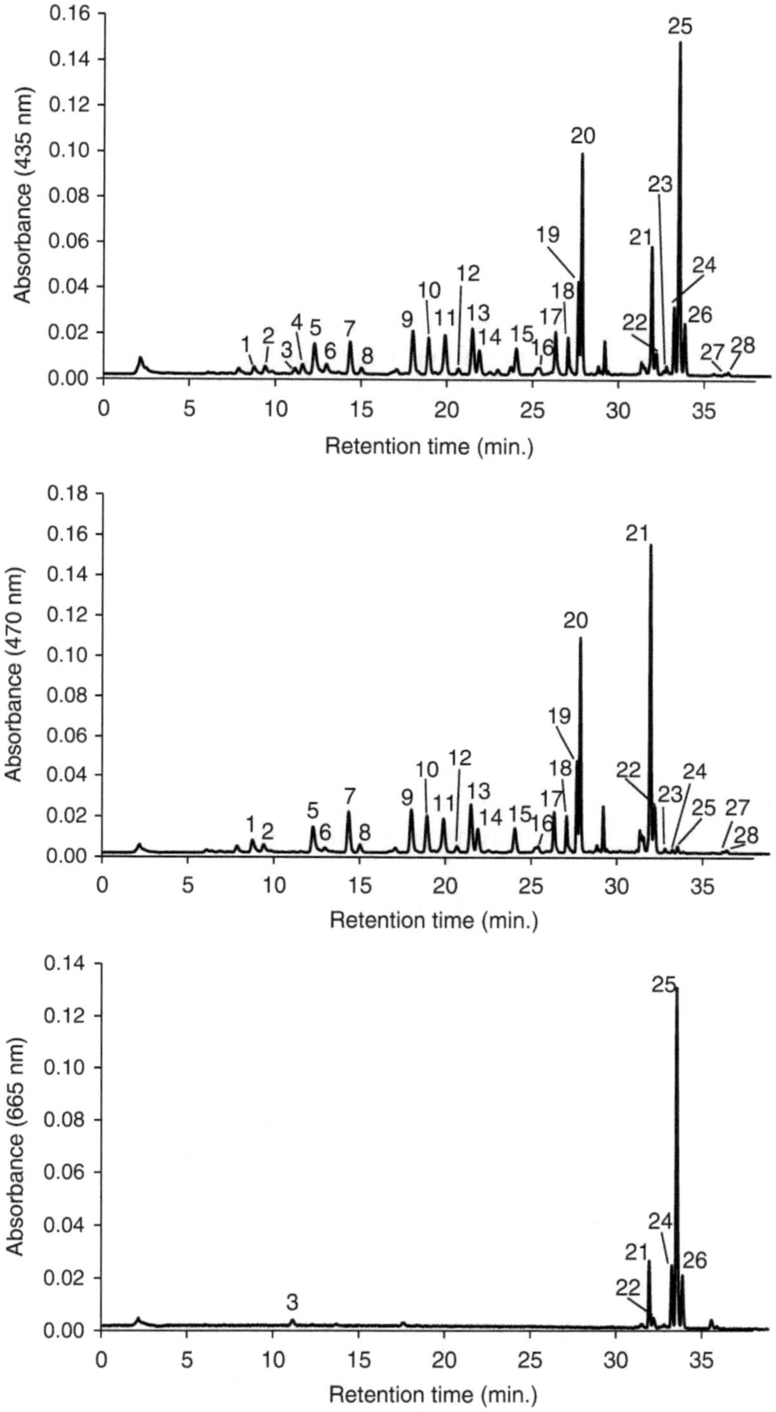

Fig. 1 High-performance liquid chromatogram of pigments from marine phytoplankton (mixed pigment standard from DHI + monovinyl (MV) Chlc_3 + Chl c_1), obtained with the method of [6] monitored at 435 nm (*upper trace*), 470 nm (*middle trace*), and 665 nm (*lower trace*). Peak identification: 1, Chl c_3; 2, MV-Chl c_3; 3, chlorophyllide *a*; 4, Mg-2,4-divinylpheopropyrin a_5 monomethyl ester (Mg-DVP); 5, Chl c_2; 6, Chl c_1; 7, peridinin; 8, peridinin-like unknown carotenoid; 9, 19′-butanoyloxyfucoxanthin; 10, fucoxanthin; 11, neoxanthin; 12, prasinoxanthin; 13, violaxanthin; 14, 19′-hexanoyloxyfucoxanthin; 15, diadinoxanthin; 16, antheraxanthin; 17, diatoxanthin; 18, alloxanthin; 19, zeaxanthin; 20, lutein; 21, divinyl (DV)-Chl *b* + Chl *b*; 22, DV-Chl *b* epimer; 23, MGDG-Chl c_2; 24, DV-Chl a; 25, Chl *a*; 26, Chl *a* epimer; 27, ß, ε-carotene; 28, ß,ß-carotene

volumes of primary standards (concentration approx. 2.5 μg/mL) and adjusting to a known volume with acetone:water (9:1, v/v). Then, prepare a series of at least six *working solutions* by diluting different volumes of the "concentrated pigment mixture" to the same final volume with acetone:water (9:1, v/v). These solutions should be prepared in different ranges of concentrations, depending on the expected concentration of pigments in the samples. Hence, different ranges of concentrations will be necessary for oligotrophic or eutrophic waters [15]. Inject each *working solution* into the HPLC system in triplicate (water must be added to each working standard just prior to injection, as for samples—*see* **step 1** of Subheading 3.2). Derive a response factor for each pigment (f_p) as the inverse of the slope of the regression line of peak area (A_p) against the weight (ng) of pigment injected (w_p) [16].

3. System calibration, internal standard method [14]: Operate as described above, but incorporate the same amount of an internal standard primary solution to each *working solution* (with growing concentrations of pigments to be analyzed in the series, while the internal standard concentration remains constant). Recommended internal standard concentration in each *working solution* should be in the middle of the range covered by the other pigments. Inject as described above. For each pigment standard, the relative response factor (f_p^{is}) can be calculated as the inverse of the slope of the regression of the ratio of areas of the pigment and internal standard (A_p/A_{is}) against the ratio of their corresponding weights in the injection (w_p/w_{is}).
4. Determining the concentration of pigments in the samples with the external standard method [14]: Use the external standardization equation to calculate Cp, the concentration of a pigment in seawater (ng/L), from its peak area (A_p), and response factor (f_p):

$$Cp = \frac{A_p \cdot f_p \cdot v_{ext} \cdot 10^3}{v_{inj} \cdot v_{filt} \cdot B}$$

where v_{ext}, v_{inj}, and v_{filt} denote the volumes of extract (mL), injected sample (μL), and filtered seawater sample (L). B is the dilution factor (<1.0), calculated as the ratio of the extract volume and the sum of extract and water added prior to injection.
5. Calculating the concentration of pigments in the samples with the internal standard method [14]:

$$Cp = \frac{A_p \cdot w_{is} \cdot f_p^{is}}{A_{is} \cdot v_{filt}}$$

where A_p is the peak area, w_{is} denotes the mass (ng) of internal standard added to the sample, and v_{filt} is the volume of filtered seawater sample (L) (*see* **Note 15**).

4 Notes

1. Filters with different nominal pore size (GF/D, GF/C) can be used to collect cells of different sizes. Fractionated samples can be achieved employing sequentially several of these filters. On the other hand, polycarbonate filters are commonly used in oceanography to size-fractionate phytoplankton. These filters can either be used directly to extract pigments (only if MeOH is used as extraction solvent (*see* **Note 2**), due to the limited chemical resistance of polycarbonate to other solvents, or they can be used to pre-filter a water sample which is then filtered on GF/F, providing information on the smaller size categories of phytoplankton (*see* ref. 17 for a recent update on filtration, storage, and extraction).
2. Different solvents have been used to extract pigments from algal cells retained on the filter (*see* ref. 17). 90 % aqueous acetone is still the most commonly used solvent, but aqueous 95 % MeOH (*see* ref. 6) is also being used with good results [18]. As methanolic extracts have been claimed to promote Chl allomerization upon prolonged storage [17], samples should be immediately injected after extraction. Sparging the sample vial with an inert gas to displace air can help reduce pigment alteration when samples have to stand in a refrigerated autosampler for several hours [18]. Dimethylformamide, alone or in combination with other solvents, has also been used with very good extraction efficiency, but many laboratories are reluctant to use this solvent because of its toxicity [17]. If the filters are extracted by grinding with a spatula followed by placing them in an ultrasonic bath, filter debris may still contain rather large fibers and centrifugation can be omitted as most debris will not pass the luer tip of the syringe barrel, thus not clogging the filter. When a sonifier probe is used to disrupt the filter, a slurry is produced so that centrifugation or prefiltration is necessary before a final filtration onto 0.22 μm filters. Filter disruption can be done directly in the syringe [2, 4].
3. The original method [6] used 0.25 M pyridine solution in the eluents, but further work showed that the same chromatographic performance can be achieved with 0.025 M. The stock solution is stable at ambient temperature for months. The working solution tends to contaminate (fungal growth has been observed). Prepare it fresh each 2 or 3 days. Purge the chromatographic tubing with MeOH when not in use. If the working solution is not to be used for several days, keep it refrigerated and refilter before the next use.
4. Photosynthetic pigments are sensitive towards light, oxygen, heat, acids, and alkalis. All operations should be done under dim light (a secure green light can be used) and keeping reasonable laboratory temperature. Perform all solvent-handling

operations in a fume hood. Wear appropriate gloves and protective glasses.

5. Highly concentrated samples (cultures, field samples from estuarine or coastal waters) can result in filter clogging. Prolonged filtration with (partially) clogged filters can promote mechanical disruption of algal cells. Filtration times (or volumes) must be adjusted in a compromise between extract concentration and sample integrity, but should not be more than 20 min.
6. Immediate freezing is best achieved with liquid nitrogen (dry shippers are convenient for this), but quick storage in a −80 °C freezer normally suffices. Both are adequate for long-term conservation of the samples. Dry ice can be an alternative for quick freezing and short-term storage, but avoid contact of the filters with vapors from dry ice (they can acidify pigments).
7. Mixing with water is necessary to match injection solvent viscosity to that of eluent, thus avoiding peak shape distortion for less retained pigments due to "viscous fingering" phenomena [19]. Losses of low-polarity pigments can occur due to their limited solubility in aqueous solvents [20]. Mixing in the injection loop minimizes these losses as any insoluble pigment will be recovered as solvent polarity decreases with the development of the gradient. MeOH:water mixtures show higher viscosities than acetone:water ones, so that the step of water addition can be omitted when MeOH:water (95:5, v/v) is used as extraction solvent and injection volumes do not exceed 100 μL.
8. When using gradients with high- or low-pressure mixing systems, some differences in retention times can be observed that are caused by the different "dwell volume" (the volume between the point of mixing and the column) in the two systems (in gradient elution, the actual composition in the gradient program is delayed in reaching the column, because it has to pass through the dwell volume). Retention times will be higher on low-pressure systems that normally have larger dwell volumes. In consequence, changes in gradient profiles and/or eluent composition can be necessary to adjust retention and resolution with different HPLC systems [6].
9. The amount of water used to achieve the retention of pigment fractions in the solid-phase extraction cartridge can vary depending on the compound and the composition of the solvent in which it is eluted from the preparative fraction. In some cases excess water causes the pigment to pass unretained through the cartridge due to the formation of micelles. If this is observed, a reduction in water content can promote retention.
10. This procedure can also be used for the preparative isolation of unknown pigments for further characterization by structural

techniques (mass spectrometry, nuclear magnetic resonance), but the co-occurrence of non-absorbing lipids usually requires additional purification.

11. Check homogeneity of spectrum at least at mid-upslope, top, and mid-downslope of each peak.
12. Different elution can be achieved by changing the stationary phase. Polymeric octadecylsilica columns (e.g., Vydac TP, The Separations Group, Hesperia, California, USA) can be used for this purpose [8].
13. Spectral checking in different regions of the peak does not always ensure peak purity. This can be the case if the impurity appears at very low concentration, if the spectrum of the impurity is very similar to that of the main pigment in the peak or if a total coelution (identical retention profile) occurs.
14. At 435 nm most pigments are detected, except pheophytin *a*, pheophorbide *a*, and their derivatives; 470 nm excludes Chl *a* derivatives from the chromatogram; and 665 nm detects only Chl *a*, pheophytin *a*, pheophorbide *a*, and their derivatives.
15. Extraction and injection volumes are not included in the equation.

References

1. Wright SW, Jeffrey SW (2006) Pigment markers for phytoplankton production. In: Volkman JK (ed) Marine organic matter, vol 2, Handbook of environmental chemistry. Springer, Heidelberg, pp 71–104
2. Wright SW (2005) Analysis of phytoplankton populations using pigment markers. http://www.antarctica.gov.au/__data/assets/pdf_file/0007/21877/ml_38651523287037_analysis20of20phytoplankton20pigments.pdf. Accessed 1 Nov 2013
3. Roy S, Llewellyn CA, Egeland E et al (eds) (2011) Phytoplankton pigments: characterization, chemotaxonomy and applications in oceanography. Cambridge University Press, Cambridge, UK
4. Jeffrey SW, Mantoura RFC, Wright SW (eds) (1997) Phytoplankton pigments in oceanography: guidelines to modern methods. UNESCO Publishing, Paris
5. Roy S, Garrido JL (2013) Pigments Liquid chromatography. In: Reedijk J (ed) Reference module in chemistry, molecular sciences and chemical engineering. Elsevier, Waltham, MA. doi:10.1016/B978-0-12-409547-2.04878-2
6. Zapata M, Rodríguez F, Garrido JL (2000) Separation of chlorophylls and carotenoids from marine phytoplankton: a new HPLC method using a reversed-phase C_8 column and pyridine-containing mobile phases. Mar Ecol Prog Ser 195:29–45
7. Van Heukelem L, Thomas CS (2001) Computer-assisted high-performance liquid chromatography method development with applications to the isolation and analysis of phytoplankton pigments. J Chromatogr A 910:31–49
8. Garrido JL, Airs R, Rodríguez F et al (2011) New HPLC separation techniques. In: Roy S, Egeland ES, Johnsen G et al (eds) Phytoplankton pigments: characterization, chemotaxonomy and applications in oceanography. Cambridge University Press, Cambridge, UK, pp 165–194
9. Roy S, Wright SW, Jeffrey SW (2011) Phytoplankton cultures for standard pigments and their suppliers. In: Roy S, Egeland ES, Johnsen G et al (eds) Phytoplankton pigments: characterization, chemotaxonomy and applications in oceanography. Cambridge University Press, Cambridge, UK, pp 653–657
10. Zapata M, Garrido JL (1991) Influence of injection conditions in reversed-phase high-performance liquid chromatography of chlorophylls and carotenoids. Chromatographia 31:589–594

11. Egeland ES (2011) Data sheets aiding identification of phytoplankton carotenoids and chlorophylls. In: Roy S, Egeland ES, Johnsen G et al (eds) Phytoplankton pigments: characterization, chemotaxonomy and applications in oceanography. Cambridge University Press, Cambridge, pp 665–822
12. Zapata M, Rodríguez F, Garrido JL et al (2004) Photosynthetic pigments in 37 species (65 strains) of Haptophyta: implications for oceanography and chemotaxonomy. Mar Ecol Prog Ser 270:83–102
13. Zapata M, Fraga S, Rodríguez F et al (2012) Pigment-based chloroplast types in dinoflagellates. Mar Ecol Prog Ser 465:33–52
14. Mantoura RFC, Repeta D (1997) Calibration methods for HPLC. In: Jeffrey SW, Mantoura RFC, Wright SW (eds) Phytoplankton pigments in oceanography: guidelines to modern methods. UNESCO Publishing, Paris, pp 407–428
15. Van Heukelem L, Hooker SB (2011) The importance of the quality assurance plan for method validation and minimizing uncertainties in the HPLC analysis of phytoplankton pigments. In: Roy S, Egeland ES, Johnsen G et al (eds) Phytoplankton pigments: characterization, chemotaxonomy and applications in oceanography. Cambridge University Press, Cambridge, pp 195–242
16. Hooker SB, Van Heukelem L (2011) A symbology and vocabulary for an HPLC lexicon. In: Roy S, Egeland ES, Johnsen G et al (eds) Phytoplankton pigments: characterization, chemotaxonomy and applications in oceanography. Cambridge University Press, Cambridge, pp 243–256
17. Pinckney JL, Millie DF, Van Heukelem L (2011) Update on filtration, storage and extraction solvents. In: Roy S, Egeland ES, Johnsen G et al (eds) Phytoplankton pigments: characterization, chemotaxonomy and applications in oceanography. Cambridge University Press, Cambridge, pp 627–635
18. Alou-Font E, Mundy CJ, Roy S et al (2013) Snow cover affects ice algal pigment composition in the coastal Arctic Ocean during spring. Mar Ecol Prog Ser 474:89–104
19. Rousseaux G, Martin M, De Wit A (2011) Viscous fingering in packed chromatographic columns: non-linear dynamics. J Chromatogr A 1218:8353–8361
20. Latasa M, van Lenning K, Garrido JL et al (2001) Losses of chlorophylls and carotenoids in aqueous acetone and MeOH extracts prepared for RPHPLC analysis of pigments. Chromatographia 53:385–391

Chapter 16

Characterization of Phlorotannins from Brown Algae by LC-HRMS

Jeremy E. Melanson and Shawna L. MacKinnon

Abstract

Phlorotannins are a class of polyphenols found in brown seaweeds that have significant potential for use as therapeutics, owing to their wide range of bioactivities. Molecular characterization of phlorotannin-enriched extracts is challenging due to the extreme sample complexity and the wide range of molecular weights observed. Herein, we describe a method for characterizing phlorotannins employing ultrahigh-pressure liquid chromatography (UHPLC) operating in hydrophilic interaction liquid chromatography (HILIC) mode combined with high-resolution mass spectrometry (HRMS).

Key words Brown seaweed, High-resolution mass spectrometry, Hydrophilic interaction liquid chromatography, Phlorotannins, Polyphenols

1 Introduction

Phlorotannins are a class of polyphenols formed by the oligomerization of 1,3,5-trihydroxybenzene, which is also known as phloroglucinol. Phlorotannins are found exclusively in brown macroalgae (seaweeds), which produce them as a defense mechanism in response to environmental stress. The molecular weight (MW) of these compounds can span as high as 100 kDa [1], and brown macroalgae can contain phlorotannins at levels ranging from 0.5 to 20 % of their dry weight [2]. However, phlorotannin levels in brown algae can vary with habitat, time of harvest, light intensity exposure, and nutrient availability in the surrounding waters [2–4].

In recent years, phlorotannins have been associated with numerous bioactivities [5–12], including antioxidant [12], antifungal [6], and anti-inflammatory activities [13]. Despite the potential of phlorotannins for use as natural health products, efforts in product development have likely been hindered by the lack of suitable analytical methods required for the characterization of extracts.

Dagmar B. Stengel and Solène Connan (eds.), *Natural Products From Marine Algae: Methods and Protocols*, Methods in Molecular Biology, vol. 1308, DOI 10.1007/978-1-4939-2684-8_16, © Springer Science+Business Media New York 2015

In many cases phlorotannins are analyzed as total phenolics, where the total contents of phenolic compounds are measured by colorimetric assays such as the Folin–Ciocalteu (FC), the Folin-Denis reagent, and 2,4-dimethoxybenzaldehyde (DMBA) [1, 14]. While these tests are reliable and simple to employ, they provide little information on the chemical composition of the phlorotannins, and this information is vital for the standardization and quality control of any potential commercial products. Therefore, instrumental methods capable of characterizing complex phlorotannin extracts would be particularly valuable.

Analytical methods for phlorotannins based on liquid chromatography (LC) have been reported [1, 15–17], but peak identification is a significant challenge as phlorotannin standards are generally not available. As a result, methods for phlorotannins taking advantage of the added specificity offered by liquid chromatography—mass spectrometry (LC-MS) have been recently described [18–21]. In particular, high-resolution mass spectrometry (HRMS) methods capable of obtaining high mass accuracy across a broad mass range are particularly attractive for complex phlorotannin extracts [18, 19]. Mass spectral data can be acquired over a broad mass range and accurate mass measurements by HRMS can be used to directly assign elemental compositions of novel structures. In addition, as HRMS data are collected in a non-targeted fashion, it allows for retrospective analysis of archived data for compounds of interest that may arise in the future.

The method described below employs HRMS in conjunction with ultrahigh-pressure liquid chromatography (UHPLC) for the characterization of phlorotannins in brown algae [18]. The method is optimized for low molecular weight phlorotannins and employs hydrophilic interaction liquid chromatography (HILIC). As instrumental settings are not tuned for individual phlorotannins in HRMS, the method is straightforward and relatively simple to implement. However, HRMS methods typically require sufficient post-acquisition data processing and analysis, and this component of the method is described in detail below.

2 Materials and Instrumentation

2.1 Solvents

1. Deionized water obtained from a Milli-Q® Advantage A10 Water Purification System or equivalent (prepared by purifying distilled water to attain a resistance of 18 MΩ at 25 °C with a TOC of <5 ppb).
2. High purity grade methanol, HPLC grade acetonitrile and dichloromethane.

2.2 Extraction and Sample Preparation Components

1. Fitz®Mill grinder (South Plainfield, NJ) or equivalent equipped with a 40-mesh size.
2. 80 % methanol in deionized water (v/v).
3. Büchner funnel and glass fiber filters (Whatman 934-AH, 42.5 mm).
4. Nitrogen flow or Genevac® EZ-2 solvent evaporator.
5. Fractionation of seaweed extracts: C_{18} Sep-Pak (6 cc, 500 mg, Waters) solid phase extraction cartridge.
6. Filtration of extract fractions prior to HILIC-HRMS analysis: Ultrafree®-MC centrifugal filters with Durapore PVDF membranes (0.45 µm pore size, Millipore).

2.3 HILIC-HRMS Components

1. Thermo Scientific Accela UHPLC equipped with a quaternary pump and autosampler, coupled to a Thermo Scientific Exactive™ bench top mass spectrometer with Orbitrap™ technology, equipped with the optional Heated Electrospray Ionization (HESI) probe (*see* **Note 1**). Xcalibur 2.1 software was used for instrument control and data acquisition.
2. Waters UPLC® BEH Amide column (100×2.1 mm; 1.7 µm).
3. Mobile phase A: 10 mM ammonium acetate at pH 9.0, prepared by dissolving 0.77 g of ammonium acetate in approximately 900 mL of deionized water, adjusting to pH 9.0 with a concentrated ammonium hydroxide solution, then diluting to 1 L with deionized water.
4. Mobile phase B: acetonitrile.
5. MS calibration solution provided by the manufacturer: 10 pmol/µL sodium dodecyl sulfate (SDS), 10 pmol/µL sodium taurocholate, 0.001 % Ultramark 1621, and 0.01 % acetic acid in 50:25:25 acetonitrile–methanol–water.

3 Methods

3.1 Preparation of Extract from Brown Seaweeds

1. Collect brown seaweeds and transport them to the lab in a cooler containing ice. Clean them of contaminants and rinse them with seawater before being frozen and stored at −20 °C.
2. Freeze-dry frozen seaweed samples until the seaweed is brittle.
3. Grind the seaweed to a fine powder using a Fitz®Mill grinder or equivalent equipped with a size 40 mesh.
4. Weigh out 2 g of dried seaweed powder into a 50 mL beaker and add 25 mL of 80 % methanol. Stir the mixture for 30 min using a magnetic stirring bar. Allow the mixture to settle for 10 min without stirring. Decant the extractant using a glass pipette. Repeat this extraction process three more times using 10 mL of 80 % methanol.

5. Filter the combined extract solutions using the Büchner funnel equipped with a glass fiber filter.
6. Evaporate the methanol under a nitrogen flow or using a Genevac® EZ-2 solvent evaporator or equivalent operated at or below 35 °C. Prepare triplicate extracts of each seaweed sample.
7. Transfer resulting aqueous filtrates into 20 mL glass scintillation vials. Partition the aqueous filtrate by adding 1 volume of dichloromethane to the vial. Mix by drawing solution up and down in a glass Pasteur pipette 15–20 times. Screw the top on each vial and let sit for 10 min to allow for phase separation.
8. Remove the bottom dichloromethane layer using a glass Pasteur pipette. Repeat partitioning procedure three more times using dichloromethane. Discard the lipid containing dichloromethane layer each time.
9. Pipette the partitioned aqueous layer into a pre-weighed 20 mL scintillation vial and evaporated any residual dichloromethane under a stream of nitrogen or using a Genevac® EZ-2 solvent evaporator or equivalent operated at or below 35 °C.
10. Freeze-dry the resulting aqueous layer to dryness. Record the weight of the vial and determine the weight of the resulting polyphenolic containing extract. Flush the vial with nitrogen and store the extract at −80 °C.

3.2 Preparation of Phlorotannin-Enriched Fraction

1. Precondition a 6 cc, 500 mg Waters Sep-Pak column cartridge with 12 mL of methanol followed by 18 mL of deionized water (*see* **Note 2**).
2. Dissolve the extract prepared in Subheading 3.1 in deionized water at a concentration of 50–150 mg/mL and sonicate for 10 min if aid in dissolution is required.
3. Using a glass Pasteur pipette, load the resulting sample on to the preconditioned column cartridge. Rinse the sample vial with 1–2 mL of high purity water and add to the column.
4. Elute the column cartridge with deionized water so that the total water volume added to the column is equal to 20 mL. Collect the resulting eluent, which contains salts and carbohydrates. Do not let the column cartridge run dry.
5. To obtain the phlorotannin enriched fraction, elute the column cartridge next with 30 mL of methanol and collect the eluent in a pre-weighed vial.
6. Evaporate the methanol from this fraction under a stream of nitrogen or using a Genevac® EZ-2 solvent evaporator or equivalent operated at or below 35 °C.
7. Freeze-dry the resulting aqueous fraction, obtain weight of fraction, purge the vial with nitrogen and store sealed extract at −80 °C until further analysis.

8. To prepare the phlorotannin-enriched fraction for HILIC-HRMS analysis, dissolve fraction in 75 % acetonitrile /25 % methanol at a concentration of 1–3 mg/mL and filter using an Ultrafree®-MC centrifugal filter (*see* **Note 3**).

3.3 HILIC-HRMS Procedure

1. Calibrate the mass spectrometer daily using the automated function for optimal sensitivity and mass accuracy according the manufacturer's recommended procedure (*see* **Note 4**) by infusing the calibration solution at a rate of 5 μL/min with a syringe pump and using a tune page optimized for this low flow rate and adjusting the HESI probe depth to position "B".
2. Create an appropriate tune page for efficient desolvation and ionization (*see* **Note 5**) at the LC flow rate of 400 μL/min using settings such as the following: spray voltage −2.7 kV, heater temperature 300 °C, capillary temperature 350 °C, sheath gas 55 psi, aux gas 18 (arbitrary units), sweep gas 0 (arbitrary units), capillary voltage −60 V, and tube lens voltage −125 V. Adjust the HESI probe depth to position "D".
3. Create an acquisition method within Xcalibur software for the both the LC and MS control, specifying the correct tune page created in **step 2**. The MS method should consist of a scan wide enough to ensure detection of all phlorotannins, such as *m/z* 150–2,000, using the "ultra-high" resolution setting (100,000, 1 Hz; *see* **Note 6**), and operating in negative ion mode. The LC program should include the flow rate of 400 μL/min, the column temperature of 30 °C, and the gradient consisting of an initial hold at 5 % mobile phase A for 1 min, followed by a linear gradient to 35 % A in 16 min, followed by re-equilibration for 5 min at 5 % A, for a total run time of 22 min.
4. After proper preconditioning of the LC column (*see* **Note 7**) at the initial mobile phase conditions, connect the column to the MS and adjust the low flow rate to 400 μL/min.
5. Complete the sample table specifying the correct acquisition method from **step 3** and using the appropriate injection volume, typically 3–10 μL, depending on sample concentration.
6. Start the sequence (*see* **Note 8**).

3.4 Data Processing

1. Following acquisition, the total ion current chromatogram (TICC) or base peak chromatogram (BPC) can be inspected for rapid visualization of the major low molecular weight phlorotannins with a degree of polymerization (DP) $\leq$10, as shown in Fig. 1. Larger phlorotannins may also be present, but will not typically be observed in the TICC (*see* **Note 9**).
2. As the TICC is only a simple representation that depicts the major sample components, masses corresponding to the individual phlorotannins must be extracted from the raw data and plotted as a function of retention time, resulting in an

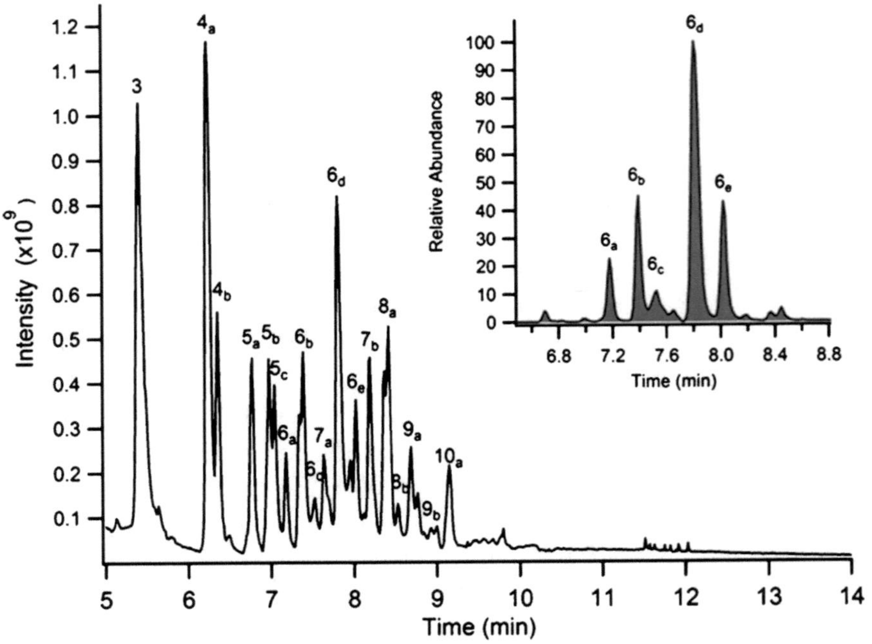

Fig. 1 TICC of a phlorotannin-enriched extract of *Fucus vesiculosus*. (inset) Extracted ion chromatogram of *m/z* 745.10463, demonstrating the number of phlorotannin isomers derived from six phloroglucinol units. Peak numbers correspond to the number of phloroglucinol units and the letters correspond to different isomers of the same degree of polymerization. For situations where multiple peaks are co-eluting, the peak assignment is based on the most intense peak in the MS spectrum. Reproduced from ref. [18] with permission from Wiley

extracted ion chromatogram (EIC). Depending on the mass accuracy and precision of the mass spectrometer, typically 5–10 ppm mass windows centered at the monoisotopic mass of the ion will be sufficient. For instance, Fig. 1 (inset) shows the result of extracting *m/z* 745.10463 with a 10 ppm mass window, corresponding to phlorotannins with six phloroglucinol units in *Fucus vesiculosus*. The various peaks at *m/z* 745.10463 correspond to structural isomers at DP = 6, caused by the various bonding types of phlorotannins (*see* **Note 10**).

3. Without knowledge of the phlorotannin profile of the extract, masses should be extracted over a wide range of phlorotannin DP. This can be accomplished manually in Xcalibur software, or in a software package designed for high-throughput automated screening such as Thermo's ToxID software or equivalent. ToxID requires only the elemental composition of the various phlorotannin species, while Xcalibur would require the exact monoisotopic masses of the ions observed. Observed masses to be screened can be calculated manually (*see* **Note 11**) or with the aid of Xcalibur software (or equivalent). Alternatively, Table 1 lists masses of observed ions for phlorotannins up to DP = 50, along with their elemental composition and molecular weight.

Table 1
List of elemental compositions, monoisotopic masses, and calculated masses for phlorotannin ions detected in negative ion mode HRMS at each degree of polymerization (DP), with the ions highlighted in bold font representing the predominant charge state detected under the experimental conditions described

DP	Elemental composition	Monoisotopic mass (Da)	Masses of observed phlorotannin ions (*m/z*)			
			[M-H]$^-$	[M-2H]$^{-2}$	[M-3H]$^{-3}$	[M-4H]$^{-4}$
1	$C_6H_6O_3$	126.03169	125.02442	62.00857	41.00329	
2	$C_{12}H_{10}O_6$	250.04774	249.04046	124.01659	82.34197	61.50466
3	$C_{18}H_{14}O_9$	374.06378	**373.05650**	186.02461	123.68065	92.50867
4	$C_{24}H_{18}O_{12}$	498.07983	**497.07255**	248.03263	165.01933	123.51268
5	$C_{30}H_{22}O_{15}$	622.09587	**621.08859**	310.04066	206.35801	154.51669
6	$C_{36}H_{26}O_{18}$	746.11191	**745.10463**	372.04868	247.69669	185.52070
7	$C_{42}H_{30}O_{21}$	870.12796	**869.12068**	434.05670	289.03537	216.52471
8	$C_{48}H_{34}O_{24}$	994.14400	**993.13672**	496.06472	330.37405	247.52872
9	$C_{54}H_{38}O_{27}$	1,118.16004	**1,117.15276**	558.07274	371.71274	278.53273
10	$C_{60}H_{42}O_{30}$	1,242.17609	**1,241.16881**	620.08076	413.05142	309.53674
11	$C_{66}H_{46}O_{33}$	1,366.19213	**1,365.18485**	682.08879	454.39010	340.54075
12	$C_{72}H_{50}O_{36}$	1,490.20818	**1,489.20090**	744.09681	495.72878	371.54476
13	$C_{78}H_{54}O_{39}$	1,614.22422	**1,613.21694**	806.10483	537.06746	402.54878
14	$C_{84}H_{58}O_{42}$	1,738.24026	1,737.23298	**868.11285**	578.40614	433.55279
15	$C_{90}H_{62}O_{45}$	1,862.25631	1,861.24903	**930.12087**	619.74482	464.55680
16	$C_{96}H_{66}O_{48}$	1,986.27235	1,985.26507	**992.12890**	661.08350	495.56081
17	$C_{102}H_{70}O_{51}$	2,110.28839	2,109.28112	**1,054.13692**	702.42219	526.56482
18	$C_{108}H_{74}O_{54}$	2,234.30444	2,233.29716	**1,116.14494**	743.76087	557.56883
19	$C_{114}H_{78}O_{57}$	2,358.32048	2,357.31320	**1,178.15296**	785.09955	588.57284
20	$C_{120}H_{82}O_{60}$	2,482.33653	2,481.32925	**1,240.16098**	826.43823	619.57685
21	$C_{126}H_{86}O_{63}$	2,606.35257	2,605.34529	**1,302.16901**	867.77691	650.58086
22	$C_{132}H_{90}O_{66}$	2,730.36861	2,729.36133	**1,364.17703**	909.11559	681.58487
23	$C_{138}H_{94}O_{69}$	2,854.38466	2,853.37738	**1,426.18505**	950.45427	712.58889
24	$C_{144}H_{98}O_{72}$	2,978.40070	2,977.39342	**1,488.19307**	991.79295	743.59290
25	$C_{150}H_{102}O_{75}$	3,102.41675	3,101.40947	1,550.20109	**1,033.13164**	774.59691
26	$C_{156}H_{106}O_{78}$	3,226.43279	3,225.42551	1,612.20912	**1,074.47032**	805.60092
27	$C_{162}H_{110}O_{81}$	3,350.44883	3,349.44155	1,674.21714	**1,115.80900**	836.60493
28	$C_{168}H_{114}O_{84}$	3,474.46488	3,473.45760	1,736.22516	**1,157.14768**	867.60894

(continued)

Table 1 (continued)

DP	Elemental composition	Monoisotopic mass (Da)	Masses of observed phlorotannin ions (m/z)			
			[M-H]$^-$	[M-2H]$^{-2}$	[M-3H]$^{-3}$	[M-4H]$^{-4}$
29	$C_{174}H_{118}O_{87}$	3,598.48092	3,597.47364	1,798.23318	**1,198.48636**	898.61295
30	$C_{180}H_{122}O_{90}$	3,722.49696	3,721.48968	1,860.24120	**1,239.82504**	929.61696
31	$C_{186}H_{126}O_{93}$	3,846.51301	3,845.50573	1,922.24922	**1,281.16372**	960.62097
32	$C_{192}H_{130}O_{96}$	3,970.52905	3,969.52177	1,984.25725	**1,322.50240**	991.62498
33	$C_{198}H_{134}O_{99}$	4,094.54510	4,093.53782	2,046.26527	**1,363.84109**	1,022.62899
34	$C_{204}H_{138}O_{102}$	4,218.56114	4,217.55386	2,108.27329	**1,405.17977**	1,053.63301
35	$C_{210}H_{142}O_{105}$	4,342.57718	4,341.56990	2,170.28131	**1,446.51845**	1,084.63702
36	$C_{216}H_{146}O_{108}$	4,466.59323	4,465.58595	2,232.28933	**1,487.85713**	1,115.64103
37	$C_{222}H_{150}O_{111}$	4,590.60927	4,589.60199	2,294.29736	**1,529.19581**	1,146.64504
38	$C_{228}H_{154}O_{114}$	4,714.62531	4,713.61804	2,356.30538	**1,570.53449**	1,177.64905
39	$C_{234}H_{158}O_{117}$	4,838.64136	4,837.63408	2,418.31340	1,611.87317	**1,208.65306**
40	$C_{240}H_{162}O_{120}$	4,962.65740	4,961.65012	2,480.32142	1,653.21185	**1,239.65707**
41	$C_{246}H_{166}O_{123}$	5,086.67345	5,085.66617	2,542.32944	1,694.55054	**1,270.66108**
42	$C_{252}H_{170}O_{126}$	5,210.68949	5,209.68221	2,604.33747	1,735.88922	**1,301.66509**
43	$C_{258}H_{174}O_{129}$	5,334.70553	5,333.69825	2,666.34549	1,777.22790	**1,332.66910**
44	$C_{264}H_{178}O_{132}$	5,458.72158	5,457.71430	2,728.35351	1,818.56658	**1,363.67312**
45	$C_{270}H_{182}O_{135}$	5,582.73762	5,581.73034	2,790.36153	1,859.90526	**1,394.67713**
46	$C_{276}H_{186}O_{138}$	5,706.75366	5,705.74639	2,852.36955	1,901.24394	**1,425.68114**
47	$C_{282}H_{190}O_{141}$	5,830.76971	5,829.76243	2,914.37758	1,942.58262	**1,456.68515**
48	$C_{288}H_{194}O_{144}$	5,954.78575	5,953.77847	2,976.38560	1,983.92130	**1,487.68916**
49	$C_{294}H_{198}O_{147}$	6,078.80180	6,077.79452	3,038.39362	2,025.25999	**1,518.69317**
50	$C_{300}H_{202}O_{150}$	6,202.81784	6,201.81056	3,100.40164	2,066.59867	**1,549.69718**

4. While it might appear that the larger phlorotannins would exceed the mass range of the mass spectrometer, larger phlorotannins can be detected as they appear as multiply charged ions under electrospray ionization conditions, gaining one negative charge for every ionized hydroxyl group (loss of H). With increasing phlorotannin size, there is greater probability of multiple charge sites. This is depicted in Table 1, with the ions highlighted in bold font representing the predominant charge state detected at each DP. For instance, [M-H]$^{-2}$ ions were the predominant ions observed from DP 14–24 (*see* **Note 12**).

5. Once the raw data are screened against the target list of phlorotannins, the resulting EIC (Fig. 1 inset) can be inspected manually to confirm the presence/absence of individual phlorotannins and to obtain peak areas suitable for relative or semi-quantitation.
6. Mass errors can be calculated to ensure confidence in results by confirming masses are detected within the instrumental specification. Typically reported in parts-per-million (ppm), the mass error (Δ_m) can be calculated by:

$$\Delta_m = \left\{\left(m_{exp} - m_{calc}\right) / m_{calc}\right\} \times 10^6$$

where m_{exp} is the mass measured experimentally (typically average m/z across chromatographic peak) and m_{calc} is the calculated monoisotopic exact mass (*see* Table 1). For instance, the DP = 6 ion was detected at m/z 745.10753 (Fig. 2a), so:

$$\Delta_m = \left\{(745.10753 - 745.10463) / 745.10463\right\} \times 10^6 = 3.9\,\text{ppm}.$$

This is within the mass accuracy tolerance of the Exactive mass spectrometer of ± 5 ppm (*see* **Note 13**) using external calibration. Specialized HRMS screening software programs such as ToxID will display all EICs, mass errors, and peak areas for easy viewing.
7. Phlorotannin hits can then be further scrutinized as needed by referring back to the mass spectral data. The mass spectra can confirm phlorotannin identity by matching the experimental isotopic pattern with the theoretical model based on the proposed elemental composition, using Xcalibur software (or equivalent; *see* **Note 14**). The isotopic pattern can also be used to confirm that the correct charge state was assigned, based on the peak spacing of the isotopes. For instance, Fig. 2b shows the mass spectrum for the phlorotannin of DP = 16. The isotopic pattern matches closely with the predicted pattern and the peak spacing of ~ 0.5 m/z is indicative of a −2 charge (*see* **Note 15**), thus confirming the structure of $C_{96}H_{66}O_{48}$.

4 Notes

1. This method could also be effectively implemented on alternative instrumentation. Any equivalent UHPLC system capable of handling sub-2-μm particle columns and pressures in excess of 1,000 bar would be sufficient. In addition, other HRMS instruments such as modern quadrupole/time-of-flight (QTOF) systems with resolving power of 20,000 FWHM or greater would be capable of resolving the isotopic patterns of the largest phlorotannins described in this method (DP = 50; $C_{300}H_{202}O_{150}$).

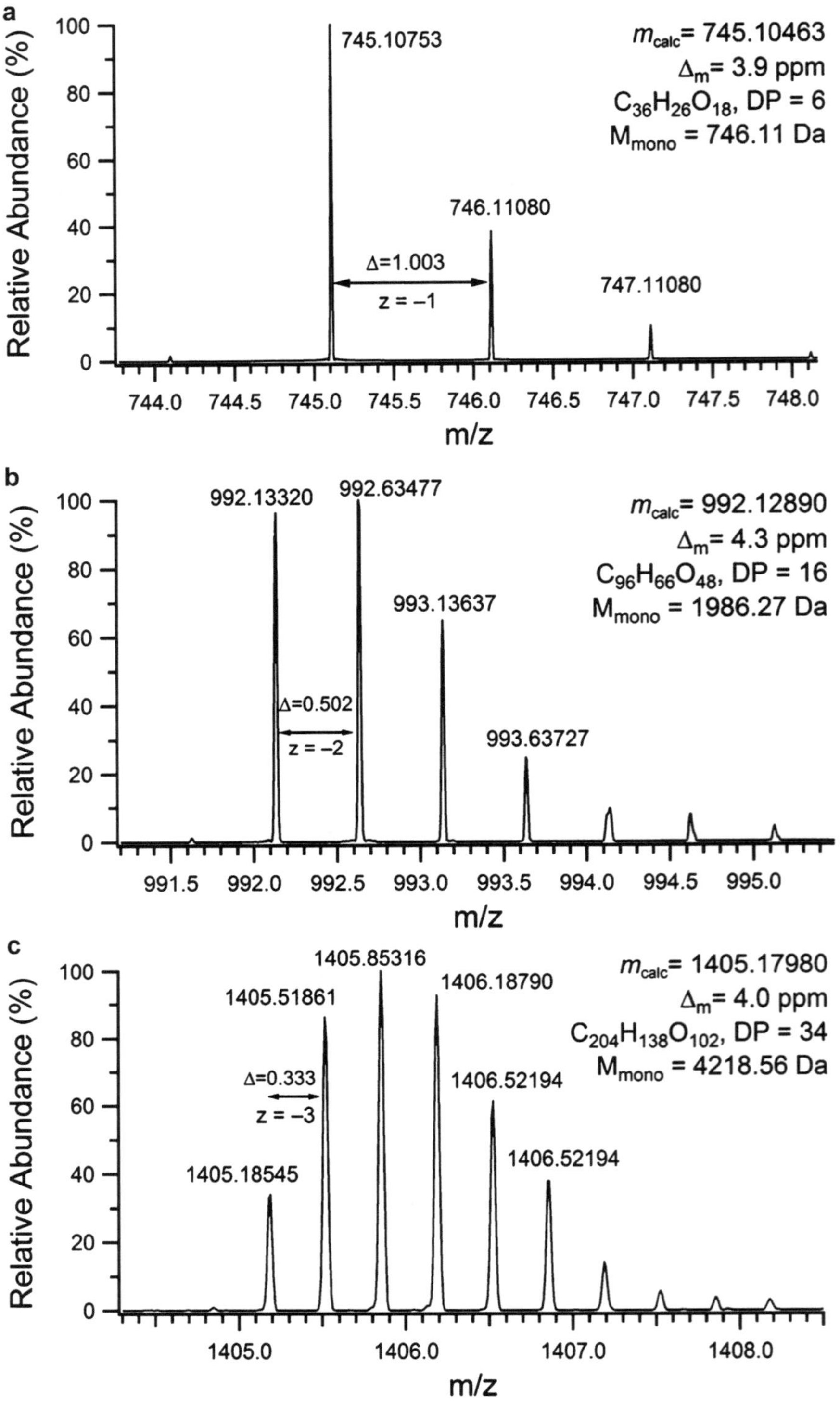

Fig. 2 Typical mass spectra of phlorotannins in *Fucus vesiculosus*: (**a**) $C_{36}H_{26}O_{18}$, DP = 6, charge state = −1; (**b**) $C_{96}H_{66}O_{48}$, DP = 16, charge state = −2; (**c**) $C_{204}H_{138}O_{102}$, DP = 34, charge state = −3. Reproduced from ref. [18] with permission from Wiley

2. It is critical that this preconditioning is conducted to ensure that the column achieves the required separation.
3. The phlorotannin enriched fractions can be verified to contain phlorotannins using proton NMR spectroscopy. To do this dissolve a portion of the phlorotannin enriched sample in methanol d_4 NMR solvent and acquire a standard proton NMR spectrum. If phlorotannins are present, peaks should be observed between 6.2 and 7.5 ppm.
4. While this method employs negative ionization mode only, performing the full calibration in both positive and negative modes is generally recommended.
5. Tuning specifically for phlorotannins is challenging as standards are generally not available. We performed tuning and optimization with an extract of *Fucus vesiculosus*. As infusing with a syringe pump at low flow rate is not useful for simulating LC conditions, tuning is best accomplished with flow injection analysis with the column removed using a representative solvent composition (i.e., mid-point of gradient). We performed rough tuning while observing the complete distribution of phlorotannins, and performed more fine tuning on an individual species such as the DP = 6 ion at m/z 745.10463.
6. As chromatographic peaks are generally quite broad in HILIC, the maximum resolution setting and slowest acquisition rate (1 spectrum/s) was used in this case and provided a sufficient number of data points across chromatographic peaks. If an alternative mode of separation was employed yielding sharper chromatographic peaks, a more rapid scan rate (lower resolution) might need to be employed to provide sufficient data points.
7. Condition the HILIC column according to manufacturer recommendations, typically beginning with an initial flush with 60 % acetonitrile for 50 column volumes (~1 h), followed by the initial conditions of 5 % mobile phase A for 20 column volumes (~30 min). Keep in mind that re-equilibration of HILIC columns between gradient runs tend to be longer than with reversed phase columns, so longer times than those described above might be necessary to achieve reproducible retention times, depending on extract complexity and concentration. Also, as HILIC columns tend to be less robust in general than reversed phase columns, it is recommended that a new column be used for this application, and used exclusively for this method.
8. Typically a blank run would be performed at the beginning, end, and in-between some injections if sample-carry over is observed or expected. A quality control sample, such as an in-house standardized extract, can be injected at specified intervals throughout the sequence to ensure method performance. We used a well-characterized extract of *Fucus vesiculosus*.

9. This LC method is optimized for low molecular weight phlorotannins and larger phlorotannins (>1,200 Da) typically generate broad peaks that elute between 10 and 14 min, many of which co-elute, and will not appear in the TICC due to their lower abundance. As the degree of polymerization increases, the number of potential isomers would presumably increase due to the variety of branching and bonding types, and isomers for these larger phlorotannins are only partially resolved and appear as broad peaks.
10. Increased retention time of phlorotannin isomers correlates with a greater number of free hydroxyl groups for the different phlorotannin bonding types, due to H-bonding with the amide stationary phase. According to ref. [16], we would expect retention times of compounds with the same degree of polymerization to increase as follows: phlorethols < fucophlorethols < fucols, based on the number of free hydroxyl groups.
11. Calculated exact masses (m_{calc}) can be derived according to:

$$m_{\mathrm{calc}} = \left\{M_{\mathrm{mono}} - \left(|z| \times 1.00728\right)\right\} / |z|$$

where M_{mono} is the monoisotopic mass of the intact phlorotannin, $|z|$ is the absolute value of the charge state of the ion, and 1.00728 is based on the monoisotopic hydrogen mass corrected for the mass due to the gain of an electron. For instance, the calculated exact mass for the $[\mathrm{M-H}]^-$ ion of $C_{36}H_{26}O_{18}$ (DP = 6) is:

$$\left\{746.11191 - \left(|-1| \times 1.00728\right)\right\} / |-1| = 745.10463.$$

12. The major charge state observed for each DP shown in Table 1 will be highly instrument-specific and will vary with experimental conditions, so this should only be used as an approximate guide.
13. In our experience, errors on negative ion mass measurements can exceed the instrument specification, especially for low-abundance ions such as some of the larger phlorotannins, so we tend to use an Δ_{m} threshold of ± 10 ppm. By contrast, if performing internal mass correction (i.e., lock-mass), then the threshold could be set as low as 2–3 ppm.
14. Note that experienced users comfortable with modeling isotopic patterns should consider screening for the M + 1 or even M + 2 isotope peak for larger phlorotannins to obtain optimal signal. For instance, the monoisotopic peak at *m/z* 1405.18545 in Fig. 2c represents less than 40 % of the intensity of the M + 2 peak at *m/z* 1405.85316. So screening for the M + 2 peak in this case would result in significantly higher signal. For simplicity, Table 1 lists only the monoisotopic ions as opposed to the most abundant isotopic peak.

15. As mass spectra display not mass, but mass-to-charge (m/z) ratio, peak spacing of the isotopic cluster is a convenient means to determine charge state. As shown in Fig. 2c, as the isotopic peak spacing is ~0.33 m/z, and isotopes will differ in mass by 1 Da, we know this correspond to a −3 ion ($z = 1/0.33$). Note that since this method employs negative ion mode, only negative ions can be detected. So peak spacing of 1 m/z indicates $z = -1$, 0.5 m/z indicates $z = 2$, 0.33 m/z indicates $z = 3$, 0.25 m/z indicates $z = 4$, 0.2 m/z indicates $z = 5$, etc.

References

1. Parys S, Rosenbaum A, Kehraus S et al (2007) Evaluation of quantitative methods for the determination of polyphenols in algal extracts. J Nat Prod 70:1865–1870
2. Ragan MA, Glombitza KW (1986) Phlorotannins, brown algal polyphenols. In: Round FE, Chapman DJ (eds) Progress in phycological research, vol 4. Biopress, Bristol, pp 129–241
3. Targett NM, Arnold TM (1998) Predicting the effects of brown algal phlorotannins on marine herbivores in tropical and temperate oceans. J Phycol 34:195–205
4. Toth GB, Pavia H (2001) Removal of dissolved brown algal phlorotannins using insoluble polyvinylpolypyrrolidone (PVPP). J Chem Ecol 27:1899–1910
5. Audibert L, Fauchon M, Blanc N et al (2010) Phenolic compounds in the brown seaweed *Ascophyllum nodosum*: distribution and radical-scavenging activities. Phytochem Anal 21:399–405
6. Lopes G, Pinto E, Andrade PB et al (2013) Antifungal activity of phlorotannins against dermatophytes and yeasts: approaches to the mechanism of action and influence on *Candida albicans* virulence factor. PLoS One 8:e72203
7. Lee SH, Jeon YJ (2013) Anti-diabetic effects of brown algae derived phlorotannins, marine polyphenols through diverse mechanisms. Fitoterapia 86:129–136
8. Kwon HJ, Ryu YB, Kim YM et al (2013) In vitro antiviral activity of phlorotannins isolated from *Ecklonia cava* against porcine epidemic diarrhea coronavirus infection and hemagglutination. Bioorg Med Chem 21:4706–4713
9. Kim H, Kong CS, Lee JI et al (2013) Evaluation of inhibitory effect of phlorotannins from *Ecklonia cava* on triglyceride accumulation in adipocyte. J Agric Food Chem 61:8541–8547
10. Kannan RRR, Aderogba MA, Ndhlala AR et al (2013) Acetylcholinesterase inhibitory activity of phlorotannins isolated from the brown alga, *Ecklonia maxima* (Osbeck) Papenfuss. Food Res Int 54:1250–1254
11. Freile-Pelegrín Y, Robledo D (2013) Bioactive phenolic compounds from algae. In: Hernandez-Ledesma B, Herrero M (eds) Bioactive compounds from marine foods: plants and animal sources. John Wiley & Sons Ltd., Oxford, pp 113–129
12. Tierney MS, Smyth TJ, Rai DK et al (2013) Enrichment of polyphenol contents and antioxidant activities of Irish brown macroalgae using food-friendly techniques based on polarity and molecular size. Food Chem 139:753–761
13. Kim AR, Shin TS, Lee MS et al (2009) Isolation and identification of phlorotannins from *Ecklonia stolonifera* with antioxidant and anti-inflammatory properties. J Agric Food Chem 57:3483–3489
14. Stern JL, Hagerman AE, Steinberg PD et al (1996) Phlorotannin-protein interactions. J Chem Ecol 22:1877–1899
15. Koivikko R, Loponen J, Pihlaja K et al (2007) High-performance liquid chromatographic analysis of phlorotannins from the brown alga *Fucus vesiculosus*. Phytochem Anal 18:326–332
16. Yanagida A, Murao H, Ohnishi-Kameyama M et al (2007) Retention behavior of oligomeric proanthocyanidins in hydrophilic interaction chromatography. J Chromatogr A 1143: 153–161
17. Kim SM, Kang SW, Jeon JS et al (2013) Determination of major phlorotannins in *Eisenia bicyclis* using hydrophilic interaction chromatography: seasonal variation and extraction characteristics. Food Chem 138:2399–2406
18. Steevensz AJ, MacKinnon SL, Hankinson R et al (2012) Profiling phlorotannins in brown macroalgae by liquid chromatography-high resolution mass spectrometry. Phytochem Anal 23:547–553

19. Ferreres F, Lopes G, Gil-Izquierdo A et al (2012) Phlorotannin extracts from Fucales characterized by HPLC-DAD-ESI-MS^n: Approaches to hyaluronidase inhibitory capacity and antioxidant properties. Mar Drugs 10:2766–2781
20. Motilva MJ, Serra A, Macià A (2013) Analysis of food polyphenols by ultra high-performance liquid chromatography coupled to mass spectrometry: an overview. J Chromatogr A 1292: 66–82
21. Tierney M, Soler-Vila A, Rai D et al (2013) UPLC-MS profiling of low molecular weight phlorotannin polymers in *Ascophyllum nodosum*, *Pelvetia canaliculata* and *Fucus spiralis*. Metabolomics 10:524–535

Chapter 17

Analysis of Betaines from Marine Algae Using LC-MS-MS

Shawna L. MacKinnon and Cheryl Craft

Abstract

Betaines are a class of quaternary ammonium compounds found in marine algae that can act as osmolytes and/or affect gene expression, and therefore improve plant tolerance to stresses such as temperature extremes, drought, and salinity when applied to agricultural crops. In humans, glycine betaine acts as a methyl donor and has been shown to protect internal organs, improve vascular risk factors, and enhance sport performance. Here we describe a sensitive LC-MS-MS method for the baseline separation and quantification of four betaines found in algae, namely, glycine betaine, δ-aminovaleric acid betaine, γ-aminobutyric acid betaine, and laminine.

Key words Glycine betaine, δ-aminovaleric acid betaine, γ-aminobutyric acid betaine, Laminine, LC-MS-MS analysis, Macroalgae

1 Introduction

Historically, brown seaweeds, such as *Ascophyllum nodosum*, have been applied directly to the soil, either freshly harvested or as dried ground seaweed meal, primarily as a soil conditioner or source of organic matter [1]. More recently, aqueous and alkaline extracts prepared from a variety of commercially available seaweeds have been used in agriculture and horticulture systems as foliar sprays, soil drenches, or often a combination of both. The beneficial effects of seaweed extract application on crop performance have been attributed to a variety of constituents, including betaines [2]. These naturally occurring quaternary ammonium compounds can act as osmolytes and/or affect gene expression, therefore improving plant tolerance to stresses such as temperature extremes, drought, and salinity [3–8].

For over 50 years glycine betaine has been added to animal feeds as a nutritional supplement [9]. In the sports medicine arena glycine betaine has been found to enhance sport performance (ergogenics) by improving muscle endurance and increasing quality of repetitions performed [10]. The bioactivity of glycine betaine

Dagmar B. Stengel and Solène Connan (eds.), *Natural Products From Marine Algae: Methods and Protocols*, Methods in Molecular Biology, vol. 1308, DOI 10.1007/978-1-4939-2684-8_17, © Springer Science+Business Media New York 2015

is associated with its ability to act as a methyl donor in the methionine cycle primarily in the human kidney and liver [9]. Other beneficial effects on human health include helping to protect cells and organs from osmolytic stress and improving heart health by improving vascular risk factors [9].

Over 13 betaine analogs have been identified in Chlorophyta (green), Heterokontophyta (brown), and Rhodophyta (red) seaweeds using thin layer chromatography and NMR techniques [11]. The isolated yield of each betaine ranged from 0 to 2 % per algal dry weight with the lowest betaine levels being reported in the Phaeophyceae. Seaweed biomass and commercial extracts prepared from *Ascophyllym nodosum*, and *Fucus* and *Laminaria* species have been found to contain glycine betaine, γ-aminobutyric acid betaine, δ-aminovaleric acid betaine, and laminine [12, 13].

Early analytical methods aimed at the quantification of betaines in plants, seaweeds, or commercial seaweed extracts included the use of thin layer chromatography [14], thin layer electrophoresis [15], and nonspecific precipitation reagents [16, 17]. The development of an HPLC-based approach to betaine analysis has been ongoing since the late 1970s [18]. Betaines do not fluoresce appreciably and can only be detected at low UV wavelengths (190–220 nm) as they lack a strong chromophoric moiety [19]. Attempts at increasing the detection limits of betaines using derivatization reagents have given unsatisfactory results because of the lack of reactivity of the betaines towards some reagents [20]. Proton magnetic resonance (^{1}H NMR) has also been developed as a method of betaine analysis for seaweed extracts and plant material [21] but has the limitation that it cannot be used to quantify the levels of betaines that are present as minor components [12].

The LC-MS-MS method described herein quantifies betaines in seaweeds and commercial seaweed extract products using a rapid sample preparation and cleanup protocol. It utilizes the sensitivity and selectivity of LC-MS-MS making it a method that is suitable for the routine analysis of glycine betaine, γ-aminobutyric acid betaine, δ-aminovaleric acid betaine, and laminine.

2 Materials and Instrumentation

2.1 Solvents and Standards

1. Ultrapure water is prepared by purifying distilled water to attain a resistance of 18 MΩ at 25 °C with a TOC of <5 ppb.
2. High purity grade methanol.
3. HPLC grade acetonitrile.
4. 0.2 % formic acid containing 5 mM ammonium formate is prepared by adding 200 μL of formic acid (90 %) and 31.5 mg of ammonium formate to 99.8 mL of ultrapure water.

5. 0.01 % formic acid containing 5 mM ammonium formate is prepared by adding 100 μL of formic acid (90 %) and 315 mg of ammonium formate to 999.9 mL of ultrapure water. After mixing the solution is filtered through a 0.2 μm filter.
6. 95 % acetonitrile–water containing 0.01 % formic acid and 5 mM ammonium formate is prepared by adding 100 μL of formic acid (90 %) and 315 mg of ammonium formate to 50 mL of ultrapure water. After mixing this solution is added to 950 mL of acetonitrile (HPLC Grade). The mixed solution is filtered using a 0.2 μm filter in preparation for HPLC use.
7. Glycine betaine and laminine (NεNεNε-trimethyl lysine). The γ-aminobutyric acid betaine and δ-aminovaleric acid betaine standards are synthesized via amine quaternization of aminobutyric acid and aminovaleric acid using a methyl iodide–potassium bicarbonate–methanol reagent [22]. Semi-purification of the desired reaction product is achieved by crystallization from methanol/ether which is repeated three times [22]. Further purification is accomplished using an ion exchange system that consists of a strong anion, weak cation and strong cation ion exchange columns arranged in a series. The procedure used to purify the betaines differs only from that reported before [23] in that Permutit Q (50–100 mesh, H^+ form) is used in place of AG50W resin (strong cation exchange resin). The eluted betaines are evaporated to dryness using a rotary evaporator and then freeze-dried. Proton-based quantitative NMR is used to determine the purity of each of the four betaine standards.

2.2 Extraction and Sample Preparation Components

1. Fractionation of seaweed extracts: C_{18} Sep-Pak (3 cc, 500 mg, Waters) solid phase extraction cartridge.
2. Filtration of samples prior to LC-MS-MS analysis: 4 mm PTFE (0.45 μm pore size) HPLC certified syringe filter.

2.3 LC-MS-MS Components

1. Betaine LC-MS-MS analysis: *Sciex* API 4000 Triple Quadrapole Mass Spectrometer (SCIEX, Streetsville, ON, Canada) coupled to an Agilent 1100 HPLC (Palo Alto, CA, USA). In our case the instrument was equipped with a Turbo IonSpray source and operated in a positive multiple reaction monitoring (MRM) mode.
2. A Waters XBridge™ HILIC column with 2.5-μm diameter particles (4.6 mm × 75 mm). The column was equipped with a Waters XBridge™ HILIC guard column (4.6 mm × 20 mm).
3. Mobile phase A consisted of 95 % acetonitrile/5 % water containing 0.01 % formic acid and 5 mM ammonium formate (v/v). Aqueous mobile phase B is composed of 0.01 % formic acid and 5 mM ammonium formate.

4. Individual betaine standard solutions are volumetrically prepared in methanol from concentrated stock solutions (*see* **Note 1**). Glycine betaine (GB) is prepared at 0.1, 1.0, and 5.0 μg/mL. Aminovaleric acid betaine (AVAB) is prepared at 1.0, 2.0, 10.0, and 20.0 μg/mL. Aminobutyric acid betaine (AVAB) is prepared at 0.2, 2.0, and 4.0 μg/mL. Lastly, laminine (LAM) is prepared at 0.2, 1.0, 2.0 μg/mL.

3 Methods

3.1 Extraction Procedure for Wet Seaweed Samples

1. Collect seaweeds, transport them to the lab in a cooler containing ice, clean them of contaminants and rinse them with seawater. Freeze the seaweeds and store them at −20 °C or colder.
2. Coarsely grind cleaned seaweed samples using liquid nitrogen and a mortar. Finely grind the coarse ground seaweed using dry ice and a grinder equipped with a size 40 mesh. Weigh out 250 mg of the finely ground seaweed into a 5 mL screw cap (Teflon lined) glass vial. Determine the dry weight of a separate wet seaweed sample by freeze-drying or drying the sample using a vacuum-drying oven.
3. Add 2 mL of methanol to each sample containing screw cap vial, screw on cap, shake and incubate in a heating block or water bath for 1 h at 78 °C. Shake the sample and continue to incubate at 78 °C for another hour. Remove incubated vial, let cool for 5 min and then remove the cap. Collect the aqueous methanol extract using a glass Pasteur pipette. Repeat this extraction protocol once more. Next, extract the seaweed pellet with 2 mL of methanol for only 1 h at 78 °C. Repeat this extraction once more.
4. Combine the four extracts and evaporate to dryness under a stream of nitrogen in a pre-weighed 20–25 mL glass vial or tube. Obtain dry weight of extract.
5. Suspend the resulting sample in 5 mL of 0.2 % formic acid with 5 mM ammonium formate using sonication to make the sample as homogeneous as possible (*see* **Note 2**).

3.2 Extraction Procedure for Dry Seaweed Samples

1. Grind dry seaweed samples (dried using an oven, a freeze dryer, etc.) using a grinder equipped with a size 40 mesh. Weigh out 100 mg of the dry seaweed powder into a 5 mL screw cap (Teflon lined) vial.
2. Grind dry seaweed extract products using a grinder equipped with a size 40 mesh. Weigh out 100 mg of the dry seaweed powder into a 5 mL screw cap (Teflon lined) vial.

3. Freeze-dry commercial liquid seaweed products, grind them using a grinder equipped with a size 40 mesh and weigh out 100 mg of the dry seaweed powder into a 5 mL screw cap (Teflon lined) vial. Determine dry weight of the product.
4. Add 2 mL of methanol to each sample containing screw cap vial, screw on cap, shake and incubate for 1 h at 78 °C in a heating block or water bath. Shake the sample and continue to incubate for another hour. Remove incubated vial from the heat, let cool for 5 min and then remove the cap. Collect the aqueous methanol extract using a glass Pasteur pipette. Repeat this extraction protocol once more. Next extract the pellet with 2 mL of methanol for 1 h at 78 °C. Repeat this extraction once more.
5. Combine the four extracts and evaporate to dryness under a stream of nitrogen in a pre-weighed 20–25 mL glass vial or tube. Obtain dry weight of extract.
6. Suspend the resulting sample in 5 mL of 0.2 % formic acid with 5 mM ammonium formate using sonication to make the sample as homogeneous as possible (*see* **Note 2**).

3.3 Preparation of Betaine-Enriched Fraction

1. Precondition a C_{18} Sep-Pak solid phase extraction cartridge, that has been fitted with a small glass wool plug on the top, by eluting it first with 5 mL of methanol, followed by 15 mL of an aqueous solution containing 0.2 % formic acid and 5 mM ammonium formate.
2. Apply the resuspended extract to the preconditioned C_{18} Sep-Pak cartridge collecting the eluent.
3. Rinse the tube twice with 5 mL of 0.2 % formic acid containing 5 mM ammonium formate. Elute the cartridge with these "rinses".
4. Combine the three cartridge eluents (~15 mL in volume) and evaporate to dryness under a stream of nitrogen.
5. Dissolve the resulting betaine containing fraction in 2.0 mL of methanol. Remove and filter 100 μL of the sample through a 4 mm PTFE (0.45 μm pore size) HPLC certified syringe filter in preparation for HPLC-MS-MS analysis. Store the remainder of the sample at −20 °C.

3.4 LC-MS-MS Procedure

1. Wash and equilibrate the Waters XBridge™ HILIC column well with 100 % solvent A making sure it is not connected to the *Sciex* API 4000 Triple Quadrapole Mass Spectrometer.
2. Open the Analyst software, build an Acquisition Method and check that the required devices are configured.
3. Under the LC Pump Method Properties create the steps (elution gradient) as in Table 1.

Table 1
Elution gradient for the LC method

Step	Total time (min)	Flow rate (μL/min)	A (%)	B (%)
0	0.00	1,000	85	15
1	7.00	1,000	85	15
2	7.01	1,000	65	35
3	15.00	1,000	65	35
4	15.01	1,000	85	15
5	20.00	1,000	85	15

Table 2
MS method for multiple reaction monitoring (MRM)

Q1 mass (Da)	Q3 mass (Da)	Dwell (ms)
118.200	58.200	200.00
146.100	87.000	200.00
159.900	101.100	200.00
188.900	83.900	200.00

4. Set the Column Oven chamber to 40 °C and the Autosampler to 4 °C with an injection volume of 1.00 μL.
5. Select the MS, create an Experiment 1 method and select the scan type as Multiple Reaction Monitoring (MRM) with positive polarity. Create Table 2.

   Set the collision gas (CAD) to 8, the curtains gas (CUR) to 20.00, both ion source gases (GS1 and GS2) to 50.00, the ion spray voltage (IS) to 5,000.00, the temperature (TEMP) to 300.00, the interface heater (ihe) to ON, the declustering potential (DP) to 60.00, the entrance potential (EP) to 10.00, the collision energy (CE) to 30.00 and the collision cell exit potential to (CPX) to 15.00. Ensure that the Q1 resolution is set to Unit and the Q3 resolution is set to Low. Save the betaine analysis method with an appropriate name.
6. Open Build Acquisition Batch giving the Set an appropriate name (usually Yr-Mo-Day). Under the Method Editor, select the betaine method that was created in 5. Select the Add Set button and then the Add Samples button so that the sample names and locations can be inputted. Input a solvent blank for the first run followed by a standard mixture that contains GB, AVAB, ABAB, and LAM so that the retention times and the

resolution of the standards can be checked. After inputting the samples, input a solvent blank followed by a series of injections that will be used for the construction of calibration curves for GB, AVAB, ABAB, and LAM. Save this batch file.

7. Attach the column to the inlet of the *Sciex* API 4000 Triple Quadrapole Mass Spectrometer.
8. Open the instrument Queue, click Ready and then, after a few minutes, click Start Sample. All samples in this queue will now be injected and run on the LC-MS-MS system.

3.5 Data Processing

1. After the runs are completed, use the Analyst software to open the Quantitate tab. Use the Quantitation Wizard to select the samples and standards for automatic integration and area determinations.
2. Copy the areas of the standards to spreadsheet/data processing software, such as Excel®, to create betaine calibration curves and to determine the betaine content in each sample on a dry and a wet sample basis.
3. Acquisition of a total ion current (TIC) chromatogram resulting from the injection of a composite betaine standard should present retention times of 3.7 min for glycine betaine (GB), 7.1 min for δ-aminovaleric acid betaine (AVAB), 8.5 min for γ-aminobutyric acid betaine (ABAB) and 9.7 min for laminine (LAM) as shown on Fig. 1 (*see* **Notes 3** and **4**).

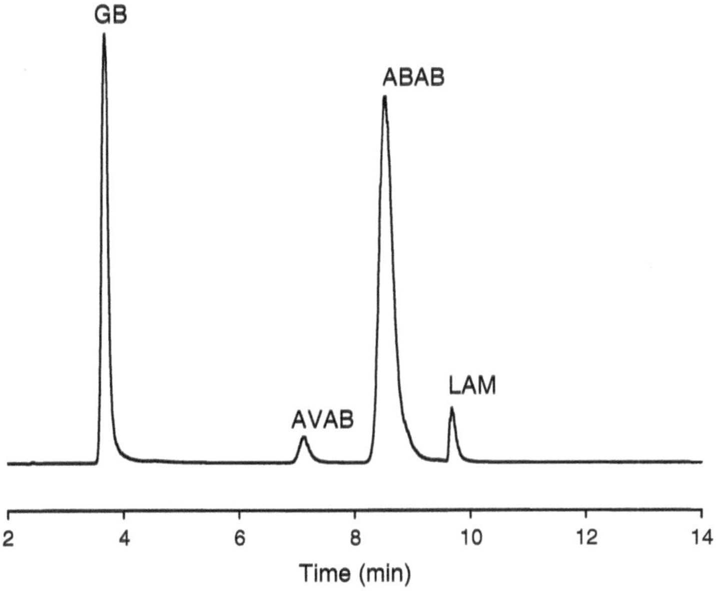

Fig. 1 Total ion current chromatogram from the LC-MS-MS analysis of a mixed betaine standard showing retention times of 3.7 min for glycine betaine (GB), 7.1 min for δ-aminovaleric acid betaine (AVAB), 8.5 min for γ-aminobutyric acid betaine (ABAB) and 9.7 min for laminine (LAM)

4 Notes

1. The percent purity obtained from quantitative NMR analysis was used in calculating the concentrations of each standard solution.
2. There may be particles in the sample that would not dissolve in the solvent. Load the sample containing the particulate onto the SPE cartridge as in Subheading 3.3. If the solvent will not move through the cartridge as a result of the particulate then break up the particulate with a glass pipette to enable solvent cartridge flow.
3. Using this method the betaine content of *Ascophyllum nodosum* was determined to vary with collection site in Nova Scotia, Canada and seasonally in the range 7.0–23 μg GB, 75–140 μg AVAB, 23–47 μg ABAB, and 2.8–8.9 μg LAM per gram dry weight of seaweed [24].
4. The betaine profile of *Saccharina latissima* was shown to be dominated by GB with AVAB, ABAB, and LAM being detected at much lower levels (Fig. 2).

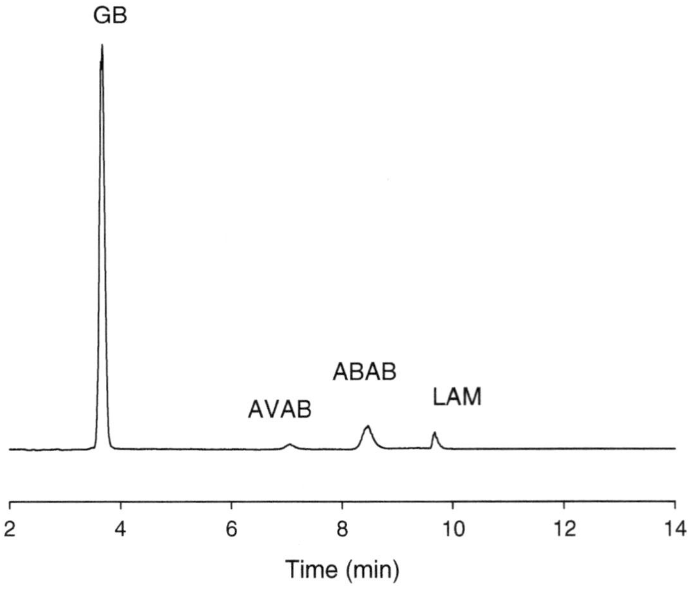

Fig. 2 Total ion current chromatogram from the LC-MS-MS analysis of *Saccharina latissima*

References

1. Temple WD, Bomke AA (1988) Effects of kelp (*Macrocystis integrifolia*) on soil chemical properties and crop response. Plant Soil 105:213–222
2. Blunden G, El Barouni MM, Gordon SM et al (1981) Extractions, purification and characterization of Dragendorff-positive compounds from some British marine algae. Bot Mar 24:451–456
3. Mason TG, Blunden G (1989) Quaternary ammonium and tertiary sulphonium compounds of algal origin as alleviators of osmotic stress. Bot Mar 32:313–316
4. Hayashi H, Alia ML et al (1997) Transformation of *Arabidopsis thaliana* with the *codA* gene for choline oxidase; accumulation of glycine betaine and enhanced tolerance to salt and cold stress. Plant J 12:133–142
5. Holmström KO, Somersalo S, Mandal A et al (2000) Improved tolerance to salinity and low temperature in transgenic tobacco producing glycine betaine. J Exp Bot 51:177–185
6. Sakamoto A, Valverde R, Alia et al (2000) Transformation of *Arabidopsis* with the *codA* gene for choline oxidase enhances freezing tolerance of plants. Plant J 22:449–453
7. Park E-J, Jeknić Z, Sakamoto A et al (2004) Genetic engineering of glycine betaine synthesis in tomato protects seeds, plants, and flowers from chilling damage. Plant J 40:474–487
8. Yang X, Wen X, Gong H et al (2007) Genetic engineering of the biosynthesis of glycine betaine enhances thermotolerance of photosystem II in tobacco plants. Planta 225:719–733
9. Craig SAS (2004) Betaine in human nutrition. Am J Clin Nutr 80:539–549
10. Hoffman JR, Ratamess NA, Kang J et al (2009) Effect of betaine supplementation on power performance and fatigue. J Int Soc Sport Nutr 6:7. doi:10.1186/1550-2783-6-7
11. Blunden G, Smith BE, Irons MW et al (1992) Betaines and tertiary sulphonium compounds from 62 species of marine algae. Biochem Syst Ecol 20:373–388
12. Tyihák E, Blunden G, Ma Y-C (1994) Quantitative estimation of betaines in commercial seaweed extracts using overpressured layer chromatography. J Appl Phycol 6:469–473
13. Blunden G, Rogers DJ, Barwell CJ (1984) Biologically-active compounds from British marine algae. In: Krogsgaard-Larsen P, Brogger CS, Kofod H (eds) Natural products and drug development, Alfred Benzon Symposium 20. Munksgaard, Copenhagen, pp 179–190
14. Müller H, Eckert H (1989) Simultaneous determination of monoethanolamine and glycine betaine in plants. J Chromatogr 479: 452–458
15. Gorham J, Coughlan SJ, Storey R et al (1981) Estimation of quaternary ammonium and tertiary sulphonium compounds by thin-layer electrophoresis and scanning reflectance densitometry. J Chromatogr 210:550–554
16. Barak AJ, Tuma DJ (1979) Simplified procedure for determination of betaine in liver. Lipids 14:860–863
17. Bao W, Gao S, Fan Z et al (1989) Isolation and identification of betaine from beet molasses and its quantitative determination. J Shenyang Coll Pharm 6:12–15
18. Dupuy P (1978) Analytical recognition of chaptalization (addition of sugar to wine after fermentation). Ann Nutr Alim 32:1123–1132
19. Zamarreño A, Cantera RG, Garcia-Mina JM (1997) Extraction and determination of glycine betaine in liquid fertilizers. J Agric Food Chem 45:774–776
20. Lever M, Bason L, Leaver C et al (1992) Same-day measurement of glycinebetaine, carnitine and other betaines in biological material. Anal Biochem 205:14–21
21. Blunden G, Cripps AL, Gordon SM et al (1986) The characterization and quantitative estimation of betaines in commercial seaweed extracts. Bot Mar 29:155–160
22. Benoiton NL, Chen FMF (1976) The synthesis of amino-acid derivatives and peptides containing N,N-dimethylamino and N,N,N-trimethylamino groups. Proc 14th Eur Peptide Symp, Wépion, Belgium, p 149–152
23. Bessieres M-A, Gibon Y, Lefeuvre JC et al (1999) A single-step purification for glycine betaine determination in plant extracts by isocratic HPLC. J Agric Food Chem 47: 3718–3722
24. MacKinnon SL, Hiltz D, Ugarte R et al (2010) Improved methods of analysis for betaines in *Ascophyllum nodosum* and its commercial seaweed extracts. J Appl Phycol 22:489–494

Chapter 18

Analysis of Marine Biotoxins Using LC-MS/MS

Bernd Luckas, Katrin Erler, and Bernd Krock

Abstract

Different clinical types of algae-related poisoning have attracted scientific and commercial attention: paralytic shellfish poisoning (PSP), diarrhetic shellfish poisoning (DSP), and amnesic shellfish poisoning (ASP). Bioassays are common methods for the determination of marine biotoxins. However, biological tests are not completely satisfactory, mainly due to the low sensitivity and the absence of specialized variations. In this context LC-MS methods replaced HPLC methods with optical detectors, allowing both effective seafood control and monitoring of phytoplankton in terms of the different groups of marine biotoxins. This chapter describes state-of-the-art LC-MS/MS methods for the detection and quantitation of different classes of phycotoxins in shellfish matrices. These classes include the highly hydrophilic paralytic shellfish poisoning (PSP) toxins. Hydrophilic interaction liquid chromatography (HILIC) has been shown to be useful in the separation of PSP toxins and is described in detail within this chapter. Another important class of phycotoxins is diarrhetic shellfish poisoning (DSP) toxins. This group traditionally comprises okadaic acid and dinophysistoxins (DTXs), pectenotoxins (PTXs), and yessotoxins (YTXs). The most recently described shellfish poisoning syndrome, azaspiracid shellfish poisoning (AZP) is caused by azaspiracids, which in turn are diarrhetic, but usually are treated separately as AZP. The last group of regulated shellfish toxins is the amnesic shellfish poisoning (ASP) toxin domoic acid, produced by species of the genus *Pseudo-nitzschia*.

Key words ASP toxins, Azaspiracids, Domoic acid, DSP toxins, LC-MS/MS determination, Okadaic acid and Dinophysistoxins, Pectenotoxins, PSP toxins, Regulated marine biotoxins, Yessotoxins

1 Introduction

Harmful algal blooms (HABs) can cause problems when algal toxins are ingested by and stored in mussels that are consumed by humans. Mussels filter approximately 20 L water/h, and during algal blooms water may contain several million algal cells per liter. Although not all algae produce toxins, it is plausible that a significant accumulation of toxins will occur in mussels. Monitoring of toxins in seafood and risk assessment for human exposure is the main task of food control as consumption of seafood contaminated with marine biotoxins may cause serious diseases. Damage to the nervous system (paralytic shellfish poisoning, PSP), the intestinal system (diarrhetic shellfish poisoning, DSP), and loss of memory

Dagmar B. Stengel and Solène Connan (eds.), *Natural Products From Marine Algae: Methods and Protocols*, Methods in Molecular Biology, vol. 1308, DOI 10.1007/978-1-4939-2684-8_18, © Springer Science+Business Media New York 2015

(amnesic shellfish poisoning, ASP) have been observed subject to the type of algal bloom [1]. Therefore, the analysis of marine food for marine biotoxins is an important task that must be conducted according to international regulations and in strict compliance with the respective restrictions [2].

1.1 Paralytic Shellfish Poisoning (PSP) Toxins

The main producers of paralytic shellfish poisoning (PSP) toxins are dinoflagellates of the genus *Alexandrium*. The toxins were named PSP toxins due to the observation that their consumption caused symptoms of poisoning in warm blooded species similar to paralytic phenomena. PSP toxins are potential neurotoxins which specifically block the excitation current in nerve and muscle cells resulting in signs of paralysis. Consequently, the development of analytical methods for the determination of poisoning caused by PSP toxins was an important task. The mouse bioassay unambiguously gives evidence of the toxic potential of a sample, since the application of higher toxin concentrations yields a shortening of the time until death of the animals [3]. However, those biological tests only reveal the total PSP toxicity of a sample expressed in MU (mouse units)/kg or in PSP/kg. Thus, the determination of individual PSP toxins was only possible after isolation and structure elucidation [4]. In 1957, a PSP toxin was isolated from *Saxidomus giganteus* (clams) from Alaska, and in 1975 the chemical structure was assigned to the so-called saxitoxin (STX) (Fig. 1). Later on, more PSP toxins were identified which were all related to saxitoxin (STX) or *N*-1-hydroxy-saxitoxin (neosaxitoxin, NEO). Today, it is convenient to distinguish between three groups of PSP toxins: carbamoyl toxins, *N*-sulfocarbamoyl toxins, and decarbamoyl toxins [5]. *N*-Sulfocarbamoyl toxins exhibit only low toxicity, whereas the toxicity of the carbamoyl and decarbamoyl toxins is significantly higher (Fig. 1).

Electrospray ionization mass spectrometry (ESI-MS) is effective for detection of the polar PSP toxins, which are quite basic and, therefore, form stable $[M+H]^+$ ions. Thus, the direct detection of underivatized PSP toxins is possible using a mass spectrometer as detector [6]. In 2007, Diener et al. [7] published the application of a ZIC-HILIC column (Dichrom) for the separation of underivatized PSP toxins including a new LC-MS/MS method which enabled the separation of all three groups of PSP toxins in a single chromatographic run. Although the limits of detection (LODs) are sometimes higher with MS detection than with fluorescence detection, the sensitivity of the MS detection is sufficient to control seafood at the regulatory limit for a PSP toxin content at 800 μg STX equivalents /kg wet weight of soft tissues [8].

1.2 Diarrhetic Shellfish Poisoning (DSP) Toxins

Since the first reports of diarrheic shellfish poisoning (DSP) in Japan, in 1978, the illness is now recognized as a threat to public health throughout the world. All DSP toxins are relatively nonpolar, with molecular weights higher than 500, and easily extractable by organic solvents. Most DSP toxins are polyether compounds

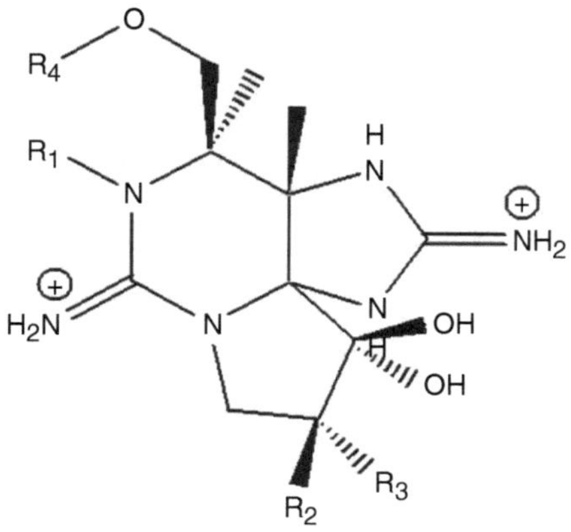

PSP toxin	group	$[M+H]^+$/ $[M+H-SO_3]^-$	R1	R2	R3	R4	toxicity factor (LD_{50} µg/kg)	MRL:
STX	carbamoyl toxins	300	H	H	H	H_2N-COO	1,0 (10)	800 mg STX-eq[2]/kg
NEO		316	OH	H	H	H_2N-COO	1,0 (65)	
GTX-1		412	OH	H	OSO_3^-	H_2N-COO	1,0 (n/a)	
GTX-2		396	H	H	OSO_3^-	H_2N-COO	0,4 (n/a)	
GTX-3		396	H	OSO_3^-	H	H_2N-COO	0,6 (n/a)	
GTX-4		412	OH	OSO_3^-	H	H_2N-COO	0,7 (n/a)	
GTX-5/ B1	N-sulfo-carbamoyl toxins	380	H	H	H	SO_3^--NH-COO	0,1 (n/a)	
GTX-6/ B2		396	OH	H	H	SO_3^--NH-COO	0,1 (n/a)	
C1		396	H	H	OSO_3^-	SO_3^--NH-COO	<0,1 (n/a)	
C2		396	H	OSO_3^-	H	SO_3^--NH-COO	0,1 (n/a)	
C3		412	OH	H	OSO_3^-	SO_3^--NH-COO	<0,1 (n/a)	
C4		412	OH	OSO_3^-	H	SO_3^--NH-COO	0,1 (n/a)	
de-STX	de-carbamoyl toxins	257	H	H	H	H	1,0 (n/a)	
de-NEO		273	OH	H	H	H	0,4 (n/a)	
de-GTX-1		369	OH	H	OSO_3^-	H	0,5 (n/a)	
de-GTX-2		353	H	H	OSO_3^-	H	0,2 (n/a)	
de-GTX-3		353	H	OSO_3^-	H	H	0,4 (n/a)	
de-GTX-4		369	OH	OSO_3^-	H	H	0,5 (n/a)	

[1]Maximum Residue Level
[2]Equivalent

Fig. 1 Chemical structures of PSP toxins and their toxicities relative to saxitoxin (STX = 1). *MRL* maximum residue level, *eq.* equivalent

with distinctive chemical structures and widely varying functional groups resulting in different toxicological and chemical characteristics [9]. Okadaic acid (OA) and dinophysistoxins (DTXs) (Fig. 2) have been classified as diarrhetic shellfish poisoning (DSP) toxins due to their production of symptoms of diarrhea [10]. Unfortunately, when pectenotoxins (PTXs) (Fig. 3) [11] and yessotoxins (YTXs) (Fig. 4) [12] were discovered and before their

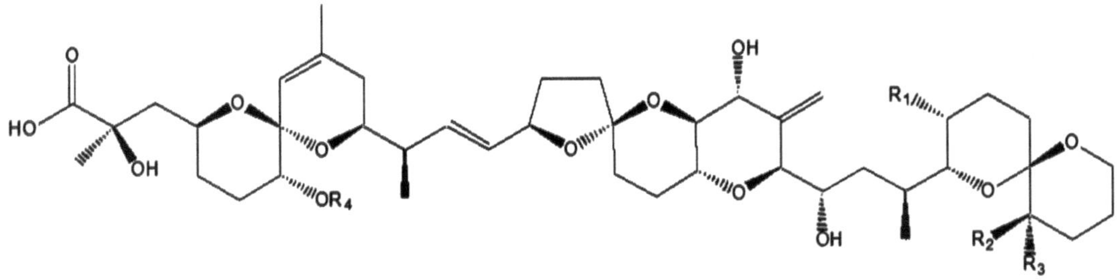

Name	R 1	R 2	R 3	R 4	m/z [M+NH4]+
OA	CH_3	H	H	H	822.5
DTX-1	CH_3	CH_3	H	H	836.5
DTX-2	H	H	CH_3	H	822.5
DTX-3 (acylated forms of DTXs)	CH_3 or H			fatty acid	-

Fig. 2 Chemical structures and $[M+NH_4]^+$ masses of okadaic acid (OA), its derivatives dinophysistoxin-1 (DTX-1), DTX-2, and DTX-3. *Note*: unlike OA, DTX-1, and DTX-2, DTX-3 are not individual compounds, but general terms for 7-acylated variants

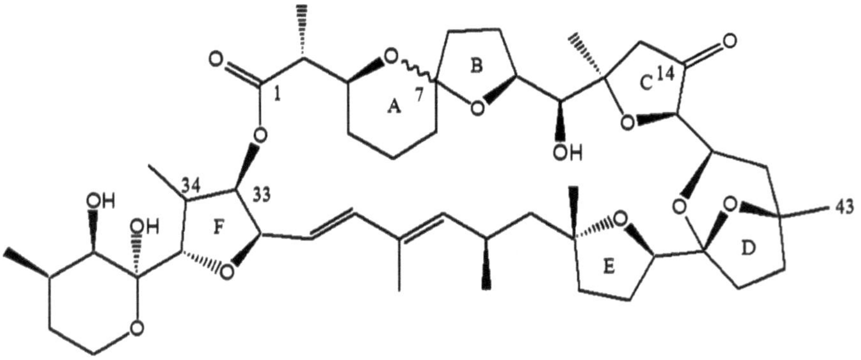

Name		R 43	R 33	R 34	m/z [M+NH4]+
Pectenotoxin-1	PTX-1	CH_2OH	-O-	H	892.5
Pectenotoxin-2	PTX-2	CH_3	-O-	H	876.5
Pectenotoxin-2 seco acid	PTX-2sa	CH_3	-OH-HO-	H	894.5
Pectenotoxin-11	PTX-11	CH_3	-O-	OH	892.5

Fig. 3 Chemical structures and $[M+NH_4]^+$ masses of pectenotoxin-1 (PTX-1), PTX-2, PTX-2 seco acid, and PTX-11

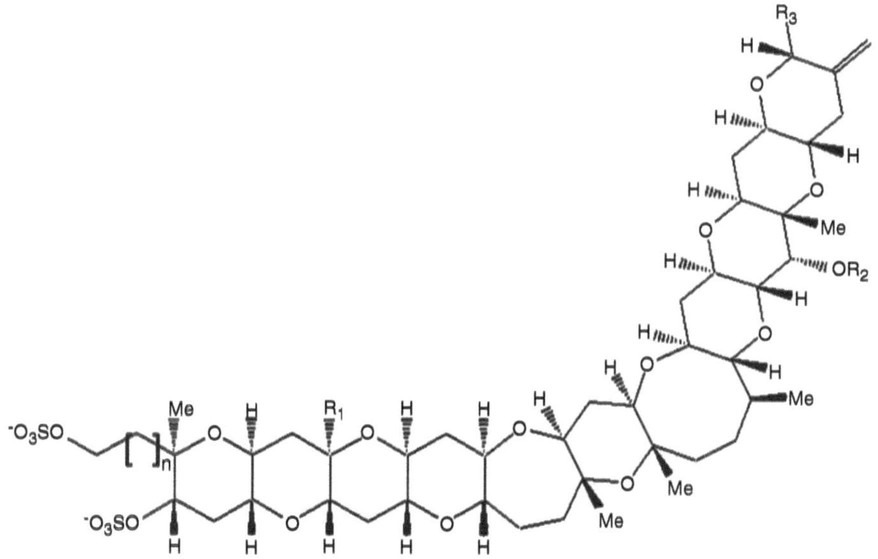

Name	n	R 1	R 2	R 3	*m/z* $[M+NH_4]^+$
YTX	1	H	H	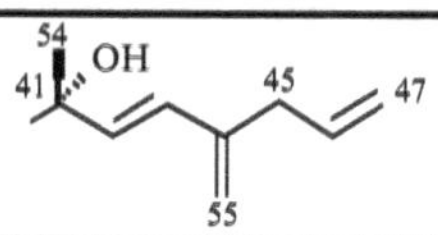	1160
45-OH-YTX	1	H	H	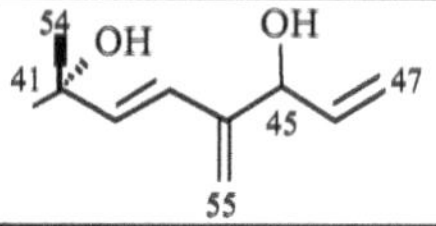	1176
1a-homo-YTX (Protoceratin I)	2	H	H	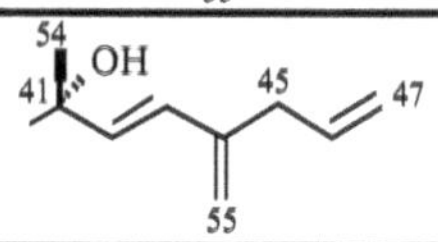	1174
45-OH homo-YTX	2	H	H	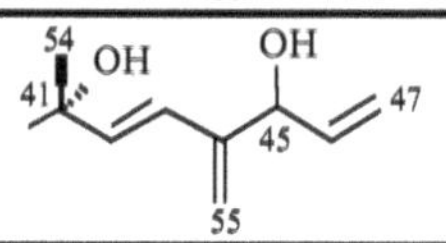	1190

Fig. 4 Chemical structures and $[M+NH_4]^+$ masses of yessotoxin (YTX), 45-OH-YTX, 1a-homo-YTX, and 45-OH-homo-YTX

toxicity was understood, they were placed into the DSP category because they often occur together with OA and DTXs and were detected by the mouse bioassay procedure first used to detect DSP toxins in lipophilic extracts generally. With the continued discovery of a series of new lipophilic toxins, such as azaspiracids (AZAs) (Fig. 5), it became clear that a better approach would be needed to categorize the toxins strictly according to their chemical classes rather than their toxic symptoms [13]. This would allow seafood safety to be regulated according to allowable levels of specific toxins (Table 1) rather than to the result of a specific assay.

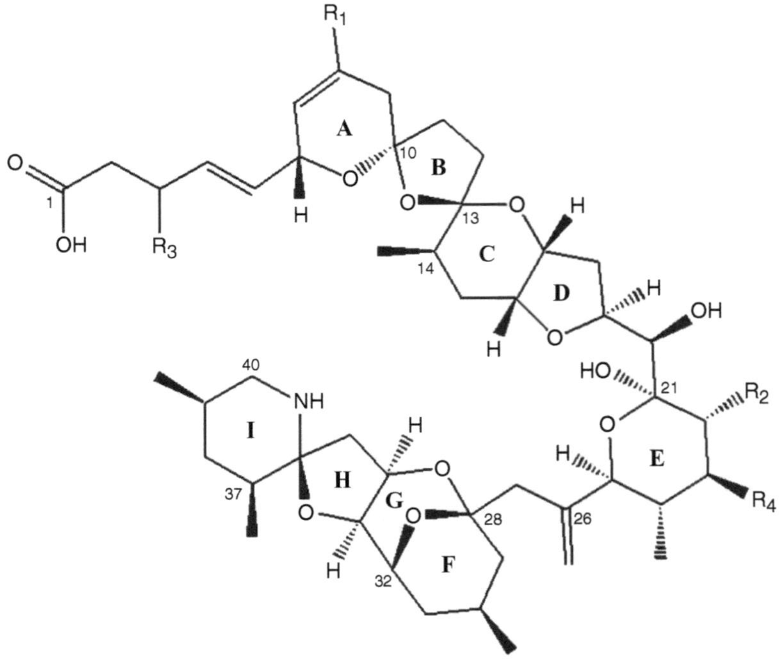

		R 1	R 2	R 3	R 4	*m/z* $[M+H]^+$
Azaspiracid-1	**AZA-1**	H	CH_3	H	H	842
Azaspiracid-2	**AZA-2**	CH_3	CH_3	H	H	856
Azaspiracid-3	**AZA-3**	H	H	H	H	828

Fig. 5 Chemical structures and $[M+H]^+$ masses of azaspiracid-1 (AZA-1), AZA-2, and AZA-3

Table 1
Regulated lipophilic toxins

Okadaic acid group	MRL	Pectenotoxin group	MRL	Yessotoxin group	MRL	Azaspiracid group	MRL
OA DTX-1 DTX-2	160 μg OA eq./kg	PTX-1 PTX-2 PTX-2sa	160 μg/kg	YTX Homo-YTX 45-OH-YTX 45-OH-homo-YTX	1 mg/kg	AZA-1 AZA-2 AZA-3	160 μg/kg

MRL Maximum Residue Level, *eq.* equivalent

In Annex II Section VII Chapter V [2] to Regulation 853/2004/EC [14], maximum levels for ASP, PSP and DSP toxins are established. Table 1 shows the maximum residue level for the different groups of DSP toxins.

1.3 Azaspiracid Shellfish Poisoning (AZP) Toxins

Amongst the known marine shellfish poisoning syndromes, azaspiracid shellfish poisoning (AZP) is the most recent one which was observed for the first time in the Netherlands in 1995. Contaminated mussels cultivated in Ireland were consumed and intoxicated at least eight people [15]. Three years later the implicated toxin was identified, isolated, structurally defined, and named azaspiracid (now called azaspiracid-1 (AZA-1)) [16]. In the following years other variants of AZA-1 were found and isolated from shellfish [17–20]. As a result, the European Union has set a regulatory limit for maximum levels of AZA-1, AZA-2, and AZA-3 in shellfish (160 μg/kg). Due to their structural characteristics AZAs were early suspected to be of dinoflagellate origin, however, the AZA-producing organism remained unknown until the isolation of *Azadinium spinosum* from the North Sea in 2007 [21].

1.4 Amnesic Shellfish Poisoning (ASP) Toxins

Domoic acid (DA) (Fig. 6) and its isomers are marine biotoxins causing amnesic shellfish poisoning (ASP) in humans. Symptoms of ASP include gastrointestinal symptoms (vomiting, diarrhea, or abdominal cramps) and/or neurological symptoms (confusion, loss of memory, or other serious signs such as seizure or coma) occurring within 24–48 h after consuming contaminated shellfish. DA is a water-soluble cyclic amino acid mainly produced by marine red algae of the genus *Chondria* and diatoms of the genus *Pseudo-nitzschia*. The first confirmed outbreak of ASP occurred in Canada in 1987 and was related to mussels affected by a bloom of the *Pseudo-nitzschia f. multiseries*. DA isomers have also been detected in shellfish in the United States and in a number of European countries. Although several isomers of DA (diastereoisomer epi-domoic acid (epi-DA) and isodomoic acids (iso-DAs)) have been identified data on the occurrence only of DA and epi-DA (expressed as sum DA) have been reported [22]. Domoic acid became a

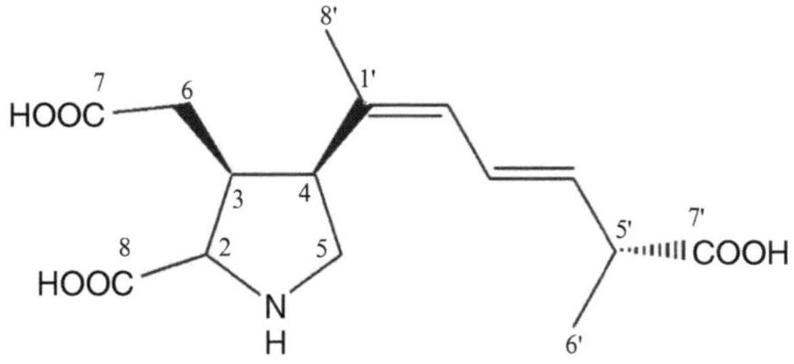

Fig. 6 Chemical structure of domoic acid (DA)

subject of interest for food control laboratories after introduction of a limit of 20 mg DA/kg flesh of mussels [14]. HPLC separation of the underivatized domoic acid on reversed-phase (RP) columns followed by UV detection at 242 nm was suggested. In addition, several methods involving MS have also been used for the determination of DA and its analogues in shellfish [23].

In this chapter, protocols for the determination of the hydrophilic PSP toxins, the lipophilic DSP toxins (including azaspiracids) and the ASP toxin domoic acid are outlined.

2 Materials

For the preparation of all extraction and mobile phase solutions use ultrapure water (18 MΩ/cm, Milli-Q; Millipore), analytical grade reagents and HPLC or LC-MS grade solvents. For the preparation of all solutions use gradient cylinder or pipettes and store all solutions at room temperature, if not indicated otherwise.

Standard stock solution of ASP, DSP, and PSP can be purchased from the Certified Reference Material (CRM) Programme of the Institute of Marine Biosciences, National Research Council, Halifax, NS, Canada. Sealed ampoules have to be stored at +4 °C (ASP) and −20 °C (DSP, PSP) until dilution.

2.1 Determination of PSP Toxins

2.1.1 Reagents

1. Extraction solvent 0.2 M hydrochloric acid: transfer 26 mL of hydrochloric acid 25 % in a 1,000 mL volumetric flask and fill up with distilled water.
2. 1 M ammonium formate: weigh 31.53 g of ammonium formate (MW 63.06 g/mol), dilute it in 200 mL of water, transfer it in a 500 mL volumetric flask and fill up with water (*see* **Note 1**).
3. 1 M formic acid: transfer 19.1 mL of formic acid 99 % in a 500 mL volumetric flask and fill up with water (*see* **Note 1**).
4. Mobile phase A: 2 mM formic acid, 5 mM ammonium formate in acetonitrile/water (80:20, v/v). Mix 5 mL of 1 M ammonium formate solution and 2 mL of 1 M formic acid with 193 mL of water and 800 mL of acetonitrile.
5. Mobile phase B: 10 mM formic acid, 10 mM ammonium formate in water. Mix 10 mL of 1 M ammonium formate solution and 10 mL of 1 M formic acid with 980 mL of water.
6. Prepare a standard mix stock solution by pipetting the volumes of the respective standard solutions given in Table 2 into a vial and add 712.6 μL of 0.03 M acetic acid (final volume: 1,000 μL).

Table 2
Original concentrations of commercially available PSTs produced by the Certified Reference Material (CRM) Programme of the Institute of Marine Biosciences, National Research Council, Canada; Volumes needed for each PST for the preparation of a standard mix stock solution; concentrations of each PST in the stock solution and final concentrations of each PST in solutions 1–4 for calibration of samples

	Original conc.	Volume for stock sol.	Conc. stock sol.	Conc. Sol. 1	Conc. Sol. 2	Conc. Sol. 3	Conc. Sol. 4
Strd.	[μM]	[μL]	[pg/μL]	[pg/μL]	[pg/μL]	[pg/μL]	[pg/μL]
STX	*65*	7,4	144,93	7,2	14,5	29,0	58,0
GTX2	*118*	16	746,52	37,3	74,7	149,3	298,6
GTX3	*39*	16	246,73	12,3	24,7	49,3	98,7
NEO	*65*	7	144,38	7,2	14,4	28,9	57,8
dcSTX	*62*	8	128,12	6,4	12,8	25,6	51,2
GTX1	*106*	132	5756,31	287,8	575,6	1151,3	2302,5
GTX4	*35*	132	1900,67	95,0	190,1	380,1	760,3
B1	*65*	12	295,93	14,8	29,6	59,2	118,4
dcGTX2	*114*	80	3212,98	160,6	321,3	642,6	1285,2
dcGTX3	*32*	80	901,89	45,1	90,2	180,4	360,8
C1	*114*	25	1354,89	67,7	135,5	271,0	542,0
C2	*35*	25	415,98	20,8	41,6	83,2	166,4

7. Prepare four dilutions from this stock solution for calibration of samples:

   Solution 1: 25 μL stock solution and 475 μL 0.03 M acetic acid.

   Solution 2: 50 μL stock solution and 450 μL 0.03 M acetic acid.

   Solution 3: 100 μL stock solution and 400 μL 0.03 M acetic acid.

   Solution 4: 200 μL stock solution and 300 μL 0.03 M acetic acid.

   The final concentrations of each PST in the four solutions are listed in Table 2 (*see* **Note 2**).

2.1.2 Materials and Instrumentation

1. Grinder or mechanical mixer (e.g., ULTRA-TURRAX).
2. 15 mL glass or plastic tubes (heat and centrifuge stable).
3. Vortex.
4. Electrical heater or water bath.
5. Ultrasonic probe or ultrasonic bath.
6. Centrifuge at least 5,000 ×*g* for 15 mL tubes.
7. 2.0 mL tubes (adapted for centrifugation).

8. Centrifuge at least 15,000 × *g* for 2.0 mL tubes.
9. Nylon or PTFE syringe micro filter with 0.45 μm pore size.
10. One way 2 mL syringe with needle.
11. 1.5 mL HPLC vials.
12. HPLC instrumentation: injection system, gradient pump, column oven (able to heat 35 °C), HILIC analytical column, and triple-quadrupole mass spectrometer (*see* **Note 3**).

2.2 Determination of DSP Toxins

2.2.1 Reagents

1. Extraction solvent: methanol–water (90:10, v/v). Mix 900 mL of methanol with 100 mL of water.
2. 1 M ammonium formate: weight 31.53 g of ammonium formate (MW 63.06 g/mol), dilute it in 200 mL of water, transfer it in a 500 mL volumetric flask and fill up with water (*see* **Note 1**).
3. Buffer stock solution: transfer 20 mL of 1 M ammonium formate solution together with 19.1 mL of formic acid 99 % in a 500 mL volumetric flask and fill up with water (*see* **Note 1**).
4. Mobile Phase A: 2 mM ammonium formate, 50 mM formic acid in water. Mix 50 mL of the buffer stock solution with 950 mL of water.
5. Mobile Phase B: 2 mM ammonium formate, 50 mM formic acid in methanol–water (95:5; v/v). Mix 50 mL of the buffer stock solution with 950 mL of methanol.
6. Solutions for the hydrolysis of dinophysistoxin-3 (DTX-3):
   - 2.5 M sodium hydroxide: weigh 10 g of NaOH and dilute it with 50 mL of water on ice. Temper the solution at room temperature, transfer it into a 100 mL volumetric flask and fill up with water.
   - 2.5 M hydrochloric acid solution: transfer 32.4 mL of 25 % HCl into a 100 mL volumetric flask and fill up with water.
7. Standard solutions: prepare a series of calibration solutions with increasing concentration of okadaic acid (OA), dinophysitoxin-1 (DTX-1), dinophysistoxin-2 (DTX-2), pectenotoxin-1 (PTX-1), pectenotoxin-2 (PTX-2), pectenotoxin-2 seco acid (PTX-2sa), yessotoxin (YTX), homo-yessotoxin (homo-YTX), 45-OH-yessotoxin (45-OH-YTX), 45-OH-homo-yessotoxin (45-OH-homo-YTX), and azaspiracid-1 (AZA-1), azapiracid-2 (AZA-2), azaspiracid-3 (AZA-3) within the range of for example 0.005–0.08 μg/mL. Dilute the standard solutions with methanol (*see* **Note 2**).

2.2.2 Materials and Instrumentation

1. Grinder or mechanical mixer (e.g., ULTRA-TURRAX).
2. Centrifuge tubes 10 or 15 mL.
3. Vortex.

4. Thermo mixer.
5. Ultrasonic probe or ultrasonic bath.
6. Centrifuge at least 5,000 × *g*.
7. Nylon or PTFE syringe micro filter with 0.2 μm pore size.
8. One way 2 mL syringe with needle.
9. 1.5 mL HPLC vials, brown glass.
10. 200 μL Micro insert for HPLC vial.
11. LC-MS/MS instrumentation: Injection system, gradient pump, column oven, analytical column C8 reversed phase, triple-quadrupole mass spectrometer (*see* **Notes 3** and **4**).

2.3 Determination of ASP Toxins

2.3.1 Reagents

1. Extraction solvent: methanol–water (50:50, v/v). Mix 500 mL of methanol with 500 mL of water.
2. 1 M ammonium formate: weigh 31.53 g of ammonium formate (MW 63.06 g/mol), dilute it in 200 mL of water, transfer it in a 500 mL volumetric flask and fill up with water.
3. Buffer stock solution: transfer 20 mL of 1 M ammonium formate solution together with 19.1 mL of formic acid 99 % in a 500 mL volumetric flask and fill up with water.
4. Mobile phase A: 2 mM ammonium formate, 50 mM formic acid in water. Mix 50 mL of the buffer stock solution with 950 mL of water.
5. Mobile phase B: 2 mM ammonium formate, 50 mM formic acid in methanol–water (95:5; v/v). Mix 50 mL of the buffer stock solution with 950 mL of methanol.
6. Standard solutions: prepare a series of calibration solutions with increasing concentration of domoic acid (DA) within the range of for example 0.2–25 μg/mL. Dilute the standard solutions with methanol–water (95:5, v/v) (*see* **Note 5**).

2.3.2 Materials and Instrumentation

1. Grinder or mechanical mixer (e.g., ULTRA-TURRAX).
2. 10 or 15 mL glass or plastic tubes (heat and centrifuge stable).
3. Vortex.
4. Electrical heater or water bath.
5. Ultrasonic probe or ultrasonic bath.
6. Centrifuge at least 5,000 × *g*.
7. Nylon or PTFE syringe micro filter with 0.45 μm pore size.
8. One-way 2 mL syringe with needle.
9. 1.5 mL HPLC vials.
10. HPLC instrumentation: injection system, gradient pump, column oven, C18 reversed phase analytical column, mass spectrometer with electro spray ionization source (*see* **Note 3**).

3 Methods

3.1 Preparation of Shellfish Sample

For representative sampling it is necessary to pool 100–150 g of tissue.

1. Clean the outside of the shellfish with fresh water, open the shell by cutting the adductor muscle, and rinse inside with fresh water to remove sand and other material.
2. Cut the meat and homogenate the tissue finely and sensitive with a suitable grinder or mechanical mixer.

3.2 Determination of PSP Toxins

3.2.1 Extraction

1. Weigh accurately 4.0 ± 0.1 g of tissue homogenate into a centrifuge tube.
2. Add 5 mL of extraction solvent, homogenize for 1 min with a vortexer, and put the sample in the ultrasonic bath for 5 min or utilize the ultrasonic probe for 1 min.
3. Close the tube and boil the mixture 15 min in an electrical heater or water bath.
4. Cool down at room temperature, centrifuge at 2,500 × *g* for 5 min, and transfer the supernatant in a 20 mL volumetric flask.
5. Repeat the extraction with 3 mL of extraction solvent two times more without heating.
6. Fill up to 20 mL with extraction solvent and transfer 1.5 mL of the extract into a 2.0 mL tube, centrifuge 10 min up to 15,000 × *g*, and filtrate the supernatant through a filter with the aid of a one-way 2 mL syringe.
7. Aliquot the filtrate into an HPLC vial and analyze the sample immediately or store it in the freezer at approximately −12 °C or lower.

3.2.2 LC-MS/MS Measurement

The determination of PSP toxins carried out by hydrophilic interaction liquid chromatography with MS/MS detection. LC conditions are listed in Table 3 and MS/MS conditions are described in Table 4. The SRM transitions for the detection of PSP toxins are listed in Table 5. The limits of detection (LOD) are relatively variable among PSP toxins and range from 5 pg on-column (S/N = 5) for GTX-3 and 7 pg for GTX-4 up to 112 pg for GTX-1. However, most PSP toxins have intermediate LODs between 10 and 50 pg on-column. Figure 7 shows the ion traces of a PSP standard mix consisting of C1/2, B1, GTX1-4, NEO, STX, dcGTX2/3, and dcSTX. The alphamers (C1, GTX1, GTX2, and dcGTX2) elute first on a HILIC column followed by the betamers (C2, GTC3, GTX4, and dcGTX3), B1, the carbamoyl toxins NEO and STX and finally by dcSTX.

Table 3
LC conditions for the determination of PSP toxins

LC conditions		
Column	ZIC-HILIC 150 mm × 4.6 mm id, 5 μm (Dichrom)	
Temperature	35 °C	
Flow	0.7 mL/min	
Injection volume	20 μL	
Run time	45 min gradient	
LC-gradient		
Time/min	A/%	B/%
0	80	20
5.0	65	35
10.0	60	40
20.0	55	45
24.0	55	45
25.0	80	20
45.0	80	20
Stop		

Table 4
MS/MS parameters for the determination of PSP toxins

MS/MS conditions on ABI-Sciex-4000 Q Trap triple-quadrupole mass spectrometer			
Mode	Positive	Temperature (°C)	650
Curtain gas (psi)	30	Nebulizer gas (psi)	70
Ion-spray voltage (V)	5,000	Interface heater	On
Auxiliary gas (psi)	70	Entrance potential (V)	10
Declustering potential (V)	66	Collision energy (V)	30
Collision gas	High	Collision cell exit potential (V)	12

3.3 Determination of DSP Toxins

3.3.1 Extraction

1. Weigh accurately 4.0 ± 0.1 g of tissue homogenate into a centrifuge tube.
2. Add 5 mL of extraction solvent, homogenize for 1 min with a vortexer, and put the sample in the ultrasonic bath for 5 min or utilize the ultrasonic probe for 1 min.
3. Centrifuge at 2,500 × *g* for 5 min and transfer the supernatant into a 20 mL volumetric flask.

Table 5
SRM transitions for selected PSP toxins

Toxin	Transition *m/z*	Toxin	Transition *m/z*	Toxin	Transition *m/z*
STX	300 > 282 300 > 204	B1	380 > 300 380 > 282	dc-STX	257 > 196 257 > 156
NEO	316 > 298 316 > 196	B2	396 > 316 396 > 298	dc-NEO	273 > 255
GTX-1	412 > 332 412 > 314	C1	396 > 316 396 > 298	dc-GTX-1	369 > 289
GTX-2	396 > 316 396 > 298	C2	396 > 316 396 > 298	dc-GTX-2	353 > 273
GTX-3	396 > 316 396 > 298	C3	412 > 332 412 > 314	dc-GTX-3	353 > 273
GTX-4	412 > 332 412 > 314	C4	412 > 332 412 > 314	dc-GTX-4	369 > 289

4. Repeat the extraction with 3 mL extraction solvent two times. The third centrifugation runs at 5,000 × *g* for 10 min.
5. Fill up to 20 mL with extraction solvent and filtrate a portion of the extract through a filter with the aid of a one-way 2 mL syringe.
6. Aliquot the filtrate into an HPLC vial and analyze the sample immediately or store it in the freezer at approximately −20 °C.

3.3.2 Hydrolysis: Transformation of DTX-3 to OA or DTX-1

1. Add to 500 μL of extract 62.5 μL of 2.5 M NaOH into an HPLC vial and mix 1 min with a vortex.
2. Heat this mixture at 75 °C for 40 min in a thermo mixer and let it shake slowly.
3. Cool down to room temperature and add 62.5 μL 2.5 M HCl and mix 1 min with a vortex.
4. Transfer the extract into a micro insert HPLC vial.
5. Analyze the hydrolyzed extract immediately and together with the unhydrolyzed samples and apply the dilution factor of the hydrolysis in the calculation of the concentrations of the present DSP toxins.

3.3.3 LC-MS/MS Measurement

The chromatographic run was divided into three periods with different mass spectrometric parameters. The respective LC conditions are listed in Table 6, the MS parameters are in Table 7 and the SRM transitions, retention times, period and LOD for each toxin are listed in Table 8. Figure 8a shows period 1 and a peak of DA, Fig. 8b shows period 2 for the cyclic imine toxins showing peaks of GYM and SPX1 and Fig. 8c shows period 3 for the

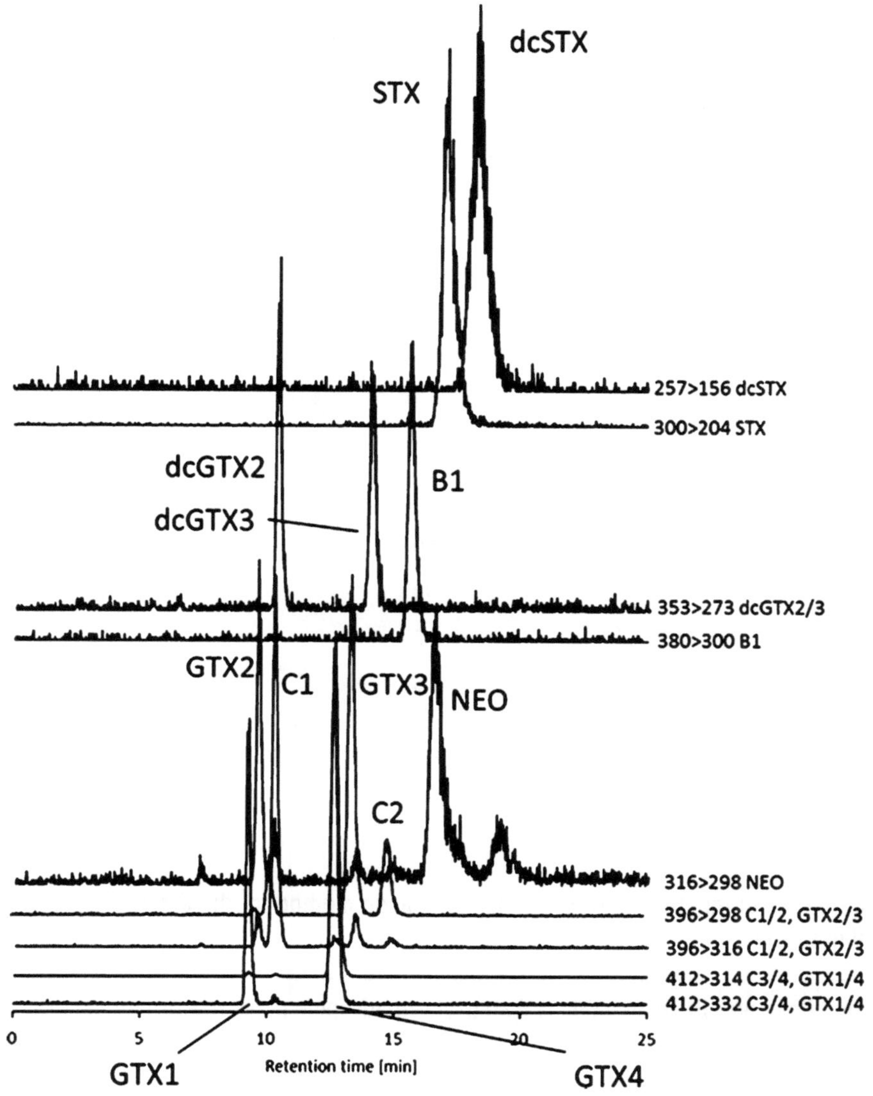

Fig. 7 Specific ion traces of a SRM chromatogram of a PSP standard mix containing C1/2, B1, GTX1-4, NEO, STX, dcGTX2/3, and dcSTX

lipophilic toxins OA and DTXs, PTXs, YTXs and AZAs. This method is also able to analyze domoic acid in period 1 and cyclic imine toxins (spirolides, gymnodimines) [24] in period 2.

3.4 Determination of ASP Toxins

3.4.1 Extraction

1. Weigh accurately 4.0 ± 0.1 g of tissue homogenate into a centrifuge tube.
2. Add 5 mL extraction solvent, homogenize for 1 min with a vortexer, and put the sample in the ultrasonic bath for 5 min or utilizes the ultrasonic probe for 1 min.
3. Close the tube and heat the mixture at 60–70 °C for 15 min in an electrical heater or water bath.

Table 6
LC conditions of the multi-toxin method for the determination of DA and lipophilic toxins

LC conditions		
Column	Hypersil BDS 120 Å 5 μm C8 reversed phase 50 mm × 2.0 + Security Guard column (Phenomenex)	
Temperature	20 °C	
Flow	0.2 mL/min	
Injection volume	5 μL	
Run time	30 min gradient	
LC-gradient		
Time/min	A/%	B/%
0	95	5
10.0	0	100
15.0	0	100
18.0	95	5
30.0	95	5
Stop		

Table 7
MS/MS parameters of the three periods of the multi-toxin method for the determination of DA and lipophilic toxins

MS/MS conditions on ABI-Sciex-4000 Q Trap triple-quadrupole mass spectrometer			
	Period 1 (0–8.75 min)	**Period 2 (8.75–11.2 min)**	**Period 3 (11.2–18.0 min)**
Curtain (psi)	20	10	10
CAD	Medium	Medium	Medium
Ion-spray voltage (V)	5,500	5,500	5,500
Temperature (°C)	275	Ambient	Ambient
Nebulizer gas (psi)	50	10	10
Auxiliary gas (psi)	50	Off	Off
Interface heater	On	On	On
Declustering potential (V)	50	50	50
Entrance potential (V)	10	10	10
Collision cell exit potential (V)	15	15	15

Table 8
SRM transitions, retention times, and period of DA and lipophilic toxins

Toxin	Transition (collision energy [V])	Retention time [min]	Period	LOD[a] [pg]
DA	312 > 266 (20) 312 > 161 (30)	7.58	1	58
OA	822 > 223 (55)	11.85	3	180
DTX-1	836 > 237 (55)	12.83	3	320
DTX-2	822 > 223 (55)	12.15	3	nd
PTX-1	892 > 213 (55)	11.33	3	nd
PTX-2	876 > 213 (55)	12.40	3	26
PTX-2sa	894 > 213 (55)	11.93	3	15
YTX	1,160 > 965 (55)	13.14	3	34
Homo-YTX	1,174 > 979 (55)	13.22	3	nd
45-OH-YTX	1,176 > 981 (55)	11.94	3	nd
45-OH-homo-YTX	1,190 > 977 (55)	11.85	3	nd
AZA-1	842 > 824 (55)	14.37	3	2
AZA-2	856 > 838 (55)	14.61	3	nd
AZA-3	828 > 810 (55)	13.90	3	nd

Nd not determined
[a]LOD defined as (S/N = 3)

4. Cool down at room temperature, centrifuge at 2,500 × *g* for 5 min, and transfer the supernatant in a 20 mL volumetric flask.
5. Repeat the extraction with 3 mL of extraction solvent two times more without heating and centrifuge at 5,000 × *g* for 10 min the last time.
6. Fill up to 20 mL with extraction solvent and filtrate a portion of the extract trough a filter with the aid of a one way 2 mL syringe.
7. Aliquot the filtrate into an HPLC vial and analyze the sample immediately or store it in the freezer approximately −12 °C or lower.

3.4.2 LC-MS/MS Measurement

The determination of DA carried out by reversed phase liquid chromatography with MS/MS detection. LC is performed on a reversed phase C18 column and LC parameters are given in Table 9. Mass spectrometric detection of DA is performed in the SRM mode with two transitions and MS parameters as listed in Table 10.

The LOD (S/N = 3) of DA is 40 pg on-column using an AB Sciex API 2000 triple quadrupole mass spectrometer. Figure 9 shows the quantifier ion trace (*m/z* 312 > 266) and the qualifier ion trace (*m/z* 312 > 161) of a DA standard.

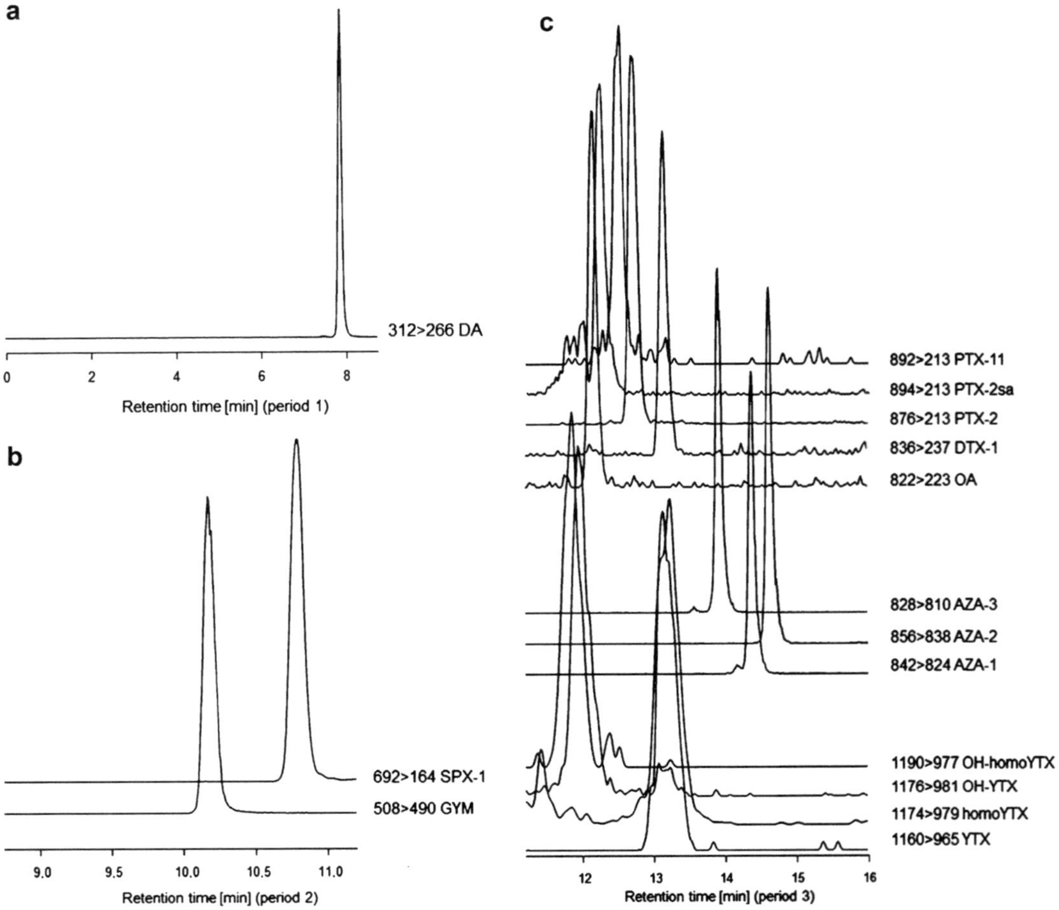

Fig. 8 SRM chromatograms of a lipophilic toxin standard mix DSP: (**a**) period 1 (0–8.5 min) for the detection of DA, (**b**) period 2 (8.5–11.2 min) for the detection of cyclic imine toxins and (**c**) period 3 (11.2–16 min) for PTXs, DTXs, AZAs, and YTXs

Table 9
LC conditions for the determination of DA

LC conditions	
Column	Luna C18 reversed phase 250 mm × 4.6 mm id, 5 μm + security guard column (Phenomenex)
Temperature	25 °C
Flow	0.8 mL/min
Injection volume	20 μL
Run time	25 min gradient
LC-gradient	

(continued)

Table 9
(continued)

Time/min	A/%	B/%
0	87	13
2.0	87	13
10.0	50	50
10.5	10	90
15.5	10	90
16.0	87	13
25.0	87	13
Stop		

Table 10
MS/MS parameters for the determination of DA

MS/MS conditions on ABI-Sciex-2000 triple-quadrupole mass spectrometer Transitions (*m/z*) for DA: 312 > 266 (quantifier), 312 > 161 (qualifier)			
Mode	Positive	Temperature (°C)	400
Curtain gas (psi)	50	Nebulizer gas (psi)	90
Ion-spray voltage (V)	5,500	Interface heater	on
Auxiliary gas (psi)	70	Entrance potential (V)	12
Declustering potential (V)	16	Collision energy (V)	20
Collision gas (V)	6	Collision cell exit potential (V)	6
Focussing potential (V)	370		

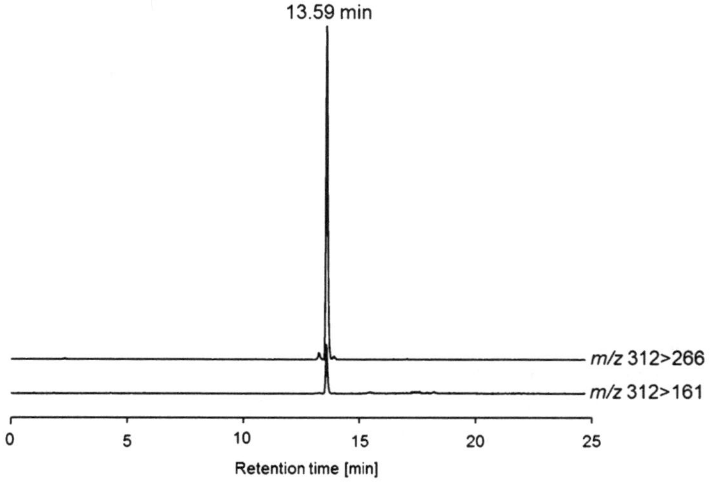

Fig. 9 Quantifier (*m/z* 312 > 266) and qualifier (*m/z* 312 > 161) of a SRM chromatogram of a DA standard

4 Notes

1. The enantiomeric pairs (e.g., GTX-1/4, GTX-2/3, C1/C2) easily convert into each other, especially under alkaline or acidic conditions as well as at ambient or higher temperatures. Even though they can be detected individually, their measured ratios do not necessarily reflect the ratios in the original sample. For this reason these enantiomeric pairs should always be given as sums.
2. Store all buffer stock solutions refrigerated at approximately 4 °C and warm up to room temperature before use.
3. Store PSP and DSP calibration solutions in the dark and in the freezer at approximately -20 °C. Let them thaw to room temperature before use.
4. The use of a guard column is recommended.
5. Store ASP solutions in the dark, refrigerated at approximately 4 °C, not more than 3 months, and do not freeze them. Warm up to room temperature before use.

References

1. Anonymous (2009) Scientific opinion of the panel on contaminants in the food chain on a request from the European Commission on marine biotoxins in shellfish: summary on regulated marine biotoxins. EFSA J 1306:1–23
2. Christian B, Luckas B (2008) Determination of marine biotoxins relevant for regulations: from the mouse bioassay to coupled LC-MS methods. Anal Bioanal Chem 391:117–134
3. Hollingworth T, Wekel MM (1990) Fish and other marine products. paralytic shellfish poisoning. biological method, final action. In: Hellrich K (ed) Official methods of analysis of the AOAC, 15th edn. AOAC, Arlington, VA, pp 881–882
4. Schantz EJ (1986) Chemistry and biology of saxitoxin and related toxins. Ann N Y Acad Sci 479:15–23
5. Luckas B, Hummert C, Oshima Y (2003) Analytical methods for paralytic shellfish poisons. In: Hallegraeff G, Anderson DM, Cembella AD (eds) Manual on harmful marine microalgae. Intergovernmental Oceanographic Commission of UNESCO, Paris, pp 191–209
6. Quilliam MA (2003) The role of chromatography in the hunt for red tide toxins. J Chromatogr A 1000:527–548
7. Diener M, Erler K, Christian B et al (2007) Application of a new zwitterionic hydrophilic interaction chromatography column for determination of paralytic shellfish poisoning toxins. J Sep Sci 30:1821–1826
8. Anonymous (2009) Scientific opinion of the panel on contaminants in the food chain on a request from the European Commission on marine biotoxins in shellfish: Saxitoxin Group. EFSA J 1019:1–76
9. Yasumoto T, Murata M, Oshima Y et al (1985) Diarrhetic shellfish toxins. Tetrahedron 41: 1019–1025
10. Anonymous (2008) Opinion of the panel on contaminants in the food chain on a request from the European Commission on marine biotoxins in shellfish: okadaic acid and analogues. EFSA J 589:1–62
11. Anonymous (2009) Scientific opinion of the panel on contaminants in the food chain on a request from the European Commission on marine biotoxins in shellfish: pectenotoxin group. EFSA J 1109:1–47
12. Anonymous (2008) Opinion of the panel on contaminants in the food chain on a request from the European Commission on marine biotoxins in shellfish: yessotoxin group. EFSA J 907:1–62
13. Anonymous (2008) Opinion of the panel on contaminants in the food chain on a request from the European Commission on marine biotoxins in shellfish: azaspiracids. EFSA J 723:1–52

14. Anonymous (2004) Regulation (EC) No. 853/2004 of the European Parliament and of the Council of 29 April 2004 laying down specific hygiene rules for food of animal origin. Off J Eur Commun. Annex III, Sect VII, Chap V, 2. L 226, p 60–61
15. McMahon T, Silke J (1996) West coast of Ireland; winter toxicity of unknown aetiology in mussels. Harmful Algae News 14:2
16. Satake M, Ofuji K, Naoki H et al (1998) Azaspiracid, a new marine toxin having unique spiro ring assemblies, isolated from Irish mussels, *Mytilus edulis*. J Am Chem Soc 120: 9967–9968
17. Ofuji K, Satake M, McMahon T et al (1999) Two analogs of azaspiracid isolated from mussels, *Mytilus edulis*, involved in human intoxication in Ireland. Nat Toxins 7:99–102
18. James KJ, Sierra MD, Lehane M et al (2003) Detection of five new hydroxyl analogues of azaspiracids in shellfish using multiple tandem mass spectrometry. Toxicon 41:277–283
19. Rehmann N, Hess P, Quilliam MA (2008) Discovery of new analogs of the marine biotoxin azaspiracid in blue mussels *Mytilus edulis* by ultra-performance liquid chromatography/tandem mass spectrometry. Rapid Commun Mass Sp 22:549–558
20. Krock B, Tillmann U, Voß D et al (2012) New azaspiracids in Amphidomataceae (Dinophyceae). Toxicon 60:830–839
21. Krock B, Tillmann U, John U et al (2009) Characterization of azaspiracids in plankton size-fractions and isolation of an azaspiracid-producing dinoflagellate from the North Sea. Harmful Algae 8:254–263
22. Anonymous (2009) Scientific opinion of the panel on contaminants in the food chain on a request from the European Commission on marine biotoxins in shellfish: domoic acid. EFSA J 1181:2–61
23. Lawrence JF, Lau BPY, Cleroux C et al (1994) Comparison of UV absorption and electrospray mass spectrometry for the high-performance liquid chromatographic determination of domoic acid in shellfish and biological samples. J Chromatogr A 659:119–126
24. Anonymous (2010) EFSA panel on contaminants in the food chain (CONTAM); scientific opinion on marine biotoxins in shellfish-Cyclic imines (spirolides, gymnodimines, pinnatoxins and pteriatoxins). EFSA J 8(6):1628, 1–39

Chapter 19

Fucoidan Analysis by Tandem MALDI-TOF and ESI Mass Spectrometry

Stanislav D. Anastyuk, Natalia M. Shevchenko, and Vladimir I. Gorbach

Abstract

The application of mass spectrometry towards the structural analysis of the most interesting sulfated biopolymers of the brown algae—fucoidans only developed relatively recently. During method development, many problems, both chemical and instrumental, have to be solved. For example, mass spectrometry has a limitation in the analysis of anionic high molecular weight (HMW) polysaccharides because of the labile nature of sulfate groups which cause the polysaccharide to desulfate rather than ionize. Thus, decomposition methods should be developed taking into account the structural features of such a complex and fragile compound. The selection of optimal instrument settings for the electrospray ionization mass spectrometry (ESIMS) and of matrix media for matrix-assisted laser desorption/ionization mass spectrometry (MALDIMS) is also required. When optimal parameters for mass spectrometric analyses are found, the application of these methods to the elucidation of structural features of fucoidans (by studying their fragments) allows researchers to rapidly obtain new and unique data, often impossible to achieve by other techniques. Herein, we describe tandem mass spectrometry of sulfated fucooligosaccharides, obtained by an autohydrolysis technique from structurally different fucoidans.

Key words Brown algae, Electrospray ionization mass spectrometry (ESIMS) fucoidan, Matrix-assisted laser desorption/ionization time-of-flight mass spectrometry (MALDI-TOF MS), Oligosaccharides, Tandem mass spectrometry (MS/MS)

1 Introduction

Mass spectrometry is a modern fundamental and applied technique. It involves the acquisition of data on the composition of a substance, its structure, physical and chemical properties, and processes of ion-molecular interactions. Electrospray ionization mass spectrometry (ESIMS) has been developed since the 1980s [1, 2]. Together with matrix-assisted laser desorption/ionization mass spectrometry (MALDIMS) [3, 4], ESIMS enables the study of high molecular weight (HMW) compounds of almost any type, including proteins, nucleotides, natural and synthetic polymers.

Structural features of sulfated polysaccharides, such as chondroitin sulfates and heparins have been successfully studied by mass

Dagmar B. Stengel and Solène Connan (eds.), *Natural Products From Marine Algae: Methods and Protocols*, Methods in Molecular Biology, vol. 1308, DOI 10.1007/978-1-4939-2684-8_19, © Springer Science+Business Media New York 2015

spectrometry because of the availability of the specific enzymes, effectively catalyzing depolymerization of the polymers [5, 6]. Since enzymes capable of effectively decomposing fucoidan [7] are not widely available, an alternative method for the soft decomposition (without the excess loss of labile sulfate groups) of fucoidans is required.

Despite of the complexity of fucoidans, two species-specific structurally different types of its main chain have been suggested [8]: a linear backbone of alternating 3- and 4-linked α-L-fucopyranose (α-L-Fuc*p*) residues with different sulfation/acetylation pattern, which was shown to be specific for the Fucales (Fig. 1a) and a linear backbone of 3-linked 2,4-disulfated α-L-Fuc*p*, which was shown to be characteristic for Laminariales (Fig. 1b). These structures have been elucidated using the "classic" methods of carbohydrate chemistry such as desulfation, Smith degradation and methylation analysis [9] and NMR.

The approach for the analysis of fucoidan (a galactofucan from the brown alga *Saccharina* (*Laminaria*) *gurjanovae*) through mass spectrometric investigation of its fragments, obtained following a solvolytic desulfation reaction was employed by our group for the first time [10]. Excess DMSO was removed with repeated co-evaporation with MeOH. The galactofucan was additionally decomposed by partial acid hydrolysis (0.2 N TFA at 60 °C). MALDI-TOF analysis of the fragments revealed galactooligosaccharides and "hybrid" oligosaccharides, composed of fucose and galactose. Unfortunately, the conditions initially selected were too severe to allow preservation of the sulfate groups (desulfation conditions) and the instrument was unable to operate in MS/MS

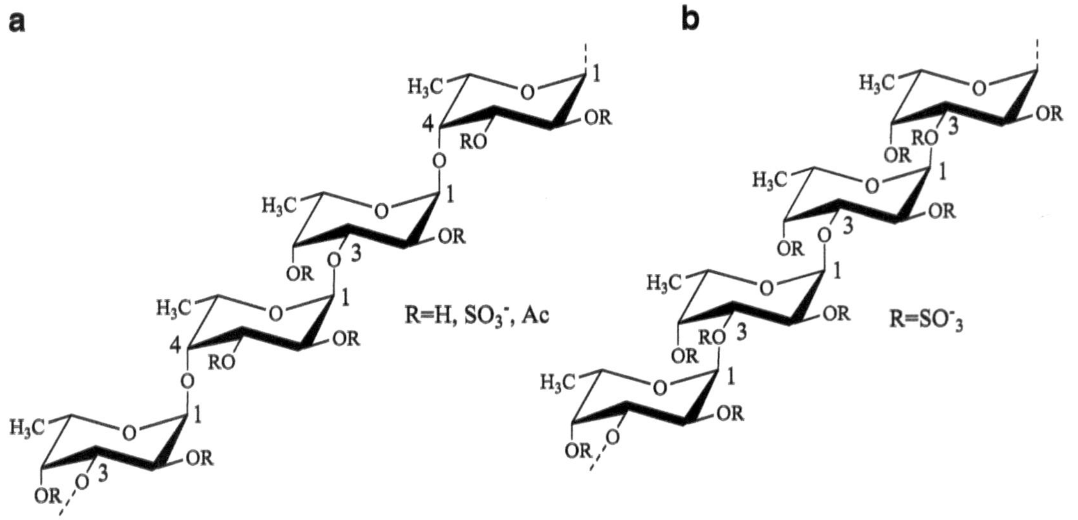

Fig. 1 Structures of main chain of fucoidans, characteristic to (**a**) Fucales, (**b**) Laminariales species of brown alga

mode to obtain structural data. Recent data [11] confirm that the fucan part of the galactofucan from *S. gurjanovae* is characteristic of Laminariales (Fig. 1b).

The first systematic mass spectrometric (ESIMS/MS) elucidation of the structural features of oligosaccharides, derived from sulfated fucan of the brown alga *Ascophyllum nodosum* by mild acid hydrolysis (0.75 mM H_2SO_4 at 60 °C) was done by [12]. The fucoidan was previously shown to contain highly branched core region with primarily 3-linked α-L-Fuc*p* residues and few α-(1→4) linkages, branch points were at position 2 of the internal 3-linked α-L-Fuc*p* residues [13]. The work [12] was based on previous investigations of the ESIMS/MS fragmentation patterns of positional isomers of sulfated fucose [14] and hexoses [15, 16] and fragments from heparin [17] and chondroitin [18]. It was confirmed that the formation of characteristic ^{0,2}A-/^{0,2}X-ions during negative-ion CID MS/MS requires an available proton at the C-3 hydroxyl group (following the nomenclature introduced by [19], Fig. 2). Hence, these observations indicated that no cross-ring cleavages during CID ESIMS/MS of 3-linked disaccharides could occur. The availability of mobile protons (at the glycosidic OH-group) is also essential [14, 20] for the production of the cross-ring cleavages, which provide information of both the sulfate position and linkage type.

The solvolytic desulfation conditions for decomposition of fucoidan were implemented on the brown alga *Fucus evanescens* [21]. This fucoidan was already shown [22] to contain a linear backbone of alternating (Fig. 1a) 3- and 4-linked α-L-Fuc*p*-2-sulfate residues. Additional sulfate occupied position 4 in a part of 3-linked fucose residues, which was elucidated by the "classic" methods of carbohydrate chemistry and NMR. We used both the MALDI-TOFMS and ESIMS/MS (with TOF analyzer, which gives great sensitivity and accuracy) techniques to observe minor

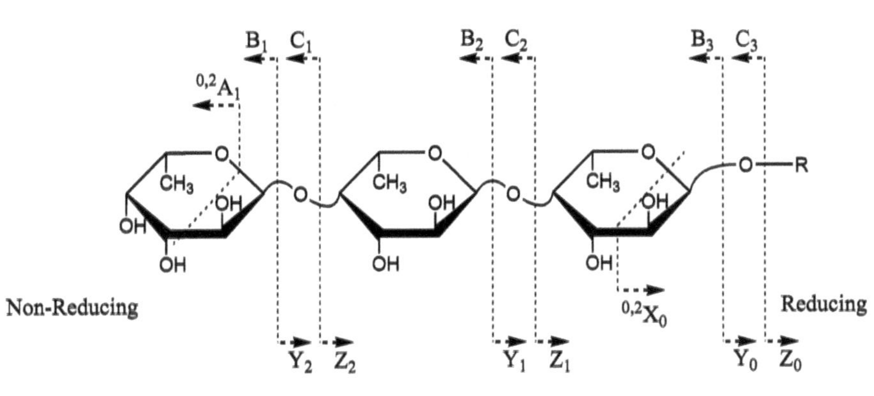

Fig. 2 Nomenclature for the fragmentation of glycoconjugates, suggested by [19]

constituents of *F. evanescens*, which were also found in monosaccharide composition by [22], but it was impossible to observe their exact positions by the classic approach. The fragments were identified by ESIMS/MS: Xyl-(1→4)-Fuc, Gal-(1→4)-Fuc, Gal-(1→4)-Gal-(1→4)-Fuc, Gal-(1→4)-Gal. Fucose, galactose, and xylose residues were shown to be mainly 2-*O*-sulfated with traces of 4-*O*-sulfation. Glucuronic acid was also found as a part of non-sulfated fucooligosaccharides: Fuc-(1→3)-GlcA, Fuc-(1→4)-Fuc-(1→3)-GlcA, Fuc-(1→3)-Fuc-(1→4)-Fuc-(1→3)-GlcA [21].

Although the solvolytic desulfation reaction produced oligosaccharides suitable for MS analysis, we were unable to change its parameters and sulfates were mostly lost. It was noticed that dry highly dialyzed fucoidan samples at high moisture (relative humidity over 80 %) spontaneously decomposed almost entirely to fucose monomer. It was clear that polysaccharide hydrolysis was initiated by the $-SO_3H$ groups of the compound as the source of acid and the term "autohydrolysis" first applied to carrageenan decomposition [23] was also used for fucan decomposition. We investigated the autohydrolysis of highly sulfated fucan samples from the brown algae *F. evanescens* and *Saccharina cichorioides* (3-linked 2,4-disulfated α-L-fucan [24]) and we found reproducible conditions for the depolymerization of these fucans with known structures, which were also determined by independent methods. The autohydrolysis must be carried out at 5 mg/mL of a polysaccharide concentration at 25–37 °C for 48–72 h that gives reproducible results and allows the absence of unwanted sulfate rearrangement effects [25], but these conditions are still sufficiently strong to cause cleavage of hybrid oligosaccharides (GalFuc, for example; [26]), which are not yielded by other techniques [22].

The results of the mass spectrometric investigation on *S. cichorioides* [27] by both tandem MALDI-TOF and ESIMS techniques were virtually the same. The derivatization through the reduction with $NaBD_4$ was performed to unambiguously assign the fragment ions from the reducing and non-reducing termini. Tandem MALDI-TOFMS technique (LIFT TOF/TOF in Bruker (Germany) implementation) was found to be more convenient for the analysis of complex mixtures of poly-sulfated monosaccharides and oligosaccharides, since only singly-charged ions are observed. But, ESIMS was found to be more sensitive and allow the observation of minor fragments containing uronic acids, which were invisible with MALDI-TOFMS, as highlighted recently in a study on the brown alga *Costaria costata* [28].

By using the MS/MS technique it is possible to unambiguously distinguish the sulfate position in a monosaccharide (except for sulfates at C-4 and at C-6 of hexoses where MS^3 experiment is required [16]), and (1→3)-, (1→2)- and (1→4)-type of linkages in disaccharides [12, 22, 26–28]. The positions of sulfate groups could be found for all linkage types with the exception of (1→3), because of the cross-ring cleavage restriction. The sulfation at C-3 is also hard to elucidate. The MS sequences of oligosaccharides with higher degrees of polymerization are much more complicated and not well enough studied. Thus, only tandem mass spectra of monosaccharides and disaccharides are presented and discussed here in detail.

2 Materials

Prepare all solutions using ultrapure water (prepared by purifying deionized water to attain a sensitivity of 18 MΩ cm at 25 °C) and reagents at least analytical grade. Prepare and store all reagents at room temperature (unless indicated otherwise).

2.1 Autohydrolysis of Fucoidans (Depolymerization Technique)

1. Fucoidan solution for depolymerization by autohydrolysis: dissolve 50 mg of the fucoidan sample in 1 mL of water.
2. Amberlite CG-120, 200–400 mesh. Use 5 mL minicolumn.
3. 0.1 M HCl: pour 50 mL of water into the volumetric flask. Add 90 μL of conc. HCl. Bring the solution volume up to 100 mL with water.
4. 2.5 % NH_4OH solution: pour 9 mL of water into the volumetric flask. Add 1 mL of 25 % aqueous ammonia (v/v).

2.2 Reduction of Oligosaccharides

1. Reducing solution: $NaBD_4$ solution: 2 mg/mL solution of $NaBD_4$ reagent (Sigma, USA) in water. Add 2 mg of $NaDB_4$ into a 2.5 mL vial. Add 1 mL of water (*see* **Note 1**).
2. Acetic acid.
3. Methanol.

2.3 Negative-Ion ESIMS of Oligosaccharides

1. Working solution for ESIMS experiment: acetonitrile–water (1:2, v/v): mix 660 μL of water with 330 μL of acetonitrile in a 2.5 mL vial.

2.4 Negative-Ion MALDI-TOFMS of Oligosaccharides

1. Working solution for ESIMS experiment: acetonitrile–water (1:2, v/v): mix 660 μL of water with 330 μL of acetonitrile in a 2.5 mL vial.
2. MALDIMS matrices: DHB/SDHB (Sigma, USA) matrix solution: 10 mg/mL solution of 2,5-dihydroxybenzoic acid (*see* **Note 2**). Add 1 mg of matrix into a 600 μL vial. Add 1 mg of L-fucose to reduce the in-source fragmentation [26]. Add 100 μL of the working solution for ESIMS. Mix with a shaker until the components are fully dissolved.
3. Arabinoosazone (phenylosazone of D-arabinose [30]) matrix solution: 10 mg/mL solution of arabinoosazone. Add 1 mg of matrix into a 600 μL vial. Add 1 mg of L-fucose. Add 100 μL of acetone–water solution (1:3, v/v). Mix with shaker until the components are fully dissolved.

3 Methods

3.1 Autohydrolysis of Fucoidans (Depolymerization Technique)

1. Set up a cation exchange minicolumn with cation exchange resin (Amberlite CG-120, 200–400 mesh).
2. Charge the resin with 10 mL 0.1 M HCl.
3. Wash with 15 mL of water.
4. Carefully load the fucoidan solution (working concentration of 5 mg/mL) for autohydrolysis onto the column.
5. Eluate oligosaccharides with 10 mL of water into a plastic tube (*see* **Note 3**).
6. Place the eluant in a temperature-controlled chamber for 48–72 h at 37 °C.
7. Neutralize with 2.5 % NH_4OH solution.
8. Evaporate under vacuum or freeze-dry.

3.2 Reduction of Oligosaccharides

1. Prepare 1 mL of the reducing solution.
2. Dissolve 1 mg of oligosaccharides in the reducing solution.
3. Leave at 4 °C overnight.
4. Add 100 μL of 100 % acetic acid.
5. Repeat **step 4** until the solution becomes neutral.
6. Evaporate under vacuum.
7. Add 1 mL of MeOH.
8. Evaporate under vacuum.
9. Repeat **steps 7** and **8** three times to remove the boric acid formed during the reaction.

3.3 Negative-Ion ESI-MS of Oligosaccharides

3.3.1 Preliminary Investigation of the Mixture Content

1. Prepare the oligosaccharides by autohydrolysis. Dissolve the sample in the working solution for ESIMS analysis at a concentration of 1 mg/mL.
2. Prepare 100–300 μL of the diluted sample in the separate vials (1:100 and 1:1,000). Use the working solution for ESIMS for dilution.
3. Set up the instrument: use the standard settings for the negative-ion mode. Set the fragmentor voltage (if present) to 160–170 V.
4. Introduce the sample into the ESIMS instrument using direct injection at a flow rate of 5 μL/min.
5. Record mass spectra. Adjust the fragmentor voltage (if present) to maximize the signal of oligosaccharides and minimize the intensities of the fragment ions at m/z 97 (sulfate anion), 225 $[FucSO_3\text{-}H_2O]^-$ and other signals, indicating the loss of water molecules (M-18).

Table 1
The (partial) composition of the autohydrolysis mixture, obtained from the fucansulfate of the brown alga *Saccharina cichorioides* [27]

ESIMS		MALDIMS	
m/z	Composition	*m/z*	Composition
225.01	$[FucSO_3\text{–}H_2O]^-$	225.0	$[FucSO_3\text{–}H_2O]^-$
231.01	$[Fuc_3(SO_3)_3]^{3-}$	739.1	$[Fuc_3(SO_3Na)_3\text{–}Na]^-$
234.01	$[Fuc_2(SO_3)_2]^{2-}$	491.0	$[Fuc_2(SO_3Na)_2\text{–}Na]^-$
243.02	$[FucSO_3]^-$	243.0	$[FucSO_3]^-$
279.70	$[Fuc_4(SO_3)_3]^{3-}$	885.3	$[Fuc_4(SO_3Na)_3\text{–}Na]^-$
298.04	$[Fuc_3(SO_3)_2\text{–}H_2O]^{2-}$	619.1	$[Fuc_3(SO_3Na)_2\text{–}H_2O\text{–}Na]^-$
307.04	$[Fuc_3(SO_3)_2]^{2-}$	637.1	$[Fuc_3(SO_3Na)_2\text{–}Na]^-$
328.38	$[Fuc_5(SO_3)_3]^{3-}$	–	Not observed
344.96	$[Fuc(SO_3Na)_2\text{–}Na]^-$	344.9	$[Fuc(SO_3Na)_2\text{–}Na]^-$
371.07	$[Fuc_2SO_3\text{–}H_2O]^-$	371.0	$[Fuc_2SO_3\text{–}H_2O]^-$
380.07	$[Fuc_4(SO_3)_2]^{2-}$	783.2	$[Fuc_4(SO_3Na)_2\text{–}Na]^-$
389.08	$[Fuc_2SO_3]^-$	389.1	$[Fuc_2SO_3]^-$

6. Try different concentrations of the sample to reach maximal signal-to-noise ratio.
7. Fill up a table with *m/z* values, charge state and a composition of the oligosaccharide mixture. An example is given in Table 1, left column.

3.3.2 CID MS/MS Mode

1. Switch into a CID MS/MS mode.
2. Set CID energy to 10 V.
3. Set the isolation window to 1.3 mass units for singlycharged ions or 4 mass units for multiply charged ions (*see* **Note 4**).
4. Equilibrate the abundance of the parent ion and the fragment ions by adjusting CID energy.
5. Record MS/MS spectra.

3.4 Negative-Ion MALDI-TOFMS of Oligosaccharides

3.4.1 Preliminary Investigation of the Mixture Content

1. Prepare 10 μL of the sample of oligosaccharides in the working solution for ESIMS experiment at a concentration of 10 mg/mL for DHB/SDHB matrix and 0.1–0.01 mg/mL for arabinoosazone matrix (*see* **Note 5**).
2. Sample introduction for DHB/SDHB matrix: mix 1 μL of the sample and 1 μL of the matrix in the vial. Apply 1 μL of the mixture onto a target. Evaporate the droplet by the air.

3. Sample introduction for arabinoosazone matrix: apply 1 μL of the matrix onto a target. Evaporate a droplet by hair drier/hot air. Apply 1 μL of the sample. Repeat evaporation step.
4. Set up the instrument: use the default settings for the negative ion reflector mode of operation (*see* **Note 6**).
5. Record mass spectra using the DHB/SDHB and arabinoosazone matrices as separate experiments. Adjust the laser power to maximize the signal of oligosaccharides and minimize the intensities of the fragment ions at *m/z* 97 (sulfate anion), 225 $[FucSO_3\text{-}H_2O]^-$ and other signals, indicating the loss of water molecules (M-18). Arabinoosazone gives better sensitivity and less fragmentation, SDHB/DHB induces fragmentation but gives wider mass range.
6. Populate a table with *m/z* values and composition of the oligosaccharide mixture. An example is given in Table 1, right column (*see* **Note 7**).

3.4.2 MS/MS Mode

1. Switch into MS/MS mode.
2. Set the laser power to minimum.
3. Select the parent ion.
4. Try to get abundant MS of the fragment ions by adjusting laser energy.
5. Record MS/MS spectra.

3.5 Interpretation of Tandem Mass Spectra (See Note 8)

3.5.1 MS/MS of Singly Sulfated Monosaccharides

1. Sulfation at C-2 of hexoses (*see* **Note 9**) (*m/z* 259), desoxyhexoses (*see* **Note 10**) (*m/z* 243) and, most likely, pentoses (*see* **Note 11**) (*m/z* 229) gives an abundant signal at *m/z* 138.9 ($^{0.2}X$) from the cross-ring cleavage. The abundant $[M\text{-}H_2O]$ signal at *m/z* [M-18] also serves as an indicator of the sulfation at C-2 [14, 16].
2. Sulfation at C-4/C-6 of hexoses gives fragment ion at *m/z* 198.9 ($^{0.2}A$) (*see* **Note 12**). Desohyhexose gives fragment ion at *m/z* 180.9. Pentose gives fragment ion at *m/z* 166.9.
3. Sulfation at C-3 prevents dissociation with the ring-opening. However, the presence of the $^{0.3}X$-type ion at *m/z* 168.9 may suggest the sulfation at C-3 [14] (*see* **Note 13**).

3.5.2 MS/MS of Sulfated Fucobioses

1. Record MS/MS of the corresponding alditol derivative (*see* **Note 14**).
2. Record MS/MS of the native (underivatized) biose ion. The Y-type ions from the cleavages of fucose residues have *m/z* 243 (*see* Figs. 2, 3, 4, 5, and 6). The additional sulfate group gives +102 Da ($+SO_3Na$—H; *see* Figs. 4 and 6; *see* **Note 15**).
3. The ions, having no shift in MWs are assumed to be C-type ions (*see* **Note 16**).

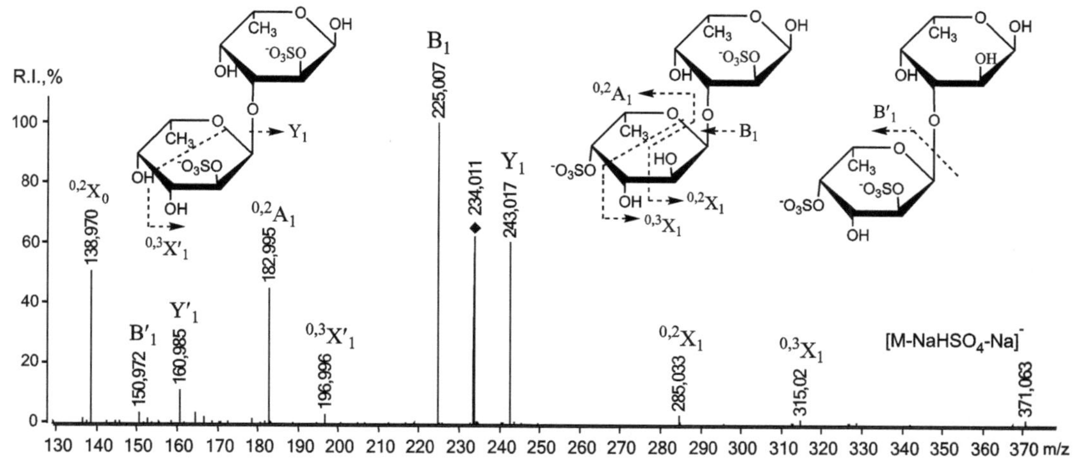

Fig. 3 Negative-ion CID ESIMS/MS of disulfated fucobiose at *m/z* 234.011. The fragment was obtained by autohydrolysis of fucoidan from brown alga *Saccharina cichorioides*. Reproduced from [27] with permission from Elsevier

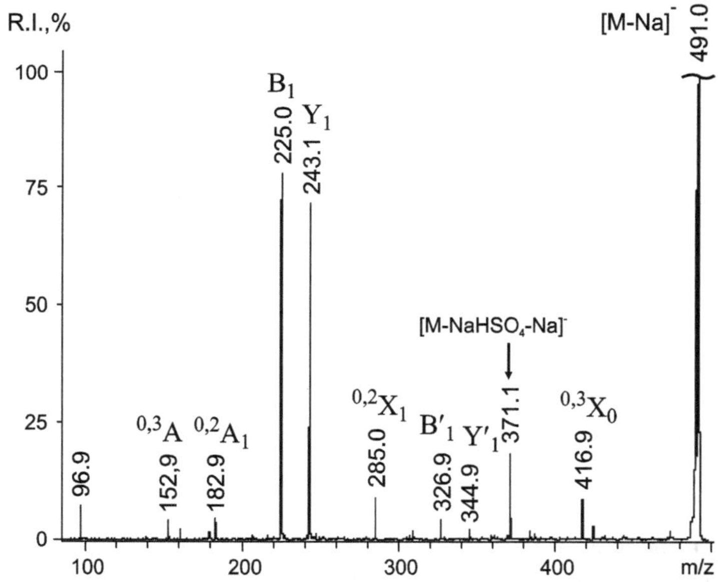

Fig. 4 Negative-ion tandem MALDI-TOFMS of disulfated fucobiose at *m/z* 491.0. The fragment was obtained by autohydrolysis of fucoidan from brown alga *Saccharina cichorioides*. Reproduced from [27] with permission from Elsevier

4. The fragment ions, having MWs lower than Y-type ions by 18, are B-type ions (*see* Figs. 3, 4, 5, and 6). Usually, they have maximum intensity if sulfate is spatially close to the glycosidic linkage, which causes easier cleavage. High intensity of B_1-type ion at m/z 225 suggests that fucose residue on non-reducing terminus is likely sulfated at C-2 (*see* Figs. 2, 3, 4, and 5).

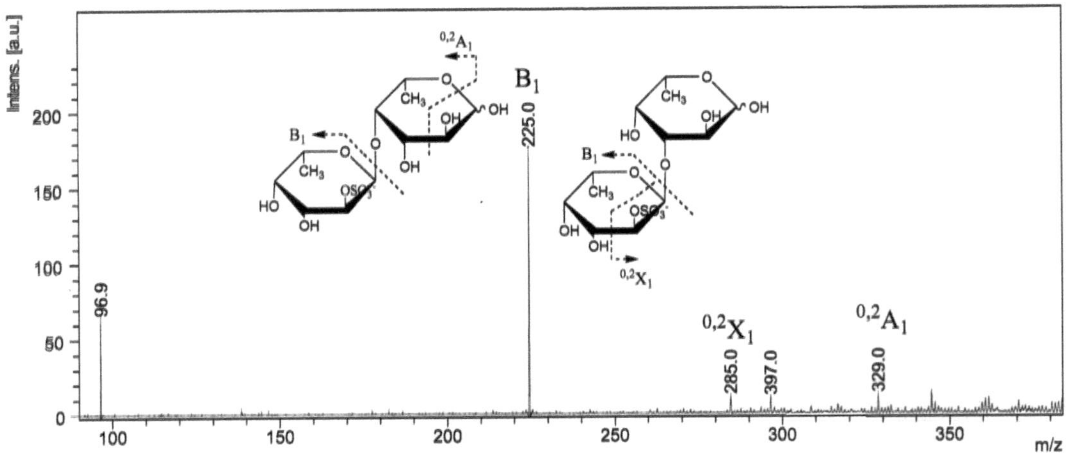

Fig. 5 Negative-ion tandem MALDI-TOFMS of the monosulfated fucobiose at *m/z* 389.1. The fragment was obtained by autohydrolysis of fucoidan from brown alga *Fucus evanescens*

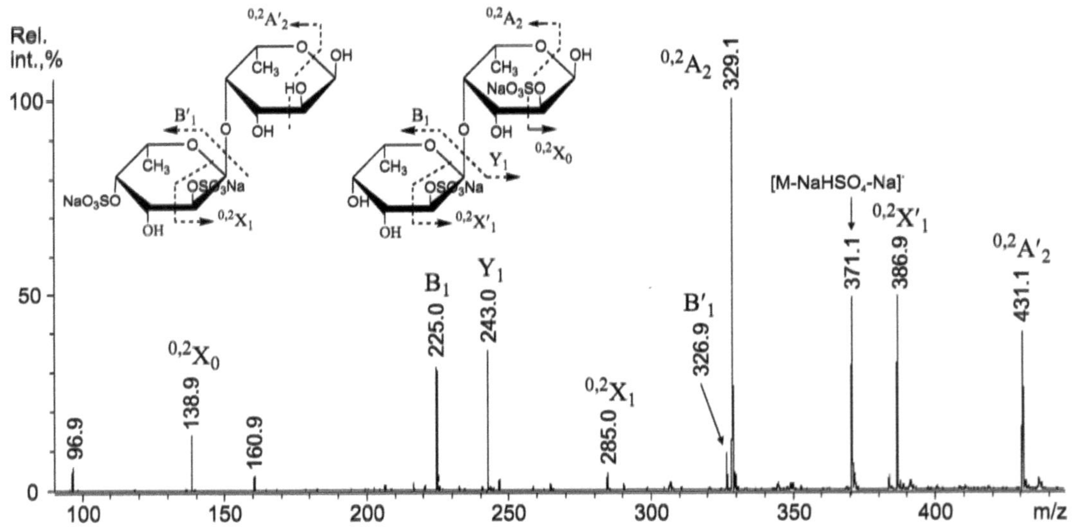

Fig. 6 Negative-ion tandem MALDI-TOFMS of the disulfated fucobiose at *m/z* 491.0. The fragment was obtained by autohydrolysis of fucoidan from brown alga *Fucus evanescens*. Reproduced from [26] with permission from Elsevier

5. Poly-sulfated oligosaccharides give [M-120] fragment ion in both tandem ESIMS and MALDI-TOFMS (*see* **Note 17**).
6. The fucoidan fragment is assumed to be (1→4)-linked if MS/MS exhibits intensive signals of ^{0,2}A-type of fragment ions (Fig. 6) and (1→4)-linked if fragment ions are weak (Figs. 2, 3, 4, and 5) (*see* **Note 18**).
7. The presence of (1→2)-type of linkage could suggest [12, 13, 28] high intensity of ^{0,2}X-type fragment ions (Fig. 7).

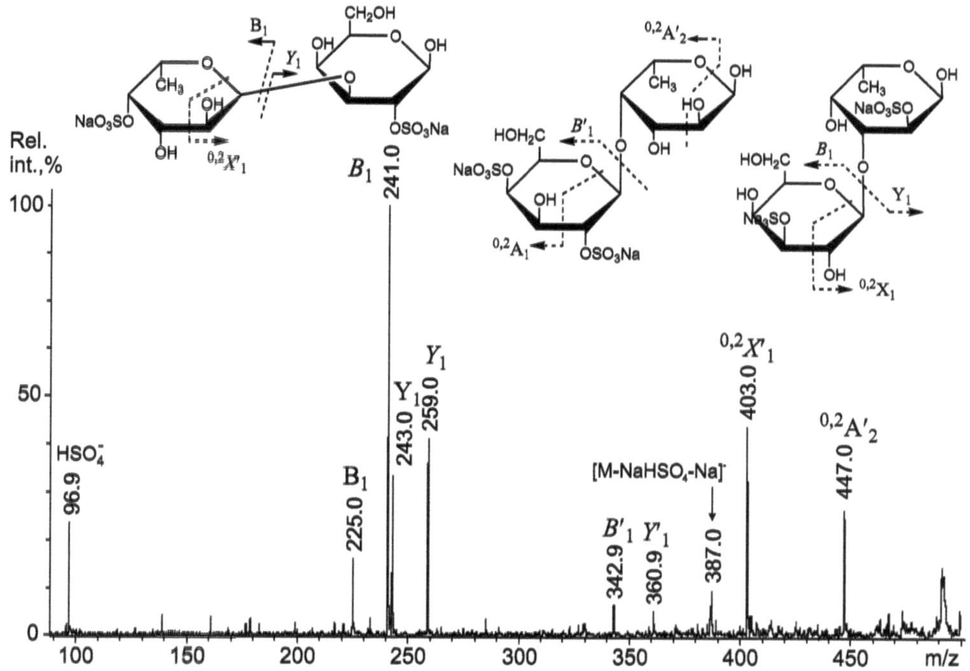

Fig. 7 Negative-ion tandem MALDI-TOFMS of the "hybrid" disulfated biose at *m/z* 507.0. The fragment was obtained by autohydrolysis of fucoidan from brown alga *Fucus evanescens*. Reproduced from [26] with permission from Elsevier

4 Notes

1. Add water very carefully as $NaDB_4$ is hygrophobic.
2. SDHB (sold as "super" DHB) is a mixture of DHB (2,5-dihydroxybenzoic acid) and 2-hydroxy-5-methoxybenzoic acid. The mixture of these organic acids gives finer crystallization than DHB alone.
3. Avoid using glass tubes. Otherwise K^+ will give unwanted signals in mass spectra.
4. MS/MS spectra of multiply charged ions must contain native isotope patterns to distinguish the charge state of the ion. Charge state = 1/(distance between ^{12}C and ^{13}C isotope MWs).
5. Arabinoosazone matrix works better in MW range of 100 Da–1.5 kDa. DHB/SDHB matrices work better in higher MW range from 500 Da to 4 kDa (or more) [29].
6. For Bruker ULTRAFLEX-III (Germany) use the following settings: accelerating voltage 25 kV; laser power 30 μJ; pulse-width 6 ns; number of shots 100; laser shot rate 66 Hz.

7. ESIMS is a more sensitive technique for minor components, especially if these components are purified. MALDI-TOFMS is more suitable for direct analyses of mixtures. Calculations of MWs; for example, disulfated fucobiose (for MALDI-TOFMS): 146 (MW of fucose) + 146 + 18 (MW of the reducing terminus) + 102 (–H + SO_3Na) + 102–23 (Na^+) = 491. ESIMS: (491–23 (remove second Na^+))/number of sulfates (2 here) = 234.
8. Use of the software for automatic calculation of MWs of the fragments is highly recommended. The interpretation is given for singly charged variants to simplify calculations.
9. Galactose (Gal), mannose (Man), glucose (Glc), etc. are hexoses ($C_6H_{12}O_6$).
10. Fucose (Fuc), rhamnose (Rha), etc. are 6-desoxyhexoses ($C_6H_{12}O_5$).
11. Xylose (Xyl) is a pentose ($C_5H_{10}O_5$).
12. The differentiation between sulfation at C-6 and C-4 requires the MS^3 technique [16].
13. The signal at *m/z* 168.9 was not observed for the oligosaccharides from *F. evanescens*, but sulfation at C-3 was found in other Fucales, e.g., *Silvetia babingtonii* by MS/MS [31] and in *Fucus serratus* [32] and *Fucus distichus* [33] by independent methods.
14. Reduction with $NaBH_4$ or $NaBD_4$ gives a +2 or a +3 Da shift in MW, respectively (if singly charged).
15. Calculation is made for singly charged "sodiated" ions for MALDI-TOFMS.
16. Y- and C-type ions have the same MWs as unfragmented by MS/MS oligosaccharides. The shift in MW allows distinguishing between them. It was shown on carrageenans [34] that the product ion spectra of $[M\text{-}Na]^-$ (where M represents the sodium salt of oligosaccharides) featured an extensive series of B- and C-type glycosidic cleavages, whereas the Y-type cleavage occurred mainly at the C-4 sulfated residues. However, in the case of fucoidans and carrageenans, fragment ions of the C-type are almost invisible and ions of the B-type are predominating [26–28]. If abundant Y-ion is observed, it could suggest C-4 sulfation of the corresponding sugar residue [34].
17. The spatially close sulfate groups may cause spontaneous in-source desulfation, appearing as $[M\text{-}NaSO_4 + H]^-$ ion (loss of 120 Da) in mass spectrum.
18. As the primary backbone of a fucoidan from *F. evanescens* was shown [22] to be composed of 2-, 2,4- and sometimes 4-sulfated alternating (1→3)- (Figs. 1a and 4) and (1→4)-linked (Fig. 5) α-L-Fuc*p* residues, the oligosaccharides were found to be "tailored" by 2-desulfation in 4-linked fucose

residues on the reducing end, probably due to the selective cleavage effect [35]. On the other hand, initially more heavily sulfated (3-linked and 2,4-disulfated) α-L-fucan from *S. cichorioides* [24] was decomposed with excess desulfation [36], but its main building block (2,4-disulfate of α-L-Fucp) and 3-linked fucooligosaccharides of the main chain were clearly observed by both tandem ESIMS and MALDI-TOFMS [27].

Acknowledgments

The work was supported by FEB RAS grant 15-II-5-016.

References

1. Aleksandrov ML, Gall LN, Krasnov NV et al (1984) Mechanism of ion formation during the electrohydrodynamic sputtering of a liquid into a vacuum. J Anal Chem USSR 39:1268–1274
2. Yamashita M, Fenn JB (1984) Electrospray ion source. Another variation of the free-jet theme. J Phys Chem 88:4451–4459
3. Hillenkamp F, Karas M (1990) Mass-spectrometry of peptides and proteins by matrix-assisted ultraviolet-laser desorption ionization. Meth Enzymol 193:280–295
4. Karas M, Hillenkamp F (1988) Laser desorption ionization of proteins with molecular masses exceeding 10000 daltons. Anal Chem 60:2299–2301
5. Harvey DJ (1999) Matrix-assisted laser desorption/ionization mass spectrometry of carbohydrates. Mass Spectrom Rev 18:349–450
6. Zaia J (2004) Mass spectrometry of oligosaccharides. Mass Spectrom Rev 23:161–227
7. Kusaykin M, Bakunina I, Sova V et al (2008) Structure, biological activity, and enzymatic transformation of fucoidans from the brown seaweeds. Biotechnol J 3:904–915
8. Cumashi A, Ushakova NA, Preobrazhenskaya ME et al (2007) A comparative study of the anti-inflammatory, anti-coagulant, anti-angiogenic, and anti-adhesive activities of nine different fucoidans from brown seaweeds. Glycobiol 17:541–552
9. Bilan MI, Usov AI (2008) Structural analysis of fucoidans. Nat Prod Comms 3:1639–1648
10. Shevchenko NM, Anastyuk SD, Gerasimenko NI et al (2007) Polysaccharide and lipid composition of the brown seaweed *Laminaria gurjanovae*. Bioorg Khim 33:96–107
11. Shevchenko N, Anastyuk S, Menshova R et al (2015) Further studies on structure of fucoidan from brown alga Saccharina gurjanovae. Carbohydr Polym 121:207–216
12. Daniel R, Chevolot L, Carrascal M et al (2007) Electrospray ionization mass spectrometry of oligosaccharides derived from fucoidan of *Ascophyllum nodosum*. Carbohydr Res 342: 826–834
13. Marais M-F, Joseleau J-P (2001) A fucoidan fraction from *Ascophyllum nodosum*. Carbohydr Res 336:155–159
14. Tissot B, Salpin JY, Martinez M et al (2006) Differentiation of the fucoidan sulfated L-fucose isomers constituents by CE-ESIMS and molecular modeling. Carbohydr Res 341:598–609
15. Karlsson NG, Karlsson H, Hansson GC (1996) Sulphated mucin oligosaccharides from porcine small intestine analysed by four-sector tandem mass spectrometry. J Mass Spectrom 31:560–572
16. Minamisawa T, Hirabayashi J (2005) Fragmentations of isomeric sulfated monosaccharides using electrospray ion trap mass spectrometry. Rapid Commun Mass Spectrom 19:1788–1796
17. Saad OM, Leary JA (2004) Delineating mechanisms of dissociation for isomeric heparin disaccharides using isotope labeling and ion trap tandem mass spectrometry. J Am Soc Mass Spectrom 15:1274–1286
18. Desaire H, Leary JA (2000) Utilization of MS3 spectra for the multicomponent quantification of diastereomeric N-acetylhexosamines. J Am Soc Mass Spectrom 11:1086–1094
19. Domon B, Costello CE (1988) A systematic nomenclature for carbohydrate fragmentations in Fab-MsMs spectra of glycoconjugates. Glycoconjugate J 5:397–409

20. Zaia J, Miller MJ, Seymour JL et al (2007) The role of mobile protons in negative ion CID of oligosaccharides. J Am Soc Mass Spectrom 18:952–960
21. Anastyuk SD, Shevchenko NM, Nazarenko EL et al (2009) Structural analysis of a fucoidan from the brown alga *Fucus evanescens* by MALDI-TOF and tandem ESI mass spectrometry. Carbohydr Res 344:779–787
22. Bilan MI, Grachev AA, Ustuzhanina NE et al (2002) Structure of a fucoidan from the brown seaweed *Fucus evanescens* C. Ag. Carbohydr Res 337:719–730
23. Ciancia M, Matulewicz MC, Stortz CA et al (1991) Room-temperature, low-field Cla spectra of degraded carrageenans. 2. On the specificity of the autohydrolysis reaction in Kappa/Iota and Mu/Nu structures. Int J Biol Macromol 13:337–340
24. Zvyagintseva TN, Shevchenko NM, Chizhov AO et al (2003) Water-soluble polysaccharides of some far-eastern brown seaweeds: distribution, structure, and their dependence on the developmental conditions. J Exp Mar Biol Ecol 294:1–13
25. Anastyuk SD, Shevchenko NM, Dmitrenok PS et al (2011) Investigation of a sulfate transfer during autohydrolysis of a fucoidan from the brown alga *Fucus evanescens* by tandem ESIMS. Carbohydr Res 346:2975–2977
26. Anastyuk SD, Shevchenko NM, Ermakova SP et al (2012) Anticancer activity *in vitro* of a fucoidan from the brown alga *Fucus evanescens* and its low-molecular fragments, structurally characterized by tandem mass-spectrometry. Carbohydr Polym 87:186–194
27. Anastyuk SD, Shevchenko NM, Nazarenko EL et al (2010) Structural analysis of a highly sulfated fucan from the brown alga *Laminaria cichorioides* by tandem MALDI and ESI mass spectrometry. Carbohydr Res 345:2206–2212
28. Anastyuk SD, Imbs TI, Shevchenko NM et al (2012) ESIMS analysis of fucoidan preparations from *Costaria costata*, extracted from alga at different life-stages. Carbohydr Polym 90:993–1002
29. Anastyuk SD, Barabanova AO, Correc G et al (2011) Analysis of structural heterogeneity of κ/β-carrageenan oligosaccharides from *Tichocarpus crinitus* by negative-ion ESI and tandem MALDI mass spectrometry. Carbohydr Polym 86:546–554
30. Chen P, Baker AG, Novotny MV (1997) The use of osazones as matrices for the matrix-assisted laser desorption/ionization mass spectrometry of carbohydrates. Anal Biochem 244:144–151
31. Anastyuk SD, Shevchenko NM, Dmitrenok PS et al (2012) Structural similarities of fucoidans from brown algae *Silvetia babingtonii* and *Fucus evanescens*, determined by tandem MALDI-TOF mass spectrometry. Carbohydr Res 358:78–81
32. Bilan MI, Grachev AA, Shashkov AS et al (2006) Structure of a fucoidan from the brown seaweed *Fucus serratus* L. Carbohydr Res 341:238–245
33. Bilan MI, Grachev AA, Ustuzhanina NE et al (2004) A highly regular fraction of a fucoidan from the brown seaweed *Fucus distichus* L. Carbohydr Res 339:511–517
34. Yu G, Zhao X, Yang B et al (2006) Sequence determination of sulfated carrageenan-derived oligosaccharides by high-sensitivity negative-ion electrospray tandem mass spectrometry. Anal Chem 78:8499–8505
35. Pomin VH, Valente AP, Pereira MS et al (2005) Mild acid hydrolysis of sulfated fucans: a selective 2-desulfation reaction and an alternative approach for preparing tailored sulfated oligosaccharides. Glycobiology 15: 1376–1385

Chapter 20

Determination of Substitution Patterns of Galactans from Green Seaweeds of the Bryopsidales

Paula Ximena Arata, Paula Virginia Fernández, and Marina Ciancia

Abstract

Sulfated and pyruvylated galactans are the major soluble polysaccharides produced by seaweeds of the Bryopsidales. Their backbones have a complex and variable pattern of substitution which, until now, has only been elucidated for a few species. Methods for determination of sulfate and pyruvic acid content, and chemical strategies to determine their position in the galactan chain are outlined here. These methods can also be applied to other sulfated and/or pyruvylated polysaccharides.

Key words Bryopsidales, Green seaweed, Methylation analysis, Polysaccharide structure, Pyruvic acid ketal, Sulfated polysaccharide, Structure of galactans, Sulfate content

1 Introduction

The Bryopsidales comprise coenocytic seaweeds which synthesize different types of sulfated polysaccharides in small quantities [1, 2]. Purification of these types of polysaccharides is a difficult task; it is usually carried out by anion exchange chromatography [3–6]. In some cases, sulfated and pyruvylated galactans can be separated from sulfated arabinans by precipitation of the latter with potassium chloride and recovery of the former from the supernatant [4]. Galactans from *Codium* species have been studied in detail [1, 3, 4, 7–10], and now galactans obtained from some other genera are under study ([5, 6]). *Codium* produces 3-linked β-D-galactans with ramifications on C-6 and terminal β-D-galactopyranose units with pyruvic acid ketal linked to *O*-3 and *O*-4 (*S*-configuration) as major structural features [3, 8–10]; galactans from *Bryopsis* are quite similar [6]. Sulfation occurs mainly on C-4 and some of the 3-linked galactose units, while other units are substituted by pyruvic acid ketals linked to *O*-4 and *O*-6 (*R*-isomer) (Fig. 1). Percentages of the different structural units, as well as degree of ramification, vary greatly with species and isolation procedure. Caribbean calcareous species of this order, including *Penicillus*

Dagmar B. Stengel and Solène Connan (eds.), *Natural Products From Marine Algae: Methods and Protocols*, Methods in Molecular Biology, vol. 1308, DOI 10.1007/978-1-4939-2684-8_20, © Springer Science+Business Media New York 2015

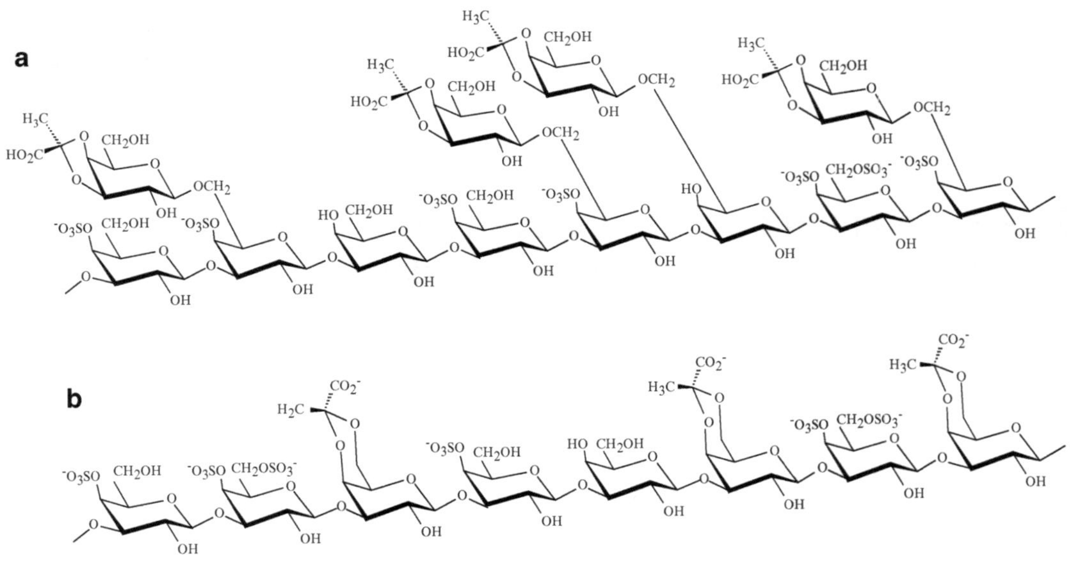

Fig. 1 A possible sequence for the galactan from *Codium vermilara* (**a**) and from *Bryopsis plumosa* (**b**)

spp., *Halimeda* spp., or *Udotea* spp., also synthesize such galactans, but the substitution pattern seems to be quite different [5]. Due to the importance of these structural characteristics in defining the biological properties of these compounds, it is of great interest to determine them and compare their activities, which may lead to the design of new drugs with special features [2, 11].

In this chapter we describe methods for the determination of sulfate and pyruvate content. In addition, determination of position of these substituents, which is usually carried out by methylation analysis and desulfation-methylation and/or depyruvylation-methylation or combinations of these procedures, is described. Although NMR spectroscopy is a very powerful tool for structural determination, spectra of these galactans are very complex, and can only be resolved with the aid of chemical methods, and usually only partial assignment is achieved. Besides, NMR spectra can be used in a simple way to determine the presence of pyruvic acid ketals and, when both ketal types are present, the ratio between them [12]. Of course these methods could be employed to determine these substituents in polysaccharides of other sources.

2 Materials

All solutions are prepared using distilled water and analytical grade reagents.

2.1 Sulfate Content

1. 1 M hydrochloric acid (HCl): pour 4.17 mL of 12 M HCl and make up to 50 mL with distilled water.

2. 3 % w/v trichloroacetic acid (TCA): dissolve 3 g in distilled water and make up to 100 mL. Store at 4 °C.
3. Barium chloride-gelatin reagent: dissolve gelatin (200 mg) (*see* **Note 1**) in 40 mL of hot double-distilled water (60–70 °C) and allow it to stand at 4 °C for at least 6 h and preferably overnight. Dissolve barium chloride (200 mg) in the semi-gelatinous fluid and leave the resultant cloudy solution standing for 2–3 h before use (*see* **Note 2**).
4. Anhydrous sodium sulfate.
5. 2 M trifluoroacetic acid (TFA): take 15.4 mL of pure TFA (density: 1.48 kg/L, MW: 114.02 g/mol) and make up to 100 mL with distilled water.
6. 1.8 mM sodium carbonate (Na_2CO_3): add 190.8 mg of Na_2CO_3 to 1 L of distilled water (MW: 105.9885 g/mol).
7. 1.7 mM sodium bicarbonate ($NaHCO_3$): add 142.8 mg of $NaHCO_3$ to 1 L of distilled water (MW: 84.004 g/mol).
8. An ion exchange chromatography system such as DIONEX DX-100 system with an AS4A column (4×250 mm), an AMMS-II micro-membrane suppressor and a conductivity detector (eluent: 1.8 mM Na_2CO_3/1.7 mM $NaHCO_3$ v/v, flow rate: 2 mL/min).
9. Spectrophotometer.

2.2 Pyruvic Acid Content

1. 0.02 M oxalic acid: dissolve 252.2 mg of oxalic acid in distilled water and make up to 100 mL (MW: 126.1 g/mol dihydrate) Store at 4 °C.
2. 0.75 M sulfuric acid (H_2SO_4): pour 2.1 mL of 17.8 M H_2SO_4 and make up to 50 mL with distilled water. Store at 4 °C.
3. 2 M HCl: pour 1.67 mL of conc. HCl and make up to 10 mL with distilled water.
4. 2,4-Dinitrophenylhydrazine (DNFH)/2 M HCl (w/v): add 10 mg of 2,4 DNFH to 2 M HCl up to 10 mL (*see* **Note 3**).
5. Ethyl acetate.
6. 10 % w/v Na_2CO_3: add 10 g of sodium carbonate (MW: 105.9885 g/mol) to distilled water and make up to 100 mL.
7. Pyruvic acid.
8. Spectrophotometer.

2.3 Methylation

1. Amberlite IR120 (H^+) resin (6 cm×0.7 cm i.d.; *see* **Note 4**).
2. 5 % triethylammonium chloride solution (w/v): add 5 g of salt (MW: 137.65 g/mol) to distilled water and make up to 100 mL.
3. Dimethyl sulfoxide (DMSO; *see* **Note 5**).
4. Sodium hydroxide (NaOH) reagent grade, ≥98 %, pellets (anhydrous).

5. Iodomethane (CH_3I), purum, ≥99.0 % (GC). Store at 4 °C.
6. Dialysis membrane of 3,500 molecular weight cutoff.

2.4 Desulfation

1. Amberlite IR120 (H^+) resin (6 cm × 0.7 cm i.d.; *see* **Note 4**).
2. 5 % w/v pyridinium chloride: add 5 g of salt (MW: 115.56 g/mol) to distilled water and make up to 100 mL.
3. Dimethyl sulfoxide (DMSO; *see* **Note 5**).
4. Pyridine (anhydrous).
5. A domestic microwave oven (1,200 W, operating at 2,450 MHz).
6. Threaded Teflon 50 mL tubes with screw caps (*see* **Note 6**).
7. Dialysis membrane of 3,500 Da molecular weight cutoff.

2.5 Removal of Pyruvic Acid Residues

1. 1 % w/v acetic acid: pour 1 mL of acetic acid and make up to 100 mL with distilled water. Store at 4 °C.
2. Saturated solution of sodium bicarbonate. Solubility in water 96 g/L (20 °C), keep adding until a precipitate is observed.
3. Dialysis membrane with 3,500 Da as molecular weight cutoff.

2.6 NMR Spectroscopy

1. Tetramethylsilane (TMS) as external reference.
2. 99.9 % deuterated water (D_2O).
3. Acetone ACS reagent, ≥99.5 %.
4. 500 MHz ^1H NMR, proton decoupled 125 MHz ^{13}C NMR spectra, and two-dimensional NMR experiments (HMQC and COSY) were recorded on a Bruker AM500 at room temperature. Parameters for ^{13}C NMR spectra were as follows: pulse angle 51.4°, acquisition time 0.56 s, relaxation delay 0.6 s, spectral width 29.4 kHz, and scans 25,000. For ^1H NMR spectra: pulse angle 76°, acquisition time 3 s, relaxation delay 3 s, spectral width 6,250 Hz, and scans 32.

3 Methods

3.1 Sulfate Content

Estimation of the ester sulfate content of polysaccharide sulfates involves acid hydrolysis, followed by determination of liberated inorganic sulfate. Sulfate can be determined turbidimetrically [13] (*see* **Note 7**). However, in the case of desulfated samples or samples in which the sulfate content is less than 5 %, the error increases greatly and the sulfate content is usually overestimated. The presence of important amounts of proteins in the sample interferes in the determination. In these cases, ion chromatography with conductometric detection should be used. An alternative would be the use of microanalysis.

3.1.1 Turbidimetric Method

1. Dissolve the polysaccharide (2–3 mg) in necessary amount of 1 M HCl in order to obtain a solution between 40 and 400 μg SO_4/mL.
2. Heat the sample at 105–110 °C for 4–5 h to hydrolyze the polysaccharide. After being cooled, mix the vial containing the hydrolyzate before opening.
3. Transfer 0.5 mL to a test tube containing 3.5 mL of 3 % TCA.
4. Add barium chloride-gelatin reagent (1 mL) and, after mixing, keep it at room temperature for 15–20 min.
5. Measure the absorbance of the solution at 360 nm using a spectrophotometer against an appropriate reagent blank (*see* **Note 8**).
6. Prepare calibration curve using anhydrous sodium sulfate (60–600 μg/mL) (*see* **Note 9**).

3.1.2 Conductometric Method

1. Heat the sample (1 mg) in 2 M trifluoroacetic acid at 120 °C for 2 h to hydrolyze the polysaccharide (*see* **Note 10**).
2. Use an ion chromatographic system such as the DIONEX DX-100 system with a micromembrane suppressor and a conductivity detector.

3.2 Pyruvic Acid Content

The percentage of pyruvic acid can be determined according to Koepsell and Sharpe [14].

1. Dissolve the polysaccharide (4–5 mg) in 1 mL of 0.02 M oxalic acid.
2. Heat the samples at 120 °C for 3 h to hydrolyze the polysaccharide. After cooling, mix them before opening.
3. Transfer 0.5 mL to a tube containing 1 mL of 0.75 M sulfuric acid, and mix.
4. Add 0.5 mL of 2,4-DNFH/HCl reagent, mix and keep at room temperature for 15–20 min.
5. Liquid-liquid extraction is needed to eliminate sub products of the reaction: add 5 mL of ethyl acetate to the test tubes, mix and then discard the lower phase. Add 3 mL of 10 % sodium carbonate solution, mix and then discard the upper phase.
6. Measure the absorbance at 380 nm using a spectrophotometer against distilled water as reagent blank.
7. Prepare a calibration curve using pyruvic acid as standard; this technique is valid between 5 and 60 μg/mL.

3.3 Methylation

Methylation analysis is an analytical method largely used for structural studies of polysaccharides. It involves methylation of the free hydroxyls in the intact polysaccharide through a bimolecular nucleophilic substitution (S_N2) reaction in alkali. This reaction, usually carried out with iodomethane as methylating agent, gives rise to methyl ethers from all the free hydroxyl groups. Cleavage of the glycosidic links by hydrolysis and derivatization of the partially methylated monosaccharides is then needed to analyze the methylation pattern by gas chromatography–mass spectrometry. Several methods have been developed to allow the monosaccharide components and positions of the glycosidic linkages to be studied in the original polysaccharide [15]. Classical techniques [16–19] have been revised and compared [20] in order to obtain a rapid method for simultaneous identification of furanosides and pyranosides in heteropolysaccharides.

1. Use the Amberlite IR 120 resin to convert the polysaccharide (5–10 mg) into the corresponding triethylammonium salt to ensure solubility in DMSO: wash the column with distilled water (50 mL), then activate the resin with the 5 % triethylammonium solution (10 mL) and wash again with distilled water (50 mL). Dissolve the sample in a small volume of distilled water (1–2 mL). Use distilled water (~10 mL) for elution of the sample. Use the phenol–sulfuric acid method [21] to ensure all sample is eluted. Freeze-dry the sample and store dry, at room temperature [18].
2. Dissolve the sample in DMSO (1 mL) to give a 5 mg/mL solution; magnetic stirring must be kept at all times (*see* **Note 11**).
3. Add 100 mg of finely powdered anhydrous NaOH to the solution (*see* **Note 12**), and stir for 60 min.
4. Add 0.5 mL of iodomethane in a cold bath maintaining the magnetic stirring. Allow the reaction to proceed at room temperature for 30 min (*see* **Note 13**).
5. Some samples need **steps 3** and **4** to be repeated to achieve permethylation.
6. Stop the reaction by adding 10 mL of water and stir for an additional interval of 10 min.
7. Recover the methylated derivative by dialysis and lyophilization (*see* **Note 14**).
8. The methylated samples can be derivatized in the same way as the parent polysaccharides for GC/MS analysis (*see* **Note 15**).

3.4 Desulfation

The reaction can be carried out by the microwave-assisted solvolytic method with DMSO described by Navarro et al. [22]. Sulfated polysaccharides are desulfated by heating their pyridinium salts dissolved in DMSO in a standard domestic microwave oven. When their pyridinium salts are generated in carefully chosen conditions, desulfation proceeds smoothly after 1 min heating, leading to a removal of 80–90 % of the original sulfate present in the case of these sulfated galactans. Some depolymerization may occur, but its effect is moderate (yields of the recovered products are 60–90 %, considering the sulfate loss), even for polysaccharides with labile bonds. Here, an in situ methylation procedure is coupled with this desulfation method in order to facilitate the retrieval of structural data.

1. Convert the polysaccharide (20–50 mg) to the pyridinium salt using the Amberlite IR 120 resin: wash the column with distilled water (50 mL), then activate it with the 5 % pyridinium chloride solution (10 mL) and wash again with distilled water (50 mL). Dissolve the sample in a small volume of distilled water (1–2 mL). Use distilled water (~10 mL) for elution of the sample. Use phenol–sulfuric acid method [21] to ensure the entire sample is eluted. Add immediately drops of pyridine to the eluate until the pH becomes neutral, and then freeze-dry the solution.
2. Dissolve the pyridinium salt of the polysaccharide in dry DMSO (10 mL) containing 2 % dry pyridine.
3. Treat the mixture for 1 min in the microwave using the Teflon tube. Total time of reaction is divided in six steps of microwave heating (10 s), followed by opening the vessel and cooling (50–60 °C) in an ice-bath to relieve the pressure (*see* **Note 16**).
4. Allow the reaction mixture to cool in an ice-bath; for in situ methylation, separate an aliquot (3 mL) and methylate it as described in Subheading 3.3.
5. Add distilled water to the remaining solution up to 25 mL and dialyze the resulting mixture against tap water and then distilled water, and freeze-dry.

3.5 Removal of Pyruvic Acid Residues

Removal of pyruvic acid residues from polysaccharides is performed according to Bilan et al. [8].

1. Dissolve the sample (25 mg) in a hydrolysis vial using 1 % acetic acid (5 mL).
2. Keep the vial at 100 °C for 4 h.
3. Allow the reaction mixture to cool and then neutralize it with a saturated solution of $NaHCO_3$.
4. Dialyze the product against tap water and then distilled water, and freeze-dry it.

3.6 NMR Spectroscopy

The samples (20 mg) were exchanged in 99.9 % D_2O (0.5 mL) four times. Chemical shifts were referenced to internal acetone (δ_{CH3} 2.175/31.1). Chemical data can be supported by NMR spectroscopy which can also detect the presence of pyruvic acid ketals in the galactan. Moreover, linkage position of pyruvic acid substituent to galactose can be clearly established. Pyruvic acid ketal in terminal 3,4-*O*-(1′-carboxy)ethylidene-D-galactopyranose (*S*-configuration) units and 3-linked 4,6-*O*-(1′-carboxy)ethylidene-D-galactopyranose (*R*-configuration) units give very distinct δ values, namely, δ_{CH3} 1.42/26.0 and 1.61/24.1, respectively, and δ for the quaternary carbon at δ 108.8 and 101.6, respectively (with reference to internal acetone, δ_{CH3} 2.12/31.2; Fig. 2). The ratio between units substituted in any of both positions can be determined by integration

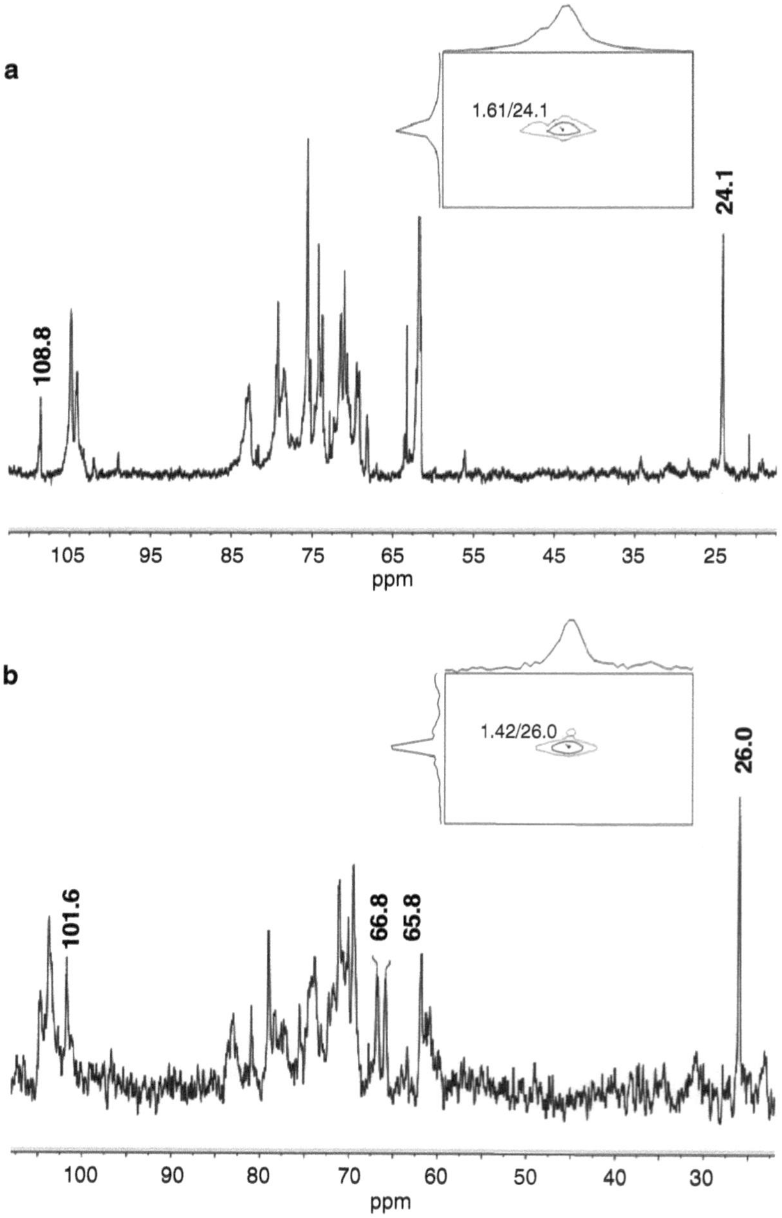

Fig. 2 ^{13}C NMR spectra and detail of the HMQC of the sulfated and pyruvylated galactans from *Codium vermilara*, where pyruvic acid mostly gives terminal 3,4-*O*-(1′-carboxy)ethylidene-D-galactopyranose units (**a**) and *Bryopsis plumosa*, where only 3-linked 4,6-*O*-(1′-carboxy)ethylidene-D-galactopyranose units were detected (**b**). The diagnostic signals for pyruvic acid ketals are indicated

of the signals at δ_H 1.42 and 1.61 in the ^{1}H NMR spectrum of the galactan [12]. In addition, this substituent produces a shift of the signal of C-6 of the 3-linked β-D-galactopyranose units to lower fields (from δ 61.8 in non-substituted residue, to δ 65.8 when this sugar is 4,6-pyruvylated) and a shift to lower fields of the signal corresponding to C-5 (from δ 75.5 in the non-substituted sugar, to δ 66.8 in the 4,6-pyruvylated unit). These peaks appear in a region of the spectrum with very few signals and are very easily distinguished (Fig. 2; [6]).

4 Notes

1. Some gelatins contain excessive amounts of SO_4^{2-} ions. Gelatin for microbiology (Merck) proved to be satisfactory in all respects.
2. This reagent (barium chloride-gelatin) can be stored at 4 °C and it is stable for about 1 week.
3. Sonication may be necessary for complete dissolution.
4. Amberlite IR120 (H^+) is a gel type strongly acidic cation exchange resin of the sulfonated polystyrene type. Any other resin with the same characteristics can be used.
5. DMSO must be anhydrous, dry by previous addition of activated molecular sieves to the bottle.
6. Closed-vessel reaction can be carried out in home-made Teflon tubes with screw caps, manufactured from 42 mm (tube) and 52 mm (cap) rods, with tube walls 5 mm thick.
7. An alternative is the colorimetric method of Terho and Hartiala [23] which allows determination of linked and inorganic sulfate separately. This is important when the polysaccharides form aggregates containing inorganic salts, which could include inorganic sulfate.
8. 1 M HCl is used as blank and for preparing calibration curve solutions.
9. The standard used is Na_2SO_4 but in the polysaccharide the sulfate content is usually in the form $NaSO_3$; therefore a correction must be done in order to express the percentage of sulfate correctly:

   % sulfate (as $NaSO_3$) = 103/142 × % Na_2SO_4 with 103 and 142 the molecular weight of $NaSO_3$ and Na_2SO_4, respectively.
10. When hydrolysis is carried out as described in Subheading 3.1.1, the mixture must be evaporated to dryness under nitrogen and redissolved in high purity water from a Milli-Q system, as HCl could interfere in the chromatogram.

11. Anhydrous conditions and good solubilization of the sample are essential to obtain a complete methylation of the polysaccharide. Mild heating and/or increasing stirring time previous to the reaction could be necessary to allow a better solubilization.
12. The NaOH pellets are crushed in order to get a fine powder, which must be added quickly to avoid absorption of the ambient humidity. It is not necessary to weigh the powdered NaOH, just add a spatula tip to the solution. Powdering increases the speed of solubilization of NaOH.
13. The reaction vial must have a needle to allow elimination of gases produced during the methylation reaction.
14. It is important to ensure that the remaining CH_3I is completely eliminated. It is clearly seen in the dialysis bag as a second liquid phase. The dialysis process may take 4–7 days, or even more.
15. If acid-labile sugars are present in the sample, a reductive hydrolysis/acetylation must be performed to the native and methylated samples [18]. Alternatively, for insoluble or only partially soluble polysaccharides or their permethylated derivatives, a strong hydrolysis could be necessary [24]. However, for galactans from the Bryopsidales, the standard procedure should be suitable [24]. The sugar mixture can be converted to the corresponding alditol acetates [24] or into the aldononitrile acetates [25].
16. All the reactions should be carried out at full microwave power. Risks of accidents are minimized by carefully observing the cooling time after each step of reaction.

Acknowledgements

This work was supported by grants from National Research Council of Argentina, CONICET (PIP 559-2009), and the National Agency for Promotion of Science and Technology, ANPCYT (PICT 2008-0500). PVF and MC are Research Members of the National Research Council of Argentina (CONICET).

References

1. Percival E, McDowell RH (1981) Algal walls – composition and biosynthesis. In: Tanner W, Loewus F (eds) Plant carbohydrates II, vol 13B. Springer, Berlin, pp 277–316
2. Fernández PV, Arata PX, Ciancia M (2014) Polysaccharides from *Codium* species: chemical structure and biological activity. Their role as components of the cell wall. In: Bourgougnon N (ed) Sea plants, vol 71, Adv Bot Res. Elsevier, Oxford, pp 253–278
3. Estevez JM, Fernández PV, Kasulin L et al (2009) Chemical and *in situ* characterization of macromolecular components of the cell walls from the green seaweed *Codium fragile*. Glycobiology 19:212–228
4. Fernández PV, Quintana I, Cerezo AS et al (2013) Anticoagulant activity of a unique sulphated pyranosic (1→3)-β-L-arabinan through direct interaction with thrombin. J Biol Chem 288:223–233

5. Arata PX, Quintana I, Canelón DJ, Vera BE, Compagnone RS, Ciancia M (2015) Chemical structure and anticoagulant activity of highly pyruvylated sulfated galactans from tropical green seaweeds of the order Bryopsidales. Carbohydr Pol 122:376–386
6. Ciancia M, Alberghina J, Arata PX et al (2012) Characterization of cell wall polysaccharides of the coencocytic green seaweed *Bryopsis plumosa* (Bryopsidaceae, Chlorophyta) from the Argentine coast. J Phycol 48:326–335
7. Ciancia M, Quintana I, Vizcarguenaga MI et al (2007) Polysaccharides from the green seaweeds *Codium fragile* and *C. vermilara* with controversial effects on hemostasis. Int J Biol Macromol 41:641–649
8. Bilan MI, Vinogradova EV, Shashkov AS et al (2007) Structure of a highly pyruvylated galactan sulphate from the pacific green alga *Codium yezoense* (Bryopsidales, Chlorophyta). Carbohydr Res 342:586–596
9. Farias EHC, Pomin VH, Valente AP et al (2008) A preponderantly 4-sulphated, 3-linked galactan from the green alga *Codium isthmocladum*. Glycobiology 18:250–259
10. Ohta Y, Lee JB, Hayashi K et al (2009) Isolation of sulphated galactan from *Codium fragile* and its antiviral effect. Biol Pharm Bull 32:892–898
11. Ciancia M, Quintana I, Cerezo AS (2010) Overview of anticoagulant activity of sulphated polysaccharides from seaweeds in relation to their structures, focusing on those of green seaweeds. Curr Med Chem 17:2503–2529
12. Bilan MI, Vinogradova EV, Shashkov AS et al (2006) Isolation and preliminary characterization of a highly pyruvylated galactan from *Codium yezoense* (Bryopsidales, Chlorophyta). Bot Mar 49:259–262
13. Dodgson KS, Price RG (1962) A note on the determination of the ester sulphate content of sulphated polysaccharides. Biochem J 84: 106–110
14. Koepsell HJ, Sharpe ES (1952) Microdetermination of pyruvic and α-ketoglutaric acids. Arch Biochem Biophys 38:443–449
15. Biermann CJ (1988) Hydrolysis and other cleavages of glycosidic linkages in polysaccharides. Adv Carbohydr Chem Biochem 46: 251–271
16. Haworth WN (1915) A new method of preparing alkylated sugars. J Chem Soc 107:8–16
17. Hakomori SJ (1964) Rapid permethylation of glycolipids and polysaccharides catalyzed by methylsulfinyl carbanion in dimethyl sulfoxide. J Biochem (Tokyo) 55:205–208
18. Stevenson TT, Furneaux RH (1991) Chemical methods for the analysis of sulphated galactans from red algae. Carbohydr Res 210:277–298
19. Ciucanu I, Kerek K (1984) A simple and rapid method for the permethylation of carbohydrates. Carbohydr Res 134:209–217
20. Sassaki GL, Gorin PAJ, Souza LM et al (2005) Rapid synthesis of partially O-methylated alditol acetate standards for GC–MS: some relative activities of hydroxyl groups of methyl glycopyranosides on Purdie methylation. Carbohydr Res 340:731–739
21. Dubois M, Gilles KA, Hamilton JK et al (1956) Colorimetric method for determination of sugars and related substances. Anal Chem 28:350–356
22. Navarro DA, Flores ML, Stortz CA (2007) Microwave-assisted desulfation of sulfated polysaccharides. Carbohydr Pol 69:742–747
23. Terho TT, Hartiala K (1971) Method for determination of the sulfate content of glycosaminoglycans. Anal Biochem 41:471–476
24. Morrison IM (1988) Hydrolysis of plant cell walls with trifluoroacetic acid. Phytochemistry 27:1097–1101
25. Stortz CA, Matulewicz MC, Cerezo AS (1982) Separation and identification of *O*-acetyl-*O*-methyl-galactonitriles by gas-liquid chromatography and mass spectrometry. Carbohydr Res 111:31–39

Chapter 21

Structural Characterization of a Hybrid Carrageenan-Like Sulfated Galactan from a Marine Red Alga *Furcellaria lumbricalis*

Youjing Lv, Bo Yang, Xia Zhao, Junzeng Zhang, and Guangli Yu

Abstract

Carrageenans are sulfated galactan isolated from marine red algae with different disaccharide forms. There are also some hybrid carrageenan-like oligomers, which are reported to possess a number of bioactivities. Here, we describe a method to study the structural characterization of a carrageenan-like sulfated galactan FB1 extracted from the red seaweed *Furcellaria lumbricalis*. We show the process of the general analysis of FB1, including the molecular weight, sulfate content, total sugar content, protein content, and 3,6-anhydrogalactose (3,6-AnG) content analyses. The fine structure identification methods, including desulfation and methylation, nuclear magnetic resonance (NMR), and electrospray ionization collision induced dissociation tandem mass spectrometry (ES-CID-MS/MS), are also described in detail.

Key words ES-CID-MS/MS, General analysis, Methylation, NMR, Sulfated galactan, Structure

1 Introduction

Carrageenans are sulfated galactan isolated from marine red algae, which are widely used as food additives due to their gel-forming ability and other rheological properties. Carrageenans are composed of linear, repeating disaccharide blocks of alternating 3-linked β-D-galactopyranose (β-Gal, unit G) and 4-linked α-d-galactopyranose (α-Gal, unit D), with the D unit often occurring as its 3,6-anhydro form (3,6-AnG, unit A) [1]. Based on the presence of a D or an A form of the 4-linked galactose and the differing sulfate contents and substitutions, carrageenans can be divided into several types. For example, κ-, ι-, and λ-carrageenans have different disaccharide building blocks of –[G4S-A]$_n$–, –[G4S-A2S]$_n$–, and –[G2S-D2S6S]$_n$–, respectively [2]. However, natural polysaccharides are rarely in the ideal form and usually consist of different biose units or copolymeric chains. The substitution of the hydroxyls in D-galactose by methyl and pyruvate groups also complicates

Dagmar B. Stengel and Solène Connan (eds.), *Natural Products From Marine Algae: Methods and Protocols*, Methods in Molecular Biology, vol. 1308, DOI 10.1007/978-1-4939-2684-8_21, © Springer Science+Business Media New York 2015

the carrageenan structure [3]. The heterogeneity of carrageenan largely depends on the algal species, growth stage, and environmental conditions. This structural heterogeneity confers a wide range of physicochemical properties and biological activities including antiviral [4, 5], antitumour [6, 7], immunomodulatory [8], and anti-Alzheimer's disease activities [9].

We obtain one hybrid carrageenan-like sulfated galactan FB1 from the hot water fraction from the red alga *Furcellaria lumbricalis* [1, 10]. The analyses to obtain basic information, such as molecular weight, sulfate content, total sugar content, crude protein content, 3,6-AnG content, and monosaccharide analyses, are carried out to describe the general properties of FB1. The precise structure of FB1 is also determined by electrospray ionization collision induced dissociation tandem mass spectrometry (ES-CID-MS/MS) technique, methylation analysis, and nuclear magnetic resonance (NMR) technique. All these analyses could also be applied to other types of carrageenan.

2 Materials

Prepare all solutions using distilled water and analytical grade reagents and prepare and store all reagents at room temperature (unless indicated otherwise).

2.1 Extraction and Purification of FB1

1. Algae were collected on the coast of Prince Edward Island, Canada, and provided by PEI Food Technology Center of Canada (Sept. 2003).
2. 60 mesh sieve.
3. 95 % ethanol.
4. 1 M sodium hydroxide (NaOH): dissolve 20.0 g of NaOH in 1,000 mL of distilled water.
5. 0.25 M potassium chloride (KCl): dissolve 18.64 g of KCl in 1,000 mL of distilled water.
6. Q-Sepharose Fast Flow ion-exchange column (5.0/10 cm; XK column, GE company; or equivalent).
7. 2 M sodium chloride (NaCl): dissolve 116.88 g of NaCl in 1,000 mL of distilled water and filter through 0.22 μm. Prepare a set of diluted NaCl solutions from the 2 M solution through a series of dilutions with distilled water to make NaCl contents as 1.5, 1.0, 0.5, and 0.2 M. Ultrasonic degas before use.
8. Sepharose 4B Fast Flow gel filtration column (XK 2.6/90 cm).
9. Dialysis membrane (cut off: 7,000 Da).
10. AKTA-FPLC (fast protein liquid chromatography) system used for polysaccharide purification through ion-exchange chromatography.

2.2 General Analysis

2.2.1 Total Sugar Content

1. Phenol solution (6 %): warm redistilled phenol in a 60 °C water bath until it melts completely. Weigh 80.0 g of phenol, add distilled water until 100 g to make an 80 % of stock solution, store at 4 °C. To get a 6 % phenol solution, dilute 15 mL of 80 % stock solution with distilled water up to 200 mL in a volumetric flask; store at 4 °C.
2. Conc. sulfuric acid (H_2SO_4; 98 %, 18.4 M).
3. Galactose (Gal) standard solutions: dissolve 10 mg of Gal in 20 mL of distilled water to make a galactose stock solution. Prepare a set of diluted sugar standards from the Gal stock solutions through a series of 1:1 dilutions with distilled water to make concentration as 0.05, 0.10, 0.15, 0.20, 0.25 mg/mL.
4. FB1 is diluted to 0.5 mg/mL with distilled water.
5. Microplate reader.

2.2.2 Sulfate Content

1. 0.5 % (w/v) gelatin aqueous buffer: dissolve 1.0 g of gelatin into 200 mL of distilled water at 60–70 °C. Store at 4 °C overnight.
2. 0.5 % (w/v) barium chloride ($BaCl_2$)-gelatin buffer: dissolve 0.5 g $BaCl_2$ into 100 mL of 0.5 % gelatin aqueous buffer to make a 0.5 % solution. Store at 4 °C overnight.
3. 8 % (w/v) trichloroacetic acid (TCA): weigh 8.0 g of TCA into a 100-mL volumetric flask; add distilled water up to the volume.
4. Potassium sulfate (K_2SO_4) standards: weigh 108.75 mg of K_2SO_4 (*see* **Note 1**) into a 100-mL volumetric flask, and add distilled water up to the volume to make the final concentration of SO_4^{2-} as 0.7768 μg/μL. Prepare a set of diluted SO_4^{2-} standards from the K_2SO_4 standard solution through a series of dilutions with distilled water to make SO_4^{2-} contents as 0.1553, 0.2330, 0.3107, 0.3884, and 0.4661 μg/μL.
5. 1 M HCl: pour 85 mL of conc. HCl to a 1,000-mL volumetric flask, add distilled water up to 1,000 mL, and mix well.
6. UV–Vis spectrophotometer.

2.2.3 Crude Protein Content

1. 4 % (w/v) sodium carbonate (Na_2CO_3) solution: add 40 g of Na_2CO_3 to 600 mL of distilled water. Transfer the solution to a volumetric flask and add distilled water up to 1,000 mL.
2. 0.2 M sodium hydroxide (NaOH) solution: add 8 g of NaOH to 600 mL of distilled water. Transfer the solution to a volumetric flask and add distilled water up to 1,000 mL.
3. 1 % (w/v) copper sulfate pentahydrate ($CuSO_4 \cdot 5H_2O$): add 10 g of $CuSO_4 \cdot 5\ H_2O$ to 600 mL of distilled water. Transfer the solution to a volumetric flask and add distilled water up to 1,000 mL.

4. 2 % (w/v) sodium potassium tartrate tetrahydrate solution: add 20 g of sodium potassium tartrate tetrahydrate to 600 mL of distilled water. Transfer the solution to a volumetric flask and add distilled water up to 1,000 mL.
5. Folin-phenol solution I:

   Reagent A: 4 % sodium carbonate solution mixed volumetric equally with 0.2 M sodium hydroxide solution.

   Reagent B: 1 % copper sulfate solution mixed volumetric equally with 2 % sodium potassium tartrate tetrahydrate solution.

   Before use, mix reagent A and B as 50:1 (v/v) (*see* **Note 2**).
6. Folin-phenol solution II: before use, dilute the commercial Folin-phenol reagent with equal volume of distilled water.
7. BSA Standards: prepare a set of diluted protein standards from the BSA standard solution (1.0 mg/mL) through a series of 1:1 dilutions with distilled water.
8. FB1 is diluted to 0.5 mg/mL with distilled water.
9. UV–Vis spectrophotometer.

2.2.4 3,6-Anhydrogalactose (AnG) Content

1. Resorcinol stock solution: dissolve 150 mg of resorcinol in 100 mL of distilled water in dark reagent bottle and store at 4 °C (*see* **Note 3**).
2. Acetal stock solution: dissolve 82 mg of acetal in 10 mL of distilled water in dark reagent bottle and store at 4 °C. Before use, it should be diluted with water (1:24, v/v) (*see* **Note 4**).
3. Resorcinol-acetal solution: combine 9 mL of resorcinol stock solution into 100 mL of concentrated HCl and 1 mL of diluted acetal solution (*see* **Note 5**).
4. D-fructose standard solutions: dissolve 10 mg of fructose in 100 mL of distilled water to make D-fructose stock solution. Prepare a set of diluted D-fructose standards from the stock solution through serial dilutions with distilled water to make concentrations as 0.005, 0.010, 0.015, 0.020, 0.025, 0.030 mg/mL.
5. FB1 is diluted to 0.5 mg/mL with distilled water.
6. UV–Vis spectrophotometer.

2.2.5 Molecular Weight Analysis

1. 0.2 M sodium sulfate (Na_2SO_4): dissolve 28.4 g of Na_2SO_4 in 1 L of distilled water and filter through 0.22 μm. Ultrasonic degas before use.
2. 5 mg/mL dextran standards and FB1 solutions: weigh 5.0 mg of each dextran standard (standards with different molecular weight: 5.9, 11.8, 22.8, 47.3, 112, 212, 404, and 788 kDa) and FB1, dissolve in 1 mL of 0.2 M Na_2SO_4, then filter through 0.22 μm.

3. PL aquagel-OH column (8 μm, 30×8 mm; Perkin Elmer, Waltham, MA, USA, or equivalent).
4. Gel permeation chromatography system with a refractive index detector.

2.3 Monosaccharide Composition Analysis

1. 80 mg/mL α-methylnaphthalene-borane (MMB): dissolve 80 mg of MMB in 1 mL of distilled water.
2. 6 and 4 M trifluoroacetic acid (TFA): to make 6 M TFA, weigh 6.84 g of TFA and add distilled water up to 10 g; for 4 M TFA, weigh 4.56 g of TFA and add distilled water up to 10 g.
3. Acetonitrile (CH_3CN) and dichloromethane (CH_2Cl_2)—chromatographic grade.
4. Inositol, acetic ether (EtOAc), acetic anhydride (Ac_2O), and perchloric acid ($HClO_4$).
5. Standard sugars: galactose, glucose, mannose, rhamnose, xylose, fucose, arabinose, and 3,6-AnG.
6. Screw-cap Teflon tubes (5 mL).
7. Fused silica capillary column DB-225 (0.25 μm, 30 m×0.32 mm; J&W Scientific, Folsom, CA, USA, or equivalent).
8. A Gas Chromatograph (GC) with a Flame-Ionization Detector (FID).

2.4 NMR Analysis Materials

1. 2 M hydrochloric acid (HCl): pour 170 mL of conc. HCl to a volumetric flask (1,000 mL), add distilled water up to 1,000 mL and mix well.
2. 1 M NaOH: add 4 g of NaOH to 60 mL of distilled water. Transfer the solution to a volumetric flask and add distilled water up to 100 mL.
3. Ethanol.
4. Deuterated water (D_2O).
5. Acetone-d6.
6. 600-MHz NMR spectrometer.

2.5 Desulfation and Methylation Analysis

1. Amberlite IR 120 (H^+) cation exchange resin.
2. Pyridine, sodium hydride (NaH), methyl iodide (CH_3I), acetic anhydride (Ac_2O), glacial acetic acid (CH_3COOH), anhydrous dimethyl sulfoxide (DMSO) (*see* **Note 6**).
3. Chlorotrimethylsilane (CTMS), chloroform ($CHCl_3$), sodium borodeuteride ($NaBD_4$), and menthol (CH_3OH)—chromatographic grade.
4. 4 M TFA: weigh 4.56 g of TFA and add distilled water up to 10 g.
5. Dichloromethane (CH_2Cl_2)—chromatographic grade.

6. 0.05 M NaOH: add 2 g of NaOH to 200 mL of distilled water. Transfer the solution to a volumetric flask and add distilled water up to 1,000 mL.
7. Sealed Teflon tube.
8. Dialysis membrane (cut off: 7,000 Da).
9. Potassium bromide (KBr) spectroscopically pure used to check the methylation process of FB1.
10. DB-225MS fused Silica Capillary Column (0.25 μm, 30 m×0.32 mm; J&W Scientific, Folsom, CA, USA, or equivalent).
11. Fourier Transform Infrared Spectrometer.
12. Gas chromatography–mass spectrometry (GC-MS) system.

2.6 Preparation and Purification of Oligosaccharides

1. 0.1 and 0.2 M H_2SO_4 solution: add slowly 1.4 mL of conc. sulfuric acid (98 %, 18.4 M) to 200 mL of distilled water. Transfer the solution to a volumetric flask and add distilled water up to 250 mL. For the 0.2 M solution: add slowly 2.8 mL of conc. sulfuric acid to 200 mL of distilled water. Transfer the solution to a volumetric flask and add distilled water up to 250 mL.
2. 2 M NaOH solution: weigh 20 g of NaOH and dissolve in 200 mL of distilled water.
3. Ethanol.
4. 0.1 M ammonium bicarbonate (NH_4HCO_3): weigh 7.9 g of NH_4HCO_3 in 800 mL of distilled water. Transfer the solution to a volumetric flask and add distilled water up to 1,000 mL. Filter through 0.22 μm and ultrasonic degas before use.
5. Superdex Peptide HR column (1.0×30 cm; Pharmacia Bioscience, Uppsala, Sweden, or equivalent).
6. High Pressure Liquid Chromatography system with a refractive index detector.

2.7 ES-MS Analysis

1. 0.01 M NaOH: weigh 0.08 g of NaOH and dissolve in 200 mL of distilled water.
2. 0.05 M sodium borodeuteride ($NaBD_4$): weigh 20.93 mg of $NaBD_4$ and dissolve in 10 mL of 0.01 M NaOH solution.
3. 50 % acetonitrile in distilled water (v/v).
4. Acetic acid (HAc)/distilled water (1:1, v/v).
5. Acetonitrile (CH_3CN)/2 mM NH_4HCO_3 (1:1, v/v): dissolve 0.0395 g of NH_4HCO_3 in 250 mL of distilled water to make a 2-mM NH_4HCO_3 solution. Mix 100 mL of CH_3CN and 100 mL of 2 mM NH_4HCO_3.
6. AG50W-X8 cation exchange resin (H^+) (Bio-Rad, Hercules, CA, USA, or equivalent).
7. Micromass Q-Tof Ultima instrument.

3 Methods

Carry out all procedures at room temperature unless specified otherwise.

3.1 Extraction and Purification of FB1

1. Pulverize the air-dried seaweed and pass it through a 60 mesh sieve.
2. Remove the fat of seaweed powder by covering the seaweed powder with 95 % EtOH at 80 °C for 4 h (three times) in a three-necked flask in a water bath. After centrifugation (4,340 × *g* 15 min), dry the residue at 35 °C.
3. Extract the dried alga (50 g) with 1,000 mL of distilled water at 25 °C in a water bath for 4 h (three times) and centrifuge (4,340 × *g*, 15 min) to remove the cold water soluble polysaccharide such as starch.
4. Extract further the alga residue with 1,000 mL of water at 85 °C in a water bath for 1 h (three times) and centrifuge (4,340 × *g*, 15 min). Combine the three supernatants. Adjust the supernatant pH to 8.0 with 1 M NaOH and precipitate it with 4 volumes of 95 % ethanol.
5. Dissolve the precipitate in distilled water again to make a 1 % (w/v) solution and add the same volume of 0.25 M KCl. Leave the solution standing for 1 h and then centrifuge (4,250 × *g*, 15 min). The insoluble polysaccharide is called FB. The extraction chart is represented in Fig. 1.

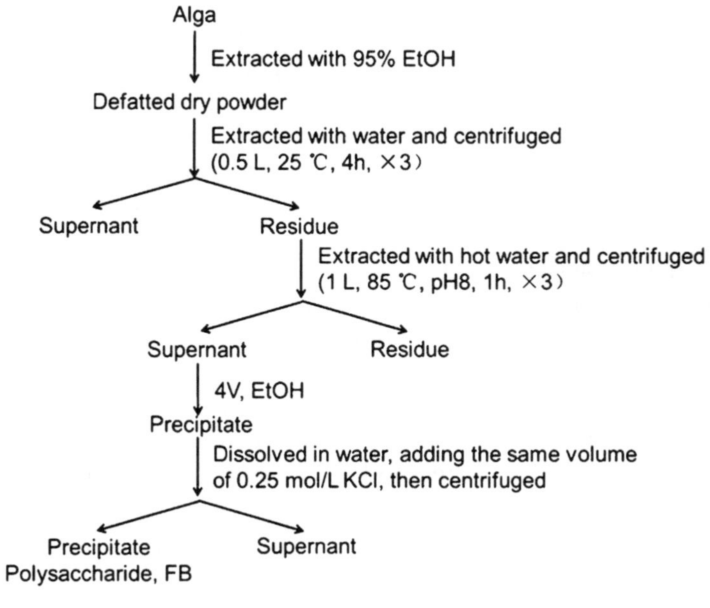

Fig. 1 The extraction chart of polysaccharides FB

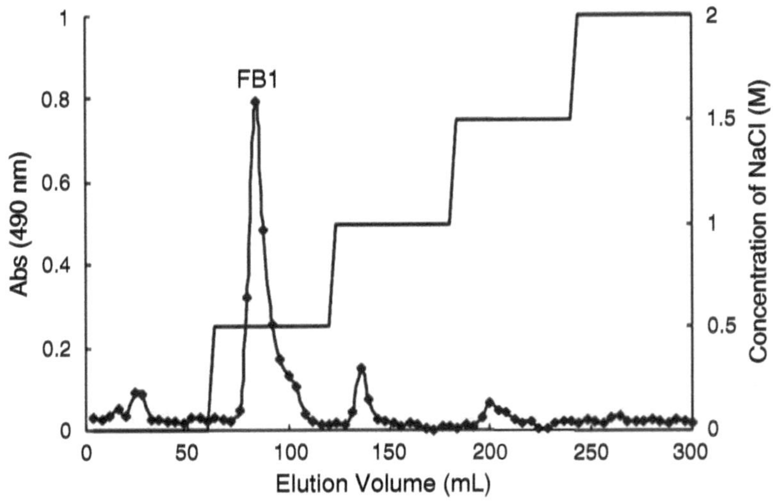

Fig. 2 Elution profile of polysaccharides FB on Q-Sepharose Fast Flow ion-exchange column. Reproduced from [1] with permission from Elsevier

6. For purification, dissolve FB (300 mg) in 50 mL of distilled water, pour it onto a Q-Sepharose Fast Flow column connected to an AKTA-FPLC system, and elute with water, followed by a stepwise addition of increasing concentrations of NaCl (0.5, 1.0, 1.5, and 2 M) at a flow rate of 120 mL/h with 500 mL (Fig. 2). Test each fraction (12 mL) for polysaccharide presence by the phenol-sulfuric acid method (*see* Subheading 3.2.1). Combine fractions eluted with the same solution of NaCl and containing polysaccharide; dialyze, concentrate, and freeze-dry the combined fractions.
7. Subject the fraction eluted by 0.5 M NaCl to gel filtration using a Sepharose 4B Fast Flow column. Elute with 500 mL of 0.2 M NaCl at a flow rate of 12 mL/h (Fig. 3). Test each fraction (1 mL) for polysaccharide presence by the phenol-sulfuric acid method (*see* Subheading 3.2.1). Dialyze the purified fraction (FB1) against distilled water and freeze-dry it.

3.2 General Analysis of the Polysaccharide FB1

The general properties of FB1, including total sugar content, sulfate content, protein content, and 3,6-AnG content, is determined by colorimetric method. The flow chart of the colorimetric method is shown in Fig. 4.

3.2.1 Total Sugar Content

The total sugar content is determined by the phenol-sulfuric acid method [11], using galactose as standard.

1. Load 200 μL of standard or sample solution (5 mg/mL) per test tube. If less sample or dilution, add distilled water to make up 200 μL.
2. Add 100 μL of 6 % phenol and 1.5 mL of concentrated sulfuric acid quickly per tube and mix.

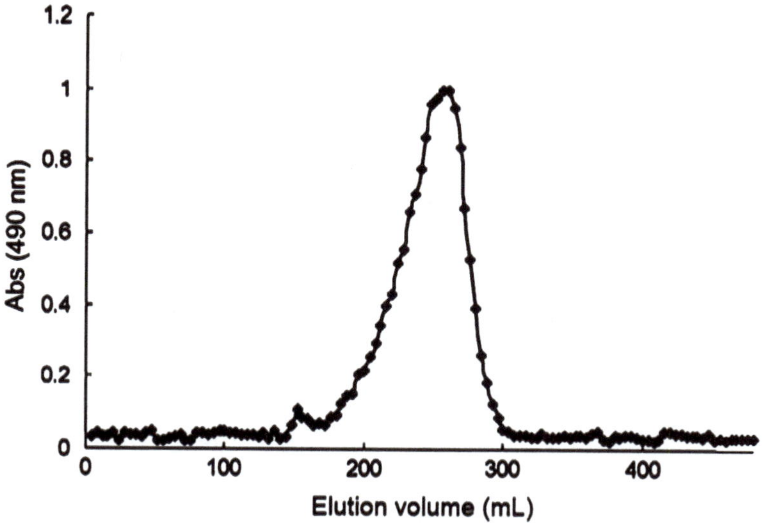

Fig. 3 Gel filtration chromatography of the polysaccharide FB1 on Sepharose 4B Fast Flow column. Reproduced from [1] with permission from Elsevier

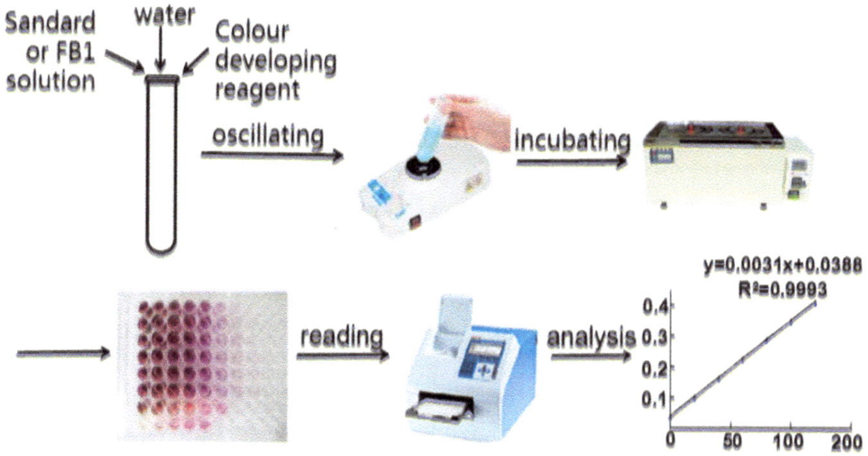

Fig. 4 Flow chart of general analysis (total sugar, sulfate, protein and 3,6-anhydrogalactose content) of polysaccharide FB1

3. Incubate the tubes at 100 °C in a water bath for 10 min.
4. Cool the tubes to room temperature (5 min) and load 200 μL of reaction solution per well on a 96-well plate.
5. Read the absorbance at 490 nm using a Microplate reader.

3.2.2 Sulfate Content

Sulfate content is determined by the $BaCl_2$-gelatin method [12].

1. Hydrolyze 5.0 mg of FB1 with 3 mL of 1 M HCl in sealed reaction vials at 105 °C in an oven for 6 h. Dry the reaction mixtures under N_2 gas and redissolve in 3 mL of distilled water.

2. Add 100 μL of hydrolyzed sample or standard in each test tube, and distilled water up to 200 μL.
3. Add 3.8 mL of 8 % TCA in each tube and 1 mL of 0.5 % $BaCl_2$-gelatin buffer.
4. Mix the tubes well and leave them standing at room temperature for 15 min.
5. Read the UV absorbance at 360 nm.

3.2.3 Crude Protein Content

The crude protein content is determined by the Lowry method [13].

1. Add 200 μL of standard or sample solutions per tube.
2. Add 1 mL of Folin-phenol solution I into each tube and mix well. Leave them standing at room temperature for 10 min.
3. Add 100 μL of Folin-phenol solution II into each tube and mix well.
4. Incubate at 37 °C in a water bath for 30 min.
5. After cooling at room temperature, read the absorbance of the tubes at 650 nm.

3.2.4 3,6-Anhydrogalactose (3,6-AnG) Content

The content of 3,6-AnG is determined by the resorcinol method using fructose as standard [14].

1. Combine 200 μL of standard or sample with 1 mL of precooled resorcinol-acetal solution in each test tube on an ice-water bath.
2. Mix well and leave standing at 20 °C in a water bath for 4 min, followed by 10 min at 80 °C in another water bath.
3. Cool the mixture in an ice-water bath for 15 min.
4. Read the absorbance at 555 nm within 15 min.

3.2.5 Molecular Weight (Mw) Analysis of FB1

1. Analyze the Mw of polysaccharide by High Performance Gel Permeation Chromatography (HPGPC), using PL aquagel-OH column. Solvent is 0.2 M Na_2SO_4 at 30 °C. The flow rate is 0.5 mL/min.
2. Calibrate with the standards (dextrans). Record the retention time of each dextran and FB1.
3. Represent the retention time as abscissa and the log value of molecular weight as ordinate to plot the standard curve and calculate the Mw of FB1 (Fig. 5).

3.3 Monosaccharide Composition Analysis

As traditional acid hydrolysis might destroy 3,6-AnG, a two-step reduction hydrolysis method is used to determine the monosaccharide composition of FB1.

1. Add 2 mg of sample, 100 μL of distilled water, 50 μL of MMB solution, and 100 μL of 6 M TFA into a screw-cap Teflon tube sequentially. Heat the mixture at 80 °C in a water bath for 30 min,

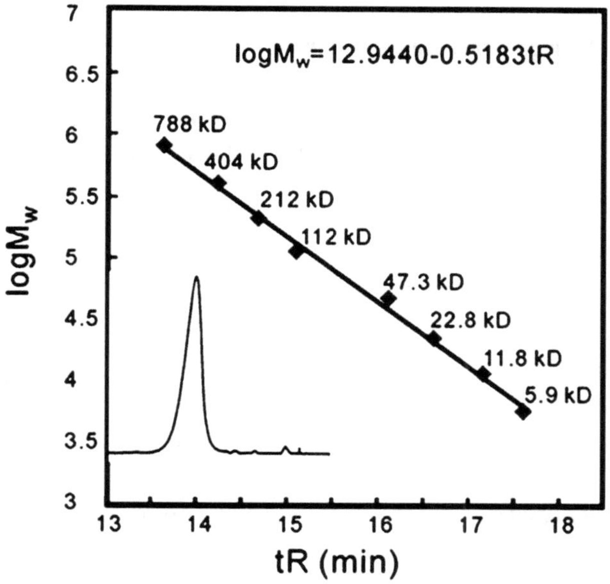

Fig. 5 Averaged molecular weight of the polysaccharide FB1 analyzed by the HPGPC method on PL aquagel-OH column. Reproduced from [1] with permission from Elsevier

then cool it to room temperature. Add 50 μL of MMB solution and stream-evaporate it under N_2 at 50 °C in a water bath.

2. Add 100 μL of distilled water and 100 μL of 4 M TFA into the tubes and hydrolyze at 120 °C in an oil bath for 1 h. After being cooled down to room temperature, add 100 μL of MMB solution and evaporate it at 50 °C in a water bath; then add 500 μL of CH_3CN and evaporate it under N_2. The sugar alditol is thus obtained.

3. Acetylate the sugar alditol with a mixture of 500 μL of EtOAc, 1.5 mL of Ac_2O, and 50 μL of $HClO_4$. Sonicate the mixture at room temperature for 10 min and leave standing for another 15 min. Add 5 mL of ultrapure water, and extract this water phase with 2 mL of CH_2Cl_2. Wash three times the CH_2Cl_2 fraction, which contains the alditol acetates, with 5 mL of water, and use the CH_2Cl_2 layer for GC analysis.

4. Analyze the alditol acetates by GC using a Fused Silica Capillary Column: the sample is detected with flame-ionization detector (FID) at 250 °C; the injector and the oven temperature are set at 250 and 210 °C, respectively. Compared with standard sugars, the composition and content of monosaccharide are determined by retention time and peak area, respectively.

The polysaccharide FB1 shows a symmetric peak on a PL aquagel-OH column, with average molecular weight of 428 kDa. The chemical composition analyses show that FB1 contained 83.6 % sugar and 15.2 % sulfate, with a small amount of crude protein

Table 1
Results of the general analyses on the polysaccharide FB1

	Total sugar content (%)	Sulfate content (%)	Protein content (%)	3,6-AnG content (%)	Monosaccharide composition (%)	MW (kDa)
FB1	83.6	15.2	0.3	39.8	Gal: 58.1 3,6-AnG: 39.8 Unknown: 2.1	428

(0.3 %) (Table 1). Monosaccharide composition analysis by GC indicates that FB1 is composed of 58.1 % Gal, 39.8 % 3,6-AnG, and 2.1 % of an unknown constituent.

3.4 NMR Analysis

In contrast to the unbranched polymeric nature of polypeptide and nucleic acid chains, carbohydrates may be branched structures, capable of substitution at several points. NMR is widely used in the carbohydrate structure analysis because it can address four essential problems: the monosaccharides present and their furanose/pyranose configurations and α/β anomeric forms; the linkage between the monosaccharides and the position of substituents; the dynamic properties of the carbohydrate molecule; the conformation or family of conformations due to the O-glycosidic bonds and hydroxy-methyl rotamers. All the four questions can be satisfactorily answered by NMR [15].

1. Dissolve 40 mg of FB1 in 8 mL of distilled water, then adjust the pH to 1.0 with 2 M HCl, and keep it for 10 min at 60 °C in a water bath (*see* **Note 7**). Neutralize the hydrolyzed FB1 with 1 M NaOH and precipitate it with 16 mL of ethanol. Dissolve the precipitate in 10 mL of distilled water and freeze-dry it.
2. Dissolve 30 mg of the hydrolyzed FB1 in 1 mL of D_2O and freeze-dry twice to replace all exchangeable protons with deuterium. Acquire the ^{1}H-NMR, ^{13}C-NMR, ^{1}H-^{1}H Correlation Spectroscopy (COSY), and Heteronuclear Multiple-Quantum Correlation (HMQC) at 20 °C using a 600-MHz NMR spectrometer. Calibrate the chemical shift value using acetone-d6 as an internal standard.
3. Assign the chemical shifts of ^{1}H and ^{13}C of the polysaccharide FB1 based on a series of ^{1}H-^{1}H COSY, and 2D ^{1}H-^{13}C (HMQC) correlation experiments. From ^{1}H-^{1}H COSY spectrum, all correlation signals could be assigned. For instance, as labeled in Fig. 6, the ring protons of unit G H1 (4.45 ppm) → H2 (3.44 ppm) → H3 (3.72 ppm) → H4 (3.98 ppm), H5 (3.54 ppm) → H6a (3.56 ppm) → H6b (3.61 ppm), and the ring protons of unit G′ H1 (4.50 ppm) → H2 (3.44 ppm) → H3 (3.86 ppm) → H4 (4.70 ppm), H5 (3.66 ppm) → H6a

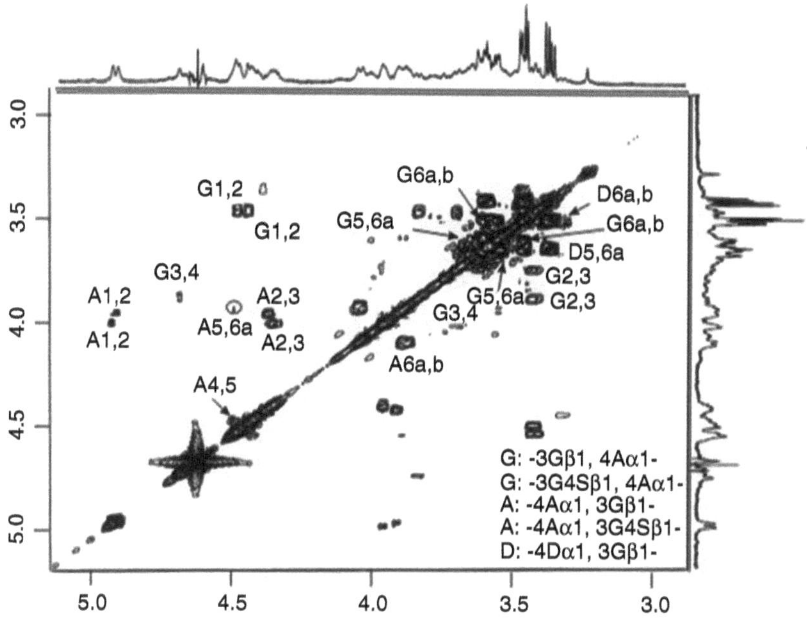

Fig. 6 ^1H-^1H COSY spectrum of the polysaccharide FB1. Reproduced from [1] with permission from Elsevier

Table 2
Correlations in ^1H-^1H COSY spectrum of the polysaccharide FB1

	Coupled H/H (ppm)				
Sugar residues	**H1/H2**	**H2/H3**	**H3/H4**	**H5/H6a**	**H6a/b**
→3G4Sβ1,4Aα1→	4.50/3.44	3.44/3.86	3.86/4.70	3.66/3.58	3.58/3.64
→3Gβ1,4Aα1→	4.45/3.44	3.44/3.72	3.72/3.98	3.54/3.56	3.56/3.61
→4Aα1,3Gβ1→	4.92/3.92	3.92/4.38	na	4.50/3.92	3.92/4.06
→4Aα1,3G4Sβ1→	4.94/3.98	3.98/4.35	na	4.49/3.90	3.90/4.07
→3Gβ1,4Dα1→	na	na	na	3.61/3.40	3.40/3.46

Reproduced from [1] with permission from Elsevier
na not assigned, *G4S* 1,3-linked-4-sulfated-galactose, *A* 1,4-linked-3,6-anhydrogalactose, *G* 1,3-linked galactose, *D* 1,4-linked galactose

(3.58 ppm)→H6b (3.64 ppm) were determined (Table 2). The proton-carbon correlation was assigned based on the HMQC spectrum (Fig. 7), which maps the correlations between carbon atoms and their directly bonded protons. Take unit G for example, the correlations of H1 with C1 at 102.5 ppm, H2 with C2 at 69.2 ppm, H3 with C3 at 79.7 ppm, H4 with C4 at 65.6 ppm, H5 with C5 at 75.0 ppm, and H6 with C6 at 60.5 ppm were clearly arisen. The proton and carbon cross-peaks of the other units could also be assigned clearly by HMQC spectrum as in Table 3.

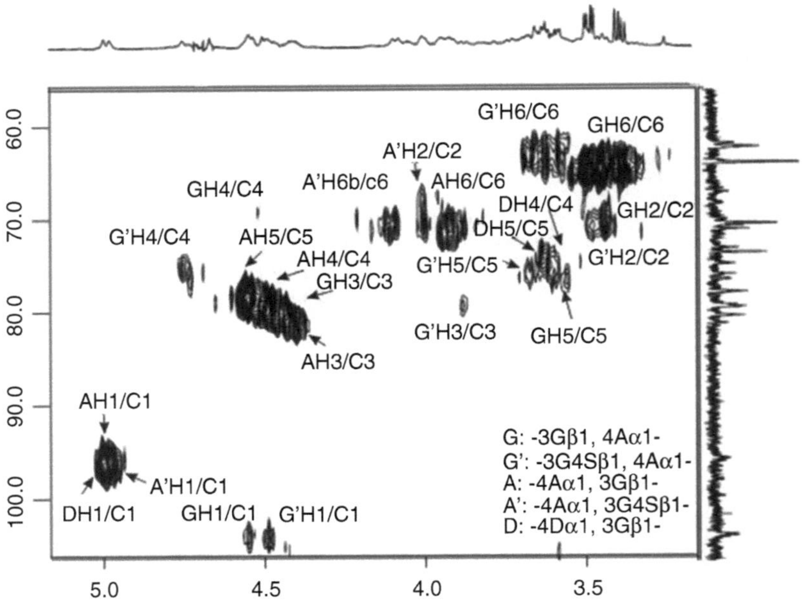

Fig. 7 HMQC spectrum of the polysaccharide FB1. Reproduced from [1] with permission from Elsevier

Table 3
Correlations revealed in HMQC spectrum of the polysaccharide FB1

	Coupled H/C (ppm)					
Sugar residues	**H1/C1**	**H2/C2**	**H3/C3**	**H4/C4**	**H5/C5**	**H6/C6**
→3G4Sβ1,4Aα1→	4.50/102.4	3.44/69.5	3.86/77.5	4.70/73.3	3.66/74.4	3.58 (3.64)/60.7
→3Gβ1,4Aα1→	4.45/102.5	3.44/69.2	3.72/79.7	3.98/65.6	3.56/75.0	3.56 (3.61)/60.5
→4Aα1,3Gβ1→	4.92/93.9	3.91/69.5	4.38/78.9	na	4.49/76.3	3.92/69.0
→4Aα1,3G4Sβ1→	4.94/94.1	3.98/68.6	4.35/79.7	4.45/77.7	4.50/76.3	4.49/76.3
→3Gβ1,4Dα1→	5.02/94.	na	na	3.61/73.3	3.64/72.1	3.40 (3.46)/62.5

Reproduced from [1] with permission from Elsevier
na not assigned, *G4S* 1,3-linked-4-sulfated-galactose, *A* 1,4-linked-3,6-anhydrogalactose, *G* 1,3-linked galactose, *D* 1,4-linked galactose

3.5 Desulfation and Methylation Analysis

Desulfation is operated as described in [16] and methylation is performed according to [17] (*see* **Note 8**). Briefly, the process is carried out as shown in Fig. 8 and outlined below.

1. Load 20 mg of FB1 on an Amberlite IR 120 cation exchange resin (20 mL). Wash the column with 20 mL of distilled water. Adjust the pH of the eluate to 9.0 with pyridine (*see* **Note 9**). Evaporate the eluate with a rotavapor and freeze-dry it to obtain the pyridinium form of FB1 (FB1-Py).

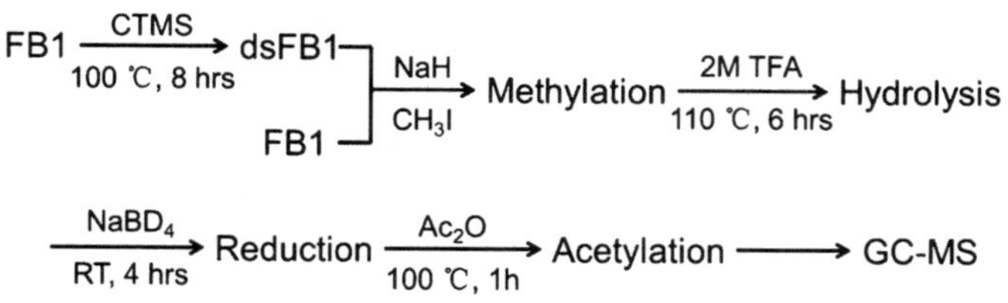

Fig. 8 Flow chart of the process of desulfation and methylation of the polysaccharide FB1

2. Dissolve the FB1-Py in 2 mL of pyridine in a sealed Teflon tube. Add CTMS until the molar ratio of CTMS:sulfate is about 400:1. Keep the tube at 100 °C in an oil bath for 8 h with a constant stirring. Wash the excess of CTMS with water, and recover the desulfated product dsFB1 by dialysis against distilled water for 3 days and then freeze-dry it.
3. Load 1 mg of FB1 and dsFB1 (*see* **Note 10**) in microtubes and add N_2 to protect samples from the air. Add 1 mL of anhydrous DMSO to dissolve the FB1 and dsFB1. Under the protection of N_2, add 100 mg of NaH powder rapidly and stir for 1 h. Wrap up the microtubes with silver paper and add 0.5 mL of CH_3I slowly (*see* **Notes 11** and **12**), stir for 2 h. Then add 0.5 mL of distilled water to end up the reaction. Extract the production with 1 mL of $CHCl_3$ twice and combine the two $CHCl_3$ layers. Wash the $CHCl_3$ layer with 2 mL of ultrapure water to remove impurities. Use 200 μL of $CHCl_3$ layer to carry out the Fourier Transform Infrared Spectrometry (FTIR) analysis (*see* **Note 13**) and dry the rest using a rotary evaporator.
4. The completion of methylation is confirmed by FTIR spectroscopy as the disappearance of the absorption of O-H stretch vibration. Grind 100–200 mg of dry KBr to powder and compress. Drop 20 μL of the $CHCl_3$ layer to the tablet and detect on the FTIR equipment.
5. Dissolve the dry production above in 80 μL of CH_3OH in 2 mL section bottles and add 200 μL of distilled water, following by 300 μL of 4 M TFA. After sealing up the bottles, put them in an oven at 110 °C for 6 h. After the reaction, remove the extra TFA under N_2 gas until dryness.
6. Add 10 mg of $NaBD_4$ into the two hydrolysis products above (from FB1 and dsFB1), 1 mL of 0.05 M NaOH to dissolve them and leave them to react at room temperature for 4 h. Add two drops of glacial acetic acid to finish the reaction and dry the liquid under N_2 gas. Add 1 mL of CH_3OH and then dry again. Repeat three times.

7. Dissolve the dried products above in 0.5 mL of pyridine in tubes and seal up the tubes. After 30 min in a water bath at 90 °C, add 0.5 mL of acetic anhydride and leave for reaction at 100 °C in an oil bath for 1 h. Dry the liquid under N_2 gas and add 1 mL of CH_2Cl_2. Wash the CH_2Cl_2 layer with 1 mL of distilled water (repeat three times) and dry in an oven at 45 °C. The partial methylated alditol acetates are thus obtained.
8. Analyze partial methylated alditol acetates by a GC-MS equipped with a DB-225MS fused silica capillary column: set the injector temperature at 220 °C; use helium as the carrier gas at a constant flow rate of 1.0 mL/min. Set the initial oven temperature at 100 °C for 4 min and increase it up to 240 °C at the rate of 58 °C/min. Set the transfer line temperature at 220 °C and the ion source temperature at 280 °C. Ions are generated by a 70 eV electron beam at a current of 2.0 mA. Acquire masses from 30 to 800 m/z at a rate of 30 spectra/s, and turn on the acceleration voltage after a solvent delay of 180 s. Analyze mass spectra of the compounds using the Complex Carbohydrate Structural Database (CCSD) of the Complex Carbohydrate Research Centre, University of Georgia (http://www.ccrc.uga.edu/).

   The native polysaccharide FB1 consisted of 1,4,5-Ac_3-2-Me-3,6-AnG, 1,3,5- Ac_3-2,4,6-Me_3-Gal, 1,4,5-Ac_3-2,3,6-Me_3-Gal and 1,3,4,5-Ac_4-2, 6-Me_2-Gal, while only a few 1,3,4,5-Ac_4-2, 6-Me_2-Gal components were found in dsFB1 (Table 4). This indicates that the sulfate group is mainly located in C4 of 1,3-Gal of FB1.

Table 4
Composition of partially methylated monosaccharides produced by methylation of the polysaccharide FB1 and its desulfated counterpart dsFB1

			Proportions (%)	
Linkage	Partially methylated alditol acetates	Main ion (m/z)	FB1	dsFB1
→4)A(1→	1,4,5-Ac_3-2-Me-3,6-AnGal	43,103,145	40.2	38.4
→3)G4S(1→	1,3,4,5-Ac_4-2,6-Me_2-Gal	117,129,305	29.7	3.2
→3)G(1→	1,3,5-Ac_3-2,4,6-Me_3-Gal	117,161,233,277	20.1	46.6
→4)D(1→	1,4,5-Ac_3-2,3,6-Me_3-Gal	112,161,189,233	7.9	9.7
→4)3MeGal(1→			2.1	2.1

A 1,4-linked-3,6-anhydrogalactose, *G4S* 1,3-linked-4-sulfated-galactose, *G* 1,3-linked galactose, *D* 1,4-linked galactose

3.6 Preparation the FB1 Oligosaccharides

1. Hydrolyze FB1 (50 mg) with 0.1 M H_2SO_4 (10 mg/mL) at 60 °C in a water bath for 2 h, followed by neutralization with 2 M NaOH. Add the same volume of ethanol and centrifuge (4,340×*g*, 15 min). Collect the supernatant (FB1S) and freeze-dry it.
2. Hydrolyze the ethanol precipitate fraction (FB1P, 10 mg) with 0.2 M H_2SO_4 (10 mg/mL) at 80 °C in a water bath for 4 h.
3. Dissolve FB1S and FB1P in 0.1 M NH_4HCO_3 (10 mg/mL). Separate the oligosaccharides (100 μL) on a Superdex Peptide HR column eluted with 0.1 M NH_4HCO_3 at a flow rate of 0.2 mL/min using a refractive index detector. Collect the different oligosaccharide fractions: 10 for FB1S1 (FB1S1-10) and 8 for FB1P1 (FB1P1-8). Vacuum concentrate these fractions at 45 °C and freeze-dry them.

   Compared with the oligosaccharides from mild acid hydrolysis of κ-carrageenan (dotted line) and FB1P (Fig. 9), those of FB1S showed a series of irregular patterns (solid line).

3.7 ES-TOF-MS Analysis

1. Mix 5 μL of each oligosaccharide fractions of FB1S (FB1S1 to 10) and FB1P (FB1P1 to 8) (2 mg/mL) with 5 μL of 50 % acetonitrile and inject them to the Micromass Q-Tof Ultima instrument to get the ES-MS spectrum. The mobile phase (CH_3CN/2 mM NH_4HCO_3, 1:1, v/v) is delivered by a syringe pump at a flow rate of 5 μL/min. The capillary voltage is maintained at 3 kV, while the cone voltage is 150 eV, depending on the size of the oligosaccharides. Use nitrogen as the desolvation and nebulizer gas at a flow rate of 500 and 50 L/h, respectively. Set the source temperature at 800 °C and the desolvation temperature at 150 °C.
2. Add 20 μL of 0.05 M $NaBD_4$ to 20 μg of the freeze-dried oligosaccharides, and carry out the reduction at 4 °C overnight. Neutralize the reaction solution to pH 7 with a solution of acetic acid/distilled water (1:1) to destroy borohydride. Load the reaction solution onto an AG50 W-X8 resin (H^+ form) and elute with distilled water until the elution becomes neutral. Vacuum concentrate and dry the eluent under N_2 (*see* **Note 14**).
3. Dissolve samples in CH_3CN/2 mM NH_4HCO_3 (1:1, v/v), typically at a concentration of 5–10 pmol/L. Inject 5 μL of each sample to the Micromass Q-Tof Ultima instrument to get the ES-CID-MS/MS spectrum. Argon is used as the collision gas at a pressure of 1.7 bar and the collision energy should be adjusted to 17–100 eV for optimal sequence information.

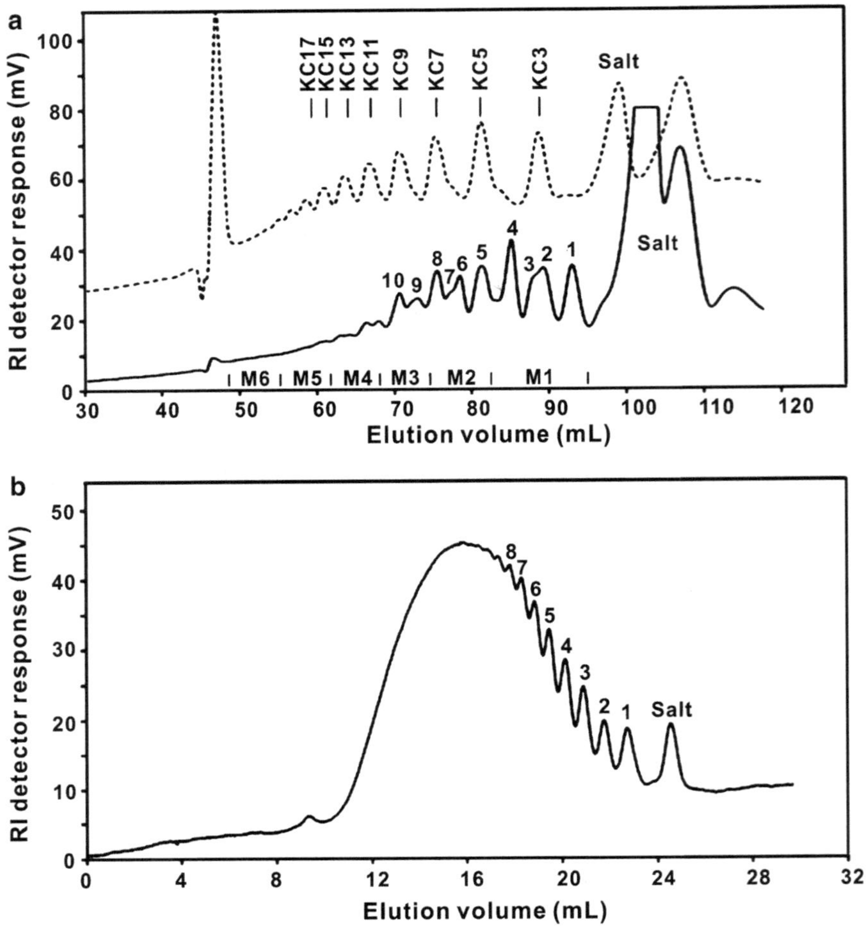

Fig. 9 Low pressure gel permeation chromatography of acid hydrolysates of the polysaccharide FB1. (**a**) FB1S is applied to a Superdex™ Peptide 10/300 GL column (*solid line*); for comparison, acid hydrolysis products from κ-carrageenan are also shown (*dotted line*). (**b**) FB1P is applied to a Superdex™ Peptide 10/300 GL column. See the text for a description of FB1S and FB1P. Reproduced from [1] with permission from Elsevier

The oligosaccharide fractions FB1S1-FB1S8 and FB1P1-FB1P6 are analyzed by negative-ion ES-MS (*see* Fig. 10, Table 5). The results show that FB1 appears to be a hybrid sulfated galactan with two major blocks, [G4S-A-G-A]*n* (75 %) and [G-D]*n* (10 %), and small building blocks of [G4S-D], [G-3MeGal], and [G4S-3MeGal]. Thus, FB1 is characterized as a copolymer of κ/β-carrageenan and nonsulfated galactan.

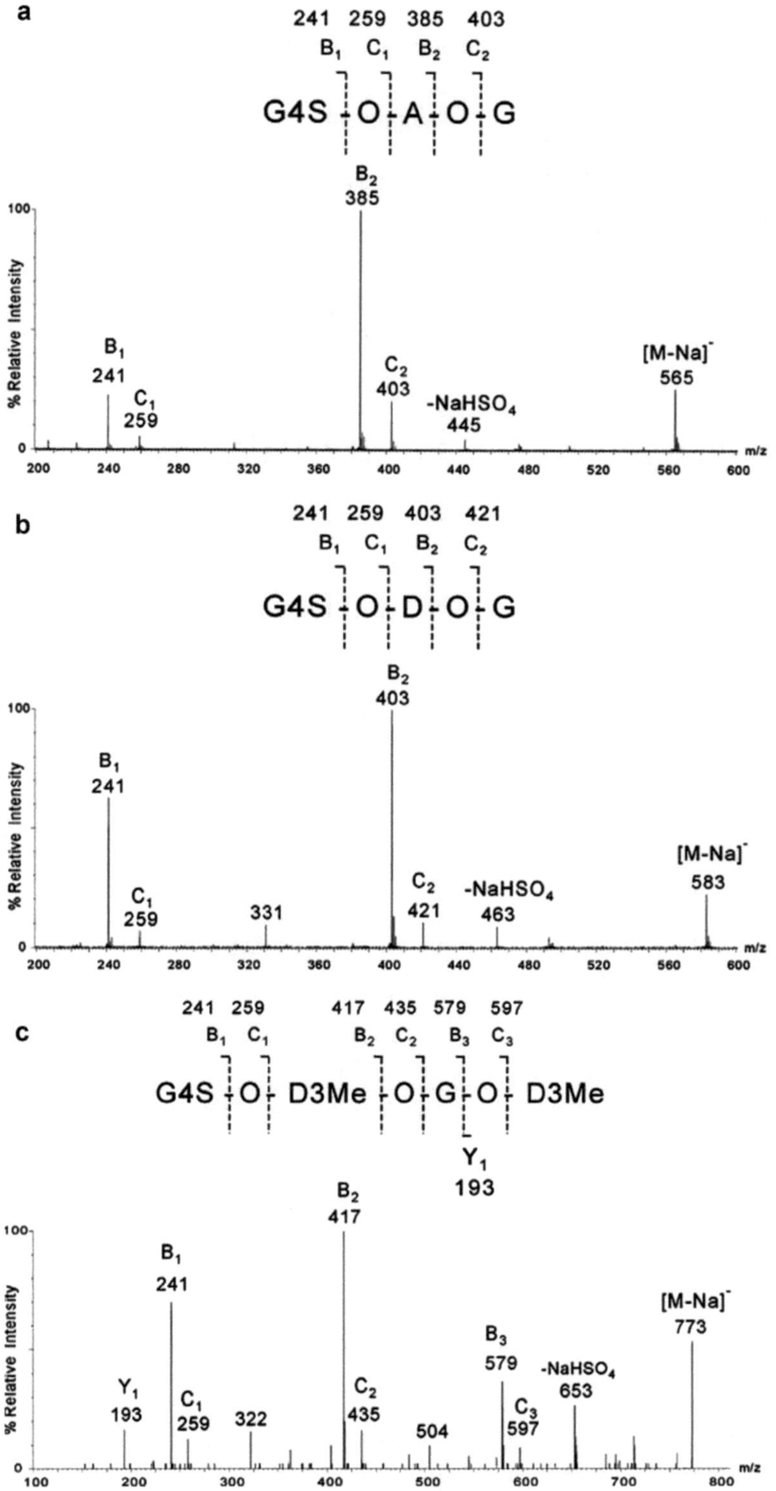

Fig. 10 Negative-ion ES-CID-MS/MS spectra of acid-hydrolyzed oligosaccharides of the polysaccharide FB1. (**a**) Sequence analysis of a trisaccharide from FB1S1; (**b**) sequence analysis of another trisaccharide from FB1S1; (**c**) sequence analysis of tetrasaccharide from FB1S3. Reproduced from [1] with permission from Elsevier

Table 5
Negative-ion ES-MS of mild acid hydrolysis products from the polysaccharide FB1

Fractions	Found ions[a] (charge)	Calculated mol mass (H form; Da)	Assignment DP[b]	Sequence	Theoretical mol mass (H form; Da)
FB1S1	565.1 (−1)	566.1	3	G4S-A-G	566.1
	583.1 (−1)	584.1	3	G4S-D-G	584.1
FB1S2	322.0 (−2)	646.0	3	G4S-A-G4S	646.1
	827.3 (−1)	828.3	5	G-D-G-D-G/D-G-D-G-D	828.3
FB1S3	745.3 (−1)	746.3	4	G-D-G-D(1S)/D-G-D-G(1S)	746.2
	773.4 (−1)	774.4	4	G4S-D3Me-G-D3Me	774.2
	907.4 (−1)	908.4	5	G-D-G-D-G(1S)/D-G-D-G-D(1S)	908.2
FB1S4	475.0 (−2)	952.0	5	G-A-G-A-G(2S)	952.2
FB1S5	343.0 (−3)	1,032.0	5	G4S-A-G4S-A-G4S	1,032.1
FB1S6	668.1 (−2) 445.0 (−3)	1,338.0	7	G-A-G-A-G-A-G(3S)	1,338.2
FB1S7	463.7 (−3)	1,394.1	8	G-D-G-D-G-D-G-D(1S) /D-G-D-G-D-G-D-G(1S)	1,394.4
FB1S8	410.1 (−4)	1,644.2	9	G-A-G-A-G-A-G-A-G(3S)	1,644.3
FB1P1	341.1 (−1)	342.1	2	G-D/D-G	342.1
FB1P2	503.1 (−1)	504.1	3	G-D-G/D-G-D	504.2
FB1P3	665.2 (−1)	666.2	4	G-D-G-D/D-G-D-G	666.2
FB1P4	827.4 (−1)	828.4	5	G-D-G-D-G/D-G-D-G-D	828.3
FB1P5	989.4 (−1)	990.4	6	G-D-G-D-G-D/D-G-D-G-D-G	990.3
FB1P6	1,151.6 (−1)	1,152.6	7	G-D-G-D-G-D-G/D-G-D-G-D-G-D	1,152.4

FB1S1-10 and FB1P1-6 denote oligosaccharide fractions derived from supernatant and precipitate, respectively, following ethanol precipitation of FB1. Reproduced from [1] with permission from Elsevier

G4S 1,3-linked-4-sulfated-galactose, *A* 1,4-linked-3,6-anhydrogalactose, *G* 1,3-linked galactose, *D* 1,4-linked galactose

[a]Major ion detected

[b]Degree of polymerization

4 Notes

1. The K_2SO_4 should be dried at 105 °C to constant weight before using.
2. When preparing the Folin-phenol buffer I, it is better to mix the reagent A and B immediately before use.
3. Use within 1 week.

4. Use within 1 month.
5. Use within 3 h.
6. NaH is used in the process of methylation and can strongly react with water to produce NaOH and H_2 under the release of a lot of heat. The strong basic and hot environment will lead to the degradation of polysaccharide and a great deal of by-product. So it is essential to make sure that all reagents are anhydrous. Different methods are used to prepare the anhydrous reagents: wash a 4 Å molecular sieve with ethanol and heat it in the muffle furnace at 400 °C for 3 h. After cooling, add the molecular sieve into the pyridine to absorb the residual water. Wash NaH powder with petroleum ether for three times and dry with N_2 before use. Add $CaCl_2$ powder into the Ch_3I to absorb the water and filter out the $CaCl_2$ before use.
7. This process of mild acid hydrolysis of FB1 is used to reduce its high viscosity.
8. All the reaction should occur in glass vessel to avoid impurities.
9. Pyridine is harmful if inhaled, swallowed, or absorbed through the skin and also has minor neurotoxic, genotoxic, and clastogenic effects. So all the operations should be done in the ventilation.
10. The sample should be dried under diminished pressure for 3 h. All the glassware should be dried in an oven at 120 °C for 12 h.
11. The NaH and DMSO could generate the $CH_3SOCH_2^-$. Before the addition of CH_3I, $CH_3SOCH_2^-$ has not disassociated all the OH^- on the carbohydrate, and if the addition of CH_3I is too fast, the excessive CH_3I will react with CH_3SOCH_2 to produce the main by-product $C_2H_5SOCH_3$. In order to decrease the side reaction and improve the yield, the weight of NaH should not be too high and the CH_3I should be added slowly, ideally drop by drop.
12. CH_3I exhibits acute toxicity for inhalation and ingestion and symptoms may include eye irritation, nausea, vomiting, dizziness, ataxia, slurred speech, and dermatitis. Make sure don't inhale or touch CH_3I and all the operations should be done in the ventilation.
13. FTIR analysis of methylation production is essential to test whether the methylation reaction is complete. The peak at 3,400–3,500/cm of –OH would disappear if the reaction is complete; otherwise the reaction should be repeated.
14. The assignment of reducing or non-reducing terminal fragments is assisted by oligosaccharide reduction and the product ion spectra of the derived alditols. The reducing terminal fragment ions in the alditols should show an increase of 2 Da from the reduction due to the opening of the saccharide ring [18, 19].

Acknowledgements

This research is supported in part by Special Fund for Marine Scientific Research in the Public Interest (201005024), National Science & Technology Support Program of China (2013BAB01B02), and the NSFC-Shandong Joint Fund for Marine Science Research Centers (U1406402). It is based on a collaborative research supported by Ocean Nutrition Canada Ltd. and the Atlantic Canada Opportunity Agency (ACOA), funded through the PEI Food Technology Center.

References

1. Yang B, Yu GL, Zhao X et al (2011) Structural characterisation and bioactivities of hybrid carrageenan-like sulphated galactan from red alga *Furcellaria lumbricalis*. Food Chem 124:50–57
2. De Ruiter GA, Rudolph B (1997) Carrageenan biotechnology. Trends Food Sci Technol 8:389–395
3. Usov AI (1998) Structural analysis of red seaweed galactans of agar and carrageenan groups. Food Hydrocolloid 12:301–308
4. Wang W, Zhang P, Hao C et al (2011) *In vitro* inhibitory effect of carrageenan oligosaccharide on influenza A H1N1 virus. Antivir Res 92:237–246
5. Kalitnik AA, Byankina Barabanova AO, Nagorskaya VP et al (2013) Low molecular weight derivatives of different carrageenan types and their antiviral activity. J Appl Phycol 25:65–72
6. Jin Z, Han YX, Han XR (2013) Degraded iota-carrageenan can induce apoptosis in human osteosarcoma cells via the Wnt/β-catenin signaling pathway. Nutr Cancer 65:126–131
7. Yuan H, Song J, Li X et al (2006) Immunomodulation and antitumor activity of kappa-carrageenan oligosaccharides. Cancer Lett 243:228–234
8. Yermak IM, Barabanova AO, Aminin DL et al (2012) Effects of structural peculiarities of carrageenans on their immunomodulatory and anticoagulant activities. Carbohyd Polym 87:713–720
9. Bravo R, Arimon M, Valle-Delgado JJ et al (2008) Sulfated polysaccharides promote the assembly of amyloid β(1-42) peptide into stable fibrils of reduced cytotoxicity. J Biol Chem 283:32471–32483
10. Yu GL, Yang B, Ren WN et al (2007) A comparative analysis of four kinds of polysaccharides purified from *Furcellaria lumbricalis*. J Ocean Univ Chin 6:16–20
11. Dubois M, Gilles KA, Hamilton JK et al (1956) Colorimetric method for determination of sugars and related substances. Anal Chem 28:350–356
12. Dodgson KS, Price RG (1962) A note on the determination of the ester sulphate content of sulphated polysaccharides. Biochem J 84: 106–110
13. Lowry OH, Rosebrough NJ, Farr AL et al (1951) Protein measurement with the Folin phenol reagent. J Biol Chem 193:265–275
14. Arsenault GP, Yaphe W (1965) Effect of acetaldehyde, acetic acid and ethanol on the resorcinol test for fructose. Anal Biochem 13:133–142
15. Weller CT (2010) Carbohydrates studied by NMR. In: Lindon J (ed) Encyclopedia of spectroscopy and spectrometry, 2nd edn. Elsevier, Amsterdam, pp 204–211
16. Kolender AA, Matulewicz MC (2004) Desulfation of sulfated galactans with chlorotrimethylsilane. Characterization of β-carrageenan by ^{1}H NMR spectroscopy. Carbohydr Res 339:1619–1629
17. Hakomori S (1964) A rapid permethylation of glycolipid, and polysaccharide catalyzed by methylsulfinyl carbanion in dimethyl sulfoxide. J Biochem 55:205–208
18. Yu GL, Zhao X, Yang B et al (2006) Sequence determination of sulfated carrageenan-derived oligosaccharides by high-sensitivity negative-ion electrospray tandem mass spectrometry. Anal Chem 78:8499–8505
19. Yang B, Yu GL, Zhao X et al (2009) Mechanism of mild acid hydrolysis of galactan polysaccharides with highly ordered disaccharide repeats leading to a complete series of exclusively odd-numbered oligosaccharides. FEBS J 276: 2125–2137

Chapter 22

Characterization of Alginates by Nuclear Magnetic Resonance (NMR) and Vibrational Spectroscopy (IR, NIR, Raman) in Combination with Chemometrics

Henrik Max Jensen, Flemming Hofmann Larsen, and Søren Balling Engelsen

Abstract

This chapter describes three different spectroscopic methods for structural characterization of the commercial important hydrocolloid alginate extracted from brown seaweed. The "golden" reference method for characterization of the alginate structure is ^{1}H liquid-state NMR of depolymerized alginate polymers using a stepwise hydrolysis. Having implemented this method, predictive and rapid non-destructive methods using vibrational spectroscopy and chemometrics can be developed. These methods can predict the M/G-ratio of the intact alginate powder with at least the same precision and accuracy as the reference method in a fraction of the time that is required to measure the alginate using the reference method. The chapter also demonstrates how solid-state ^{13}C CP/MAS NMR can be used to determine the M/G ratio on the intact sample by the use of multivariate chemometrics and how this method shares the characteristics of the solid-state non-destructive IR method rather than its liquid-state counterpart.

Key words ^{13}C CP/MAS NMR spectroscopy, Chemometrics, ^{1}H NMR spectroscopy, Infrared spectroscopy, IR, Multivariate curve resolution, MCR, Partial hydrolysis, Partial least squares regression, PLSR, Sodium alginate

1 Introduction

Alginates are extracted from brown seaweeds (Phaeophyceae) in which the polysaccharides function as an important structural component of the thalli. The alginate polysaccharide is a linear polymer with repeating units of 1,4-linked β-D-mannuronic acid (M) and α-L-guluronic acid (G) arranged in a blockwise pattern along the polymer chain as homopolymeric (MM or GG) or heteropolymeric (MG) regions. The natural amount of M- and G-units as well as block structure depends on the seaweed species, geographical location, season, vegetative phase and the collected part of the algae species. In the algae the M/G ratio is controlled by the mannuronan C-5 epimerase [1]. Sodium alginate is the

Dagmar B. Stengel and Solène Connan (eds.), *Natural Products From Marine Algae: Methods and Protocols*, Methods in Molecular Biology, vol. 1308, DOI 10.1007/978-1-4939-2684-8_22, © Springer Science+Business Media New York 2015

most common form of commercial alginate and is widely used in numerous applications, for example to modify viscosity and texture of foods, soothing and gelling in over-the-counter pharma products for heartburn and GERD (Gastro Oesophageal Reflux Disorder), drug delivery vehicle, cell encapsulation, dental impressions and in a variety of industrial applications within the paper and textile industry. In food systems sodium alginate is added in low amounts and by interactions with calcium in the food a cross-linked rigid gelling network is formed which results in a heat-stable gel, e.g., in the jam of pastries.

The viscosity of alginate hydrocolloids depends on the average molecular weight, whereas the gelling properties are affected by the distribution of the M- and G-units and on counter ions present. A low M/G-ratio typically results in a brittle gel, while a high M/G-ratio results in a softer and more flexible gel. The focus of the current chapter is on sodium alginates and therefore the most important structural parameter in relation to the gelling properties is the M/G-ratio.

The reference method for determination of the M/G ratio is ^{1}H liquid-state Nuclear Magnetic Resonance (NMR) spectroscopy [2–4] of 0.1–0.5 % (w/v) aqueous alginate solutions. The practical implementation of this method and the feasibility of replacing this method with rapid vibrational spectroscopic methods (infrared, near infrared, or Raman spectroscopy) for high-throughput quality control is the aim of this chapter.

In order to record ^{1}H liquid-state NMR spectra, the alginate samples must be solubilized and liquefied in the best way possible. This is achieved by depolymerization to reduce the molecular weight and by recording the spectra at elevated temperatures to reduce the viscosity. The depolymerization is performed by a stepwise mild partial hydrolysis using mineral acids [2–4]. This somewhat laborious method is required in order to avoid complete hydrolysis and subsequent possible epimerization leading to erroneous M/G ratios. For the procedure described here, a high temperature is required (≥353 K) to record the ^{1}H NMR spectrum. The elevated temperature serves two purposes, firstly the decreased sample viscosity results in narrower line-widths and secondly, the water resonance is moved away from the spectral region of interest due to waters temperature dependent chemical shift. Data acquisition using water suppression has been tested, but has so far resulted in misleading M/G-ratios [5, 6]. It has been suggested by Chhatbar et al. [7] to use microwaves for assisting the breakdown of the polymer. Another approach has been described by Li et al. [8] and Lundqvist et al. [9] in which alginate lyases are used to degrade alginate down to oligosaccharides, followed by chromatographic purification and 2D NMR and Electron Spray Ionization (ESI) Mass spectrometry (MS) characterization of the oligomer fractions.

Besides the M/G ratio, information about the block-structure of the sodium alginate can also be obtained from the ^{1}H NMR

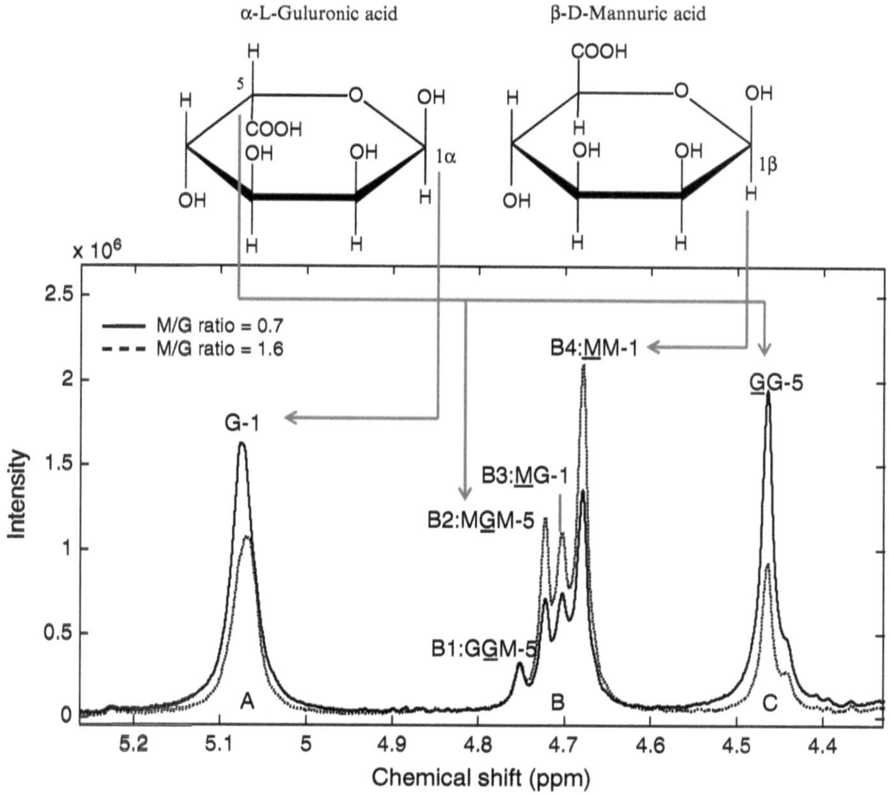

Fig. 1 Liquid-state ^{1}H NMR spectra recorded at 14.1 T (^{1}H Larmor frequency: 600 MHz) of alginates. Only the spectral region 4.35–5.25 ppm is displayed. *Solid line*: M/G-ratio = 0.7, *broken line*: M/G-ratio = 1.6. *Arrows* indicate the assignment of peaks

spectra by carefully integrating the H1 and H5 protons of the uronic acids (*see* Fig. 1) using the equations deduced by Grasdalen [3]. Both diade (MM, GM, GG) and triade (e.g., MGM) information can be obtained by this approach. Beside the pioneering work using ^{1}H liquid-state NMR [10] also have shown that ^{13}C liquid-state NMR spectra can be used for monomeric and block-structure information of partly depolymerized alginate. However, the lower sensitivity and thus longer analysis time of this technique make it less attractive for routine analysis.

It has been shown, on a limited sample material, that intact polymer analysis (M/G ratio) can be performed on alginates in solution using a standard diffusion edited NMR pulse sequence [6]. In this approach the large difference in water and polymer-sizes, hence different rates of diffusion, is utilized to suppress water. Solid state ^{13}C Cross Polarization-Magic Angle Spinning (CP-MAS) NMR techniques can also provide M/G ratio of the intact polymer [11] in combination with multivariate curve resolution [12, 13]. In contrast to the ^{1}H liquid-state NMR analysis, this method is not sensitive to the presence of small amounts of calcium ions.

The two intact polymer NMR methods are reporting only on the M/G ratio but have so far not block information.

The NMR methods mentioned above are both time-consuming and labor-intensive (hydrolysis). In order to facilitate time-effective analyses of alginates, faster analytical methods are desirable. To fulfil these requirements, vibrational spectroscopic techniques such as infrared (IR), near infrared (NIR), and Raman spectroscopy have been tested [14] and by multivariate calibration to ^{1}H liquid-state NMR their performance in determination of the M/G-ratio has been evaluated. Partial Least Squares Regression (PLSR) [15] is employed to develop the prediction methods.

Establishment of a statistically sound PLSR calibration model for determination of the M/G-ratio from vibrational spectra requires a fairly large calibration sample-set. The large sample-set is used to build a robust cross-validated model based on the reference values of M/G ratio (obtained from NMR spectroscopy on hydrolysed samples) and associated vibrational spectra (obtained on solid state samples) so that the model then can be used to rapidly predict the M/G ratio from a vibrational spectrum of a new sample. It has previously been demonstrated that 80–100 sodium alginate samples are sufficient to obtain a high quality prediction model that can predict the M/G ratio from vibrational (IR, NIR, Raman) spectra of the intact powder samples [14].

In this chapter an example is presented implementing the PLSR prediction model based on alginate IR spectra. This analysis reduces the time of analysis of the M/G ratio from several hours to a few minutes per sample, which facilitates much more frequent quality control (QC). The drawback of this type of rapid PLSR model is that it may not be universal to other types of alginate samples than those included in the calibration set. Under normal conditions the model covers only the M/G-ratio range included for the dependent variable and the alginate types included in the calibrations. If raw material samples of alginate are not included in the model these samples cannot be expected to be predicted by a final product QC model due to the presence of other polysaccharides and protein in the raw material. This is not a major problem as such PLSR models can be extended with new samples on a later stage. In other words, the PLSR model sometimes can, but cannot be expected to, extrapolate outside the variations included in the calibration sample set.

2 Materials

Below the preparation of the most important chemicals is described. Please make sure to use the type of water (Milli-Q, D_2O, etc.) indicated as this is really important for liquid-state NMR samples.

2.1 Partial Hydrolysis

1. 96 % ethanol.
2. 0.1 M HCl: slowly add 20.5 mL of 37 % HCl to 100 mL of Milli-Q water. Adjust the final volume of solution to 250 mL with Milli-Q water (1 M HCl). Dilute 10 mL of this 1 M HCl solution to 100 mL with Milli-Q water.
3. 0.1 M NaOH: weigh out 10 g of NaOH pellets and dissolve in 100 mL of Milli-Q water. Adjust the final volume of solution to 250 mL with Milli-Q water (1 M NaOH). Dilute 10 mL of this 1 M NaOH solution to 100 mL with Milli-Q water.
4. Deuterium oxide (D_2O; >99 % D).
5. Deuterium oxide 99.9 % (D_2O).
6. 0.1 % 3-(trimethylsilyl)-propionic-2,2,3,3-d4 acid sodium salt (TSP-d4). Dissolve 10 mg TSP-d4 in 10 mL of 99.9 % D_2O.
7. 1.0mg/mLTTHA(triethylenetetramine-*N,N,N′,N″,N‴,N‴*-hexaacetic acid): weigh out 25 mg TTHA in a 25 volumetric flask, dissolve TTHA in D_2O, and fill to the mark.
8. Heat-resistant glassware for hydrolysis incl. flasks.
9. Reflux condenser.
10. 5 mm (outer diameter) NMR tubes.
11. Freeze-drying equipment or other equivalent drying equipment.

2.2 NMR Spectroscopy

1. NMR spectrometer equipped with a liquid-state probe having at least a proton channel and deuterium lock channel. The probe should be temperature controlled, having an upper limit of at least 90 °C. The magnetic field strength should be at least 4.7 T (corresponding to a proton resonance frequency of 200 MHz), but preferred above 9.4 T (proton resonance frequency above 400 MHz).
2. NMR spectrometer with a magnetic field of 7.1–16.4 T and a double-resonance probe equipped for 4 mm (o.d.) rotors to acquire ^{13}C CP-MAS NMR spectra.

2.3 Vibrational Spectroscopy (Example: IR Spectroscopy)

1. FT-IR (Fourier Transform InfraRed) Spectrometer equipped with an ATR (Attenuated Total Reflectance) unit (FT-IR-ATR).

2.4 PLSR Model Building and Multivariate Curve Resolution (MCR)

1. Software for NMR, IR spectral analysis, integration, deconvolution (e.g., typical instrument vendor software) and software for multivariate data analysis/chemometrics (like Unscambler (www.camo.com), SIMCA (www.umetrics.com), LatentiX (www.latentix.com), PLS-toolbox (www.eigenvector.com).

3 Methods

The method described here is the optimal method to obtain quantitative data through ^1H NMR (liq) of partial hydrolysed sodium alginate samples and is a condensation of best practises from relevant literature [2–4, 10, 13, 14].

3.1 Partial Hydrolysis (See Note 1)

The partial hydrolysis must be performed in a well-ventilated fume-hood at room temperature. Due to handling of acid/base and boiling water protective eyewear and suitable gloves must be worn. The procedure can be scaled if needed.

1. Weigh out 100 mg of Alginate powder to prepare a 0.1 % solution, wet the Alginate powder with 1–2 mL ethanol (96 %) to break the surface tension and add 100 mL of Milli-Q water while stirring (*see* **Note 2**).
2. Adjust the pH of the solution to 5.6 using HCl (1 and 0.1 M, *see* **Note 3**) and reflux the mixture at 100 °C for 1 h while stirring.
3. After cooling to room-temperature, adjust the pH to 3.8 with HCl (1 and 0.1 M, *see* **Note 2**).
4. Reflux the sample at 100 °C for not more than 30 min and stop the hydrolysis by cooling on ice.
5. Place the sample on a magnetic stirrer at ambient temperature and adjust the pH to 7–8 using NaOH (1 and 0.1 M, *see* **Note 3**).
6. Freeze-dry the sample overnight and redissolve it in 5 mL of D_2O followed by freeze-drying to remove residual H_2O/HDO from the sample.
7. Dissolve 10 mg of the sample in 1 mL 99.9 % D_2O (with 0.1 % TSP-d4) and pipette 700 μL hereof to a 5 mm NMR tube.
8. Add four drops of the TTHA chelator solution (*see* **Note 4**) to the NMR tube, close the cap of the tube and mix by shaking the NMR tube.

3.2 NMR Spectroscopy

3.2.1 ^1H Liquid-State NMR Spectroscopy

1. Set the temperature of the probe to 353 K and wait until the temperature is stable.
2. Insert the NMR tube containing the partly hydrolysed sodium alginate (Subheading 3.1) into the NMR probe and wait for 10 min to obtain temperature equilibrium.
3. Select a single pulse experiment, set the relaxation delay to 2 s, the lock solvent to D_2O, the spectral width to 15 ppm, the acquisition time to 4 s and the number of scans to 64 or 128.
4. Optimize the receiver gain and record a spectrum.
5. In the subsequent Fourier transformation use no zero filling and an exponential line broadening of 1.0 Hz.

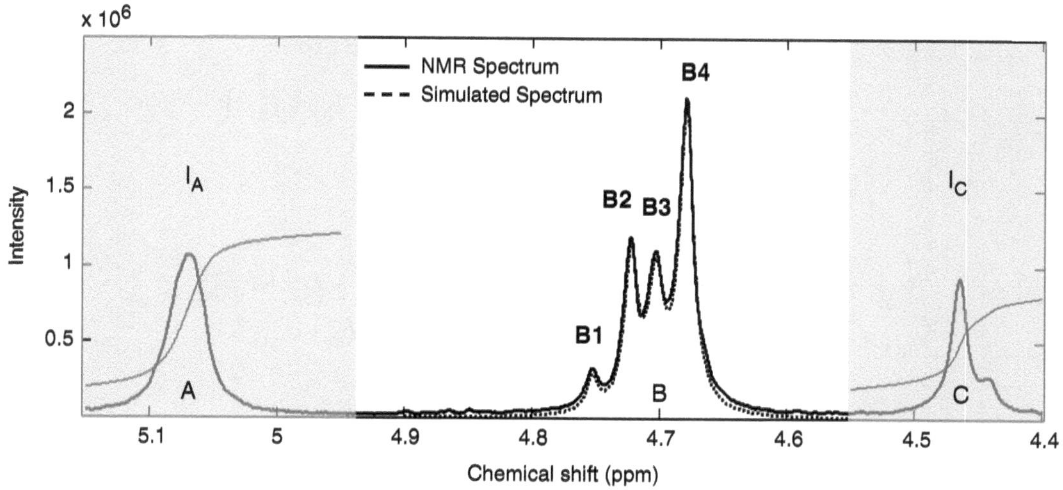

Fig. 2 Experimental (*solid line*) and deconvoluted (*broken line*) liquid-state ^{1}H NMR spectrum of the spectral region 4.40–5.15 ppm for a sample with M/G-ratio of 1.6 recorded at 14.1 T (^{1}H Larmor frequency: 600 MHz). The peaks A (*left region*) and C (*right region*) are integrated (I_A and I_C respectively) while integrals for peaks B1, B2, B3, and B4 are obtained by deconvolution (*dotted line*)

6. Perform automatic phase correction and base line correction. Check the result by visual inspection and improve if needed.
7. Carefully reference the TSP-d4 resonance to 0.000 ppm.
8. Integrate the following regions: A (4.96–5.18 ppm), B (4.57–4.82 ppm), and C (4.38–4.55 ppm) to obtain the corresponding integrals: I_A, I_B, and I_C (Fig. 2a) or go to next **step 9**.
9. If the instrumentation and spectral resolution allows deconvolution of peaks B1–B4 this can be performed to replace I_B, hereby obtaining diade, triade information as shown in Fig. 2b and formulas below. Deconvolution of NMR peaks can be performed in most NMR software packages. Please consult the manual for your specific software. Typical parameters for deconvolution are a chemical shift range (careful chemical shift referencing is mandatory), peak picking (peak maxima), line-width and expected peak shape of the simulated spectrum (Lorentizian, Gaussian, or a mixture).
10. Calculations of M/G-ratio and Diade/Triade block information from the liquid-state ^{1}H NMR spectra:

    If the spectral conditions do not allow for resolution of resonances B1, B2, B3, and B4 the calculation of the M/G ratio is performed by integration of regions A, B, and C (*see* Fig. 2a):

$$M/G = (I_B + I_C - I_A)/(I_A)$$

    If the spectral resolution allows the integration of peaks A and C and deconvolution of integrals of resonances B1–B4 (*see* Fig. 2b) the following relations are used to calculate the M/G-ratio, diade and triade fractions:

$$G = 0.5 \times (I_A + I_C + 0.5 \times (I_{B1} + I_{B2} + I_{B3}))$$
$$M = I_{B4} + 0.5 \times (I_{B1} + I_{B2} + I_{B3})$$
$$M / G \text{ ratio} = M / G$$

$$GG = 0.5 \times (I_A + I_C - 0.5 \times (I_{B1} + I_{B2} + I_{B3}))$$
$$MG = GM = 0.5 \times (I_{B1} + I_{B2} + I_{B3})$$
$$MM = I_{B4}$$

$$GGM = MGG = I_{B1} \times 0.5 \times (I_{B1} + I_{B2} + I_{B3}) / (I_{B1} + I_{B2})$$
$$MMG = I_{B2} \times 0.5 \times (I_{B1} + I_{B2} + I_{B3}) / (I_{B1} + I_{B2})$$
$$GGG = GG - GGM$$

Sequence fraction information can then be calculated as follows:

$$F_G = G / (M + G); F_M = M / (M + G)$$
$$F_{GG} = GG / (M + G); F_{MM} = MM / (M + G); F_{GM} = F_{MG} = MG / (M + G)$$
$$FGGG = GGG / (M + G); FMGM = MGM / (M + G); FGGM = FMGG = GGM / (M + G)$$

*Solid-State ^{13}C CP MAS NMR Spectroscopy (See **Note 5**)*

The procedure for recording ^{13}C CP MAS NMR spectra is described for spectrometers of magnetic field of 7.1–16.4 T using a double-resonance probe equipped for 4 mm (o.d.) rotors. At higher fields different spin-rates and higher rf-field strengths may be required to avoid overlap between isotropic resonances and spinning side bands and to obtain efficient cross-polarization and ^{1}H decoupling. Before running these experiments the magic angle should be carefully adjusted—e.g., by a ^{79}Br MAS NMR experiment on KBr.

1. Fill the rotor with alginate powder and insert the end-cap.
2. Insert the rotor into the probe and start spinning using a spin-rate of 14 kHz. Please check that the spinning sidebands from the carbonyl carbon are not located in the spectral region 60–110 ppm.
3. Select a CP-MAS experiment and set the recycle delay to 2 s, the contact time to 2 ms, the spectral width to 500 ppm, the acquisition time to 10 ms and the number of scans to 1,024. High power ^{1}H decoupling using either TPPM or SPINAL-64 at an rf-field strength of at least 62.5 kHz should be employed. Likewise both ^{1}H and ^{13}C rf-field strengths should be at least 62.5 kHz during cross polarization.
4. Optimize receiver gain and record a FID.

5. Prior to Fourier transformation zero filling to twice the number of acquired data points should be performed. Apply no line broadening.
6. Perform automatic phase and baseline correction. Visually inspect the spectrum and perform manual phase or baseline correction if necessary.

3.3 Vibrational Spectroscopy (Example: IR Spectroscopy)

Samples for vibrational spectroscopy are alginate powder used without any preparation (*see* **Note 5**). By using a diamond attenuated total reflectance (ATR) accessory, solid state sampling is very fast and reproducible, but gives less intense absorption intensity (few bounces in the ATR) than can be achieved by using classical transmission spectroscopy based on KBr pellet technology. The method reported here is described for a FT-IR spectrometer equipped with a single bounce diamond ATR accessory [14]. Representative FT-ATR-IR alginate spectra are shown in Fig. 3.

1. Clean and dry the diamond ATR crystal and record a background spectrum with no sample.
2. Place 1–2 mg of sample on the ATR crystal—just enough to cover the crystal surface. Apply pressure (by the pressure arm) to obtain maximal contact (highest absorbance) and record an IR spectrum using a resolution of 4 cm^{-1} and 16 scans in the range 4,000–650 cm^{-1} (*see* **Note 6**).

3.4 PLSR Model Building

The calibration models for the determination of the M/G-ratio are obtained by the same procedure for all the vibrational spectroscopic methods [14]. Only the procedure described for the IR data is described here.

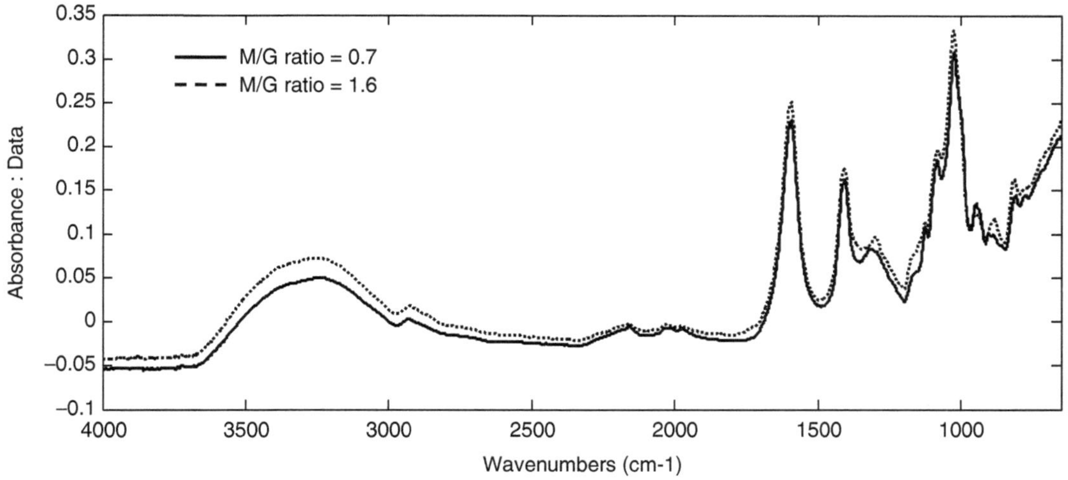

Fig. 3 IR spectra of sodium alginate samples with M/G ratios of 0.7 (*solid line*) and 1.6 (*broken line*), respectively

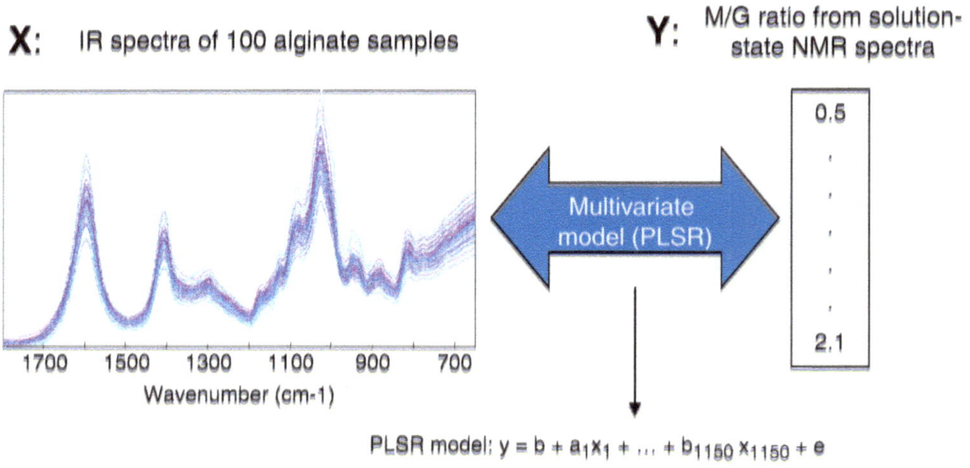

Fig. 4 Scheme demonstrating the correlation between the 1,150 wavenumbers (1,800–650 cm^{-1}) of the IR spectra of 100 samples of sodium alginate (X-matrix) with 100 M/G ratios (y-vector) as determined by ^1H liquid-state NMR

The chemometric calibration model is built from a sample-set from which both IR spectra are recorded (denoted variable) and a reference M/G ratio measured by ^1H NMR (liq.) calculated as described in Subheading 3.2.1 (dependent variable denoted the *y*-variable) (Fig. 4). The IR spectra will be contained in a matrix, **X**, with the dimensions *samples*× *wavenumbers* and the reference values from NMR will be contained in a vector with the dimension *samples*× 1. Any chemometric software and some instrument vendor software packages can handle the procedure described here.

The sample-set has to be chosen to cover all the variation (season, batch, etc.) that the final prediction model is expected to cover in a high throughput setup. The number of samples in a minimal setup is typically about 100. From the data-set a subset of samples (approx. 25 % of the total) is selected as a validation test-set. The remaining data-set is used as a calibration-set for building the calibration model. After optimization the test-set is used to validate the model. The validation test-set should cover approximately the same M/G range, though selected randomly. True triplicate (or more) measurements are preferred to cover NMR integration variation as well as variations due to a small sampling size from one large sodium alginate batch. This critical sampling issue will not be further addressed here.

It is recommended to inspect the spectral calibration-set using unsupervised models like Principal Component Analysis (PCA) [16]. The model is composed of a number of Principal Components (PCs) with associated scores and loadings. Normally, few PCs are required to explain the entire systematic variation in the spectral data-set—in this case the number of PCs used was less than 6 for all models (NMR, Raman, NIR, and IR) and 4 for the IR model [14]. Inspection of the PCA-model scores-plot is done to reveal outliers or subclasses within the data-set (*see* **Note 7**) and furthermore to test different preprocessing tools for the IR spectra (**X**-variables). When using the ATR sampling device, preprocessing (second derivatives or multiplicative scatter correction (MSC) [17]; *see* **Note 8**) of the data is normally used in order to remove scatter effects and hence optimize the PCA model towards the relevant chemical information. Inspection of the PCA-model loadings can reveal which of the **X**-variables contributes to any observed classification/outlier.

Beware when transferring the independent variable, i.e., the M/G ratio into the chemometric software that alignment of each *y*-variable must be correctly aligned with the sample ID as loaded with the spectral **X**-variables. The response *y*-variable should be mean centered prior to PLSR modeling. A supervised PLSR-model is built using the full spectral range and evaluated using the validation test-set. For model comparisons a plot of ***predicted*** *y* versus ***measured*** *y* as well as the parameters r^2 (model linearity) and RMSEP (Root Mean Squared Error of Prediction) are used to evaluate the model (Fig. 5). For a parsimonious and robust model, the number of PCs and the prediction error should be as low as possible explaining as much of the variation as possible with the squared correlation coefficient, r^2, as close to 1.00 as possible. The prediction model can be optimized using variable selection. This can optimally be performed by using interval based PLSR (as iPLS; [18]) for

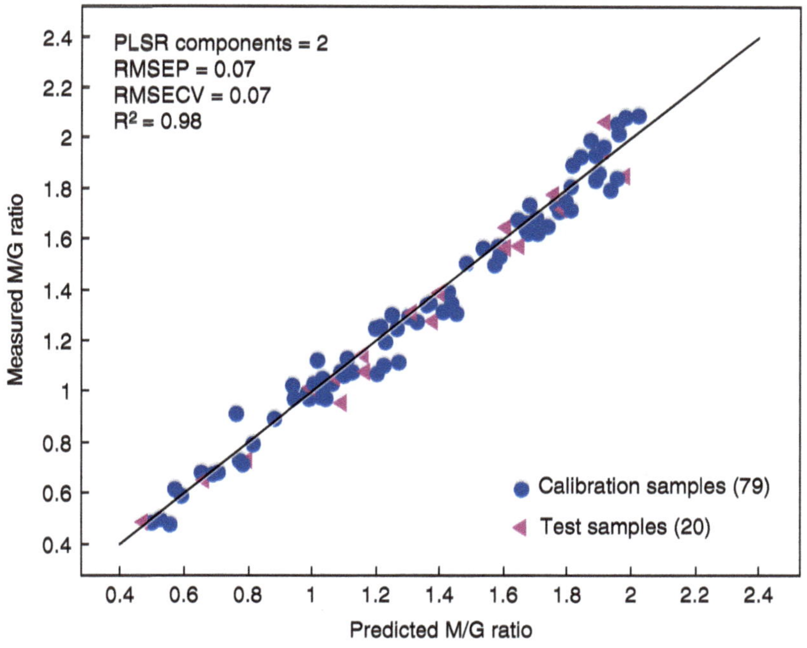

Fig. 5 Result of the PLSR prediction model preprocessed using second derivative and MSC corrected spectra. The model uses 79 samples for calibration and 20 samples for test set validation. Two PLS components are required to describe 98 % of the **y**-variance and the prediction error is 0.07 M/G-ratio. RMSEP: Root Mean Squared Error of Prediction; RMSECV: Root Mean Squared Error of Cross Validation

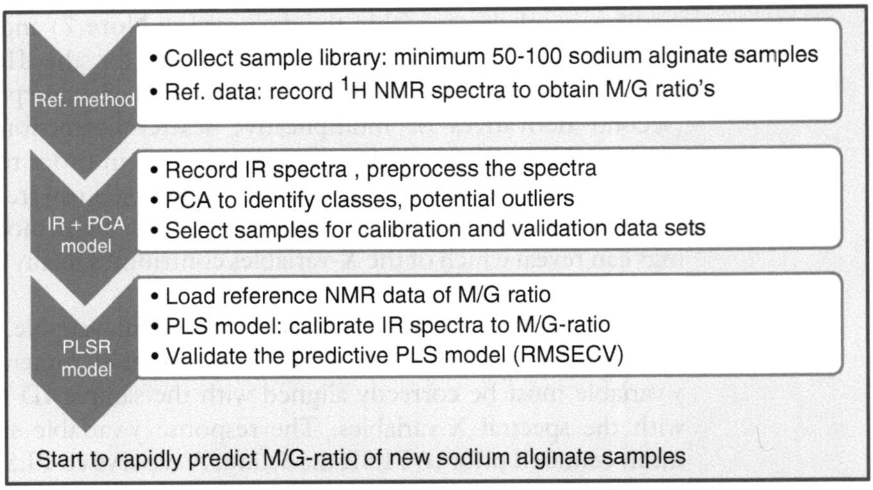

Fig. 6 Summary of the workflow for building a PLSR model of sodium alginate based on IR spectra and the M/G ratio determined by ^{1}H liquid-state NMR

narrowing the spectral range with the highest correlation to the M/G-ratio. In case of the IR spectra to M/G ratio the optimum was found to be the broad spectral range of 650–1,800 cm^{-1}.

In Fig. 6 is depicted a summary of the PLSR-model building workflow.

Maintenance of the PLSR prediction model is the last step to be mentioned in this section. Following the implementation of a PLSR-model in the laboratory or factory it is important to consider the following issues:

- Check the sampling method and the instrument performance by frequent use of reference samples with known *y*-values.
- Implement a warning method for outlying samples that do not fit the model well or is *y*-extremes. Outlying samples are important because they contain variation not observed in the calibration samples: highly impure samples, new biological origin samples, fraud samples, etc. (often a sample-to-model distance measure—Mahalanobis or similar—is an integral part of the chemometric software).
- If new sample types are to be included in the model: test if they are outliers in the existing model and, if they are, then extend the PLSR model by including a larger set of the new sample types. Revalidate the quality of the new PLSR model.

3.5 MCR/Multivariate Curve Resolution

The ^{13}C CP-MAS spectra of alginates are shown in Fig. 7 with a number of well-defined resonances based on liquid-state ^{13}C NMR assignments [10]. Due to overlapping resonances in integration are ruled out for quantification of individual carbon sites and attempts

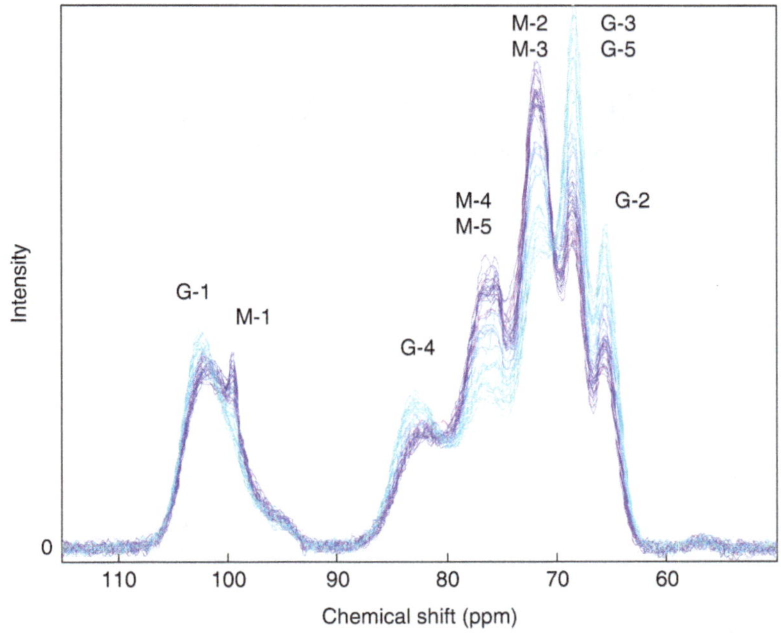

Fig. 7 Solid state ^{13}C CP-MAS NMR spectra of alginates recorded at 11.7 T

to deconvolve the resonances using a standard deconvolution approach resulted unambiguously in strong biases and nonlinear effects. However, the unsupervised multivariate method Multivariate Curve Resolution (MCR) [19, 20] was able to provide a deconvolution of the alginate spectra into just two components: a poly-mannuronic like spectrum and a poly-guluronic like spectrum [13].

MCR is similar to PCA, but without the PCA-constraint that the PCs must be orthogonal. In the ideal case MCR can provide loadings which are spectra of the pure components in the mixture and related scores which are good estimations of the concentrations of the mixture components. The MCR model is shown in Fig. 8. Just like in the case of PCA and PLSR, only a few components should be required in order to make a good parsimonious multivariate model. In contrast to PCA, MCR requires a guess for S (pure spectra) in order to initiate the algorithm. MCR is in practise often implemented using Alternating Least Squares (ALS), which is slowly converging; therefore it is often implemented with non-negativity constraints on both the concentrations and the spectra when applied to spectroscopic data (*see* **Note 9**).

Beware that the MCR solutions are generally not unique (more than one solution) which is due to the so-called rotational ambiguity. Hence, for any specific MCR model, uniqueness must be assessed before the solution can be assumed to be providing estimates of real chemical analytes [20].

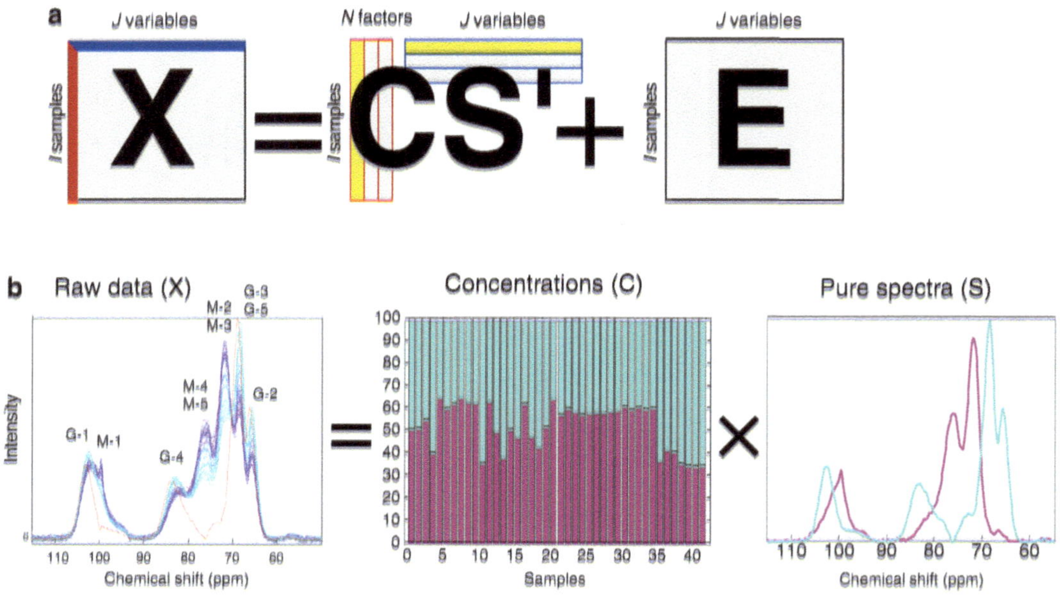

Fig. 8 Multivariate curve resolution ^{13}C CP/MAS spectra of alginate. (**a**) The MCR model and (**b**) the application to ^{13}C CP/MAS spectra of alginate

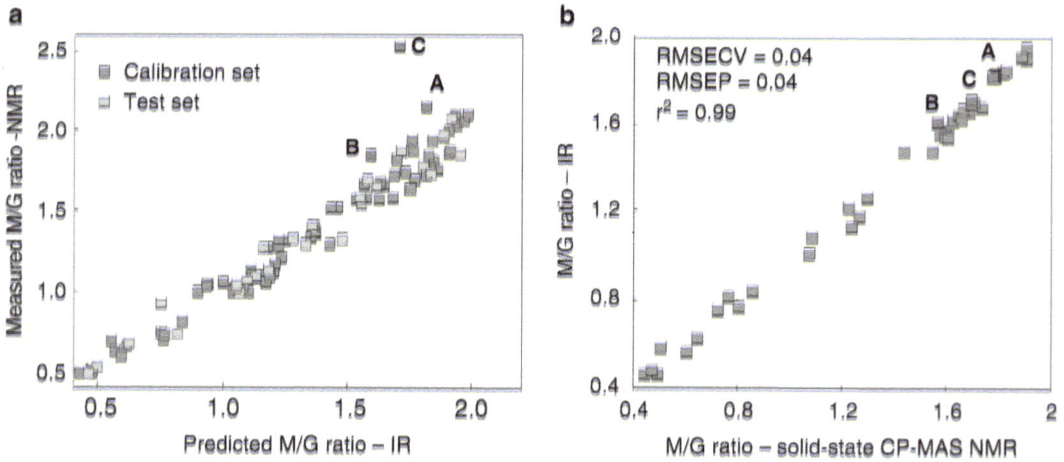

Fig. 9 The prediction performance of (**a**) the IR model vs. the ^{1}H liquid-state NMR reference measurement and (**b**) the solid-state ^{13}C CP/MAS NMR method vs. the IR model. Three outlier samples with high calcium contents (w/w %: **a**: 1.1, **b**: 1.4, and **c**: 2.4) are not well predicted by the ^{1}H liquid-state NMR method but the two solid-state methods (IR and ^{13}C CP-MAS NMR) are in perfect agreement

However, in the case of ^{13}C CP-MAS spectra it was demonstrated that the two component MCR generated scores could be directly translated into the M/G ratio of the investigated alginates [13]. Figure 9 shows the result of the M/G prediction from the MCR model compared to the prediction performance of the IR model and it shows an almost perfect match ($r^2 = 0.99$). The two solid state methods compare well which is not always the case when they are compared to the solution NMR method. In cases of high calcium content the solution state reference method underestimate the G content because the junction zone guluronic egg-box fragments become rigid and thus not visible to the solution NMR methods. This can, and should, normally be overcome by using a chelating agent in the reference method, but for the solid state methods high calcium contents represents no problem.

4 Notes

1. Typically six samples can be hydrolysed per day per person.
2. The use of a heat-resistant pyrex conical flasks and a hot plate with magnetic stirring is recommended as is the use of reflux-condensers during the reflux period.
3. Start with the 1 M solution. When the pH is close to 5.6 adjust with the 0.1 M solution. Notice that the solution reacts very slowly.
4. The chelating agent THAA is used to react with Calcium ions in the sodium alginate in order to minimize line-broading due to network formation between. EDTA is less appropriate due chemical shift overlap with the protons under investigation.
5. Examples of applications can be found in refs. 11–13.
6. The particle size may have an influence on the intensity of the spectrum obtained. The intensity of the spectrum also depends on how much sample and pressure (by the pressure arm) that is applied. Alternative the sample can be milled and sieved to obtained uniform particle size of the samples. Multiplicative scatter correction (MSC) might be able to compensate for any observed drift of the baseline. This has to be evaluated in a pretest after evaluation of the range of particle size in the sample set at hand.
7. The reason for outlying samples could typical due to impurities not identified during the partial hydrolysis for NMR or a sodium alginate powder with a particle size larger than the other samples. Large particle size might give rise to low IR absorbance due to decrease in the contact surface with the ATR accessory.
8. Preprocessing of the IR spectra can be performed batch-wise in the chemometric software packages as well as in some instrument vendor software. Second derivative spectra, Multiplicative Scatter Correction (MSC) and Standard Normal Variate (SNV) are improving baseline differences. In ref. [14] Extended Inverted Scatter Correction (EISC) or Extended

Multiplicative Scatter Correction [21] was evaluated as the most promising method. If these methods are available in the chemometric software package this is recommended, otherwise MSC/SNV can be an excellent choice [17].

9. The MCR algorithm can be obtained from academic sources (http://www.mcrals.info/) and is implemented in several of the aforementioned commercial chemometrics software packages.

Acknowledgements

The work has been supported by a grant to Tina Salomonsen for an Industrial Ph.D. and by a grant from the strategic research council to the MicroPAT project under the InSPIRe (Danish Industry-Science Partnership for Innovation and Research in Food Science) consortium. The Faculty of Science and The Ministry of Science and Technology is acknowledged for a grant to the NMR metabolomics infrastructure.

References

1. Aarstad O, Strand BL, Klepp-Andersen LM et al (2013) Analysis of G-block distributions and their impact on gel properties of in vitro epimerized mannuronan. Biomacromolecules 14:3409–3416. doi:10.1021/bm400658k
2. ASTM (2012) Active standard ASTM F2259, p 5
3. Grasdalen H (1983) High-field, H-1-NMR spectroscopy of alginate – sequential structure and linkage conformations. Carbohyd Res 118:255–260. doi:10.1016/0008-6215(83)88053-7
4. Grasdalen H, Larsen B, Smidsrod O (1979) NMR study of the composition and sequence of uronate residues in alginates. Carbohyd Res 68:23–31. doi:10.1016/s0008-6215(00)84051-3
5. Salomonsen T, Jensen HM, Larsen FH et al (2009) The quantitative impact of water suppression on NMR spectra for compositional analysis of alginates. In: Guðjónsdóttir M, Belton P, Webb G (eds) Magnetic resonance in food science. Royal Society of Chemistry, London, pp 12–19. doi:10.1039/9781847559494-00012
6. Vilen EM, Klinger M, Sandstrom C (2011) Application of diffusion-edited NMR spectroscopy for selective suppression of water signal in the determination of monomer composition in alginates. Magn Reson Chem 49:584–591. doi:10.1002/mrc.2789
7. Chhatbar M, Meena R, Prasad K et al (2009) Microwave assisted rapid method for hydrolysis of sodium alginate for M/G ratio determination. Carbohyd Polym 76:650–656. doi:10.1016/j.carbpol.2008.11.033
8. Li LY, Jiang XL, Guan HS et al (2011) Preparation, purification and characterization of alginate oligosaccharides degraded by alginate lyase from *Pseudomonas* sp. HZJ 216. Carbohyd Res 346:794–800. doi:10.1016/j.carres.2011.01.023
9. Lundqvist LCE, Jam M, Barbeyron T et al (2012) Substrate specificity of the recombinant alginate lyase from the marine bacteria *Pseudomonas alginovora*. Carbohyd Res 352:44–50. doi:10.1016/j.carres.2012.02.014
10. Grasdalen H, Larsen B, Smidsrod O (1981) C-13-NMR studies of monomeric composition and sequence in alginate. Carbohyd Res 89: 179–191. doi:10.1016/s0008-6215(00)85243-x
11. Sperger DM, Fu S, Block LH et al (2011) Analysis of composition, molecular weight, and water content variations in sodium alginate using solid-state NMR spectroscopy. J Pharm Sci 100:3441–3452. doi:10.1002/jps.22559
12. Salomonsen T, Jensen HM, Larsen FH et al (2009) Alginate monomer composition studied by solution- and solid-state NMR – a comparative chemometric study. Food Hydrocolloid 23:1579–1586. doi:10.1016/j.foodhyd.2008.11.009

13. Salomonsen T, Jensen HM, Larsen FH et al (2009) Direct quantification of M/G ratio from ^{13}C CP-MAS NMR spectra of alginate powders by multivariate curve resolution. Carbohyd Res 344:2014–2022. doi:10.1016/j.carres.2009.06.025
14. Salomonsen T, Jensen HM, Stenbaek D et al (2008) Chemometric prediction of alginate monomer composition: a comparative spectroscopic study using IR, Raman, NIR and NMR. Carbohyd Polym 72:730–739. doi:10.1016/j.carbpol.2007.10.022
15. Wold S, Martens H, Wold H (1983) The multivariate calibration-problem in chemistry solved by the PLS method. Lect Notes Math 973:286–293
16. Hotelling H (1933) Analysis of a complex of statistical variables into principal components. J Educ Psychol 24:417–441
17. Rinnan A, van den Berg F, Engelsen SB (2009) Review of the most common pre-processing techniques for near-infrared spectra. Trend Anal Chem 28:1201–1222. doi:10.1016/j.trac.2009.07.007
18. Nørgaard L, Saudland A, Wagner J et al (2000) Interval partial least squares regression (i PLS): a comparative chemometric study with an example from the near infrared spectroscopy. Appl Spectrosc 54:413–419. doi:10.1366/0003702001949500
19. Lawton WH, Sylvestre EA (1971) Self modeling curve resolution. Technometrics 13:617–633. doi:10.2307/1267173
20. Engelsen SB, Savorani F, Rasmussen MA (2013) Chemometric exploration of quantitative NMR data. eMagRes 2:267–278. doi:10.1002/9780470034590.emrstm1304
21. Martens H, Nielsen JP, Engelsen SB (2003) Light scattering and light absorbance separated by extended multiplicative signal correction. Application to near-infrared transmission analysis of powder mixtures. Anal Chem 75:394–404. doi:10.1021/ac020194w

Chapter 23

Imaging and Identification of Marine Algal Bioactive Compounds by Surface Enhanced Raman Spectroscopy (SERS)

Mats Josefson, Alexandra Walsh, and Katarina Abrahamsson

Abstract

Surface-enhanced Raman spectroscopy (SERS) is a highly selective technique that can be used for imaging of single algae cells. In contrast to normal Raman spectroscopy, SERS utilizes light interaction with colloidal gold or silver particles working as antennas to match the sensitivity of fluorescence measurements. Furthermore, SERS enables a more profound picture of not only the analyte of interest but also the present biological matrix without the need for additional fluorescence labelling. The introduction of an internal standard in the form of a thiol self-assembled monolayer (SAM) on the colloidal gold or silver particles can be used to normalize the SERS response that otherwise would also depend on the locations of the colloid particles in the microscope image.

In light of the vast amounts of data that is generated in each spectrum and the large variance in enhancement signal, multivariate analysis is necessary for accurate evaluation. This can be done by the use of transposed orthogonal projections to latent structures (T-OPLS), where the variations of properties in the reference spectra, *Y* table, and the variation in spectra, *X* table, are correlated.

Key words Macroalgae, Microalgae, Resonance Raman spectroscopy, SAM, SERS, T-OPLS

1 Introduction

Chemical imaging with Raman spectroscopy can be used to identify and determine the spatial distribution of individual compounds. Optical spectroscopic imaging techniques balance the qualitative, quantitative, and spatial information gained of compounds in a sample. Fluorescence and Raman imaging are by far the most common techniques, where fluorescence imaging has been superior in sensitivity and quantitative information gained, but lacks selectivity and the ability to positively identify single compounds. Thus, the role of Raman spectroscopy in imaging has so far mostly been limited to solid-state analysis and studies of concentrated samples. Recently, the technique has been applied to biological systems such

Dagmar B. Stengel and Solène Connan (eds.), *Natural Products From Marine Algae: Methods and Protocols*, Methods in Molecular Biology, vol. 1308, DOI 10.1007/978-1-4939-2684-8_23, © Springer Science+Business Media New York 2015

as studies of the distribution of biomolecules on algal surfaces and within single algal cells [1–3].

The introduction of surface enhanced Raman spectroscopy (SERS) has cancelled out the difference in sensitivity between fluorescence and Raman spectroscopy, but retains the innate ability of the latter to identify single compounds. As a bonus, Raman spectroscopy not only gives information about the identity of compounds but also about their local environment, which can be used to study interaction between the compounds and their environment.

The major difficulties with SERS originate from the lack of control over the nanoscale surface of the substrate used for enhancement, interaction between substrate and environment, and also interactions between colloids when colloidal substrates are used [4, 5]. The variation in enhancement resulting from these processes causes the signal obtained to be more related to the amount of enhancement than concentration of the compound studied. On a macroscale this variation is less of a problem, since the average enhancement is quite stable but on microscale or nanoscale mechanisms for compensation of the enhancement are necessary. Labelling of substrates by internal standards, such as the thiol self-assembled monolayers (SAMs), is the most versatile of the approaches to control surface enhancement [4, 5].

Multivariate evaluation of data is a necessity due to the amount of information available in each spectrum and image in combination with low signal to noise levels, but this brings its own set of challenges. The large variation in surface enhancement inherent in SERS combined with internal standard to control enhancement results in a ratiometric multivariate model, which is sensitive to variation in the spectral background. Surface enhanced background and interfering processes, such as fluorescence, are easily handled on a spectrum-to-spectrum basis when each solution can be tweaked but are considerably more difficult when the evaluation is automated.

One way to address this is orthogonal partial least squares (OPLS). Standard OPLS is a mathematical projection method for one X and one Y table. OPLS delivers the projection of the largest variation in X that is correlated to the variation in Y. Standard OPLS is used by placing measured spectra on rows in the X-table. The collective profile of one spectrum is also called an observation. The Y-table will contain properties paired with each observation. Examples of these properties are concentrations of the chemical components present in the paired spectrum. In other words, an OPLS model is a multivariate regression model where the variation of the properties in Y is correlated with the multivariate spectral variation in X.

The main feature of OPLS is that it provides the projection in two parts: the predictive variation p that is correlated with Y and the orthogonal variation po that is present above the noise level but is uncorrelated to Y. To use standard OPLS, we would need more than one Raman spectrum with known concentrations to build a multivariate model. This is not practically possible for biological matrices in general. However, with transposed orthogonal projections to latent structures (T-OPLS) it is possible to utilize the two part separation to describe spectral variation of known reference spectra in the first predictive part (*p*) and the remaining variation of spectra without separate reference in the second orthogonal part (*po*).

In T-OPLS, the standard OPLS algorithm [6] is used, but spectral data is arranged column wise. The result of this arrangement is that content of loadings and scores has switched compared to normal OPLS. The predictive loadings now contain relative measures for the amount of compound present for each reference spectrum. The scores contain spectral profiles as interpreted by the model. In SERS with internal standard there will be at least two references, one for the analyte and one for the internal standard, and there will be two sets of predictive loadings p_1 and p_2. T-OPLS has been used to study the distribution of carotenoids in algae and to estimate the distribution of a biologically active compound (bromoheptanone) on the surface of a read alga [2, 3].

With a reference spectrum of the compound of interest it is possible to use T-OPLS to map the correlation of only the specified reference or set of references even if the sample contain other major spectral variation. A simpler example without SERS and internal standard is shown in Fig. 1a where the carotenoid distribution of *Dunaliella* is shown from a hyperspectral resonance Raman image after the processing by T-OPLS. Additionally it is possible to map other chemical compounds not present as a reference in subsequent images that are separated from the referenced compound (Fig. 1b, c).

T-OPLS can be used to put the question: how much of the pattern in this reference spectrum is present in a set of spectra acquired from one or more hyperspectral Raman images?

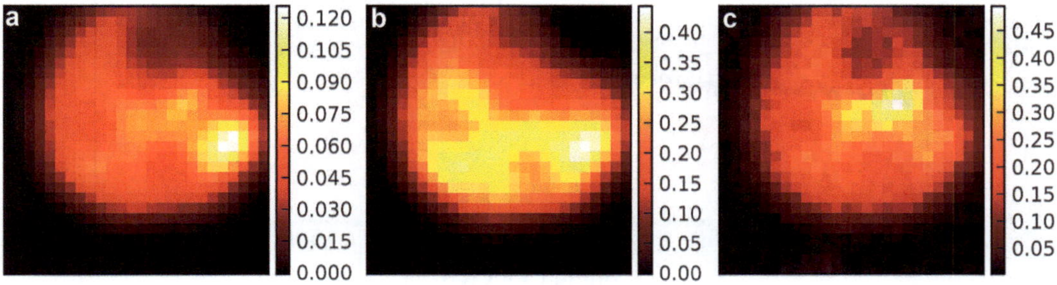

Fig. 1 Reconstructed images of carotenoid variation in *Dunaliella* from a T-OPLS treated hyperspectral Raman image. (**a**) The predictive loading correlating with beta-carotene, (**b**) the first orthogonal loading describing one or more carotenes not correlating with beta-carotene, (**c**) the second orthogonal loading describing another variation around 1,520 cm^{-1} indicating the presence of a third slightly dissimilar Raman carotenoid profile

In this example, the purpose of T-OPLS is to map or correlate all image spectra with one reference spectrum for the analyte. There is no requirement to have spectra of all chemical components in the image, but the spectral response should be linear. Spectral data should contain non-correlated variations of the amount of chemical components in the acquired spectra to yield a good result. The natural random variation in hyperspectral images is usually sufficient to provide non-correlated spectral variations. The OPLS model fails to separate chemical components if they have the same proportions in every part of the image.

The T-OPLS t_1 and t_2 contains the combined spectral features of the analyte and the internal standard, if present. These profiles should be similar to the references and can be used to verify the OPLS model. The score spectra t_1 and to_1 for the *Dunaliella* T-OPLS model are shown in Fig. 2. The t_1 contains a reproduction of the reference spectra from model data. The to_1 contains the joint spectrum of other carotenoids, not present in the reference but still present in the alga *Dunaliella*. The main difference is small but can be seen as a shift in the largest peak just above 1,500 cm^{-1}.

In this chapter we present one example of how to evaluate resonance Raman spectra with T-OPLS in order to identify the distribution of natural products in complex matrices such as algae. In the end of the chapter, the authors have listed notes that account for their personal experiences surrounding experimental setup and multivariate data analysis.

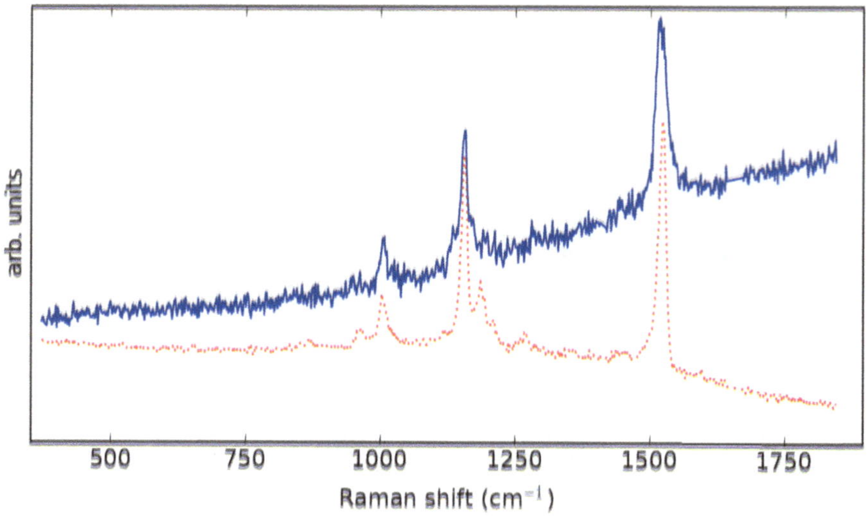

Fig. 2 Predictive (*solid line*, t_1) and first orthogonal (*dotted line*, to_1) score spectra in arbitrary units (offset and scaled to compare spectral peak positions). The score spectra are from the T-OPLS model of *Dunaliella* aimed to describe the variation of carotenoids from the hyperspectral Raman image. The main difference in the dominant peak is that the maximum is at 1,518 cm^{-1} for the predictive part and at 1,522 cm^{-1} for the orthogonal part

2 Materials

2.1 Colloid Coating Components

1. 10 mM MBN stock solution: weigh 1.35 mg 4-mercapto benzonitrile (MBN). Transfer to 15 mL Falcon tube and dissolve in 10 mL ethanol. Store at room temperature.
2. 1 mM MBN coating solution: dilute 1 mL of MBN stock solution with 10 mL ethanol. Store at room temperature in a Falcon tube.
3. Gold colloids (60 nm) suspended in water (BBI Solutions, UK) (*see* **Note 1**). Store at 6–8 °C.

2.2 Algal Preparation

1. Any algal sample.
2. Lab-Tek chamber slides. Select well size according to the dimensions of algal samples.
3. Glass coverslips.

2.3 Instrumentation

1. Spectra collection: confocal Raman spectrometer, e.g., Labram, with an inverted confocal microscope equipped with a 100× objective.
2. Detector: charged couple device (CCD).
3. Laser: 785 nm diode, power 40 mW (*see* **Notes 2** and **3**).

2.4 Data-Processing Software

1. Conversion tool: tool to convert from microscope hyperspectral image format to a general formats, such as HDF5 or MATLAB.
2. Multivariate analysis: SIMCA-P 13.0 (MKS Umetrics AB, Sweden) or alternative software that has OPLS.
3. Image presentation: routines from Matplotlib [7] were used with Python 2.7.

3 Method

Algae preparation and measurement procedures are to be carried out during the same day, at room temperature. Multivariate data analysis can be carried out at any point in time (*see* **Notes 4** and **5**).

3.1 Coating of Colloids

1. Add 150 μL of colloidal solution and 150 μL of MBN coating solution into each 1.5 mL Eppendorf tube and leave to coat for at least 48 h at 6–8 °C. Longer storage does not affect the quality of the self-assembled monolayer [8].
2. After the coating process, centrifuge the tube for 5 min at 2,040 × *g*. Remove carefully the supernatant and resuspend the pellet in 300 μL Milli-Q water. Repeat the centrifuging procedure. Wash thus three times. After the last washing, remove as much of supernatant as possible and preserve the colloidal pellet. The pellet should be used directly after washing.

3.2 Preparation of Algae for Measurement

1. Spread out algal tissue sample in a Lab-Tek glass chamber.
2. Resuspend the colloidal pellet in 50 μL Milli-Q to be spread over the algal sample, which then is covered by a coverslip to keep it fixated under the microscope [3].
3. Prepare at least one coverslip with algae as blank (no colloids) and one coverslip with non-coated colloids, and then optional amount of coverslips with MBN-coated colloids (*see* **Note 6**).

3.3 Instrument Parameters

1. Set the spectral range at 251–1,999 cm^{-1} (*see* **Note 7**).
2. Measure each spectrum three times with acquisition time of 3 s (*see* **Note 8**).
3. Scan the algae in *xy* plane in 1 μm steps and mean area of 10×10 μm^2. This created 2D maps of the sample area. The algae can also be mapped in layers in *z* axis with 1 μm separation between each layer [3].

3.4 Data Treatment and Transposed OPLS Calculations

1. Convert microscope hyperspectral data to a general format. For the exemplified equipment, in-house software to unfold hyperspectral TIFF images from the instrument was used. If spectra are acquired during many days, check that the wavenumber scales are consistent (*see* **Note 9**).
2. Unfold spectra, i.e., take converted spectra from the image and put them in a two-dimensional table. Index each spectrum with its image number and *x* and *y* coordinates for the original image pixel. If more than one physical layer is acquired from the sample, keep the image layer in the index as well. These operations can be carried out in MATLAB (The MathWorks, Inc., Natick, MA, USA) or in Microsoft Excel.
3. Remove spikes caused by cosmic rays from the spectra. Inspect each spectrum using MATLAB, Simca-P, or other plotting tool. Keep a log of all spike positions and then remove corresponding spectral values from the dataset, e.g., directly in the multivariate software (*see* **Note 10**).
4. Put the image spectra, reference spectrum for the analyte, and the internal standard reference spectrum for MBN as columns in the software for OPLS calculations (*see* **Note 11**).
5. Make a PCA model with centered but unscaled spectra to check the presence of outliers. Here additional spectra with undetected spikes from cosmic rays may turn up.
6. Set the image spectra as *x*-variables (columns) and the reference spectra as *y*-variables. Make an OPLS model with unit variance scaling for X and Y. Observe the score spectra (*see* **Note 12**). Are these similar to the reference spectra but maybe with a bit more noise? Then we are on the right track (*see* **Notes 13–15**).

7. Plot the predictive loadings for the analyte and the internal standard (MBN) as *Y*-related profiles [9].
8. With acceptable output from the T-OPLS modelling, take the predictive loadings for the analyte and the internal standard. With two references there will be two sets of predictive loadings p_1 and p_2, one for the analyte and one for the internal standard. The maximum readings of p_1 and p_2 indicate the maximum presence of the analyte and internal standard respectively.
9. Export the *Y*-related profiles and orthogonal loadings from the multivariate software for further calculations.
10. Scale the *Y*-related profiles p_1 and p_2 to be in the range 0–1 where the zero level is defined either by well-defined baselines or sample blanks (*see* **Note 16**). Divide the analyte signal from each pixel with the MBN signal from the corresponding pixel. In this way the irregular positioning of the colloid particles in the micro scale and the variations in colloid response depending on colloid aggregation will be accounted for.
11. Reconstruct the image resulting from the division using the indices for the image, image layer, and pixel coordinates. The resulting images show relative measures of the analyte over the mapped surface.

4 Notes

1. The size of the colloid spheres varies between 57 and 63 nm according to manufacturer [3].
2. Let the laser run at least 15 min before proceeding to measurements.
3. Other laser wavelength can be used as well, such as 514 nm but longer wavelength lasers, e.g., 785 nm, are often preferred if there are problems with fluorescence.
4. It is easy to get lost among all spectral files! When acquiring spectra, make sure to assign each measurement as detailed and unique file name as possible.
5. It is recommended to acquire adequate basic training in multivariate data analysis before attempting to finalize results based on the OPLS calculations.
6. Although the method we have chosen to present here is successful in accumulating colloidal matter inside living cells [10] in order to achieve enhancement, the colloids cannot penetrate the algal cell wall. Thus, all SERS enhancements that could be measured are produced from colloids on the surface of algae. However, another group has described a method for growing gold nanoparticles directly inside the algae [11], but this approach eliminates the use of self-assembly monolayers as internal standards.

7. Observe the amount of orthogonal OPLS components needed to make the model and inspect the corresponding score spectra. If many orthogonal components are generated but with noisy score profiles without spectral shape, it is highly probable that we over-fit the data. If there are few orthogonal components that contain distinct spectral features we have a better model. Also check the statistics from cross validation and follow advice for regular OPLS modelling.
8. If you need two grating settings in a microscope to cover the full wavenumber range, the measurements can be much quicker if you restrict to use only one grating and adapt the range to distinct features in the algal spectra and also of the MBN spectrum. This also facilitates the multivariate evaluation since we do not get a splice between the regions of the two grating settings.
9. Longer exposure times can produce photobleaching of algal pigments.
10. If individual spectra have a large contribution of fluorescence they may need to be removed completely but will render empty pixels in the final image reconstruction.
11. The T-OPLS method works well also in situations when an internal standard is not needed, provided that the analyte has a spectrum under the measurement conditions. One example is the resonance Raman image of *Dunaliella* given as an example above (*see* Figs. 1 and 2). In this case the spectra were not unit variance scaled but only centered in the T-OPLS model. Here, the predictive loadings were used directly without the transformation to a *Y*-related profile, rescaling to a 0–1 scale, or division.
12. You may use Pareto scaling in order to not to over emphasize spectral baseline variation.
13. If there are large features of MBN also in the orthogonal score spectra of the single *y* T-OPLS model for the MBN, this is an indication that the data are not fully linear [12] and another wavenumber range should be attempted.
14. The most favorable case is when the predictive loadings have a baseline in the bottom and signal excursions towards higher magnitudes. With smaller signal-to-noise ratio and without sufficient spectral separation, the loadings will contain more negative spikes. In this case, try to select a wavenumber range with unique similar size peaks for both the analyte and the internal standard. Alternatively, make one separate T-OPLS model for the analyte and one for the internal standard. With separate models, different wavenumber ranges can be attempted for the two T-OPLS models (*see* Fig. 3).

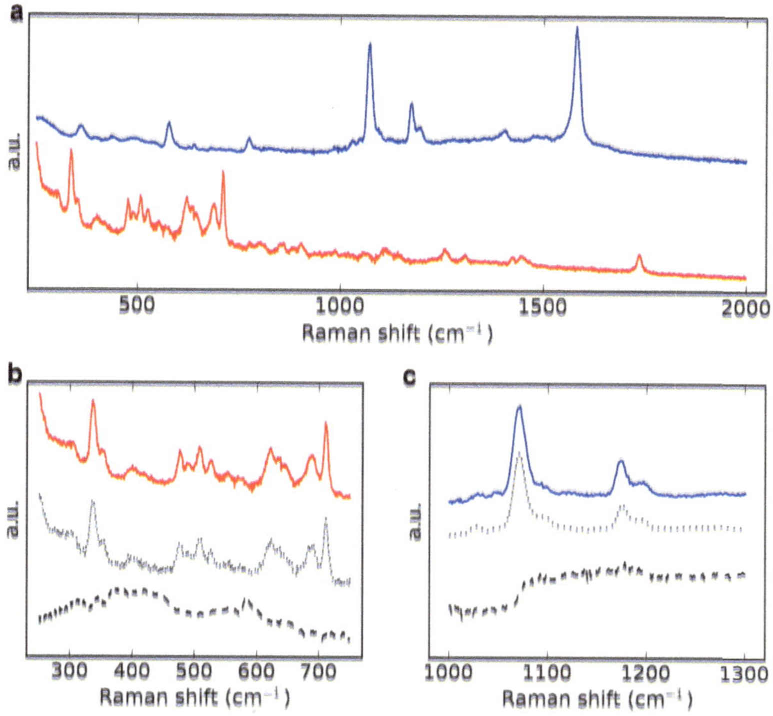

Fig. 3 (**a**) Full range reference spectra of 1,1,3,3-tetrabromo-2-heptanone (*lower*) and MBN (*upper*). Actual optimized ranges used in ref. [3]; (**b**) the reference (*solid*), predictive (*dotted*), and first orthogonal (*interrupted*) scores spectra of 1,1,3,3-tetrabromo-2-heptanone; (**c**) the internal standard reference (*solid*), predictive (*dotted*), and first orthogonal (*interrupted*) scores spectra of MBN. T-OPLS models with separate ranges were used for the analyte and the internal standard to avoid an MBN contribution in the orthogonal loadings. N.B. first orthogonal scores spectrum in (**c**) does not show any MBN peaks

15. All multivariate models are sensitive to frequency/wavenumber drift. This means that it essential to keep track of the wavenumber calibration during each measurement day, e.g., through the use of a wavelength standard. If the frequency calibration changes with time and data from different measurement occasions are used together, all spectral data used together must have a common *x*-scale. This can be achieved by re-interpolation of spectra that have a misfit in relation to spectra taken at a different time.
16. The minimum readings of p_1 and p_2 should be defined by images of the same type of sample but with the internal standard absent and, if possible, by a corresponding image with the analyte absent. Alternatively, the analyte baseline can be identified by a line plot of p.

References

1. Huang YY, Beal CM, Cai WW et al (2009) Micro-Raman spectroscopy of algae: composition analysis and fluorescence background behaviour. Biotechnol Bioeng 105:889–898
2. Abbas A, Josefson M, Abrahamsson K (2011) Characterization and mapping of carotenoids in the algae *Dunaliella* and *Phaeodactylum* using Raman and target orthogonal partial least squares. Chemometr Intell 107:174–177
3. Abbas A, Josefson M, Nylund G et al (2012) Chemical images of marine bio-active compounds by surface enhanced Raman spectroscopy and transposed orthogonal partial least squares (T-OPLS). Anal Chim Acta 737:37–44
4. Lorén A, Engelbrektsson J, Eliasson C et al (2004) Internal standard in surface-enhanced Raman spectroscopy. Anal Chem 76:7391–7395
5. Lorén A, Engelbrektsson J, Eliasson C et al (2004) Self-assembled monolayer coating for normalization of surface enhanced spectra. Nano Lett 4:309–312
6. Trygg J, Wold S (2002) Orthogonal projections to latent structures (O-PLS). J Chemometrics 16:119–128
7. Hunter JD (2007) Matplotlib: A 2D graphics environment. Comput Bus 9:90–95
8. Schlenoff JB, Ming L, Ly H (1995) Stability and self-exchange in alkanethiol monolayers. J Am Chem Soc 117:12528–12536
9. Trygg J (2004) Prediction and spectral profile estimation in multivariate calibration. J Chemometr 18:166–172
10. Eliasson C, Lorén A, Engelbrektsson J et al (2005) Surface-enhanced Raman scattering imaging of single living lymphocytes with multivariate evaluation. Spectrochim Acta 61:755–760
11. Sicard C, Brayner R, Margueritat J et al (2010) Nano-gold biosynthesis by silica-encapsulated micro-algae: a "living" bio-hybrid material. J Mater Chem 20:9342–9347
12. Stenlund H, Johansson E, Gottfries J et al (2009) Unlocking interpretation in near infrared multivariate calibrations by orthogonal partial least squares. Anal Chem 81:203–209

Chapter 24

In Vitro Protocols for Measuring the Antioxidant Capacity of Algal Extracts

Owen Kenny, Nigel P. Brunton, and Thomas J. Smyth

Abstract

In the last decade a large amount of research has been directed at targeting algal resources for biologically active molecules. High-throughput in vitro antioxidant assays are routinely used to screen for biologically active compounds present in algal extracts when the requirement is to identify samples for progression to more detailed biological scrutiny. Whilst a myriad of antioxidant assays have been developed, this present chapter aims to give step-by-step practical guidance on how to carry out some of the most popular and biologically relevant assays at the bench.

Key words Algal extracts, Antioxidants, Fluorescence assays, Preventative oxidation assays, Radical scavenging assays, Redox assays

1 Introduction

Macroalgae are a potentially abundant, sustainable, and underutilized source of compounds with useful biological properties [1–4]. Screening for biologically active compounds in a large number of algal species and extracts usually involves the initial use of high-throughput assays to eliminate samples with limited activity. Compounds with antioxidant activity often have other useful biological properties related to their ability to scavenge free radicals such as antimicrobial [5], anti-inflammatory [6], and antihypertensive [7] activities, and for this reason researchers often use high-throughput in vitro antioxidant activity assays as their initial screen. Unfortunately there is no single agreed upon method for assessing antioxidant activity, and over time, a myriad of techniques have been developed, each using different free radicals, probes and protocols [8]. In simplest terms, antioxidant activity can be defined as the ability of a compound to scavenge free radicals from a substrate when present at a concentration lower than the substrate [9]. Therefore, most antioxidant assays measure the ability of a compound to scavenge either stable free radicals or those induced or

Dagmar B. Stengel and Solène Connan (eds.), *Natural Products From Marine Algae: Methods and Protocols*, Methods in Molecular Biology, vol. 1308, DOI 10.1007/978-1-4939-2684-8_24, © Springer Science+Business Media New York 2015

produced as part of the assay. In addition, in vitro antioxidant assays can be divided into those that employ a nonbiological substrate and those that measure the ability of the compound to protect a biological substrate from attack by a free radical [10]. Whilst assays involving a biological substrate can give very useful information on the behavior of antioxidant species in foods or other biological systems, they are often much more expensive and time consuming than their nonbiological counterparts. For this reason most high-throughput antioxidant assays employ a nonbiological substrate. Screening for biologically relevant samples is often carried out as the next step.

Therefore, the present chapter concentrates mostly on describing detailed protocols for a range of high-throughput antioxidant assays employing nonbiological substrates as the probe. These assays can again be divided into two categories: Hydrogen atom transfer assays (HAT) measure the capacity of an antioxidant to quench free radicals by hydrogen donation [11]. The oxygen radical absorbance capacity (ORAC) assay is an example of this type of assay. Single electron transfer (SET) methods measure the capacity of a potential antioxidant to transfer one electron to a compound. There are many examples of this type of assay including the ferric ion reducing antioxidant power (FRAP), the Trolox equivalent antioxidant capacity (TEAC), and 2, 2-diphenyl-1-picrylhydrazyl (DPPH) assays. It should be noted that HAT assays are the preferred choice as hydrogen donation is the route followed in vivo [12].

Whilst it is important to keep in mind the mechanism by which an assay operates, the present chapter concentrates on the practical aspects of assay employment and, thus, divides the descriptions of the protocols into four categories, which relate to this aspect. The first category covers radical scavenging assays in which the end point is monitored by either the disappearance or generation of a distinct chromophore. Redox assays measure the ability of the antioxidant to scavenge free radicals from a compound that can exist in more than one oxidation state. In most cases the redox species is a transition metal. Fluorescence assays use the ability of the antioxidant to protect a fluorescent compound from free radical attack [13]. Thus the end-point is usually measured as a disappearance of fluorescence. Finally, a number of high-throughput assays operate by monitoring the ability of the antioxidant to protect an oxidatively labile compound from oxidation [14]. Most of these assays use a biologically relevant oxidatively labile probe such as an unsaturated fatty acid or a carotenoid. While a large number of in vitro antioxidant assays have been developed, it is beyond the scope of this chapter to discuss them all. Rather, the aim here is to provide detailed practical guidance for researchers intending to carry out the most commonly used assays applied to algal species and extracts.

2 Materials

Following the extraction of metabolites from macroalgal samples, of which there are a number of different methodologies to choose from, extracts need to be suitably prepared for the screening of their antioxidant activity. Subsequently, organic solvent extracts should be dried under nitrogen or with the aid of a rotary evaporator, while aqueous extracts should be freeze-dried. Where appropriate, analytical grade reagents and deionized water should be used for the preparation of solutions and/or reagents. Reagents should be prepared freshly and stored as outlined in brackets below. Consult material safety data sheets (MSDS) for disposal of reagents and adhere to guidelines specific to each laboratory for disposal of reagents and organic solvents.

2.1 DPPH Microtiter Assay

1. DPPH solution (0.12 M): weigh 23.5 mg of DPPH (2,2-diphenyl-1-picrylhydrazyl 1,1-diphenyl-2-picrylhydrazyl radical, $C_{18}H_{12}N_5O_6$) reagent into a 100 mL volumetric flask. Add 50 mL of methanol and dissolve (*see* **Note 1**). Adjust to 100 mL using methanol. Dilute this solution (0.6 M) with methanol (1:5, v/v) to obtain the 0.12 M DPPH solution (*see* **Note 2**) (fridge, weekly).
2. Trolox® stock solution (20 mM): weigh 5 mg of Trolox® into a 2 mL eppendorf tube. Add 1 mL of methanol and dissolve (daily).
3. Methanol.

2.2 ABTS Assay

1. ABTS solution (7 mM): weigh 36 mg of ABTS (2,2′-azino-bis(3-ethylbenzothiazoline-6-sulphonic acid), $C_{18}H_{18}N_4O_6S_4$) into a 10 mL volumetric flask. Add 5 mL of distilled water and dissolve. Adjust to the mark with distilled water (daily).
2. Potassium persulfate solution (2.45 mM): weigh 4.41 mg of potassium persulfate ($K_2S_2O_8$) into a 10 mL volumetric flask. Add 5 mL of distilled water and dissolve. Adjust to the mark with distilled water (room temperature, weekly).
3. Trolox® stock solution (20 mM): weigh 5 mg of Trolox® into a 2 mL eppendorf tube. Add 1 mL of ethanol and dissolve (daily).
4. Ethanol.

2.3 DMPD Assay

1. DMPD solution (100 mM): weigh 209 mg of DMPD (*N,N*-Dimethyl-*p*-phenylenediamine dihydrochloride, $(CH_3)_2NC_6H_4NH_2 \cdot 2HCl$) into a 10 mL volumetric flask. Add 5 mL of deionized water and dissolve. Adjust to the mark with deionized water (daily).
2. Ferric chloride solution (0.05 M): weigh 8.1 mg of ferric chloride ($FeCl_3$) into a 2 mL eppendorf tube. Add 1 mL of deionized water and dissolve (daily).

3. 1 M acetic acid: prepare by adding 5.75 mL of concentrated glacial acetic acid (17.4 M) to 94.25 mL of deionized water (room temperature, monthly).
4. Acetate buffer (0.1 M, pH 5.3): weigh 8.2 g of sodium acetate anhydrous ($C_2H_3NaO_2$) into a 1 L beaker. Add 800 mL of deionized water and dissolve. Adjust to pH 5.3 using 1 M acetic acid. Transfer to a 1 L volumetric flask and adjust to the mark with deionized water (room temperature, monthly).
5. Trolox® stock solution (20 mM): weigh 5 mg of Trolox® into a 2 mL eppendorf tube. Add 1 mL of methanol and dissolve (daily).
6. Methanol.

2.4 Hypochlorus Acid Scavenging Assay

1. Potassium phosphate buffer (50 mM, pH 6.6) containing 5 mM EDTA and 20 mM sodium borohydrate: weigh 6.8 g of potassium phosphate (KH_2PO_4), 756.6 mg of sodium borohydride ($NaBH_4$), and 1.46 g of EDTA into a 1 L beaker. Add 800 mL of distilled water and dissolve. Adjust to pH 6.6. Transfer to a 1 L volumetric flask and adjust to the mark with distilled water (room temperature, weekly).
2. Hypochlorus acid (HOCl) solution prepare immediately before use. Prepare 1 % solution of NaOCl by adding 1 g to 100 mL of deionized water. Prepare 0.6 M sulfuric acid by adding 16.8 mL of concentrated sulfuric acid (18 M) to 83.2 mL of deionized water. Adjust the 1 % solution of NaOCl to pH 6.2 with 0.6 M sulfuric acid (daily).
3. TNB solution (1 mM): weigh 39.6 mg of 5,5′-Dithiobis(2-nitrobenzoic acid) (DTNB, $(\text{-}SC_6H_3(NO_2)CO_2H)_2$) into a 100 mL volumetric flask. Adjust to 100 mL with the potassium phosphate buffer (50 mM, pH 6.6; *see* **Note 3**) (daily). Incubate for 30 min at 37 °C prior to use.
4. Lipoic acid standard solution (1 mM): weigh 10.3 mg of lipoic acid ($C_8H_{14}O_2S_2$) into a 50 mL volumetric flask. Adjust to mark with ethanol (daily).
5. Ethanol.

2.5 FRAP Assay

1. 1 M acetic acid: Prepare by adding 5.75 mL of concentrated glacial acetic acid (17.4 M) to 94.25 mL of deionized water (room temperature, monthly).
2. 1 M sodium hydroxide: prepare by adding 4 g of sodium hydroxide to 100 mL of deionized water (room temperature, monthly).
3. Acetate buffer (pH 3.6): weigh 3.1 g of sodium acetate anhydrous ($C_2H_3NaO_2$) into a 1 L beaker. Add 800 mL of distilled water followed by 20 mL of 1 M acetic acid. Adjust the pH

accordingly using 1 M acetic acid or 1 M sodium hydroxide. Transfer contents to a 1 L volumetric flask and fill to the mark with distilled water (room temperature, monthly).

4. 40 mM HCl: add 20 mL of distilled water to a 100 mL volumetric flask. Add 400 μL of 1 M HCl and mix. Adjust to 100 mL with distilled water (room temperature, monthly).
5. TPTZ solution (10 mM): weigh 31.3 mg of 2,4,6-tris(2-pyridyl)-1,3,5-triazine (TPTZ, $C_{18}H_{12}N_6$) in a 10 mL volumetric flask. Add 5 mL of the HCl solution and dissolve. Adjust to the mark with the HCl solution (daily).
6. Iron (III) chloride solution (20 mM): weigh 54.1 mg of Iron (III) chloride hexahydrate ($FeCl_3{\cdot}6H_2O$) into a 10 mL volumetric flask. Add 5 mL of distilled water and dissolve. Adjust to the mark with distilled water (daily).
7. Trolox® stock solution (20 mM): weigh 5 mg of Trolox® into a 2 mL eppendorf tube. Add 1 mL of methanol and dissolve (daily).
8. Methanol.

2.6 Total Phenolic Assay by Folin–Ciocalteu Reagent

1. Sodium carbonate (20 % w/v): weigh 20 g of sodium carbonate (Na_2CO_3) into a 500 mL beaker. Add 100 mL of distilled water to the beaker and stir using a magnetic stirrer with heating hotplate at 40 °C until Na_2CO_3 is fully dissolved (room temperature, monthly).
2. Folin–Ciocalteu Reagent® (FCR): store FCR bought from manufacturer in darkness and use as a neat solution (room temperature).
3. Gallic acid stock solution (30 mM): weigh out 5 mg of gallic acid into a 2 mL eppendorf tube. Add 1 mL of methanol and dissolve. Alternatively phloroglucinol could be used as a standard.
4. Methanol.

2.7 CUPRAC Assay

1. Copper (II) chloride solution (10 mM): weigh 426.2 mg of cupric (II) chloride dihydrate ($CuCl_2{\cdot}2H_2O$) into a 250 mL volumetric flask. Add 100 mL of distilled water and dissolve. Adjust to the mark with distilled water and store in the dark until required (daily).
2. Ammonium acetate buffer (1 mM, pH 7): weigh 19.27 g of ammonium acetate ($C_2H_3O_2NH_4$) into a 250 mL volumetric flask. Add 100 mL of distilled water and dissolve. Adjust to the mark with distilled water (room temperature, monthly).
3. Neocuproine (Nc) solution (7.5 mM): weigh 39 mg of neocuproine ($C_{14}H_{12}N_2$) into a 25 mL volumetric flask. Add 10 mL of ethanol and dissolve. Adjust to the mark with ethanol and store in the dark until required (daily).

4. Trolox® stock solution (20 mM): weigh 5 mg of Trolox® into a 2 mL eppendorf tube. Add 1 mL of ethanol and dissolve.
5. Ethanol.

2.8 ORAC Assay

1. Phosphate buffer (75 mM, pH 7.4): prepare by adding 0.238 g of monosodium phosphate and 1.556 g of disodium phosphate to 100 mL of deionized water and adjust the pH to 7.4 if required (room temperature, monthly).
2. Fluorescein (FL) stock solution (8.16 mM): weigh 271 mg of fluorescein salt ($C_{20}H_{10}Na_2O_5$) into a 100 mL volumetric flask. Add 50 mL of the phosphate buffer and dissolve. Adjust to the mark with phosphate buffer (*see* **Note 4**) (daily).
3. AAPH solution (153 mM): weigh 415 mg of 2,2′-Azobis(2-methylpropionamidine) dihydrochloride (AAPH; $C_8H_{20}Cl_2N_6$) into a 10 mL volumetric flask. Add 5 mL of the phosphate buffer and dissolve. Adjust to the mark with the phosphate buffer (*see* **Note 5**) (daily).
4. Trolox® stock solution (20 mM): weigh 5 mg of Trolox® into a 2 mL eppendorf tube. Add 1 mL of the phosphate buffer and dissolve (daily).
5. Methanol.

2.9 HORAC Assay

1. Phosphate buffer (75 mM, pH 7.4): prepare by adding 0.238 g of monosodium phosphate and 1.556 g of disodium phosphate to 100 mL of deionized water and adjust the pH to 7.4 if required (room temperature, monthly).
2. Fluorescein (FL) stock solution (1.345 mM): weigh 22.5 mg of fluorescein salt in a 50 mL volumetric flask. Add 20 mL of the phosphate buffer and dissolve. Adjust to the mark with the phosphate buffer (daily).
3. H_2O_2 solution (1.1 M): dilute 30 % H_2O_2 solution (8.8 M) by adding 11.4 mL of H_2O_2 to 88.6 mL of deionized water (daily).
4. Cobalt solution: weigh 15.7 mg of cobalt(II) fluoride tetrahydrate and 20 mg of picolinic acid into a 10 mL volumetric flask. Add 5 mL of distilled water and dissolve. Adjust to the mark with distilled water (daily).
5. Gallic acid stock solution (40 mM): weigh out 6.8 mg of gallic acid into a 2 mL eppendorf tube. Add 1 mL of the phosphate buffer and dissolve (daily).
6. Methanol.

2.10 Carotenoid Bleaching Assay

1. β-carotene solution (372 μM): weigh 5 mg of β-carotene ($C_{40}H_{56}$) into a 25 mL volumetric flask. Add 10 mL of chloroform and dissolve. Adjust to the mark with chloroform (daily).

2. β-carotene reagent: weigh 20 mg of linolenic acid ($C_{18}H_{30}O_2$) and 200 mg of Tween® 40 into a 100 mL beaker. Add 1 mL of the β-carotene solution to the beaker. Evaporate chloroform under N_2 gas. Add 50 mL of ultrapure water to the beaker and dissolve using a sonicator. Transfer the emulsion to a 50 mL volumetric flask and adjust to the mark with ultrapure water (daily).
3. Trolox® stock solution (20 mM): weigh 5 mg of Trolox® into a 2 mL eppendorf tube. Add 1 mL of ethanol and dissolve.
4. Ethanol.

2.11 Ferric Thiocyanate (FTC) Method

1. 3.5 % HCl: add 3.5 mL of concentrated hydrochloric acid to 96.5 mL of deionized water (room temperature, monthly).
2. Ferrous chloride solution (20 mM): weigh 50.7 mg of ferrous chloride ($Fe(III)Cl_2$) into a 20 mL volumetric flask. Add 10 mL of 3.5 % HCl and dissolve the $Fe(III)Cl_2$. Adjust to the mark using 3.5 % HCl (daily).
3. Ammonium thiocyanate solution (0.3 M): weigh 228 mg of ammonium thiocyanate (NH_4SCN) into a 10 mL volumetric flask. Add 5 mL of distilled water and dissolve. Adjust to the mark with distilled water (daily).
4. Linolenic acid solution (2.5 mM): weigh 35.1 mg of linoleic acid into a 50 mL volumetric flask. Add 20 mL of ethanol and dissolve. Adjust to the mark with ethanol (daily).
5. α-Tocopherol and/or butylated hydroxytoluene (BHT): prepare 50 μg/mL solution by adding 5 mg to 100 mL of deionized water (daily).
6. Phosphate buffer (20 mM, pH 7): prepare by adding 0.117 g of monosodium phosphate and 0.31 g of disodium phosphate to 100 mL of deionized water and adjust pH if required (room temperature, monthly).
7. 75 % ethanol in water (v/v): prepare by adding 75 mL of ethanol to 25 mL of deionized water (room temperature, monthly).
8. Ethanol.

2.12 Thiobarbituric Acid (TBA) Assay

1. Trichloroacetic acid solution (0.2 M): weigh 3.268 g of trichloroacetic acid (TCA) into a 100 mL volumetric flask. Add 50 mL of distilled water and dissolve. Adjust to the mark with distilled water (room temperature, monthly).
2. Thiobarbituric acid solution (6.7 mM): weigh 96.6 mg of thiobarbituric acid (TBA) into a 100 mL volumetric flask. Add 50 mL of distilled water and dissolve. Adjust to the mark with distilled water (room temperature, monthly).
3. Linolenic acid solution (2.5 mM): weigh 35.1 mg of linoleic acid into a 50 mL volumetric flask. Add 20 mL of ethanol and dissolve. Adjust to the mark with ethanol (daily).

4. α-Tocopherol and/or butylated hydroxytoluene (BHT): prepare 50 μg/mL solution by adding 5 mg to 100 mL of deionized water (daily).
5. Phosphate buffer (20 mM, pH 7): prepare by adding 0.117 g of monosodium phosphate and 0.31 g of disodium phosphate to 100 mL of deionized water and adjust pH if required (room temperature, monthly).
6. Ethanol.

3 Methods

3.1 DPPH Microtiter Assay

The diphenylpicrylhydrazyl (DPPH) assay was first reported by Blois [15] and measures the decoloration of the DPPH radical in the presence of antioxidant radical scavengers. The DPPH radical is considered a stable free radical due to the delocalization of the unpaired electron [16]. This radical chromogen absorbs strongly within the visible spectrum between 515 and 528 nm. However, when a compound with the potential to transfer an electron, such as an antioxidant, reacts with the DPPH radical it becomes reduced to diphenylpicrylhydrazine (pale yellow) and the absorption of DPPH between 515 and 528 nm is reduced [17] due to the loss of its electron paramagnetic resonance (EPR) free radical signal [15, 18]. The steric accessibility of an antioxidant compound to the DPPH radical plays an important role in the overall functionality of the assay. Small molecules, such as phenolic compounds, will have better access to the radical site in comparison to larger molecules [19]. This can lead to a lag time for the reaction of larger molecules with DPPH, while smaller molecules may appear to have a higher antioxidant capacity due to the faster color change [19]. In addition, it is important that the reaction occurs at a neutral pH, otherwise the ionization equilibrium of any phenolic compounds present in the extract may be affected, resulting in a change to the rate of reaction [20]. The DPPH assay is one of the most frequently used methods for evaluating the scavenging antioxidant capacity of extracts because of its simplicity, cost, speed and reproducibility. The following protocol has been broadly based on that described by Goupy et al. [21] with minor changes to accommodate the use of a microtiter approach.

1. Trolox® standard curve: dilute 50 μL of Trolox® stock solution in 950 μL of methanol. From this solution (1 mM) prepare dilutions for the following concentrations, 40, 36, 32, 28, 24, and 0 μM of Trolox using methanol.
2. Weigh 5 mg of dried algal extract into 2 mL eppendorf tubes. Add 1 mL of methanol to each eppendorf tube (5 mg/mL) and dissolve. Dilute extracts accordingly to create a suitable

concentration range (*see* **Note 6**). The initial highest concentration should not be above 1 mg/mL. Extracts should be diluted to concentrations that will inhibit the DPPH radical by between 20–80 %. This may require some trial and error, which can be minimized by testing various concentrations of each extract.

3. Pipette 100 μL of sample/Trolox® standard/methanol blank into corresponding wells on a 96-well flat-bottom microplate in triplicate.
4. Measure the neat absorbance of each well at 515 nm, correcting the path length of the UV–VIS microplate spectrophotometer for 100 μL (*see* **Note 7**).
5. Using a multichannel pipette, add 100 μL of DPPH solution to each well of the microplate. The final volume of each well is 200 μL.
6. Replace the lid and incubate the microplate in darkness for 30 min at room temperature (*see* **Note 8**).
7. Measure the absorbance of each well at 515 nm using an UV–VIS microplate spectrophotometer. Correct the path length for 200 μL.
8. Calculate the antioxidant capacity of samples as the percentage inhibition (% *I*) of the DPPH radical at 515 nm (*see* Eq. 1).

$$\%I = \left[\left(A_0 - A_1\right) / A_0\right] \times 100 \qquad (1)$$

where A_0 = absorbance of DPPH control reaction minus the absorbance of the methanol blank after 30 min. A_1 = absorbance of the sample reaction minus the neat sample absorbance after 30 min.

9. Alternatively, calculate the antioxidant capacity of samples as the concentration required to cause a 50 % inhibition in the initial absorbance (IC_{50}) of the DPPH radical at 515 nm (*see* Eq. 2).

$$IC_{50} = C_1 - \left[\left(C_1 - C_2\right) \times PI_1 - 50\right] / \left(PI_1 - PI_2\right) \qquad (2)$$

where PI_1 (%) = sample dilution with absorbance immediately above the 50 % absorbance of DPPH control reaction (including neat sample absorbance) minus the absorbance of the methanol blank after 30 min (A_0). PI_2 (%) = sample dilution with absorbance immediately below the 50 % absorbance of A_0. C_1 and C_2 are the dilution factors of the sample leading to PI_1 and PI_2.

3.2 ABTS Assay

The ABTS assay (2,2′-azino-bis(3-ethylbenzothiazoline-6-sulfonic acid) is a radical scavenging assay that was first described by Miller et al. [22] to measure the amount of the radical cation $ABTS^{\cdot+}$ formed in the presence of metmyoglobin and H_2O_2 with or without the addition of an antioxidant sample. The method has since been adapted to measure the decrease in the absorbance of a preformed $ABTS^{\cdot+}$ solution in the presence of an antioxidant compound. The $ABTS^{\cdot+}$ chromophore (blue/green) is generated through the reaction between ABTS and potassium persulfate. The decolorization of $ABTS^{\cdot+}$, in the presence of an antioxidant, can be measured at 734 nm. The ABTS method has a distinct advantage over other radical scavenging assays (i.e., DPPH and DMPD) in that the ABTS cation is soluble in both aqueous and organic solvents [19]. Therefore, the ABTS method can easily be adapted to measure both lipophilic and hydrophilic algal extracts. It is recommended [23] to dissolve non polar extracts in dichloromethane, medium polar extracts in ethanol and polar extracts in water. In addition, the ABTS cation is not affected by the ionic strength of compounds and so can be used over a wide pH range [24].

1. Trolox® standard curve: dilute 5 μL of Trolox® stock solution in 995 μL of ethanol. From this solution (0.1 mM), prepare the following dilutions using ethanol; 3, 6, 9, 12, and 15 μM.
2. To form the $ABTS^{\cdot+}$ radical, pour 10 mL of ABTS solution into a 50 mL conical flask. Add 10 mL of potassium persulfate solution. Mix contents using a glass rod. Transfer the mixture to a 20 mL conical flask and store in darkness at room temperature for 12–16 h (*see* **Note 9**).
3. Weigh 5 mg of each algal extract into eppendorf tubes in triplicate. Add 1 mL of ethanol to each eppendorf (5 mg/mL) and dissolve. Dilute extracts accordingly to create a suitable concentration range to ensure a concentration is obtained that reduces the absorbance due to the ABTS radical below 50 %.
4. Dilute the $ABTS^{\cdot+}$ solution with ethanol (or the same solvent used for dissolution of the algal extracts) to an absorbance of 0.7 ± 0.02 AU, measured using a UV–VIS spectrophotometer at 734 nm (*see* **Note 10**).
5. Pipette 990 μL of $ABTS^{\cdot+}$ solution into a 1.5 mL eppendorf tube. Add 10 μL of sample/Trolox® standard/ethanol blank into the eppendorf tube. Repeat in triplicate.
6. Vortex each eppendorf tube and incubate for 6 min at 30 °C (*see* **Note 11**).
7. Measure the absorbance of each sample using a UV–VIS spectrophotometer at 734 nm (*see* **Note 12**).
8. Calculate the antioxidant capacity of samples as the percentage inhibition (% *I*) of the ABTS radical at 734 nm (*see* Eq. 1).

$$\%I = \left[\left(A_0 - A_1\right) / A_0\right] \times 100 \quad (1)$$

where A_0 = absorbance of ABTS control reaction minus the absorbance of the ethanol blank after 6 min. A_1 = absorbance of the sample reaction minus the absorbance of the ethanol blank after 6 min.

9. Alternatively, calculate the antioxidant capacity of samples as the concentration required to cause a 50 % inhibition in the initial absorbance (IC_{50}) of the ABTS radical at 734 nm (*see* Eq. 2).

$$IC_{50} = C_1 - \left[(C_1 - C_2) \times PI_1 - 50 \right] / (PI_1 - PI_2) \qquad (2)$$

where PI_1 (%) = sample dilution with absorbance immediately above the 50 % absorbance of ABTS control reaction minus the absorbance of the methanol blank after 6 min (A_0). PI_2 (%) = sample dilution with absorbance immediately below the 50 % absorbance of A_0. C_1 and C_2 are dilution factor of the sample leading to PI_1 and PI_2.

3.3 DMPD Assay

The DMPD (*N,N*-dimethyl-*p*-phenylenediamine) method for measuring the radical scavenging capacity of antioxidants was first described by Fogliano et al. [25] for monitoring the antioxidant capacity of wines. Similar to the ABTS assay, this test measures the decoloration of the preformed DMPD radical cation (purple) in the presence of an antioxidant. The DMPD radical cation has a maximum absorbance at 505 nm and can be measured using a UV–VIS spectrophotometer. The method described below is based on that described by Mehdi and Rizvi [26] with minor adjustments for microtiter use. The original method described by Fogliano et al. [25] investigated the potential use of several oxidant species (hydrogen peroxide, sodium hypochlorite, azinobis-2-amidinopropane (ABAP), copper chloride and ferric chloride) to generate the DMPD cation. However, ferric chloride was shown to form the most stable DMPD cation. The DMPD cation (purple) is formed through the oxidation of DMPD (colorless) using ferric chloride. Acetate buffer is used in this reaction in order to lower the pH, thus preserving the solubility of ferric iron, helping to drive the deprotonation of DMPD and the formation of the phenolate cation.

1. Trolox® standard curve: dilute 50 μL of Trolox® stock solution in 950 μL of deionized water. From this solution (1 mM), prepare the following dilutions using deionized water: 8, 16, 24, 32, 40, and 48 μM.
2. To form the $DMPD^{\cdot+}$ radical, add 98.8 mL of acetate buffer into a 100 mL volumetric flask. Add 1 mL of DMPD solution to the volumetric flask and mix. Add 200 μL of ferric chloride solution to the volumetric flask and mix (*see* **Note 13**).
3. Weigh 5 mg of each algal extract into eppendorf tubes. Add 1 mL of methanol to each eppendorf (5 mg/mL) and dissolve. Dilute extracts accordingly with deionized water to create a suitable concentration range (*see* **Note 14**).
4. Pipette 20 μL of sample/Trolox® standard/deionized water blank into corresponding wells (in triplicate) in a clear flat-bottom 96-well microplate.

5. Using a multichannel pipette, add 180 μL of DMPD radical into each well. The final volume of each well is 200 μL.
6. Cover with plate lid and incubate in darkness for 15 min at room temperature (*see* **Note 15**).
7. Measure the absorbance of each well at 505 nm using a UV–VIS microplate spectrophotometer, corrected to the path length for 200 μL.
8. Calculate the antioxidant capacity of extracts as the percentage inhibition (% *I*) of the DMPD radical at 505 nm (*see* Eq. 1).

$$\%I = \left[\left(A_0 - A_1 \right) / A_0 \right] \times 100 \quad (1)$$

where A_0 = absorbance of DMPD control reaction minus the absorbance of the water blank after 15 min. A_1 = absorbance of the sample reaction minus the neat sample absorbance and the absorbance of the water blank after 15 min.

9. Alternatively, calculate the antioxidant capacity of samples as the concentration required to cause a 50 % inhibition in the initial absorbance (IC_{50}) of the DMPD radical at 505 nm (*see* Eq. 2).

$$IC_{50} = C_1 - \left[\left(C_1 - C_2 \right) \times PI_1 - 50 \right] / \left(PI_1 - PI_2 \right) \quad (2)$$

where PI_1 (%) = sample dilution with absorbance immediately above the 50 % absorbance of DMPD control reaction (including the neat sample absorbance) minus the absorbance of the water blank after 15 min (A_0). PI_2 (%) = sample dilution with absorbance immediately below the 50 % absorbance of A_0. C_1 and C_2 are the dilution factors of the sample leading to PI_1 and PI_2.

3.4 Hypochlorus Acid Scavenging Assay

The hypochlorus acid (HOCl) scavenging assay is a colorimetric method first described by Ching et al. [27] to measure the antioxidant capacity of samples via the inhibition of HOCl induced oxidation of thionitrobenzoic acid (TNB). HOCl is formed in vivo from the catalytic action of the enzyme myeloperoxidase (MPO) [28], which could lead to oxidative damage of biomolecules [29, 30]. TNB is used as a thiol substrate and is the product of the reaction between 5,5-Dithiobis(2-nitrobenzoic acid) (DTNB) and free thiols. The synthesis of the TNB solution is based on a similar procedure described by Aune and Thomas [31]. HOCl acts as an oxidant and causes the decolorization of TNB, which absorbs strongly at 412 nm. Antioxidants inhibit the actions of HOCl allowing their antioxidant potential to be measured. The following protocol has been adapted from Fernandes et al. [32].

1. Prepare a lipoic acid standard curve using the following dilutions of lipoic acid solution with ethanol; 40, 50, 60, 70, 80, and 90 μM.

2. Weigh 5 mg of algal extract into an eppendorf tube. Add 1 mL of ethanol and dissolve. Dilute to the following concentrations in ethanol: 100, 150, 200, 250, and 500 μg/mL. Extracts should be diluted to concentrations that will inhibit the reduction of TNB by between 20 and 80 %. This may require some trial and error, which can be minimized by testing various concentrations of each extract.
3. Pipette 120 μL of sample/lipoic acid standard/ethanol blank into corresponding wells (in triplicate) of a clear flat-bottom 96-well microplate.
4. Using a multichannel pipette, add 15 μL of TNB solution to each well. Measure the absorbance of each well using a UV–VIS microplate spectrophotometer at 412 nm, correcting the path length to 135 μL.
5. Add 65 μL of HOCl solution to each well so the final concentration of HOCl solution is 25 μM. The final volume of each well is 200 μL.
6. Cover with lid and incubate at 37 °C for 5 min.
7. Measure the absorbance of each well at 412 nm using a UV–VIS microplate spectrophotometer, correcting the path length to 200 μL (*see* **Note 16**).
8. Calculate the antioxidant capacity of extracts as the percentage inhibition (% *I*) of TNB oxidation at 412 nm (*see* Eq. 1).

$$\%I = \left[\left(A_0 - A_1\right) / A_0\right] \times 100 \quad (1)$$

where A_0 = Initial absorbance of TNB, before the addition of HOCl, minus the absorbance of the ethanol blank. A_1 = Absorbance of TNB after the addition of HOCl minus the neat sample absorbance and the absorbance of the ethanol blank after 5 min.

3.5 FRAP Assay

The FRAP assay is a simple and inexpensive colorimetric method that was originally designed to measure the antioxidant capacity of biological fluids such as plasma, saliva, tears, urine and cerebrospinal fluid [33]. In this assay, the trispyridyltriazine (TPTZ) reagent acts as a chromogenic ligand that forms a complex with the in excess ferric ion to yield [Fe (III)]-TPTZ [33]. In the presence of an antioxidant, the [Fe (III)]-TPTZ complex is reduced via electron transfer to a ferrous [Fe (II)]-TPTZ complex. At a low pH (3.6), the latter complex forms an intense blue color, which can be read spectrophotometrically at 593 nm. The FRAP assay has since been widely used in the antioxidant evaluation of natural products [34].

1. Trolox® standard curve: dilute 200 μL of Trolox® stock solution in 800 μL of methanol. From this solution (4 mM), prepare the following dilutions in methanol; 80, 160, 240, 320, 400 and 480 μM.

2. Weigh 5 mg of each algal extract into 2 mL eppendorf tubes. Add 1 mL of methanol to each eppendorf (5 mg/mL) and dissolve. Dilute extracts accordingly to create a suitable absorbance range.
3. Pipette 20 μL of sample/Trolox® standard/methanol blank into corresponding wells on a 96-well flat-bottom microplate in triplicate.
4. To make the FRAP reagent, add 10 mL of TPTZ solution to 10 mL of Iron(III) chloride solution in a 250 mL conical flask. Pour 100 mL of acetate buffer (pH 3.6) into the conical flask. Cover the top of the conical flask with Parafilm® and heat for 5 min at 37 °C using a water bath, while continuously swirling (*see* **Notes 17** and **18**).
5. Using a multichannel pipette, add 180 μL of the preheated FRAP reagent to each well of the microplate. The final volume of each well is 200 μL.
6. Cover with lid and incubate the plate in darkness for 40 min at 37 °C (*see* **Note 19**).
7. Measure the absorbance of each well using a UV–VIS microplate spectrophotometer at the wavelength of 593 nm. Correct the path length of the spectrophotometer to 200 μL (*see* **Note 20**).
8. Calculate the results as Trolox® equivalents (TE) using the slope of the standard curve.

3.6 Total Phenolic Assay by Folin–Ciocalteu Reagent

The total phenolic assay by Folin–Ciocalteu reagent (FCR) is a fast, simple and reliable colorimetric method for measuring the antioxidant capacity derived from the phenolic content of a particular sample. The original method was developed by Folin [35] for tyrosine analysis but was later improved by Singelton and Rossi [36] to analyze the total phenol content of wines. Chemically, the assay is based on the reduction of the FCR in the presence of a phenolic antioxidant. FCR consists of both tungsten [W(VI)] and molybdenum [Mo(VI)] oxides [37]. In the presence of phenolic antioxidants, these compounds are reduced to form a greenish-blue phosphate-tungsten [W(IV)] and phosphate-molybdenum [Mo(V)] complexes, which can be measured using a UV–VIS spectrophotometer at 765 nm. The following protocol is modelled on that described by Singelton and Rossi [36] with some modifications. There are a number of non-phenolic organic substances as well as sugars, aromatic amines, and organic acids [19], which can cause interferences with this assay. In addition, a number of inorganic substances have also been noted to interfere with this assay [38, 39]. This may lead to an overestimation in the phenolic content of an extract and should be taken into account when analyzing extracts.

1. Gallic acid standard curve: dilute 200 μL of gallic acid stock solution in 800 μL of methanol. From this solution (6 mM), prepare the following dilutions in methanol; 0.6, 0.12, 0.3, 0.6, 0.9, and 1.2 mM.
2. Weigh 5 mg of each algal extract into eppendorf tubes in triplicate. Add 1 mL of methanol to each eppendorf (5 mg/mL) and dissolve. Dilute extracts accordingly with methanol to create a suitable concentration range.
3. Pipette 100 μL of sample/gallic acid standard/methanol blank into 1.5 mL eppendorf tubes.
4. Pipette 100 μL of methanol followed by 100 μL of neat FCR into all eppendorf tubes.
5. Add 700 μL of Na_2CO_3 solution to each eppendorf tube, close, and vortex (*see* **Note 21**).
6. Place all eppendorf tubes in darkness for 20 min at room temperature.
7. Centrifuge all eppendorf tubes at 10,000 × *g* for 3 min (*see* **Note 22**).
8. Transfer the supernatant of each eppendorf tube to 1 mL plastic cuvettes and measure the absorbance of each sample at 765 nm using a UV–VIS spectrophotometer (*see* **Note 23**).
9. Repeat the assay in triplicate for all samples.
10. Calculate the results as gallic acid equivalents (GAE) using the slope of the standard curve.

3.7 CUPRAC Assay

The CUPRAC assay, which was developed by Apak et al. [40], is a colorimetric method similar to the FRAP assay but instead measures the reduction of copper (Cu). In this assay, neocuproine (Nc) acts a chromogenic ligand that forms a complex with Cu(II) to give a [Cu(II)]-Nc chromogenic oxidizing reagent. At pH 7, this reagent can become reduced in the presence of antioxidants to form an orange-yellow [Cu(I)]-Nc chromophore, which exhibits an absorption maximum at 450 nm. The redox potential of Cu(II)-Nc complex reduction to Cu(I)-Nc is higher (0.6 eV) than that of Cu(II) to Cu(I) alone (0.153 eV). This has been shown to result in more rapid and efficient oxidation of polyphenols [41]. In addition, the redox potential of Cu(II)-Nc/Cu(I)-Nc reaction is lower than that of the Fe(III)-TPTZ/Fe(II)-TPTZ reaction (0.771 eV) in the FRAP assay (*see* Subheading 3.5). As a result, the reaction is faster and more selective thus allowing for the measurement of thiol oxidants [42] but not reducing sugars and citric acid, which can cause interferences in the FRAP assay [40]. A crucial limitation for a number of antioxidant assays (e.g., DPPH, FRAP, and DMPD assays) is their inability to adapt to accurately measure the antioxidant capacity of both lipophilic and hydrophilic extracts.

In the case of the CUPRAC method, the solubility of the reagents used means the assay can be easily adapted for either eventuality by changing the solvent [40, 41]. The following protocol is based on that described by Apak et al. [40] with some modifications.

1. Trolox® standard curve: dilute 50 μL of Trolox® stock solution in 950 μL of ethanol. From this solution (1 mM), prepare the following dilutions using ethanol; 10, 20, 30, 40, 50, and 60 μM.
2. Weigh 5 mg of each algal extract into 2 mL eppendorf tubes in triplicate. Add 1 mL of ethanol to each eppendorf (5 mg/mL) and dissolve. Dilute extracts accordingly with ethanol to create a suitable concentration range.
3. Add 50 μL of sample/Trolox® standard/blank (ethanol) to corresponding wells on a clear flat-bottom 96 well microplate.
4. Using a multichannel pipette, add 50 μL Cu(II) chloride solution, 50 μL ammonium acetate buffer and 50 μL of Nc solution to each well. The final volume of each well is 200 μL.
5. Incubate for 30 min, in darkness, at room temperature (*see* **Note 24**).
6. Measure the absorbance of each well at 450 nm using a UV–VIS microplate spectrophotometer (*see* **Note 25**).
7. Calculate the antioxidant capacity of samples as Trolox® equivalents (TE) using the slope of the Trolox® standard curve.

3.8 Oxygen Radical Absorbance Capacity (ORAC) Assay

The ORAC assay is a fluorimetric assay used to measure the chain-breaking capacity of an antioxidant against peroxyl radicals by hydrogen atom transfer. In this assay, AAPH (2,2′-azobis-(2-amidinopropane hydrochloride)) is used to generate peroxyl radicals, which react with a fluorescent probe and cause it to decay into a nonfluorescent product over time. However, in the presence of an antioxidant, the rate of decay of the fluorescent probe is decreased due to the hydrogen atom transfer to peroxyl radicals. The ORAC method was developed by Cao et al. [43] and is based on Ghiselli et al. [13] and Glazer [44], who described the effects of free radicals and antioxidants on fluorescent probes. The original ORAC method [43] used the protein β-phycoerythin (B-PE) as the fluorescent probe to measure the antioxidant response in the presence of peroxyl radicals. The present method describes instead uses Fluorescein (FL) because B-PE was found to undergo a 53 % decrease in intensity due to photo-bleaching after exposure to excitation by light [45] and can also cause nonspecific protein binding by polyphenols [46]. The following protocol is based on the high-throughput ORAC assay described by Huang et al. [45].

1. Trolox® standard curve: dilute 100 μL of Trolox® stock solution in 900 μL of phosphate buffer. From this solution (1 mM), prepare the following dilutions using phosphate buffer: 5, 10, 20, 30, 40, and 50 μM.
2. Weigh 5 mg of each algal extract into 2 mL eppendorf tubes. Add 1 mL of methanol to each eppendorf (5 mg/mL) and dissolve. Dilute extracts accordingly using phosphate buffer to create a suitable concentration range.
3. Pipette 100 μL of FL stock solution into 9.9 mL of phosphate buffer and vortex. Dilute 10 μL of this solution (81.6 μM) into 9.99 mL of phosphate buffer to obtain a working concentration (81.6 nM) of FL solution.
4. Pipette 25 μL of sample/Trolox® standard/phosphate buffer blank into corresponding wells on a black flat-bottom 96-well microplate. Add 150 μL of FL solution (81.6 nM) to each well. Cover the plate and incubate in a preheated (37 °C) plate shaker at 20 rpm for 10 min.
5. Add 25 μL of AAPH solution to each well. The final volume of each well is 200 μL.
6. Measure the absorbance of each well every minute for 35 min using a microplate fluorometer at the excitation and emission wavelengths of 495 and 515 nm (*see* **Note 26**).
7. Calculate the net area under the curve (AUC) of each well and express the results as ORAC values (*see* Eq. 3).

$$\text{ORAC value} = \left(\text{AUC}_{\text{S}} / \text{AUC}_{\text{T}}\right) \times \left(\left[\text{Trolox}^{®}\right] / [\text{Sample}]\right) \quad (3)$$

where AUCs = AUC of the sample minus the AUC of the phosphate buffer blank after 35 min. AUC_{T} = AUC of Trolox® minus the AUC of the phosphate buffer blank after 35 min.

3.9 HORAC Assay

The HORAC assay is a fluorometric method that has been developed by Ou et al. [46] to measure the hydroxyl radical scavenging capacity of antioxidants. Cobalt (II) fluoride is used as an oxidant to generate hydroxyl radicals in a Fenton-like reaction involving hydrogen peroxide. Fluoroscein (FL) is used as a fluorescent probe that, in the presence of hydroxyl radicals, begins to decay resulting in a loss of fluorescence over time. Antioxidants can prevent the decay of FL by reducing hydroxyl radicals. The following protocol is based on the HORAC assay described by Huang et al. [45].

1. Gallic acid standard curve: dilute 250 μL of gallic acid stock solution in 750 μL of phosphate buffer. From this solution (10 mM), prepare the following dilutions using phosphate buffer: 0.1, 0.2, 0.4, 0.6, and 0.8 mM.
2. Weigh 5 mg of each algal extract into 2 mL eppendorf tubes. Add 1 mL of methanol to each eppendorf (5 mg/mL) and

dissolve. Dilute extracts accordingly using phosphate buffer to create a suitable concentration range.

3. Pipette 50 μL of FL stock solution into 9.95 mL of phosphate buffer and vortex. Further pipette 100 μL of this solution (6.7 μM) into 9.9 mL of phosphate buffer to obtain a working concentration (670 nM) of FL solution (*see* **Note 4**).
4. Pipette 10 μL of sample/gallic acid standard/phosphate buffer blank into each well of a black flat-bottom 96-well microplate.
5. Add 180 μL of preheated (37 °C) FL solution (670 nM) and 10 μL of H_2O_2 solution to each well.
6. Measure the absorbance of each well at the excitation and emission wavelengths of 495 and 515 nm using a microplate fluorometer. Then add 10 μL of cobalt solution to each well (*see* **Note 27**).
7. Measure the absorbance of each well at the excitation and emission wavelengths of 495 and 515 nm after 0.5 min and then every minute thereafter for 35 min (*see* **Note 28**).
8. Calculate the net area under the curve (AUC) of each well and express the results as HORAC values (*see* Eq. 4).

$$\text{HORAC value} = \left(\text{AUC}_{\text{S}} / \text{AUC}_{\text{GA}}\right) \times \left(\text{molarity of gallic acid} / [\text{Sample}]\right) \tag{4}$$

where AUCs = AUC of the sample minus the AUC of the phosphate buffer blank after 35 min. AUC_{GA} = AUC of gallic acid minus the AUC of the phosphate buffer blank after 35 min.

3.10 Carotenoid Bleaching Assay

The β-carotene bleaching assay is a decolorization method that was first described by Macro [47] to measure the decay (bleaching) of a β-carotene probe over time. The bleaching of carotenoids, such as β-carotene, can occur in a number of ways: autoxidation, oxidation by environmental conditions (light and heat) or by preformed peroxyl radicals [19]. In the method described by Koleva et al. [48], an aqueous emulsion of linoleic acid and β-carotene is discolored by the radicals generated by the spontaneous oxidation of the fatty acid, which is promoted by thermal induction. The reduction in the intense orange color over time is measured at 470 nm. However, in the presence of an antioxidant the decay of β-carotene is prevented to some degree depending on the strength of the antioxidant. In this assay, β-carotene acts as highly oxidizable lipophilic substrate. Consequently, an emulsion is formed upon the addition of water when making the β-carotene reagent. The lipophilic nature of β-carotene makes it highly functional in measuring the antioxidant capacity of lipophilic extracts but less sensitive to hydrophilic extracts. As a result, crocin, which is a hydrophilic carotenoid, is often used to measure the antioxidant capacity of hydrophilic extracts [49]. Both substrates share the same operative conditions and requirements and so can be carried out simultaneously. The following protocol has been adapted by that described by Koleva et al. [48] with some modifications.

1. Trolox® standard curve: from Trolox® stock solution, prepare the following dilutions using ethanol: 0.4, 0.8, 1.2, 1.6, and 2.0 mM.
2. Weigh 5 mg of each algal extract into 2 mL eppendorf tubes in triplicate. Add 1 mL of ethanol to each eppendorf (5 mg/mL) and dissolve. Dilute extracts accordingly to create a suitable concentration range.
3. Pipette 25 μL of sample/Trolox® standard/ethanol blank into corresponding wells on a clear flat bottomed 96-well microtiter plate. Repeat this action in triplicate.
4. Using a multichannel pipette, add 175 μL of preheated (55 °C) β-carotene reagent to each well.
5. Measure the absorbance of each well immediately ($t = 0$) using a UV–VIS microplate spectrophotometer at 470 nm. Cover with lid and incubate the plate for 120 min at 55 °C (*see* **Note 28**).
6. After incubation, measure the absorbance of each well at 470 nm ($t = 120$ min) (*see* **Note 29**).
7. The results can be expressed as a percentage inhibition (% I) relative to the control sample (*see* Eq. 5). Extracts should be diluted to concentrations that will inhibit the oxidation of the carotenoid substrate by between 20 and 80 %. This may require trial and error but can be minimized by testing various concentrations of each extract.

$$\%I = \left[\left(A_0 - A_1 \right) / A_0 \right] \times 100 \tag{5}$$

where A_0 = absorbance of β-carotene control reaction ($t_0 - t_{120}$) minus the absorbance of the ethanol blank ($t_0 - t_{120}$). A_1 = absorbance of the sample reaction ($t_0 - t_{120}$) minus the absorbance of the ethanol blank ($t_0 - t_{120}$).

3.11 Ferric Thiocyanate (FTC) Method

The FTC method was developed by Wagner et al. [50] to measure the amount of lipid hydroperoxides (LOOH) formed in the initial stages of lipid oxidation. In the method described by Kikuzaki and Nakatani [51], linolenic acid is used as a substrate for lipid oxidation. The measurement of LOOH by UV–VIS spectrometry is difficult due to the lack of suitable chromophores [52]. As a result, the determination of LOOH production requires the use of more sophisticated chromatographic (i.e., GC) [53] and spectroscopic (i.e., NMR) [54] techniques. This can be overcome by instead measuring the oxidation of iron, induced by the presence of LOOH, from an Fe(II)-thiocyanate complex to an Fe(III)-thiocyanate complex [55], resulting in an intense colored (red) endpoint. The rate of lipid peroxidation is measured daily using a UV–VIS spectrophotometer at 500 nm until the controls reach maximum absorbance. In the presence of an antioxidant, lipid oxidation becomes impeded and the endpoint of the reaction with ferric chloride is inhibited. The following protocol has been adapted from the method described by Kikuzaki and Nakatani [51] and is generally carried out in conjunction with the Thiobarbituric (TBA) assay (*see* Subheading 3.12).

1. Weigh 5 mg of extract into a 2 mL eppendorf tube. Add 1 mL of ethanol and dissolve. Dilute extracts accordingly to create a suitable concentration range. Use α-Tocopherol and/or butylated hydroxytoluene (BHT) as positive controls.
2. Pipette 1 mL of sample (as prepared in **step 1**)/positive control/blank (ethanol) into a 10 mL test tube. Add 1.025 mL of linoleic solution, 2 mL of phosphate buffer and 975 μL of distilled water to each tube so that the final volume of each tube is 5 mL.
3. Incubate in darkness at 40 °C for 24 h (*see* **Note 30**).
4. After incubation, take 10 μL of the mixture and add it to 970 μL of 75 % ethanol, 10 μL of ammonium thiocyanate solution and 10 μL of ferrous chloride solution in a 2 mL eppendorf tube. The final volume of each eppendorf tube is 1 mL. Put original mixture back in the incubator.
5. After 3 min, measure the absorbance of each sample at 500 nm using a UV–VIS spectrophotometer.
6. Repeat the assay in triplicate for all samples and average for best results.
7. Repeat **steps 4** and **5** every 24 h until 1 day after the control has reached its maximum absorbance value and express results in AU at 500 nm (*see* **Note 31**). Alternatively, the antioxidant capacity of an extract can be calculated as the percentage inhibition of Fe(II)-thiocyanate oxidation by LOOH radicals at 500 nm (*see* Eq. 6).

$$\%I = \left[\left(A_0 - A_1\right) / A_0\right] \times 100 \tag{6}$$

A_0 = absorbance of Fe(III)-thiocyanate control reaction on the final day following saturation of Fe(III)-thiocyanate detection. A_1 = absorbance of sample reaction on the final day following saturation of Fe(III)-thiocyanate detection by the blank reaction.

3.12 Thiobarbituric Acid (TBA) Assay

The color of TBA was first [56] observed to change (red) in the presence of animal tissue; this color change was later attributed to the formation of a red complex between the oxidation products of unsaturated fatty acids and TBA [57, 58]. The TBA assay, described by Kikuzaki and Nakatani [51], measures the amount of malondialdyehyde (MDA) formed from the gradual decomposition of lipid peroxides over time. MDA reacts with TBA to form a red MDA-TBA complex that can be measured using a UV–VIS spectrophotometer at 532 nm. In the presence of an antioxidant, the initial lipid peroxide production is slowed, and therefore, the amount of MDA available to react with TBA is lower compared to the blank control. The following protocol is based on the method described by Kikuzaki and Nakatani [51] and is generally performed in conjunction with the FTC assay (*see* Subheading 3.11).

1. Weigh 5 mg of algal sample into a 2 mL eppendorf tube. Add 1 mL of ethanol and dissolve. Dilute 1 in 5 with ethanol (1 mg/mL). Use α-Tocopherol and/or butylated hydroxytoluene (BHT) as positive controls.
2. Pipette 1 mL of sample (as prepared in **step 1**)/positive control/blank (ethanol) into a 10 mL test tube. Add 1.025 mL of linoleic acid solution, 2 mL of phosphate buffer, and 975 μL of distilled water to each tube so that the final volume of each tube is 5 mL.
3. Incubate in darkness at 40 °C until the final day of the FTC assay (*see* Subheading 3.11).
4. After incubation, take 240 μL of the mixture and add it to 480 μL of TCA solution and 480 μL of TBA solution in a 2 mL eppendorf tube. The final volume of each eppendorf tube is 1.2 mL. Put original mixture back in the incubator.
5. Place each eppendorf tube in boiling water for 10 min (*see* **Note 32**).
6. Cool eppendorf tubes and centrifuge at 1,000 × *g* for 20 min.
7. Measure the absorbance of each tube at 532 nm using a UV–VIS spectrophotometer. Extracts should be diluted to concentrations that will inhibit the production of MDA by between 20 and 80 %. This may require trial and error but can be kept to a minimum by testing various concentrations of each extract.
8. Express the results as AU at 532 nm and compare to the blank reaction. Alternatively, the antioxidant capacity of an extract can be calculated as the percentage inhibition (% *I*) of MDA production at 532 nm (*see* Eq. 7).

$$\%I = \left[\left(A_0 - A_1 \right) / A_0 \right] \times 100 \tag{7}$$

where A_0 = absorbance of MDA production in the control reaction on the final day. A_1 = absorbance of MDA production in sample reaction on the final day.

4 Notes

1. DPPH is a hydrophobic compound that can only be dissolved in organic solvents, which is an obvious limitation for measuring the antioxidant capacity of hydrophilic antioxidants [59]. Above a certain level of water the DPPH reagent coagulates and becomes inaccessible to antioxidant molecules [60]. Studies by Stasko et al. [61] have shown a ratio of 1:1 methanol/water (v/v) as being an appropriate solvent mixture limit before coagulation, thus allowing for the antioxidant capacity of both hydrophilic and hydrophobic antioxidants to be measured.
2. The DPPH stock solution can react with light and oxygen. In order to prevent degradation cover the volumetric flask with aluminum foil and place in darkness. If stored correctly, the DPPH stock solution is stable for up to 1 week.
3. The concentration of sodium borohydrate plays an important role in the formation of TNB. Ching et al. [27] reported that at 20 mM sodium borohydrate, a maximum yield of between 60 and 70 % TNB is formed. This is compared with just 5 % using the original method of Winterbourn [28].
4. FL is a pH sensitive probe and so the pH of the phosphate buffer (pH 7.4) should be monitored carefully [26]. The FL stock solution should be kept refrigerated (4 °C) in darkness and is stable under these conditions for several months [45].
5. AAPH is used to generate peroxyl radicals in this assay through thermal decomposition. Therefore, the AAPH solution (153 mM) should be kept on ice throughout the assay to prevent premature decomposition. This solution is stable for 8 h after which it must be discarded [45].
6. When using water-based extracts, a centrifugation step may be required to remove insoluble material.
7. At 515 nm the neat sample absorbance of an extract or compound may cause interference when measuring the decoloration of the DPPH radical. For example, carotenoids are a group of intensely colored compounds that can absorb at this wavelength [62]. This absorbance value must be added to the absorbance of DPPH in order to obtain a true measurement of radical scavenging activity.
8. The incubation reaction time can differ from study to study and represents one of the main setbacks in standardizing results of the DPPH assay. Depending on the lag time of various antioxidant compounds the reaction time may need to be increased in order for the complete reaction to occur. Since DPPH is a stable radical, the reaction time can be

prolonged to account for reaction lag time. Depending on the antioxidant compounds present, the DPPH assay can take between 20 min and 6 h [17].

9. The ABTS cation (blue) is formed through the deprotonation of ABTS (colorless) sulfonate groups by potassium persulfate. The reaction between ABTS and potassium persulfate is stoichiometrically incomplete (ratio 1:0.5) and as a result the ABTS solution is not fully oxidized, leading to the formation of an unstable ABTS cation [23]. The instability of the ABTS cation means that it is not ready to be used immediately and must be left in darkness for at least 12 h in order to stabilize. The initial instability of the ABTS cation is often viewed as a drawback to this method.
10. The ABTS cation solution is concentrated and must be diluted before use. ABTS is soluble in both aqueous and organic solvents. The maximum absorbance of the ABTS cation solution should be 0.7 ± 0.02 AU [23]. Dilute the ABTS cation solution in the solvent corresponding to that used to dissolve the extracts. The reaction between ABTS cation solution and extracts is carried out at 30 °C. Therefore, the adjustment of ABTS cation solution must also be carried out at 30 °C.
11. The incubation time for this reaction can vary depending on the type of antioxidants present in extracts. Many phenolic compounds have a redox potential lower than that of the ABTS cation [19]. This means the endpoint of the reaction can be measured after 6 min for phenolic rich extracts [23]. In the case of slower reacting antioxidants, 6 min may not be sufficient to observe a complete reaction.
12. The ABTS cation has a number of absorption maxima and thus can be measured at wavelengths of 415, 645, 734, and 815 nm. Due to the probable interference of neat extracts at lower wavelengths, 734 nm is often the preferred absorption maximum used for measuring the decoloration of the ABTS cation in the presence of natural product extracts.
13. After formation, the $DMPD^{\cdot+}$ should be left in darkness until use. The $DMPD^{\cdot+}$ is stable for up to 12 h.
14. DMPD is only soluble in aqueous solvents. The measurement of hydrophobic antioxidants can greatly decrease the sensitivity and reproducibility of the assay [63]. Extracts should be diluted to concentrations that will inhibit the DMPD cation by between 20 and 80 %. This may require trial and error but can be kept to a minimum by testing various concentrations of each extract.
15. The reaction of the $DMPD^{\cdot+}$ and an antioxidant sample is rapid and forms a very stable endpoint. This is seen as a distinct advantage of the assay [25]. However, if the incubation time is increased, DMPD, in the presence of oxidants, can

produce free radicals leading to a gradual increase of color intensity over time [26].

16. In the absence of a strong antioxidant, TNB becomes oxidized by HOCl and reforms DTNB. The antioxidant capacity of an extract is measured by the decrease of TNB absorbance at 412 nm. Alternatively, the antioxidant capacity of an extract can be measured by the increase of DTNB at 325 nm. However, this alternative is seldom used as the TNB solution used in the reaction may contain preformed DTNB, which may result in high blank absorbances.
17. The reduction of the Fe(III)-TPTZ complex is a nonspecific reaction. Therefore, any compound present in an extract that has a redox potential less than that of the Fe(III)-TPTZ complex (0.771 eV) will cause its reduction and the formation of the Fe(II)-TPTZ complex. Furthermore, the reaction mechanism in the FRAP method is based on electron transfer and is therefore unable to measure the antioxidant capacity of radical quenching antioxidants.
18. It is recommended that the FRAP reagent is prepared freshly every day. When not in use it is best to keep the formed FRAP reagent incubated at 37 °C. The solubility of the iron-TPTZ complexes is vital to this reaction and so must be carried out under acidic conditions (pH 3.6). Acetate buffer is used to lower the pH so that iron remains soluble [19]. This is often seen as a limitation of the FRAP assay as it does not mimic physiological pH.
19. Selecting the correct incubation time for the FRAP assay is a crucial variable when attempting to measure the antioxidant capacity of an extract accurately. Compounds that have a redox potential lower than that the Fe(III)-TPTZ complex (0.771 eV) will cause a reduction. Certain antioxidants such as polyphenols may react quite rapidly with the FRAP reagent, while others may react more slowly [64]. Therefore, a practical compromise must be made between both scenarios. The 40 min incubation time described in this assay is only a suggestion and can be altered depending on the antioxidant compounds present in each sample.
20. The absorbance of the Fe(II)-TPTZ complex formed as a result of antioxidant mediated reduction of the Fe(III)-TPTZ complex should range between 0.2 and 1.8 ± 0.02 AU when measured at 593 nm. If this is not the case, extracts should be either concentrated or diluted accordingly. This may require trial and error but can be minimized by testing various concentrations of each extract.
21. Sodium carbonate is used in this reaction to increase the alkalinity of the mixture (approx. pH 10). Under alkaline conditions the phenolic compounds become deprotonated forming a

phenolate ion, which then reacts faster with the FC reagent (acidic) than it would under acidic conditions.

22. The addition of sodium carbonate causes the reaction mixture to become cloudy due to the precipitation of sodium phosphate. After the incubation time has elapsed, the samples must be centrifuged in order to remove the precipitate from solution. Since the endpoint of the reaction is quite stable, time can be taken to perform this task.

23. When measuring the absorbance (765 nm) of the reaction endpoint in the presence of a standard or sample, the value should range between 0.2 and 1.8 ± 0.02 AU. If this is not the case extracts should be either concentrated or diluted accordingly. This may require trial and error but can be minimized by testing various concentrations of each extract.

24. Crude extracts will generally contain a range of different unknown antioxidant compounds. In the CUPRAC assay several small phenolic compounds have been shown [40] to quickly reduce the Cu(II)-Nc complex, while other larger molecules act much more slowly. The reaction time can be altered to ensure the reaction is complete. Alternatively, the temperature can be increased to 50 °C [40].

25. The absorbance of the Cu(I)-Nc complex formed as a result of antioxidant mediated reduction of the Cu(II)-Nc complex should range between 0.2 and 1.8 ± 0.02 AU when measured at 450 nm. If this is not the case, extracts should be either concentrated or diluted accordingly. This may require some trials and error can be minimized by testing various concentrations of each extract.

26. Sample reactions that show a rapid decay in fluorescence signal are too weak and may need to be concentrated. On the other hand sample reactions that show little decay in fluorescence signal over 35 min are too strong and should be diluted.

27. In the reaction mixture, the Co(II) fluoride solution reacts with H_2O_2 in a Fenton-like reaction to produce hydroxyl radicals [65]. In the original method described by Ou et al. [46], Cu(I) and Fe(II) were also considered for hydroxyl radical production. However, these metals were shown to give unstable production of hydroxyl radicals as they are either prone to oxidation by air or prevent fluorescence decay at pH 7.4 [46]. Co(II) fluoride is also not soluble in water at this pH and forms a complex with picolinic acid (PA). The Co(II)-PA complex has demonstrated adequate fluorescence decay curves under these experimental conditions [46].

28. The control of temperature throughout this assay is crucial for reproducibility. The oxidation of the carotenoid substrate depends on heat. If the heat is not constant, oxidation will occur at different rates between wells on the plate.

29. β-Carotene exhibits a maximum absorbance at 470 nm. However, at this wavelength in crude extracts other pigments can also absorb quite strongly causing interference with accurately measuring their antioxidant capacity. Therefore, it is not recommended to use strongly absorbing extracts at high concentrations with this assay. Furthermore, if crocin is being used as the carotenoid substrate, then measure the absorbance using a UV–VIS spectrophotometer at 450 nm.
30. At this stage of the assay, linolenic acid is auto-oxidized by heat (40 °C) resulting in the formation of LOOH radicals. If an antioxidant is present, LOOH radicals become scavenged and are therefore not available to oxidize Fe(II)-thiocyanate in the next step of the reaction.
31. The endpoint of the reaction is detected spectroscopically by the saturation of the Fe(III)-thiocyanate chromophore at 500 nm. This normally occurs between 8 and 10 days after the initial reaction. Consequently, the FTC assay has been widely criticized for its extensive reaction time. Results are generally expressed in AU at 500 nm and compared directly to the blank control or positive controls (α-Tocopherol and BHT).
32. The production MDA requires the oxidation of fatty acids containing at least three methylene-interrupted double bonds. In the assay, linolenic acid is broken down to form lipid peroxides that are further broken down to MDA after boiling in acidic conditions.

References

1. Blunt JW, Copp BR, Keyzers RA et al (2013) Algal natural products. Nat Prod Rep 30: 237–323
2. Sasso S, Pohnert G, Lohr M et al (2012) Microalgae in the postgenomic era: a blooming reservoir for new natural products. FEMS Microbiol Rev 36:761–785
3. Harnedy PA, FitzGerald RJ (2012) Bioactive peptides from algal processing waste and shellfish: a review. J Funct Foods 4:6–24
4. Wijesinghe W, Jeon YJ (2012) Enzyme-assistant extraction (EAE) of bioactive components: a useful approach for recovery of industrially important metabolites from seaweeds: a review. Fitoterapia 83:6–12
5. Konaté K, Hilou A, Mavoungou JF et al (2012) Antimicrobial activity of polyphenol-rich fractions from *Sida alba* L. (*Malvaceae*) against cotrimoxazol-resistant bacteria strains. Ann Clin Microbiol Antimicrob 11:5
6. Ghiringhelli F, Rebe C, Hichami A et al (2012) Immunomodulation and anti-inflammatory roles of polyphenols as anticancer agents. Anticancer Agents Med Chem 12:852–873
7. Alu'datt MH, Ereifej K, Abu-Zaiton A et al (2012) Anti-oxidant, anti-diabetic, and anti-hypertensive effects of extracted phenolics and hydrolyzed peptides from barley protein fractions. Int J Food Prop 15:781–795
8. Carocho M, Ferreira IC (2013) A review on antioxidants, prooxidants and related controversy: natural and synthetic compounds, screening and analysis methodologies and future perspectives. Food Chem Toxicol 51: 15–25
9. Dangles O (2012) Antioxidant activity of plant phenols: chemical mechanisms and biological significance. Curr Org Chem 16:692–714
10. Patras A, Yuan YV, Costa HS et al (2013) Antioxidant activity of phytochemicals. In: Tiwari BK, Brunton NP, Brennan CS (eds) Handbook of plant food phytochemicals: sources, stability and extraction. Wiley, Oxford, pp 452–472

11. Huang D, Ou B, Prior RL (2005) The chemistry behind antioxidant capacity assays. J Agric Food Chem 53:1841–1856
12. Schaich KM (2006) Developing a rational basis for selection of antioxidant screening and testing methods. Acta Hort (ISHS) 709:79–94
13. Ghiselli A, Serafini M, Maiani G et al (1995) A fluorescence-based method for measuring total plasma antioxidant capability. Free Radic Biol Med 18:29–36
14. Ruiz-Larrea MB, Leal AM, Liza M et al (1994) Antioxidant effects of estradiol and 2-hydroxyestradiol on iron-induced lipid peroxidation of rat liver microsomes. Steroids 59:383–388
15. Blois MS (1958) Antioxidant determinations by the use of a stable free radical. Nature 26:1199–1200
16. Molyneux P (2003) The use of stable free radical diphenylpicrylhydrazyl (DPPH) for estimating antioxidant activity. J Sci Technol 26:211–219
17. Brand-Williams W, Cuvelier ME, Berset C (1995) Use of a free radical method to evaluate antioxidant. LWT Food Sci Technol 28:25–30
18. Papariello GJ, Janish MAM (1966) Diphenylpicrylhydrazyl as an organic analytical reagent in the spectrophotometric analysis of phenols. Anal Chem 38:211–214
19. Prior RL, Wu X, Schaich K (2005) Standardised methods for the determination of antioxidant capacity and phenolics in foods and dietary supplements. J Agric Food Chem 53:4290–4302
20. MacDonald-Wicks LK, Wood LG, Garg ML (2006) Methodology for the determination of biological antioxidant capacity *in vitro*: a review. J Sci Food Agric 86:2046–2056
21. Goupy P, Hugues M, Boivin P et al (1999) Antioxidant composition and activity of barely (*Hordem vulgare*) and malt extracts and of isolated phenolic compounds. J Sci Food Agric 79:1625–1634
22. Miller NJ, Diplock AT, Rice-Evans C et al (1993) A novel method for measuring antioxidant capacity and its application to monitoring the antioxidant status in premature neonates. Clin Sci 84:407–412
23. Re R, Pellegrini N, Proteggente A et al (1999) Antioxidant activity applying an improved ABTS radical cation decolourisation assay. Free Radic Biol Med 26:1231–1237
24. Lemanska K, Szymusiak H, Tyrakowska B et al (2001) The influence of pH on the antioxidant properties and the mechanisms of antioxidant action of hydroxyflavones. Free Radic Biol Med 31:869–881
25. Fogliano V, Verde V, Randazzo G et al (1999) Method for measuring antioxidant activity and its application to monitoring the antioxidant capacity of wines. J Agric Food Chem 47: 1035–1040
26. Mehdi MM, Rizvi SI (2013) *N,N*-dimethyl-p-phenylenediamine dihydrochlorite-based method for the measurement of plasma oxidative capacity during human aging. Anal Biochem 436:165–167
27. Ching TL, de Jong J, Bast A (1994) A method for screening hypochlorous acid scavengers by inhibition of the oxidation of 5-thio-2-nitrobenzoic acid: application to anti-asthmatic drugs. Anal Biochem 218:377–381
28. Winterbourn CC (1985) Comparative reactivities of various biological compounds with myeloperoxidase-hydrogen peroxide-chloride and similarity of the oxidant to hypochlorite. Biochim Biophys Acta 840:204–210
29. Pullar JM, Winterbourn CC, Vissers MC (1999) Loss of GSH and thiol enzymes in endothelial cells exposed to sublethal concentrations of hypochlorous acid. Am J Physiol 277:1505–1512
30. Hampton MB, Kettle AJ, Winterbourn CC (1998) Inside the neutrophi phagosome; oxidants, myeloperoxidase and bacterial killing. Blood 92:3007–3017
31. Aune TM, Thomas EL (1977) Accumulation of hypothiocyanite ion during peroxidise-catalyzed oxidation of thiocyanate ion. Eur J Biochem 80:209–214
32. Fernandes E, Toste SA, Lima JLFC et al (2003) The metabolism of sulindac enhances its scavenging activity against reactive oxygen and nitrogen species. Free Radic Biol Med 35:1008–1017
33. Benzie IFF, Strain JJ (1996) The ferric reducing ability of plasma (FRAP) as a measurement of "antioxidant power": the FRAP assay. Anal Biochem 47:633–636
34. Stratil P, Klejdus B, Kuban V (2006) Determination of total content of phenolic compounds and their antioxidant activity in vegetables – evaluation of spectrophotometric methods. J Agric Food Chem 54:607–616
35. Folin O (1927) Tyrosine and tryptophan determinations in proteins. J Biol Chem 73:672
36. Singelton VL, Rossi JA (1965) Colorimetry of total phenolics with phosphomolybdic-phosphotungstic acid reagents. Am J Enol Vitic 16:144–158
37. Singh S, Singh RP (2008) In vitro methods of assay of antioxidants: an overview. Food Rev Int 24:392–415

38. Box JD (1983) Investigation of the Folin-Ciocalteu phenol reagent for the determination of polyphenolic substances in natural water. Water Res 17:511–525
39. Peterson GL (1979) Review of the Folin phenol protein quantitation method of Lowery, Rosebrough, Farr and Randall. Anal Biochem 18:201–220
40. Apak R, Guclu K, Ozyurek M et al (2004) Novel total antioxidant capacity index for dietary polyphenols and vitamins C and E, using the cupric iron reducing capability in the presence of neocuproine: CUPRAC method. J Agric Food Chem 52:2970–2981
41. Apak R, Guclu K, Ozyurek M et al (2005) Total antioxidant capacity assay of human serum using copper(II)-neocuproine as chromogenic oxidant: the CUPRAC method. Free Radic Res 39:949–961
42. Karadag A, Ozcelik B, Saner S (2009) Review of methods to determine antioxidant capacities. Food Anal Method 2:41–60
43. Cao GH, Alessio H, Cutler RG (1993) Oxygen-radical absorbance capacity assay for antioxidants. Free Radic Biol Med 14: 303–311
44. Glazer AN (1990) Phycoerythin fluorescence-based assay for reactive oxygen species. Methods Enzymol 186:161–168
45. Huang D, Ou B, Hampsch-Woodill M et al (2002) High-throughput assay of oxygen radical absorbance capacity (ORAC) using a multichannel liquid handling system coupled with a microplate fluorescence reader in 96-well format. J Agric Food Chem 50:4437–4444
46. Ou B, Hampsch-Woodill M, Flanagan J et al (2002) Novel fluorometric assay for hydroxyl radical prevention capacity using fluoroscein as the probe. J Agric Food Chem 50:2772–2777
47. Macro G (1968) A rapid method for the evaluation of antioxidants. J Am Oil Chem Soc 45:594–598
48. Koleva II, van Beek TA, Linssen JPH et al (2002) Screening of plant extracts for antioxidant activity: a comparative study on three testing methods. Phytochem Anal 13:8–17
49. Bors W, Michel C, Saran M (1984) Inhibition of the bleaching of the carotenoid crocin a rapid test for the quantifying antioxidant activity. Biochem Biophys Acta 796:312–319
50. Wagner CD, Clever HL, Peters ED (1947) Evaluation of the ferrous thiocyanate colorimetric method. Anal Chem 19:980–982
51. Kikuzaki H, Nakatani N (1993) Antioxidant effects of some ginger constituents. J Food Sci 58:1407–1410
52. Chan HWS, Coxon DT (1987) Autoxidation of unsaturated lipids. Academic, London
53. van Kuijk FJGM, Thomas DW, Stephens RJ et al (1990) Gas chromatography-mass spectrometry assays for lipid peroxides. Methods Enzymol 186:388–398
54. Frankel EN, Neff WE, Weisleder D (1990) Determination of methyl linoleate hydroperoxides by ^{13}C nuclear magnetic resonance spectroscopy. Methods Ezymol 186:380–387
55. Kolthoff IM, Medalia AI (1951) Determination of organic peroxides by reaction with ferrous ion. Anal Chem 23:595–603
56. Kohn NJ, Liversedge M (1944) On a new metabolite whose production by brain is inhibited by apomorphine, emetine, ergotamine, epinephrine and menadione. J Pharmacol Exp Ther 82:292–300
57. Bernheim F, Bernheim MLC, Wilbur KM (1947) The reaction between thiobarbituric acid and the oxidation products of certain lipids. J Biol Chem 174:247–264
58. Wilbur KM, Bernheim F, Shapiro OW (1949) The thiobarbituric acid reagent as a test for the oxidation of unsaturated fatty acids by various agents. Arch Biochem 24:305–313
59. Arnao MB (2000) Some methodological problems in the determination of antioxidant activity using chromogen radicals: a practical case. Trends Food Sci Technol 11:419–421
60. Magalhaes LM, Segundo MA, Reis R et al (2008) Methodological aspects about *in vitro* evaluation of antioxidant properties. Anal Chim Acta 613:1–19
61. Stasko A, Brezova V, Biskupic S et al (2007) The potential pitfalls of using 1,1-diphenyl-2-picrylhydrazyl to characterise antioxidants in mixed water solvents. Free Radic Res 41: 379–390
62. Noruma T, Kikuchi M, Kawakami Y (1997) Proton-donative antioxidant activity of fucoxanthin with 1,1-diphenyl-2-picrylhydrazyl (DPPH). Biochem Mol Biol Int 42:361–370
63. Gulcin I (2012) Antioxidant activity of food constituents: an overview. Arch Toxicol 86: 345–391
64. Pulido R, Bravo L, Saura-Calixto F (2000) Antioxidant activity of dietary polyphenols as determined by a modified ferric reducing/antioxidant power assay. J Agric Food Chem 48:3396–3402
65. Heckman RA, Espenson JA (1979) Kinetics and mechanism of oxidation of cobalt(II) macrocycles by iodine, bromide and hydrogen peroxide. Inorg Chem 18:38–43

Chapter 25

Disk Diffusion Assay to Assess the Antimicrobial Activity of Marine Algal Extracts

Andrew P. Desbois and Valerie J. Smith

Abstract

Marine algae are a relatively untapped source of bioactive natural products, including those with antimicrobial activities. The ability to assess the antimicrobial activity of cell extracts derived from algal cultures is vital to identifying species that may produce useful novel antibiotics. One assay that is used widely for this purpose is the disk diffusion assay due to its simplicity, rapidity, and low cost. Moreover, this assay gives output data that are easy to interpret and can be used to screen many samples at once irrespective of the solvent used during preparation. In this chapter, a step-by-step protocol for performing a disk diffusion assay is described. The assay is particularly well suited to testing algal cell extracts and fractions resulting from separation through bioassay-guided approaches.

Key words Antibacterial, Antibiotic, Bactericidal, Bioassay-guided fractionation, Natural products

1 Introduction

Marine algae are an untapped source of natural products with antimicrobial activities, and relatively few studies have systematically screened these polyphyletic organisms for potential new antibiotics (e.g., [1–3]). Various assays are used to detect antimicrobial activity in cell extracts from microorganisms; one of the most popular of these methods is the disk diffusion assay [4]. This simple assay relies on the diffusion of compounds contained in the cell extract from a paper disk into agar that has been seeded with a particular target microbe. The compounds in the cell extract diffuse from the disk into the agar and their concentrations reduce exponentially with distance. If the compounds are antimicrobial and present at sufficient concentrations to prevent microbial growth, a clear zone appears around the disk after the plate has been incubated to allow the development of a microbial lawn of growth on the agar. As a result, zone size corresponds approximately to the total antimicrobial activity in the cell extract and the output data are considered

Dagmar B. Stengel and Solène Connan (eds.), *Natural Products From Marine Algae: Methods and Protocols*, Methods in Molecular Biology, vol. 1308, DOI 10.1007/978-1-4939-2684-8_25, © Springer Science+Business Media New York 2015

to be semiquantitative. Moreover, the results of disk diffusion studies are easy to interpret and obtained typically within 24–48 h.

The disk diffusion assay is an inexpensive approach suitable for assessing the antimicrobial actions of relatively high numbers of cell extracts, such as might be undertaken in an early screening program. In addition, this assay is ideal to bioassay-guided fractionation approaches during subsequent studies aimed at separating and identifying the active compound in the cell extracts [4]. An extract screen would likely generate tens of samples for testing, whereas a typical fractionation by solid-phase extraction and high-performance liquid chromatography could generate hundreds of fractions, and the disk diffusion assay is well suited to these medium- and high-throughput experiments. In addition, it is possible to test extracts and samples prepared in any solvent because this is evaporated prior to the commencement of the test. Often this can prevent the use of liquid-based approaches to assess antimicrobial activity of samples prepared in toxic solvents. One drawback of the disk diffusion assay is its inability to differentiate between growth inhibitory and microbicidal activities, and subsequent tests have to be performed to elucidate this distinction. Furthermore, the assay is regarded as semiquantitative and caution has to be exerted when comparing the activities of different samples because many factors can influence the apparent antimicrobial activity of a sample, including the concentrations, molecular sizes, solubilities, and polarities of the active compounds.

The aim of this chapter is to provide a step-by-step method for performing the disk diffusion assay to assess the antimicrobial activity of algal cell extracts and fractions from bioassay-guided separations. Additional notes are provided to clarify any steps, offer tips and "know-how," and explain the underlying theory.

2 Materials

The disk diffusion assay can be performed in any general microbiology laboratory with the purchase of few additional pieces of equipment or consumables. Thus, it is expected that the laboratory has workspace suited to performing aseptic technique and access to standard infrastructure, such as incubators and an autoclave. Importantly, as this assay can be performed with human pathogens, it is vital to adhere to all local, national, and international safety precautions and procedures associated with working with these hazardous microorganisms and disposing of contaminated waste materials (e.g., in the United Kingdom, *see* [5]).

The consumables and equipment required to perform this assay are:

1. "Blank" 6 mm diameter antibiotic susceptibility paper disks.
2. Absolute ethanol.

3. Distilled/deionized water.
4. Aluminum foil.
5. Sterile cotton wool swabs.
6. Sterile Petri dishes (9 cm diameter; polystyrene).
7. Agar media suitable for growing the target microbes of interest.
8. Pipette and sterile tips suitable for transferring up to 20 μL of liquid.
9. Forceps (metal or sterile disposable plastic).
10. Blue roll, ruler, black marker pen, and calculator.

3 Methods

3.1 Preparation of Materials

1. Autoclave the paper disks, water, agar, forceps (if using metal ones), and tips at 121 °C for 15 min (*see* **Note 1**). Most other equipments are either provided sterile from the manufacturer or do not require sterilization.
2. Agar for the assay should be prepared according to manufacturers' instructions. Once poured into Petri dishes (depth of approximately 4 mm), the plates should be left to dry with lids on at room temperature for at least 24–48 h (*see* **Note 2**).
3. Cover a piece of rigid cardboard or plastic (approximately 20 × 30 cm) in aluminum foil, as the paper disks rest on this board during the loading and drying steps (*see* **Note 3**).
4. An inoculum of the target bacterium or fungus is required and this should come from an agar plate with multiple colonies or lawn growth (*see* **Note 4**).

3.2 Preparation of the Disks

1. Spray the foil-covered board with absolute ethanol and wipe clean with blue roll. Repeat this step with sterile water (*see* **Note 5**).
2. Once the board surface has dried completely, place the paper disks on the board and arrange in rows using clean forceps (*see* **Note 6**).
3. Load 20 μL of sample onto the center of the first paper disk using a pipette. Repeat this process by using a separate disk for each sample (*see* **Note 7**).
4. Let the paper disks to dry completely at room temperature (*see* **Note 8**).
5. Reload the paper disks with further sample (20 μL each time) to increase if needed the total quantity of sample to be tested (*see* **Note 9**).
6. Prepare a positive control paper disk to contain a quantity of a compound known to give a particular zone size against the target microbe (*see* **Note 10**).

7. Prepare negative control paper disks loaded with volumes equal to the maximum quantities of solvents loaded onto any of the test disks (*see* **Note 11**).
8. It is desirable, but not always necessary, to have control paper disks for each agar plate because it may not be possible to fit all the disks onto a single plate.
9. Let all of the loaded paper disks and corresponding control disks dry completely at room temperature.

3.3 Preparation of the Plate

1. Swab a visible quantity of microbial growth from the inoculum agar plate with the use of a sterile cotton wool swab. It is recommended that this material is derived from more than one colony (e.g., 3–5 colonies) (*see* **Note 12**).
2. Starting from the center of the Petri dish, streak the microbes on the cotton wool swab across the plate in multiple rapid strokes until reaching the top of the plate (Fig. 1). It is important that the agar surface is evenly covered with the microbial inoculum (*see* **Note 13**).
3. Turn the plate 180 ° and repeat the streaking process exactly as before (Fig. 1).
4. Turn the plate 90 ° and, starting from the center of the Petri dish, again streak across the plate until reaching the top (Fig. 1).

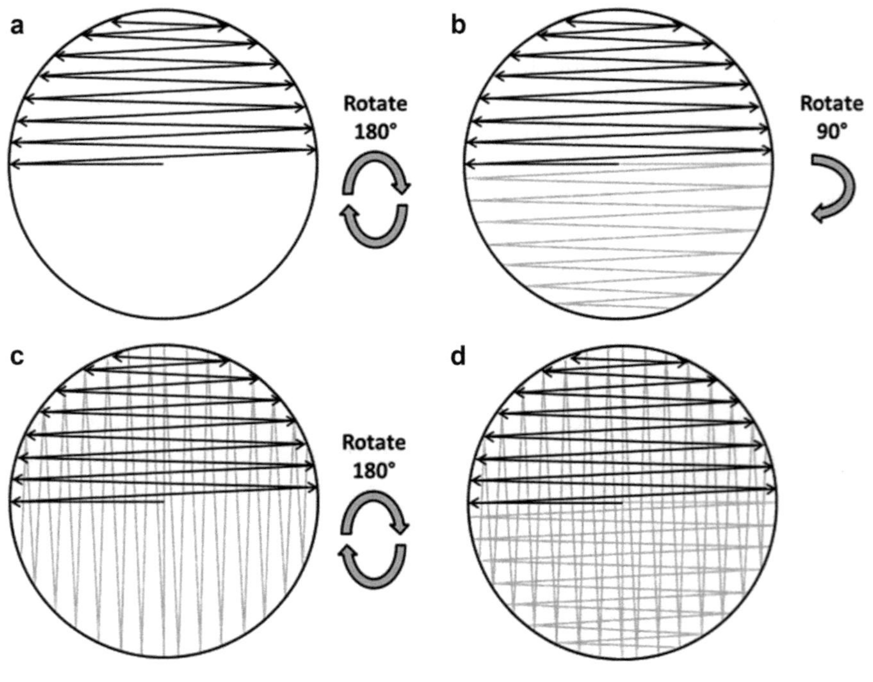

Fig. 1 Schematic representation of how to spread the microbial inoculum evenly across the agar plate with a sterile cotton wool swab (steps (**a**) to (**d**)). In reality, the lines should be virtually overlapping such that the whole surface of the agar comes into contact with the head of the swab

Finally, turn the plate 180 ° and for the final time repeat the streaking process exactly as before (Fig. 1) (*see* **Note 14**).

5. Label the underside of the plate to indicate where each disk should be positioned. The disks should be positioned in a regular pattern with plenty of space between them (*see* **Note 15**).
6. Place the loaded paper disks and corresponding controls with sterile forceps onto the agar surface in positions matching their labels and press down gently on each disk to ensure that it is attached to the surface (*see* **Note 16**).
7. Invert the agar plates and incubate at the optimum growth temperature for the target microbe (*see* **Note 17**).
8. The plates should be incubated for a period sufficient to allow the development of a lawn of growth, which is typically 24–48 h for most microbial species (*see* **Note 18**).

3.4 Reading the Plate

1. Read the results of the assay by holding the agar plate up to a light or placing it onto a dark background such as black paper. The presence of a clear zone around a paper disk indicates that antimicrobial compounds have diffused into the agar from the disk in sufficient concentrations to prevent microbial growth (*see* **Note 19**).
2. Mark the x- and y-plane edges of each clear zone with a black marker pen.
3. With the aid of a ruler, measure the x- and y-plane diameters of each clear zone to the nearest millimeter (*see* **Note 20**).
4. Calculate the mean radius for the clear zone from the two diameter measurements using the formula $r = \left(d_1 + d_2\right) / 4$.
5. Calculate the approximate total area of the clear zone (A) using the equation $A = \pi r^2$, where r is the mean radius of the clear zone (*see* **Note 21**).
6. Finally, calculate the area of the paper disk and subtract this value from each of the total area of the clear zone values to give an estimate of the antimicrobial activity in each sample (*see* **Note 22**).

4 Notes

1. It is recommended to autoclave batches of 100–200 paper disks in separate glass universal bottles.
2. It is important to use dry agar plates for the assay because moisture on the agar surface will influence the diffusion of compounds contained in the test sample from the paper disk. Nevertheless, if the agar is especially dry or cracking then the plates should be discarded.

3. The board can be reused after washing with ethanol and water but to prevent contamination of later tests it is probably advisable to prepare a fresh board for each experiment.
4. It is desirable that the inoculum agar is the same as that being used for the disk diffusion assay plate, as this will reduce any lag growth effects caused by adaptation to new growth conditions.
5. The foil-covered board should be cleaned thoroughly to ensure that it is free from any contaminating compounds that could be absorbed into the paper disks.
6. Do not use fingers to move the paper disks because there are antimicrobial compounds on the skin that could be absorbed into the disk and influence the outcome of the assay.
7. If the sample to be tested is less than 20 μL, it is advisable to increase the sample volume to 20 μL so that it will be evenly distributed throughout the paper disk, which should ensure uniform diffusion in all directions through the agar.
8. Different solvents will take varying times to evaporate from the paper disk. The disks should be allowed to dry away from drafts created by equipment or workers because these can cause the disks to move, which risks them becoming mixed up. Though the volumes of solvents being dried are very small, please consider whether there is a need to dry the paper disks in an extraction hood. The drying step ensures that the paper disk is free from potentially toxic solvents prior to testing. Furthermore, it is important to ensure that the disks are dried completely, as residual solvent may prevent further loading of the disk or influence the diffusion of the sample on the agar plate.
9. It is not unusual to repeat the reloading step up to five times, which would then assess the antimicrobial effects of the compounds in 100 μL of the sample.
10. Preliminary experiments should be performed to ascertain suitable positive controls for each target microbe. A zone size with a diameter of approximately 15 mm is recommended.
11. Often multiple negative control paper disks are required as the test samples may be constituted in a variety of solvents.
12. As this assay is semiquantitative there is no need to use a precise standardized inoculum. The assay can be performed for Gram-positive and Gram-negative bacteria, as well as fungi and other microbes that divide to form lawns of growth on agar plates.
13. Generally it is easier to spread "up" the plate, but be careful not to press too hard during streaking because this may damage the agar surface. If the agar is damaged, for example by tearing, it should be discarded and a fresh plate prepared.
14. Ensure that the whole surface of the agar is covered with the inoculum as evenly as possible. It will not be possible to see the coverage of the microbe clearly because this should be a

very fine film of cells; however it should be possible to see where the swab has contacted the agar.

15. There should be room for up to 25 paper disks on each agar plate and a paper grid or template can be prepared to aid with spacing.
16. When applying gentle force to the top of each disk be careful to ensure that the disk does not move horizontally as this may create a small clear zone after incubation. Never try to replace a disk once it has contacted the surface.
17. There is no need to seal the agar plates with tape or products such as Parafilm M®.
18. Though distinct lines may be apparent in the lawn (corresponding to where the swab was streaked across the surface), the lawn should be reasonably uniform across the entire agar surface.
19. There should be no clear zone around the negative control paper disks but there should be a visible clear zone of approximately the expected size (close to 15 mm) around the positive control disk.
20. If a clear zone has apparently been restricted in size by the edge of the plate, then calculate its diameter by taking the radius from the side that can be read.
21. Area is a better measure for potency than simple diameter or radius measurements due to the two-dimensional nature of the diffusion.
22. Zone sizes will vary according to various properties of the compound(s) responsible for the antimicrobial activity, including concentration, molecular size (larger compounds will diffuse more slowly), solubility, and polarity. This assay is semi-quantitative and caution is advised when comparing zone sizes, particularly when the nature of the antimicrobial compound(s) responsible for the activity is unknown. It is worth remembering that in samples containing a mixture of chemical constituents there is no way of distinguishing a high-potency compound at low concentration or a low-potency compound at high concentration as both scenarios could give identical zone sizes. Moreover, there is inherent plate-to-plate variability caused by other factors, such as the size of inoculum used for each plate and the thickness of the agar (since the compounds will diffuse in three dimensions a shallow agar will produce a larger clear zone than a deeper agar). Medium composition, such as pH and ion concentrations (especially divalent cations such as calcium and magnesium), will also influence the sizes of the clear zones. Finally, agarose can be used in place of agar to set the plates, as the former is less chemically complex and less charged, meaning it generally demonstrates fewer physicochemical interactions with biomolecules such as proteins.

References

1. Reichelt JL, Borowitzka MA (1984) Antimicrobial activity from marine algae: results of a large-scale screening programme. Hydrobiologia 116–117:158–168
2. Kellam SJ, Walker JM (1989) Antibacterial activity of marine microalgae in laboratory culture. Br Phycol J 24:191–194
3. Guedes AC, Barbosa CR, Amaro HM et al (2011) Microalgal and cyanobacterial cell extracts for use as natural antibacterial additives against food pathogens. Int J Food Sci Technol 46:862–870
4. Desbois AP, Lebl T, Yan L et al (2008) Isolation and structural characterisation of two antibacterial free fatty acids from the marine diatom, *Phaeodactylum tricornutum*. Appl Microbiol Biotechnol 81:755–764
5. Advisory Committee on Dangerous Pathogens (2005). Biological agents: managing the risks in laboratories and healthcare premises. Health and Safety Executive, UK. http://www.hse.gov.uk/biosafety/biologagents.pdf. Accessed 12 Nov 2013

Chapter 26

Screening of a Marine Algal Extract for Antifungal Activities

Graciliana Lopes, Paula B. Andrade, and Patrícia Valentão

Abstract

Over the past few years algal extracts have become increasingly interesting to the scientific community due to their promising biological properties. Phlorotannin extracts are particularly attractive partly due to their reported antifungal activity against several yeast and dermatophyte strains.

The micromethod used for the evaluation of the minimum inhibitory concentration (MIC) and the minimum lethal concentration (MLC) represents an effective and solvent-saving procedure to evaluate the antifungal activity of algae extracts. Here we describe the micromethod for determining the MIC and the MLC of algal extracts by using the example of a purified phlorotannin extract of brown algae.

Key words Algae, Antifungal, Micromethod, Minimum inhibitory concentration (MIC), Minimum lethal concentration (MLC), Phlorotannins

1 Introduction

Over the past few decades, marine organisms have been the source for the discovery of many new molecules with a broad spectrum of biological activities [1, 2]. Amongst these, algae hold a prominent place, not only because of their chemical and structural diversity, but also because of their historical use in food and traditional medicine since ancient times [1].

Despite the interest stimulated by most algal groups investigated, brown algae hold a prominent position as they are the only group of algae capable of producing phlorotannins, a class of metabolites with unique features [2, 3]. These polyphenols are polymers of phloroglucinol (1,3,5-trihydroxybenzene), biosynthesized through the acetate-malonate pathway in the Golgi apparatus in the perinuclear area of the cell [3]. In brown algae, phlorotannins can be found as part of the cell membranes and stored in physodes, having as main function the defense of the organism against oxidative stress, pathogens, and herbivores [4]. Regardless of the interesting bioactivities displayed by whole extracts of algae

Dagmar B. Stengel and Solène Connan (eds.), *Natural Products From Marine Algae: Methods and Protocols*, Methods in Molecular Biology, vol. 1308, DOI 10.1007/978-1-4939-2684-8_26, prepared with different solvents [5, 6], phlorotannin extracts have been the focus of many studies. In addition to their important natural ecological function, phlorotannins display several biological activities that can be important for the pharmaceutical industry. Thus, compared to several drugs currently used in therapeutics, phlorotannins show reduced undesirable side effects and microorganism resistance [7]. The wide array of biological properties reported for phlorotannins has been attributed to both extracts and isolated compounds. Amongst the most studied biological activities, anti-inflammatory [5, 8], antioxidant [5, 8–13], antibacterial [8, 12–14], and antifungal [8, 15, 16] can be highlighted. Amongst these, antifungal activity has been the least explored and the studies that are available are only relatively recent.

The search for new antifungal agents is an emergent need, not only due to the increment of fungal infections, but also because of the recent increase in resistance to antifungal agents [17]. Additionally, the increment in using antifungal drugs has increased the number and variety of fungal resistance. Resistant microorganisms can cause morbidity and mortality, entailing high costs for the health care systems [18]. Therefore, substantial attention has been given to natural products with antifungal properties, stimulating the search for therapeutic alternatives.

Phlorotannins seem to be a good alternative for the treatment of fungal infections, especially those caused by dermatophytes [8, 15, 16]. To our knowledge, there is only one study concerning the antifungal activity of isolated phlorotannins [16]. Lee et al. isolated dieckol from the brown algae *Ecklonia cava* Kjellman and evaluated its antifungal activity against the dermatophyte *Trichophyton rubrum*. This study showed the fungicidal activity of dieckol, possibly due to its ability to promote the loss of cytoplasmic membrane integrity of the dermatophyte. A screening of the antifungal activity of phlorotannins was performed by our research group [8, 15] who evaluated the antifungal activity of phlorotannin purified extracts over a wide range of yeast, dermatophytes, and *Aspergillus* sp. [15]. The results demonstrate that dermatophytes were the most sensitive to phlorotannins confirming previous results [8, 15, 16]. The phlorotannin extracts studied were fungistatic for yeast and inactive against *Aspergillus* sp. [8, 15]. An assessment of the mechanism of antifungal action was recently performed and the influence of phlorotannin extracts on the germ tube production by the yeast *Candida albicans* ATCC 10231 was also assessed [15].

This chapter describes the methodology to assess the antifungal activity of an algal extract. This screening for the antifungal activity comprises the determination of the minimum inhibitory concentration (MIC) and minimum lethal concentration (MLC) against different types of microorganisms. The methodology described here consists in a micromethod performed in a flat-bottomed 96-well plate, modified according to the Clinical and Laboratory Standards Institute (CLSI) reference protocols [19, 20]. It is solvent saving and represents a valuable improvement to avoid trailing effects generally observed with *C. albicans* [21]. This method can be applied to any algal extract; in this chapter, a purified phlorotannin extract from brown seaweeds is used as an example.

2 Materials

Use non-sterile deionized water (unless indicated otherwise) and analytical grade reagents to prepare the solutions. Store all reagents and cultures at 4 °C (unless specified otherwise). Follow the waste disposal regulations concerning reagents and microbial cultures.

2.1 Determination of the MIC

1. Yeast cells (*Candida albicans* ATCC 10231 and *Candida krusei* ATCC 6258), dermatophyte (*Trichophyton rubrum*), and *Aspergillus* sp. (*Aspergillus fumigatus*), stored in broth medium with 20 % glycerol, at −70 °C, and subcultured in Sabouraud dextrose agar (SDA) with chloramphenicol before preparing the suspensions.
2. RPMI-1640 broth (with L-glutamine, without bicarbonate, and with phenol red as pH indicator) (Biochrom AG, Berlin, Germany): To prepare 500 mL of culture medium, weigh 5.2 g of RPMI and 17.27 g of 3-(*N*-morpholino) propanesulfonic acid (MOPS) and transfer the powders to a 500 mL cylinder (*see* **Note 1**). Add water to a volume of 400 mL, mix, and adjust the pH with NaOH (*see* **Note 2**). Make up to 500 mL with water, sterilize by filtration through a sterile 0.22 μm membrane, under aseptic technique, and store in the refrigerator.
3. RPMI-1640 with 5 % DMSO: To prepare 5 mL of this solution, pipette 4,975 μL of RPMI-1640 to a sterile Falcon, with a 5,000 μL volumetric pipette, under aseptic technique. Add 25 μL of DMSO, vortex and store.
4. Fluconazole (Pfizer): To prepare the stock solution (2.048 mg/mL) weigh 2.5 mg of fluconazole into an Eppendorf. Dissolve in 1,220 μL of water and sterilize by filtration through a sterile 0.22 μm membrane, under aseptic technique. Divide in two aliquots of 250 μL and one aliquot of 220 μL and store at −80 °C.
5. 0.85 % NaCl solution: To prepare 100 mL of solution, weigh 850 mg of NaCl and transfer to a beaker. Dissolve by gradually adding 100 mL of water. Transfer to a glass bottle and sterilize by autoclaving for 15 min at 120 °C. Leave the bottle at room temperature to cool down and then store it in the refrigerator.

2.2 Determination of the MLC

1. SDA (Bio-Mérieux, Mary L'Étoile, France): To prepare 500 mL of culture medium weigh 22.75 g of SDA and transfer the powder to a 1 L Erlenmeyer. Add 250 mL of water, shake gently to dissolve the powder, and add the remaining water. Close the Erlenmeyer with cotton and sterilize by autoclaving at 120 °C for 15 min. Leave the Erlenmeyer at room temperature for 5 min and then put it in a water bath at 50 °C for 15 min.
2. Remove the medium from the water bath and distribute it through sterile Petri dishes. Use a volume of approximately 18–20 mL for each dish. Close the dishes and leave them at room temperature until complete solidification. Invert the dishes and store at 4 °C (*see* **Note 3**).

3 Methods

3.1 Preparation of Phlorotannin Extract for the Screening of Antifungal Activity

1. Add 500 μL of DMSO to the dry phlorotannin extract (*see* **Note 4**).
2. Transfer the solution to a 5 mL volumetric flask and complete the volume with water (*see* **Note 5**).
3. Sterilize the solution by filtration with a sterile 0.22 μm size pore membrane, under aseptic technique.
4. Keep the solution in the freezer (−80 °C) until use.

3.2 Yeast and Spore Suspension Preparation

Prepare cells or spore suspensions from recent cultures of the different strains of fungi on SDA with chloramphenicol and dilute them with RPMI-1640 broth in order to achieve the desired colony-forming units (CFU)/mL. The cell and spore suspensions must be freshly made, and should be used at least within the 2 h that follow their preparation.

3.2.1 Yeast Suspension (C. albicans ATCC 10231 and Candida krusei ATCC 6258)

1. Prepare a yeast suspension in an API® tube and adjust the turbidity to 0.5 MFA in a densitometer.
2. Make a 1:50 dilution with RPMI-1640, in an Eppendorf (v/v; 10 μL yeast suspension + 490 μL RPMI-1640), followed by a 1:20 dilution with RPMI-1640, in a Falcon (v/v 250 μL of the 1:50 dilution + 4750 μL of RPMI-1640) and vortex.

3.2.2 Dermatophyte Spore Suspension (T. rubrum)

1. Prepare a spore suspension in a Falcon with 2 mL of sterile 0.85 % NaCl solution, vortex, and count the number of spores in a Neubauer chamber (*see* **Note 6**).
2. Adjust the suspension between 20 and 60 spores/quadrant, adding more spores or diluting with sterile 0.85 % NaCl if necessary.
3. Make a 1:100 dilution in RPMI-1640 (v/v; 50 μL of spore suspension + 4,950 μL RPMI-1640) and vortex.

3.2.3 Aspergillus sp. (A. fumigatus) Spore Suspension

1. Prepare a spore suspension in a Falcon with 2 mL of sterile 0.85 % NaCl solution, vortex, and count the number of spores in a Neubauer chamber (*see* **Note 6**).
2. Adjust the suspension to 200–250 spores/quadrant, adding more spores or diluting with sterile 0.85 % NaCl if necessary.
3. Make a 1:50 dilution in RPMI-1640 (v/v; 100 μL of spore suspension + 4,900 μL RPMI-1640) and vortex.

3.3 Determination of the MIC (See Fig. 1) [19, 20]

All the procedures must take place under aseptic conditions.

1. Make a 1:2 (v/v) dilution of the phlorotannin extract in RPMI-1640 (*see* **Note 7**).
2. Make 11 serial dilutions of the first solution, using RPMI-1640 with 5 % DMSO (*see* **Note 8**).
3. In a sterile 96-well plate, add 100 μL of each extract dilution to each well, in duplicate (*see* **Note 9**).
4. Add 100 μL of the testing microorganism suspension (yeast, dermatophyte, or *Aspergillus* sp.) to each well (*see* **Note 10**).
5. Include a positive control (mix 100 μL RPMI-1640 + 100 μL microorganism suspension) and sterility controls (200 μL of all reagents used in the assay, i.e., RPMI-1640 and 0.85 % NaCl) in each assay (*see* **Note 11**), according to the scheme shown in Fig. 1. If there is no growth on the positive control, or if there is growth on the wells corresponding to the sterility control, the MIC assay cannot be validated.
6. Include a methodology control using a known strain (*C. krusei* ATCC 6258) with a known antifungal drug (fluconazole), for which the value/range of MIC is known (e.g., 16 < MIC < 128 for *C. krusei*). The control is performed along with samples (*see* Fig. 1):
   - Prepare serial dilutions of fluconazole stock solution (2.048 mg/mL) using RPMI-1640: add 250 μL of RPMI-1640 to an aliquot of 250 μL of the fluconazole stock solution (1:2 dilution, v/v); make serial dilutions from the previous one in RPMI-1640.
   - Apply the fluconazole serial dilution in the 96-well plate, using the same procedure as for the extract.
   - Apply 100 μL of *C. krusei* ATCC 6258 suspension (after vortex) on each well.
   - Incubate along with *C. albicans* ATCC 10231 and read the MIC.
7. Incubate the plates at 35 °C in a humidified oven (put a beaker inside the oven with deionized water to avoid mediums drying), without agitation, for 48 h for *C. albicans* ATCC 10231, *C. krusei* ATCC 6258 and *A. fumigatus*, and at 30 °C, for 96 h, for *T. rubrum*.

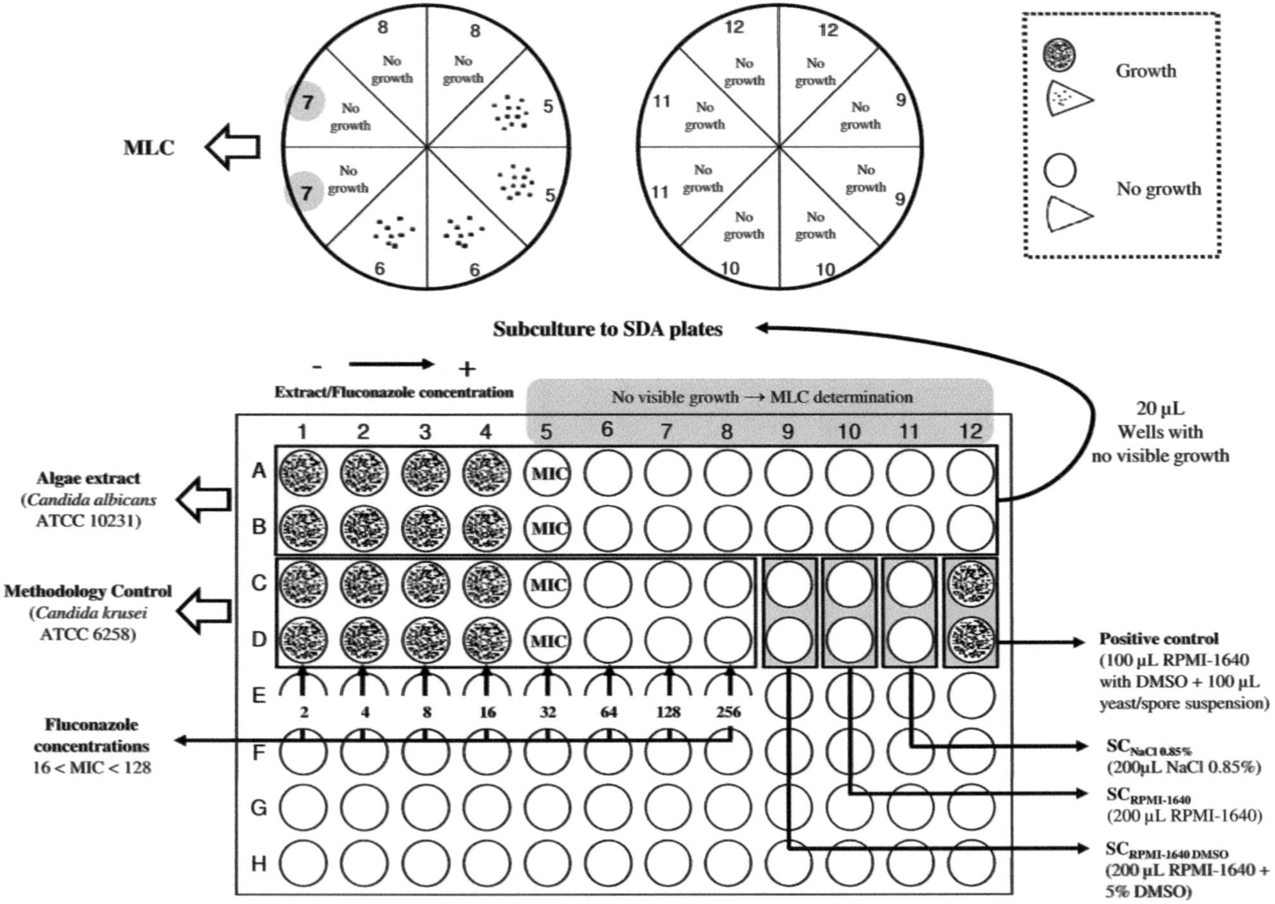

Fig. 1 Schematic representation of the micromethod for the evaluation of the MIC of algal extract. $SC_{NaCl\ 0.85\%}$ sterility control of the 0.85 % NaCl solution, $SC_{RPMI\text{-}1640}$ sterility control of the culture medium, $SC_{RPMI\text{-}1640\ DMSO}$ sterility control of the culture medium with 5 % DMSO, *SDA* Sabouraud dextrose agar, *MIC* minimum inhibitory concentration, *MLC* minimum lethal concentration

8. Read the MIC after the incubation period: The MIC value corresponds to the lowest concentration resulting in 100 % of growth inhibition (*see* **Note 12**).

3.4 Determination of the MLC (See Fig. 1) [22]

1. Remove 20 μL from all wells showing no visible growth on the 96-well plate (*see* **Note 13**) to an SDA Petri dish: divide each Petri dish into eight sections; mark each section with the extract concentration found in the well to be transferred; apply the 20 μL in the center of each division, as a drop (*see* **Note 14**).
2. Close the Petri dish and leave it at room temperature until completely dry (*see* **Note 15**).
3. Invert the Petri dish and incubate for 48 h at 35 °C for yeast and *Aspergillus* sp. and during 96 h at 30 °C for dermatophytes.
4. Read the MLC after the incubation time: the MLC corresponds to the lowest concentration showing 100 % of growth inhibition (*see* **Note 16**).

4 Notes

1. The presence of water at the bottom of the cylinder is important to prevent the powders from sticking to the glass. Close the cylinder with the stopper and shake it vigorously. RPMI-1640 and MOPS will easily dissolve.
2. First, use a 5 N NaOH solution, to reduce the difference between the starting pH and the required pH. Thereafter, a 1 N NaOH can be used until the required pH is reached, since a solution with a lower ionic strength avoids a sudden rise in the pH above the required one.
3. Petri dishes have to be filled with SDA at around 50 °C—to assure that the culture medium is still liquid, but not too hot to melt the dishes. If SDA is still hot when the Petri dish is closed, it will lead to the formation of condensation water in the cover. For this reason, and to avoid the condensed water to drop into the culture medium, as soon as SDA is solidified, dishes have to be inverted and stored.
4. Phlorotannin extracts are normally not completely soluble in water. These compounds need at least 10 % of methanol or DMSO to be dissolved. DMSO is preferred as methanol easily evaporates, making it more difficult to keep concentration of the solution constant.
5. After transferring the solution, add small amounts of water to the Erlenmeyer (in order to remove all the compounds) and transfer it to the volumetric flask, until the volume is complete.

6. For counting spores use the four lateral quadrants of the Neubauer chamber. These quadrants are divided into 16 smaller squares, corresponding to those used in biochemistry for the counting of leukocytes. Each side of the Neubauer chamber should be filled with 10 μL of spore suspension (previously vortexed). The number of spores corresponds to the mean of the spores in each of the four quadrants.
7. Make all the dilutions in Eppendorfs, using 500 μL of extract + 500 μL of RPMI-1640. The phlorotannin extract has a final DMSO concentration of 10 %. The 1:2 dilution (v/v) with RPMI-1640 gives a solution with 5 % DMSO.
8. The serial dilutions are prepared in RPMI-1640 with 5 % DMSO in order to assure that all samples used for the screening of antimicrobial activity have the same DMSO concentration. Use 500 μL of the previous solution in addition to 500 μL of RPMI-1640 with 5 % DMSO. The final DMSO concentration in the wells will be 2.5 %, because the extract will be further diluted with the microorganism suspension in RPMI-1640.
9. Vortex each phlorotannin (or other compound dilution) before adding it to the well. Start with the most diluted sample, from left to right, according to Fig. 1.
10. Vortex the microorganism suspensions for 1 min before adding them to the wells. Microorganisms tend to sediment out. In case of testing more than one phlorotannin or other compound or extract, the microorganism suspension should be vortexed between extract additions. After adding 100 μL of microorganism suspension to the wells containing 100 μL of the extract, the extract concentration will suffer a 1:2 dilution (v/v). The results will be relative to the final concentration in the well and not to the extract concentration in the Eppendorf.
11. A positive control and a sterility/negative control must always be performed with the assay, these being the first wells to be read. The positive control assures that the microorganisms are growing well while the sterility control assures that the reagents are well sterilized and that no contaminations occurred during the manipulation. The assay can only be validated if all the controls give the expected result.
12. Growth evaluation is made with the naked eye, evaluating the presence/absence of turbidity (growth), and by comparison with the positive control.
13. Mix the content of each well with a pipette before transferring it to the SDA dish. Start from the well with the highest extract concentration, using the same tip, under aseptic conditions. If it is not possible to perform the MLC assay after determining

the MIC, the 96-well plates can be kept closed in the refrigerator and the MLC can be determined within the following 24 h.

If there is no visible growth, this does not mean that the microorganism is dead. It can be kept alive in the well, but had lost the capacity to growth in contact with the tested compounds. With this assay we can evaluate the fungistatic capacity of the extracts, by the determination of their MIC. On the other hand, in order to determine whether the extracts have also fungicidal activity, the content of each well where no growth has occurred has to be transferred to a new culture medium where it will be no longer in contact with the extracts. This subculture of the 96-well plate suspension will make clear whether the microorganisms are viable or not.

14. Do not spread the liquid and keep the dish on the horizontal until completely dry. If the Petri dish is moved during this procedure, the transferred volumes can flow, mix, and hinder or preclude the reading of the results.
15. If it is taking too long for the sample drops to dry, the Petri dish can be opened and carefully moved around the Bunsen flame, avoiding liquid flow. A gentle warming will allow the drops to dry faster.
16. For an extract with fungicidal activity, the MLC value is normally equal or differs from one or two serial dilutions from MIC. If the range from MIC and MLC is higher than this, the extract cannot be considered fungicidal, but only fungistatic.

References

1. Mohamed S, Hashim SN, Rahman A (2012) Seaweeds: a sustainable functional food for complementary and alternative therapy. Trends Food Sci Technol 23:83–96
2. Blunt JW, Copp BR, Munro MHG et al (2012) Marine natural products. Nat Prod Rep 29:144–222
3. Shibata T, Kawaguchi S, Hama Y et al (2004) Local and chemical distribution of phlorotannins in brown algae. J Appl Phycol 16:291–296
4. Koivikko R, Loponen J, Honkanen T et al (2005) Contents of soluble, cell-wall-bound and exuded phlorotannins in the brown algae *Fucus vesiculosus*, with implications on their ecological functions. J Chem Ecol 31:195–212
5. Kim AR, Shi TS, Lee MS et al (2009) Isolation and identification of phlorotannins from *Ecklonia stolonifera* with antioxidant and anti-inflammatory properties. J Agric Food Chem 57:3483–3489
6. Genovese G, Leithner S, Minicante S et al (2013) The Mediterranean red alga *Asparagopsis taxiformis* has antifungal activity against *Aspergillus* species. Mycoses 56: 516–519
7. Mayer AMS, Rodríguez AD, Taglialatela-Scafati O et al (2013) Marine pharmacology in 2009-2011: marine compounds with antibacterial, antidiabetic, antifungal, anti-inflammatory, antiprotozoal, antituberculosis, and antiviral activities; affecting the immune and nervous systems and other miscellaneous mechanisms of action. Mar Drugs 11:2510–2573
8. Lopes G, Sousa C, Silva LR et al (2012) Can phlorotannins purified extracts constitute a novel pharmacological alternative for microbial infections with associated inflammatory conditions? PLoS One 7:e31145
9. Ferreres F, Lopes G, Gil-Izquierdo A et al (2012) Phlorotannin extracts from Fucales characterized by HPLC-DAD-ESI-MSn: approaches to hyaluronidase inhibitory capacity and antioxidant properties. Mar Drugs 10:2766–2781

10. Li Y, Qian ZJ, Ryu B et al (2009) Chemical components and its antioxidant properties *in vitro*: an edible marine brown alga, *Ecklonia cava*. Bioorg Med Chem 17:1963–1973
11. Parys S, Kehraus S, Krick S et al (2010) *In vitro* chemopreventive potential of fucophlorethols from the brown alga *Fucus vesiculosus* L. by antioxidant activity of selected cytochrome P450 enzymes. Phytochemistry 71:221–229
12. Devi KP, Suganthy N, Kesika P et al (2008) Bioprotective properties of seaweeds: *in vitro* evaluation of antioxidant activity and antimicrobial activity against food borne bacteria in relation to polyphenolic content. BMC Complement Altern Med 8:38
13. Cox S, Abu-Ghannam N, Gupta S (2010) An assessment of the antioxidant and antimicrobial activity of six species of edible Irish seaweeds. Int Food Res J 17:205–220
14. Eom SH, Kim YW, Kim S-K (2012) Antimicrobial effect of phlorotannins from marine brown algae. Food Chem Toxicol 50:3251–3255
15. Lopes G, Pinto E, Andrade PB et al (2013) Antifungal activity of phlorotannins against dermatophyte and yeast: approach to the mechanism of action and influence on *Candida albicans* virulence factor. PLoS One 8:e72203
16. Lee MH, Lee KB, Oh SM et al (2010) Antifungal activities of dieckol isolated from the marine brown alga *Ecklonia cava* against *Trichophyton rubrum*. J Korean Soc Appl Biol Chem 53:504–507
17. Richardson MD, Warnock DW (2012) Fungal infection: diagnosis and management, 4th edn. Wiley-Blackwell, Oxford, UK
18. Ghannoum MA, Rice LB (1999) Antifungal agents: mode of action, mechanisms of resistance and correlation of these mechanisms with bacterial resistance. Clin Microbiol Rev 12:501–517
19. Clinical and Laboratory Standard Institute (CLSI) (2008) Reference method for broth dilution antifungal susceptibility testing of filamentous fungi: approved standard, 2nd ed. M38-A2, Wayne, PA
20. Clinical and Laboratory Standard Institute (CLSI) (2008) Reference method for broth dilution antifungal susceptibility testing of yeasts: approved standard, 3rd ed. M27-A3, Wayne, PA
21. Coenye T, Vos MD, Vanderbosch D et al (2008) Factors influencing the trailing endpoint observed in *Candida albicans* susceptibility testing using the CLSI procedure. Clin Microbiol Infect 14:495–497
22. Fairhead VA, Amsler CD, McClintock JB et al (2005) Variation in phlorotannin content within two species of brown macroalgae (*Desmarestia anceps* and *D. menziesii*) from the Western Antarctic Peninsula. Polar Biol 28: 680–686

Chapter 27

Protocol for Assessing Antifouling Activities of Macroalgal Extracts

Claire Hellio, Rozenn Trepos, R. Noemí Aguila-Ramírez, and Claudia J. Hernández-Guerrero

Abstract

The development of novel environmentally friendly antifouling (AF) solutions is a very active field in fundamental and applied research. An attractive option in producing such material resides in biomimetic studies: living organisms have evolved well-adapted structures and materials over geological times through natural selection. In this chapter, we explain the experimental procedure to be followed for the preparation of macroalgal extracts and to assess their AF efficiency towards key species. All bioassays described here have the advantage of being fast, reliable, and standardized.

Key words Antifouling, Bacteria, Barnacles, Bioassay, Macroalgae, Marine extracts, Microalgae, Mussels

1 Introduction

The process of biofouling (accumulation of organisms on immersed surface) is a phenomenon that affects severely numerous productive sectors (aquaculture, oil, shipping) by causing damage to platforms, structures, and ship hulls, besides generating huge economic losses [1, 2]. Antifouling (AF) coatings are used in order to avoid the colonization of surfaces by biofoulers and, consequently, the high costs relative to transport delays, hull repairs, cleaning of desalination units, and biocorrosion (estimated at 150 billion USD per year) [3]. Given the toxicity of AF paint formulations (incorporating copper and biocides) and the fact that bacteria can rapidly develop resistance to biocides [4], it is crucial to work continuously on the development of new solutions. With these objectives, there is nowadays an intensive research effort towards the search for environmentally friendly alternative solutions based on biomimetics and linking chemical ecology to marine biotechnology [5]. A promising area is the discovery and screening of marine natural products from macroalgae. These organisms, which are non-motile

Dagmar B. Stengel and Solène Connan (eds.), *Natural Products From Marine Algae: Methods and Protocols*, Methods in Molecular Biology, vol. 1308, DOI 10.1007/978-1-4939-2684-8_27, © Springer Science+Business Media New York 2015

and often have soft body, synthesize bioactive compounds for the protection against external environmental pressures such as predation and competition or to prevent the settlement of other organisms on their surfaces [5–10]. However, a major difficulty for marine natural chemists is to distinguish which organisms produce the bioactive compounds: the basibiont (host) or the epibionts (e.g., surface-associated microflora) [11]. Since chemical extraction processes do not discriminate whether compounds originate from algae or from its associated microflora, a key step to be made immediately prior to extraction and isolation of molecules is the deletion of epibionts from the algae studied. Within chemical, ecological, and biotechnological contexts, the clear identification of source organisms becomes crucial when symbiotic interactions occur. The protocols commonly used vary according to the moisture content of macroalgae (*see* Subheading 3). The advantage of the proposed methodologies relies on the allowance for significant decontamination of algal covers without altering the surfaces. The succeeding steps consist of chemical extraction (using several solvents), followed by bioactivity assessment against key colonizers: bacteria, microalgae, macroalgae, and invertebrates. A previous study [12] has emphasized that the chance of finding new compounds displaying AF activity is higher when sampling and testing organisms from the same environment.

One of the first steps in the chemistry of marine natural products consists of obtaining organic extracts from organisms that are assumed to contain bioactive compounds. Selection of the extracting solvent thus depends on the research objective; yet some aspects must be considered, such as solvent polarity and high capacity of solvation for the substance to be extracted. Regarding extraction of polar nature compounds (such as phenolic compounds), methanol is a good extracting solvent, while hexane or chloroform allow the extraction of compounds with non-polar properties. When the nature of the bioactive extract is unknown, it is recommended to use mixtures of solvents of different polarity, or else use a polar solvent (ethanol) and conduct the further extract partitioning with solvents of decreasing polarity. The hexane dipping method [13] enables extraction of surface molecules and provides very suitable tools for the rapid extraction of potential bioactive compounds.

Bioactivity screening of extracts using lab-based assays is essential to evaluate new compound/formulation effectiveness [14] (however, when a bioactive compound is detected, field tests must be performed in order to obtain a more empirical data). For each bioassay it is important to cautiously choose positive controls using AF compounds and/or biocides that have previously been used. The commercial AF product Sea-Nine 211 (Dow), a herbicide used in AF paint formulations, which acts as PSII electron transport inhibitor, can be used as a positive control [15].

For the assessment of bioactivity against microfouling organisms (bacteria and microalgae), it is important to assess both the inhibition of cell adhesion and growth inhibition. The proposed method has the advantages that it uses only small amounts of compounds, providing a very rapid bioactivity assessment, and is not subject to any seasonal constraints.

Most sessile marine organisms produce planktonic propagules referred to as larvae for invertebrates and spores for algae [14]. Adhesion and settlement are crucial events in the fouling process; thus their inhibition provides AF protection for surfaces. The main constraint for assessing AF efficiency towards these organisms is that there is seasonality in the release of spores and larvae (all-year long for tropical species, twice a year for temperate species, and once a year for cold species) [3]. Standardized protocols have only been developed for particular species and do not represent the real biodiversity of marine foulers in natural assemblage. For the purpose of this chapter, we have chosen to select the most common species used in AF tests: *Ulva* sp. (macroalga) and *Balanus* sp. (barnacle), and the most common bioassays (based on the inhibition of settlement). However, it is important to state that in order to obtain more information on the mode of action of bioactive compounds, further bioassays targeting organism behavior should be performed (*see* [14] for review) but these will not be described here. When research laboratories are not located near any coast or if researchers have no easy access to spores or larvae, an alternative method is to assess the efficiency of compounds using biochemical assays, such as targeting enzymes involved in adhesive production (e.g., phenoloxidase).

2 Materials

2.1 Microflora Removal from Macroalgal Surfaces (Prior Chemical Extraction)

1. Sterilized forceps.
2. Sterile artificial seawater.
3. 40 and 50 % ethanol (in distilled water, v/v).
4. 1 % Sodium hypochlorite (NaOCl) in 40 or 50 % aqueous ethanol (w/v).

2.2 Preparation of Macroalgae Extracts

1. Filter paper (Whatman No. 3).
2. Mortar and pestle.
3. Round-bottom flasks.
4. Separatory funnel and glass stopper.
5. Pasteur pipette.
6. Acetone:methanol (1:1, v/v) analytical grade.
7. Ethyl ether analytical grade.

8. *n*-Butanol analytical grade.
9. Hexane analytical grade.
10. Rotary evaporator and pump.

2.3 Antibacterial Bioassays (Inhibition of Adhesion and Growth)

1. Bacterial strains (*see* **Note 1**).
2. Glass vials, 20 mL (for solution and strain inoculation).
3. Polypropylene MicroWell plates, 96-well (0.35 mL) round bottom with lid.
4. Spectrophotometer disposable cuvette (10 mL).
5. Marine bacterial medium: Sterile Marine Broth or artificial seawater with 0.5 % peptone (w/v).
6. Methanol analytical grade.
7. 0.3 % aqueous crystal violet (1 % aqueous crystal violet solution dissolved in water, (v/v) for adhesion measurement only).
8. 1 mg/mL stock solutions of controls (known AF such as Sea-Nine 211, Dow).
9. Autoclave.
10. Spectrophotometer.
11. UV sterilization cabinet.
12. Incubator.

2.4 Antimicroalgal Bioassay (Inhibition of Adhesion and Growth)

1. Microalgal strains (*see* **Note 2**).
2. GF/F filter (glass microfiber Whatman, 0.7 μm).
3. Glass vials, 20 mL.
4. Black polystyrene MicroWell plates, 96-well (0.35 mL) round bottom with lid.
5. f/2 culture media ([16]; *see* **Note 3**).
6. Methanol analytical grade.
7. Room with constant light (fluorescent lamp 100 W, irradiance 140 μmol/m^2/s).
8. Autoclave.
9. Microplate reader measuring fluorescence.
10. Chlorophyll *a* from spinach (C5753 Sigma, extinction coefficient at 663 nm in methanol: 77 L/g/cm).
11. 1 mg/mL stock solutions of controls (known AF such as Sea-Nine 211, Dow).

2.5 Experiments Assessing the Inhibition of Ulva sp. Settlement and Germination

1. Plastic tray containing artificial seawater to place fresh algae immediately after collection.
2. Polystyrene well plates, 24-well (1 mL) flat bottom with lid.
3. Polypropylene MicroWell plates, 96-well (0.35 mL) round bottom with lid.

4. Sterile von Stosch medium ([17, 18]; *see* **Note 4**).
5. 10 % formalin diluted with artificial seawater (v/v).
6. Autoclave.
7. Incubator equipped with lamps (140 μmol/m²/s).
8. Light microscope.
9. Hemocytometer.
10. 1 mg/mL stock solutions of positive controls (known AF such as Sea-Nine 211, Dow).

2.6 Experimentation Assessing the Settlement Inhibition of Barnacle Larvae

1. Plankton net.
2. Nitex plankton netting (150 mm) or equivalent.
3. 24-Well Iwaki microplates or equivalent.
4. Artificial seawater (*see* **Note 5**).
5. UV sterilization cabinet (Scie-plas G/E UVSC or equivalent).
6. Dissecting microscope.
7. 1 mg/mL stock solutions of positive controls (known AF such as Sea-Nine 211, Dow).
8. Incubator.

2.7 Inhibition of Blue Mussel Mytilus edulis Settlement

1. Spectrophotometer.
2. Phenoloxidase (PO) or tyrosinase (EC1.14.18.1).
3. 10 mM L-dopa or catechol prepared in ultrapure water.
4. Phosphate buffer 50 mM (pH 6.8): Two stock solutions are required: weigh 71.6 g of dibasic sodium phosphate ($Na_2HPO_4*12H_2O$) and pour in ultrapure water up to 1 L (0.2 M; solution A); weigh 27.6 g of monobasic sodium phosphate ($NaH_2PO_4*H_2O$) and pour in ultrapure water (0.2 M; solution B). To prepare 1 L of working buffer add 245 mL of solution A and 255 mL of solution B, fill up to 1 L with ultrapure water, and adjust the pH to 6.8.
5. 10 μg/L Biocide TBTO prepared in ultrapure water.

3 Methods

3.1 Determination of the Moisture Content of Macroalgae

1. Collect fresh material (six batches or specimens depending of algae size) and gently press them between sheets of absorbent paper to remove excess moisture.
2. Measure wet weights.
3. Dry macroalgal tissue for 48 h at 70 °C in an oven.
4. Measure dry weights.

5. Determine moisture content of each specimen using the following equation:

$$\text{Moisture content} = \left[\left(\text{wet weight} - \text{dry weight}\right) / \text{wet weight}\right] \times 100$$

6. Calculate average moisture content of all replicates.

3.2 Removal of Microflora from Macroalgal Surfaces (Prior to Chemical Extraction)

1. After collection, immerse macroalgal fronds twice for 5 min in sterile artificial seawater.
2. Dry fragments of fronds between sheets of absorbent paper to remove excess moisture.
3. Choose **step 4** or **5** depending on the moisture content of macroalgae (determined in Subheading 3.1).
4. For algae with moisture content ≥80 % immerse fronds in a solution containing 50 % ethanol with 1 % NaOCl for 30 s.
5. For algae with moisture content <80 % immerse fronds in a solution containing 40 % ethanol with 1 % NaOCl for 60 s.

3.3 Preparation of Macroalgal Extracts

3.3.1 Total Extract Preparation

1. Weigh 50 g of dry algal tissue.
2. Cut the specimens into small pieces or pulverize them in a blender.
3. Place them in a glass container, add 250 mL of acetone-methanol 1:1 to cover the dry material, and leave to macerate for 1 h.
4. Then in a mortar with a pestle, grind well the dry material to allow the extraction of organic compounds. Filter the solution, and repeat this step three to four times with the pellet. Record the amount of solvent used.
5. Mix all filtered solutions and evaporate to dryness under reduced pressure in the rotary evaporator (36 °C) to get the crude extract. To obtain the weight of this dried extract, first weigh the round-bottom flask empty and subtract it to the weight after evaporation, of the flask with the dried extract.
6. Resuspend the crude extract in 50 mL of distilled water and place it in the separatory funnel. Add 50 mL of ethyl ether.
7. Seal it with the glass stopper and shake it vigorously for 2 min with periodic venting to release pressure excess (Fig. 1). Subsequently, place the separatory funnel at room temperature for 30 min to allow separation of the two phases (water and ether; *see* **Note 6**).
8. Separate the two phases (*see* **Note** 7) and repeat the liquid/liquid extraction twice with the aqueous phase (*see* **Note 8**).
9. Concentrate the ether phase under reduced pressure to obtain the ether extract.
10. Extract the aqueous phase once more with 50 mL of butanol. Place the aqueous phase and butanol in the separatory funnel and repeat **steps** 7 and **8**.

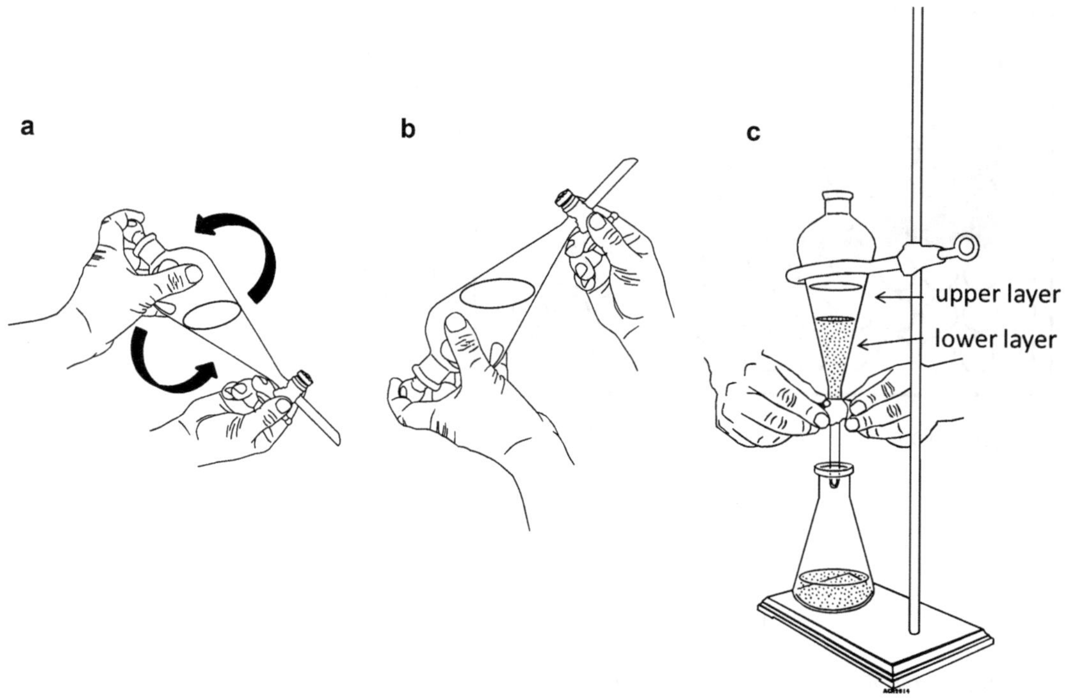

Fig. 1 Illustration of liquid-liquid extraction. (**a**) Vigorous shaking; (**b**) venting to release excess pressure; (**c**) separation of the two phases

11. Evaporate butanol phase under reduced pressure in the rotary evaporator (41 °C) to obtain the butanol extract (*see* **Note 9**).
12. Record the extract weights and calculate yields (expressed as percentage) for each extract as follows:

$$\text{Yield}\,(\%) = [\text{dry weight of the extract} \,/\, \text{dry weight of algal tissue}] \times 100$$

13. Store extracts in vials or round-bottom flasks at −20 °C until bioactivity assays. The extracts can be kept for 1 year without losing activity.

3.3.2 Surface Extracts (Hexane Dipping Method) [13]

1. Macerate the fresh algal tissue with hexane (1 L/kg alga wet weight) in dark conditions between 10 and 30 s.
2. Filter the solution and evaporate it to dryness under reduced pressure.

3.3.3 Extract Solutions

1. Calculate the amount of extract required for all bioassays (*see* **Note 10**).
2. Dissolve the extract in methanol (used as carrier solvent) to adjust concentration to 1 mg/mL.
3. Dilute this stock solution appropriately to get the following extract concentrations: 0.01, 0.1, 1, 10, and 50 μg/mL.

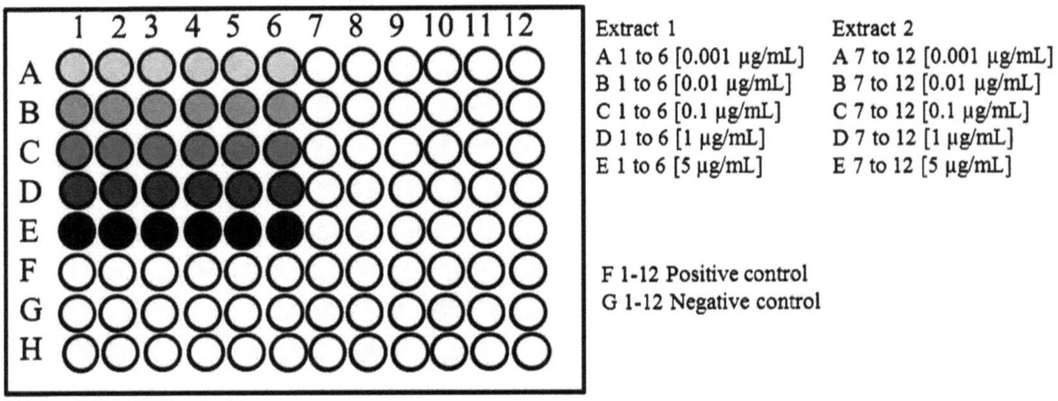

Fig. 2 Example of the distribution on a microplate of different concentrations of extracts to be tested and controls

A representation of extract and control distribution on a microplate is presented in Fig. 2.

3.4 Antibacterial Bioassays [15]

3.4.1 Growth Inhibition

1. Place 100 µL of each extract solutions and controls (known AF compounds, methanol only) in 6 wells of the 96-well plate for each of the bacterial strains. Leave wells to dry by evaporation at room temperature for 12 h (use different plates for each bacterial strain in order to avoid cross-contamination; *see* **Note 1**).
2. Seal the plates and sterilize them in a UV sterilization cabinet (Scie-plas G/E UVSC or equivalent) for 30 min.
3. Prior to the experimental work, cultivate marine bacteria in sterile seawater (autoclave 121 °C, 15 min) enriched with 0.5 % neutralized bacteriological peptone or in marine broth.
4. Dilute the bacterial cultures according to the Amsterdam method ([19], *see* **Note 11**) for a cell density of 2×10^8 cells/mL. To check the density, measure the absorbance of a tube previously inoculated with the testing strain (culture of 24 h) at 620 nm using a spectrophotometer.
5. Remove the lid of the plates under aseptic conditions.
6. Add 100 µL of the microorganisms-inoculated medium (concentration of 2×10^8 cells/mL) in each well (containing extracts and controls).
7. Seal the plates and incubate them at 20 °C for 48 h.
8. After incubation, measure the absorbance at 620 nm with a plate reader. Use the medium as a blank. Minimum inhibition concentration (MIC) value is defined as the lowest concentration that produces a reduction in growth.

3.4.2 Adhesion Inhibition

1. Prepare microplates and inoculate them as stated above (*see* **steps 1–7**).
2. After 48-h incubation, empty the wells and rinse them with 100 μL of sterile seawater to remove the non-attached cells. Dry the wells in air at room temperature.
3. Stain the remaining bacterial biofilm with 100 μL of 0.3 % crystal violet (*see* **Notes 12** and **13**).
4. Measure the absorbance at 595 nm (*see* **Note 14**). Use the medium as a blank. MIC is defined as the lowest concentration that produces a reduction in adhesion. The effective concentration (EC_{50}) is obtained for 50 % adhesion rate reduction.

3.5 Antimicroalgal Bioassay

3.5.1 Adhesion Inhibition

1. Place 100 μL of each extract solution and control (known AF compounds, methanol only) in 6 wells of the 96-well plate for each of the microalgal strains. Leave wells to dry by evaporation at room temperature for 12 h (use different plates for each microalgal strain in order to avoid cross-contamination).
2. Seal the plates and sterilize them in a UV sterilization cabinet (Scie-plas G/E UVSC or equivalent) for 30 min.
3. Cultivate microalgal strains using f/2 media (*see* **Notes 2** and **15**) and maintain them at 20 °C and constant light (incident irradiance: 140 μmol/m^2/s).
4. Estimate microalgal culture biomass through determination of chlorophyll *a* (Chl *a*) concentration [20]: Collect 5 mL of microalgal culture on a GF/F filter (Whatman) and transfer the filter immediately to a vial containing 5 mL of methanol analytical grade. Keep the vial in the dark at 4 °C for 30 min. Measure the fluorescence of the pigment extract (PolarSTAR Optima BMG Labtech or equivalent; excitation: 485 nm, emission: 645 nm). Determine the Chl *a* concentration using a calibration curve with spinach Chl *a*.
5. After the determination of Chl *a* concentrations for each microalgal species, dilute algal solutions in order to prepare aliquots of microalgal suspension with a starting concentration of 0.1 mg Chl *a* /L.
6. Remove the lid of the plates under aseptic conditions.
7. Add 100 μL of this suspension to each well of previously coated microplates containing the extract solutions and controls under aseptic conditions.
8. Incubate the inoculated microplates for 5 days at 20 °C in constant light (incident irradiance: 140 μmol/m^2/s).
9. After 5 days, remove the medium using a pipette and quantify the pigment concentration with the fluorimetric method [20]: Measure the fluorescence (PolarSTAR Optima BMG Labtech or equivalent; excitation: 485 nm, emission: 645 nm) and use

the medium as a blank. MIC is defined as the lowest concentration that produces a reduction in adhesion. EC_{50} are obtained for 50 % adhesion rate reduction.

3.5.2 Growth Inhibition

1. Prepare black microplates and inoculate them as stated above (*see* Subheading 3.5.1, **steps 1–7**).
2. After 5 days of incubation, centrifuge the microplates at 2,380*g* for 10 min at 4 °C (Beckman Coulter Allegra 25R or equivalent) and empty the plates.
3. Add 100 μL of 100 % methanol analytical grade in each well to liberate Chl *a*.
4. Quantify the pigment concentration employing the fluorimetric method [20]: Measure the fluorescence (PolarSTAR Optima BMG Labtech or equivalent; excitation: 485 nm, emission: 645 nm) and use the medium as a blank. MIC is defined as the lowest concentration that produces a reduction in adhesion. EC_{50} are obtained for 50 % growth rate reduction.

3.6 Inhibition of Settlement and Germination of Macroalgal Spores

1. Collect the macroalga *Ulva* sp. (e.g., *U. intestinalis*) from the shore.
2. Examine the algae to identify areas of discoloration which correspond to fertile area, usually around the edges.
3. Wash selected fertile areas in filtered seawater and then blot them using absorbent paper. Cut these areas into small pieces (2–3 cm length).
4. Place these macroalgal pieces in individual wells on 28-well plates and seal them with parafilm.
5. Leave the pieces to dry for 6 h at controlled temperature in a growth room (20 °C, constant light 140 μmol/m^2/s).
6. Add 2–3 drops of von Stosch medium in each well after drying.
7. The green coloration of the media indicates the presence of spores; this is confirmed using a light microscope (x20 magnification; *see* **Note 16**).
8. Count the spores using a hemocytometer and determine the number of spores per each 50 μL using the average of spore counts that results from a total of ten counts.
9. Dilute the solution using von Stosch medium at the concentration of 140 spores/μL approximately.
10. Transfer 100 μL of extract solutions and controls into 96-well plates.
11. Add 50 μL of spore solution to each well.
12. Seal the plates with parafilm and keep them in the dark for 2 h.

13. Then empty the plates by inversion and refill all wells with 100 μL of von Stosch medium.
14. Leave the plates for 5 days in a growth room (20 °C, 140 μmol/m^2/s under constant light).
15. Fix then the spores by adding 50 μL of 10 % formalin to the wells for avoiding further germination.
16. Count the spores from three fields of view for each well using an inverted microscope. Count also the number of spores settled or germinated: spores are considered to have germinated if two or more cells and cell wall are visible.
17. Express the results as EC_{50} values (concentrations leading to 50 % inhibition of the total number of spores settled or germinated spores).

3.7 Inhibition of Barnacle Larvae Settlement [12]

The most common studied species are *Balanus amphitrite* (tropical species) and *Balanus improvisus* (temperate and cold waters).

1. Collect barnacle larvae from the field using a planktonic mesh (*see* **Note 17**).
2. After collection, prevent the settlement by maintaining the larvae in Nitex plankton net (150 mm) (which is an unfavorable surface for settlement) for 24 h at 14 °C.
3. Carry out settlement assays in 24-well Iwaki microplates (or equivalent).
4. Prepare microplates as follows: Place 100 μL of each extract solution and controls in 6 wells of the plate and keep them under a fume hood for 12 h to allow the evaporation of the carrier solvent.
5. Seal the plates and sterilize them in a UV sterilization cabinet (Scie-plas G/E UVSC or equivalent) for 30 min.
6. Fill the wells with artificial seawater and add one cyprid (barnacle larva) per well.
7. Incubate the microplates in the dark at 14 °C for *Balanus improvisus* and at 28 °C for *Balanus amphitrite* for 48 h.
8. Perform each experiment with two batches of larvae.
9. After incubation, examine the physical state of each larva under a dissecting microscope. Larvae are classified as (1) dead (floating cyprids with extended thoracopods with no movement, or floating cyprids that do not respond to a light touch produced by a metal probe); (2) settled (permanently attached or metamorphosed individuals); and (3) swimmers.
10. For each extract, calculate EC_{50} values (concentration of extract which results in a 50 % inhibition of settlement compared to the blank).

3.8 Inhibition of the Blue Mussel *Mytilus edulis* Settlement [21]

Inhibition of mussel settlement is measured spectrophotometrically by recording phenoloxidase activity (enzyme involved in byssus thread synthesis).

1. Extract phenoloxidase (PO) from fresh mussels [21] (*see* **Note 18**) or buy mushroom tyrosinase (EC1.14.18.1) from Sigma (a previous study has demonstrated the correlation of activity of these two enzymes) [21].
2. Use 10 mM L-dopa or catechol in 50 mM phosphate buffer (pH 6.8) as substrate.
3. Incubate the enzyme at 25 °C with the substrate and algae extracts (*see* **Note 19**).
4. Prepare two controls: one positive control with the biocide TBTO (10 μg/mL) and one negative control with buffer only.
5. Calculate the phenoloxidase activity from the increment of absorbance readings at 475 nm from 30 s to 1 min after incubation. One unit of enzyme activity is defined as the amount of enzyme that catalyzes the formation of 1 μmol of dopachrome per minute under the experimental conditions.
6. Run the test in triplicate.
7. Express the results as the inhibition percentage of phenoloxidase activity compared to a control that contains only buffer.

4 Notes

1. Marine bacteria can be ordered from ATCC. The most common strains used in antifouling studies are *Halomonas aquamarina* ATCC 14400, *Pseudoalteromonas elyakovii* ATCC 700519, *Polaribacter irgensii* ATCC 700398, *Shewanella putrefaciens* ATCC 8071, *Roseobacter litoralis* ATCC 49566, *Vibrio harveyi* ATCC 14126, *Vibrio proteolyticus* ATCC 15338, *Vibrio harveyii* ATCC 35084, *Vibrio aestuarianus* ATCC 35048, *Vibrio natriegens* ATCC 14048, and *Pseudoalteromonas citrea* ATCC 29720.
2. Marine phytoplanktonic species can be ordered from several algal culture collections, including Algobank, SAG (Goettingen) and CCAP (Scotland). The most common strains used in antifouling studies are *Exanthemachrysis gayraliae* AC 15 (Algobank code), *Cylindrotheca closterium* AC 170, *Pleurochrysis roscoffensis* AC 32, *Porphyridium purpureum* AC 122, *Halamphora coffeaeformis* AC 713, *Lotharella globosa* AC132, *Rhodosorus marinus* AC 119, and *Pleurochrysis carterae* AC 1.
3. The recipe of the f/2 medium can be found in [16]. Stock solutions of $NaNO_3$, NaH_2PO_4, trace metals, and vitamin can

be prepared in advance and stored at 4 °C. The vitamin stock solution is light sensitive; it is advised to keep it covered with foil.

4. The recipe of the von Stosch medium is very well described in [17], which is modified from the original recipe [18].
5. If your laboratory does not have access to natural seawater, you can use artificial seawater.
6. When the separatory funnel is placed on the support ring for the separation of phases, the glass stopper must be removed to avoid excess pressure.
7. When separating the two phases during the liquid-liquid extraction, it must be considered that the lower phase is water and the upper one is either ether or butanol.
8. If an emulsion is created when separating the two phases (aqueous and ether), sodium chloride (NaCl) must be used to break emulsion. Most organic compounds are poorly soluble in aqueous solutions saturated with strong electrolytes.
9. To achieve a better evaporation of butanol, water at a ratio of 1:1 (v/v) must be added.
10. When the stock solutions and dilutions (to adjust the different concentrations of each sample to be tested: 0.01, 0.1, 1, 10, and 50 μg/mL) are prepared, you must take into consideration the evaporation of the solvent and prepare more of the required volume. This volume will depend on the number of strains to be tested and the number of bioassays to be undertaken.
11. The Amsterdam method [19] allows the absorbance measurement at 620 nm of a bacterial suspension to estimate its concentration following a standardized method. Table 1 displays how to prepare the bacterial solution for the bioassay (in order to start all the bioassays with a bacterial suspension of 2×10^8 cellules/mL). The procedure to follow is as follows: (a) measure the OD of the bacterial stock solution at 620 nm; (b) if the OD is <0.2 and the cell number is too low it is advised to place the culture back in the incubator and to repeat the operation 24 h later; (c) if the OD is >0.65, the bacterial solution should be diluted in sterile media in order to reach an OD value between 0.2 and 0.65; (d) if the OD is between 0.2 and 0.65, then use the table below to prepare the bacterial solution that will be used for the inoculation of the MicroWell plates: the corresponding volume of bacterial culture (in μL) should be added to sterile culture medium for a final volume of 10 mL.
12. It is important to wear goggles, gloves, and protective clothing while working with crystal violet. Stains can be removed using ethanol in case of spillage.

Table 1
Amsterdam table [19]

OD at 620 nm	Volume (μL)
0.20	50
0.25	40
0.30	33
0.35	29
0.40	25
0.45	22
0.50	20
0.55	18
0.60	16
0.65	15

Volume (μL) of bacterial suspension to add to a sterile medium to get a final bacterial concentration of 2×10^8 cells/mL in the inoculum (final volume 10 mL) according to the measurement of the absorbance of the bacterial suspension at 620 nm

13. Crystal violet is harmful to aquatic organisms and may cause long-term adverse effects in the aquatic environment. Do not discharge into drains or water courses or onto soil.
14. It is recommended to gently shake the microplates after staining.
15. If you encounter problems during microalgal culture (e.g., insufficient growth of microalgae), try using seawater previously stored for a couple of days in darkness instead of fresh seawater for medium preparation.
16. If you encountered difficulty in obtaining spores, try avoiding **step 6** and after **step 7** place the microplates in the fridge for 2 h. The temperature shock can enhance sporulation.
17. Depending on the age of larvae collected from the field, they may settle very quickly in the bucket. To avoid this, you can place the larvae in cold water (around 6 °C) and keep them refrigerated (using a cool box) until arrival in the laboratory.
18. Phenoloxidase can be extracted and purified for mussel's foot. The procedure involves a series of purification steps [21]: (1) precipitation with acetic acid at pH 5; (2) ultrafiltration; (3) DEAE-Sepharose chromatography; and (4) Sephacryl S-300 HR chromatography. Pure enzyme aliquots can be kept frozen for 12 months.
19. Phenoloxidase activity is significantly affected by temperature. Thus, in order to avoid variability in the results, it is important to work in a controlled-temperature room (18–23 °C).

Acknowledgements

Prof Claire Hellio and Mrs Rozenn Trepos wish to thank the LEAF project which is funded by the European Union inside the seventh framework programme FP7 (2007-2013) under grant agreement no. 314697. Prof R. Noemí Aguila-Ramírez is grateful to CONACYT for the scholarships grant, and together with Prof. Claudia J. Hernández-Guerrero wish to thank their institution and incentives for research (EDI and COFAA).

References

1. Stowe S, Richards J, Tucker A et al (2011) Review. Anti-biofilm compounds derived from marine sponges. Mar Drugs 9:2010–2035
2. Li YX, Wu HX, Xu Y et al (2013) Antifouling activity of secondary metabolites isolated from Chinese Marine organisms. Mar Biotechnol (NY) 15:552–558
3. Maréchal JP, Hellio C (2009) Challenges for the development of new non-toxic antifouling solutions. Int J Mol Sci 10:4623–4637
4. Mudryk ZJ (2002) Antibiotic resistance among bacteria inhabiting surface and subsurface water layers in estuarine lake Gardno. Pol J Environ Stud 11:401–406
5. Chapman J, Hellio C, Sullivan T et al (2014) Bioinspired synthetic macroalgae: examples from nature for antifouling applications. Int Biodeter Biodegrad 86:6–13
6. Proksch P (1994) Defensive roles for secondary metabolites from marine sponges and sponge-feeding nudibranchs. Toxicon 32: 639–655
7. Wulff JL (2006) Rapid diversity and abundance decline in a Caribbean coral reef sponge community. Biol Conserv 127:167–176
8. Dobretsov S, Abed RM, Teplitski M (2013) Mini-review: inhibition of biofouling by marine microorganisms. Biofouling 29:423–441
9. Mieszkin S, Callow ME, Callow JA (2013) Interactions between microbial biofilms and marine fouling algae: a mini review. Biofouling 29:1097–1113
10. da Gama B, Plouguerné E, Pereira RC (2014) The antifouling defense mechanisms of marine macroalgae. In: Jacquot J, Gadal P, Bourgougnon N (eds) Advances in Botanical Research No 71, Sea plants, 1st edn. Elsevier, Amsterdam, pp 413–440
11. Kientz B, Thabard M, Cragg S et al (2012) A new method for removing microflora from macroalgal surfaces: an important step for natural product discovery. Bot Mar 54:457–469
12. Maréchal JP, Hellio C (2011) Antifouling activity against barnacle cypris larvae: do target species matter (*Amphibalanus amphitrite* versus *Semibalanus balanoides*)? Int Biodeter Biodegrad 65:92–101
13. De Nys R, Dworjanyn SA, Steinberg PD (1998) A new method for determining surface concentrations of marine natural products on seaweeds. Mar Ecol Prog Ser 162:79–87
14. Dahms H, Hellio C (2009) Laboratory bioassays for screening marine antifouling compounds. In: Hellio C, Yebra D (eds) Advances in marine antifouling coatings and technologies. Woodhead Publishing, Cambridge, pp 275–307
15. Trepos R, Cervin G, Hellio C et al (2014) Antifouling compounds from the sub-arctic ascidian *Synoicum pulmonaria*: synoxazolidinones A and C, pulmonarins A and B, and synthetic analogues. J Nat Prod 77:2015–20113. doi:10.1021/np5005032
16. Guillard RRL, Ryther JH (1962) Studies of marine planktonic diatoms. I. *Cyclotella nana* Hustedt and *Detonula confervacea* Cleve. Can J Microbiol 8:229–239
17. Andersen RA (2005) Algal culturing techniques. Academic, London, p 578
18. Guiry MD, Cunningham EM (1984) Photoperiodic and temperature responses in the reproduction of north-eastern Atlantic *Gigartina acicularis* (Rhodophyta: Gigartinales). Phycologia 23:357–367
19. Amsterdam D (1996) Susceptibility testing of antimicrobials in liquid media. In: Lorian V (ed) Antibiotics in laboratory medicine. Williams & Wilkins, Baltimore, pp 52–111
20. Chambers L, Hellio C, Stokes K et al (2011) Investigation of *Chondus crispus* as a potential source of new antifouling agents. Int Biodeter Biodegrad 65:939–946
21. Hellio C, Bourgougnon N, Gal YL (2000) Phenoloxidase (EC 1.14. 18.1) from the byssus gland of *Mytilus edulis*: purification, partial characterization and application for screening products with potential antifouling activities. Biofouling 16:235–244

INDEX

A

Algae (macroalgae and microalgae)
- algae and agri-horticulture 11–12
- algae and cosmetics 12–13
- algae and food 5, 11
- algae and health 13
- algal biotechnology 24–28
- algal breeding 24–25
- algal cultivation 14–24
- algal diversity 4–5
- algal harvesting 13, 14
- current applications of algae 5–13
- genetic modification 25–27
- synthetic biology and molecular pharming 27
- synthetic chemistry 27–28

Algal secondary metabolites
- acetogenins 44–46
- alkaloids 51–52
- alkylated phenyl 47–48
- bromophenols 49
- MAAs 52
- oxylipin 44–46
- peptides 50–51
- pheromones 45
- phlorotannins 48–49
- quinones 48
- steroids 43–44
- terpenes 41–43

Alginates
- infra-red spectroscopy 351, 355
- multivariate curve resolution 351, 358–361
- partial hydrolysis 351, 352
- partial least squares regression model 351, 355–358
- resonance magnetic nuclear spectroscopy 351–355

B

Betaines
- betaine enriched fraction 269, 271
- liquid chromatography–tandem mass 267–274
- solid–liquid extraction 268–271
- spectrometry 269–273

Bioactivity of algal compounds
- antifouling activity
  - antibacterial bioassay 424, 428–429
  - antimicroalgal bioassay 424, 429–430
  - inhibition of barnacle settlement 425, 431
  - inhibition of macroalgal settlement and germination 424–425, 430–431
  - inhibition of mussel settlement 425, 432
  - preparation of extract 423–428
- antifungal activity
  - minimum inhibitory concentration 413, 415–416
  - minimum lethal concentration 414, 416
  - preparation of extract 413–415
- antimicrobial activity
  - disk preparation 404–406
  - plate preparation and reading 405–407
- antioxidant capacity
  - ABTS assay 377, 383–385
  - β-carotene bleaching assay 380–381, 392–393
  - CUPRAC assay 379–380, 389–390
  - DMPD assay 377–378, 385–386
  - DPPH assay 377, 382–383
  - FRAP assay 378–379, 387–388
  - FTC method 381, 393–394
  - HORAC assay 380, 391–392
  - hypochlorus acid scavenging assay 378, 386–387
  - ORAC assay 380, 390–391
  - TBA assay 381–382, 394–395
  - total phenolic assay by Folin–Ciocalteu reagent (FCR) 379, 388–389

C

Carbohydrates
- spectrophotometric assays 77–78, 82–84

E

Ecology of algal secondary metabolites
- allelopathy 59
- anthropogenic impacts 61–62
- biotic interactions 61
- defensive ecology 55–59
- geographical variations 52–54
- life history stages 61
- sensory ecology 54–55
- trophic interactions 60

Dagmar B. Stengel and Solène Connan (eds.), *Natural Products From Marine Algae: Methods and Protocols*, Methods in Molecular Biology, vol. 1308, DOI 10.1007/978-1-4939-2684-8, © Springer Science+Business Media New York 2015

Extraction of algal compounds
enzyme-assisted extraction 146–149
microwave-assisted extraction 153–156

F

Fucoidan
autohydrolysis of fucoidan 303, 304
electrospray ionization mass spectrometry 303–305
matrix-assisted laser desorption/ionization time-of-flight mass spectrometry (MALDI-TOFMS) 303, 305–309
reduction of oligosaccharides 303, 304

G

Galactan
3,6-anhydrogalactose content 328, 334
desulfation 316, 318–319, 329–330, 338–340
electrospray ionisation collision induced dissociation tandem mass spectrometry 330, 341–344
methylation analysis 315–318, 329–330, 338–340
molecular weight analysis 328, 334
nuclear magnetic resonance spectroscopy 316, 319–321, 329, 336–338
purification using high performance gel permeation chromatography 330, 341
pyruvic acid content 315, 317
solid–liquid extraction 326, 331–332
sulfate content 314–317, 327, 333–334
total sugar content 327, 332–333

L

Lipids and fatty acids
fractionation 177, 179–180
gas chromatography–flame ionisation detector 176, 182–186
solid–liquid extraction 176–179
thin layer chromatography 177–178, 180–182

M

Marine biotoxins
amnesic shellfish poisoning toxins
description 283–284
extraction 287, 291–293
liquid chromatography–tandem mass spectrometry 287, 293–295
azaspiracid shellfish poisoning toxins
description 283
diarrhetic shellfish poisoning toxins
description 278–283
extraction 286, 289–290
liquid chromatography–tandem mass spectrometry 287, 290–291
paralytic shellfish poisoning toxins
description 278
extraction 284, 288
liquid chromatography–tandem mass spectrometry 285–286, 288–289
Mycosporine-like amino acids
liquid chromatography–mass spectrometry 122–127
solid–liquid extraction 121–124

N

Nuclear magnetic resonance spectroscopy
classical 194–195, 199–202
HRMAS 194, 196–199

O

Oxylipin
enrichment and partial purification 163, 165–166
liquid chromatography–mass spectrometry 163–164, 166–169
solid–liquid extraction 163–165

P

Phenolic compounds, phlorotannins
hydrophilic interaction liquid chromatography–high resolution mass spectrometry 255, 257–261
liquid–liquid purification of phlorotannins 133–135
molecular size separation of phlorotannins 133, 135–136
nuclear magnetic resonance spectroscopy 133, 137–139
phlorotannin enriched fraction 255–257
radical scavenging activity of phlorotannins 133, 137
solid–liquid extraction 132–134, 255–256
spectrophotometric assays 75–79, 84–88, 133, 136–137
Phycobiliproteins
electrophoresis 111, 113–115
R-phycoerythrin purification 110–111, 113
solid–liquid R-phycoerythrin extraction 110, 112
spectrophotometric assays 79, 93–94
Pigments (chlorophylls and carotenoids)
high performance liquid chromatography 242–248
solid–liquid extraction 242, 244
spectrophotometric assays 79, 88–93
Proteins
enrichment 105
solid–liquid protein extraction 104–105
spectrophotometric assays 76–77, 80–82

S

Surface-enhanced Raman spectroscopy
algal preparation 369, 370
data treatment............ 369–371
Raman spectrometer............ 369, 370

T

Terpenes
flash chromatography 227, 228, 232, 235
fractionation by column chromatography............ 212–214
gas chromatography–mass spectrometry 227, 230–231
nuclear magnetic resonance spectroscopy............ 213–217, 227, 231–232
purification by high performance liquid chromatography............ 213, 214
solid–liquid extraction 212, 213, 227–229
thin layer chromatography............ 213, 227, 229–230

MIX
Papier aus verantwortungsvollen Quellen
Paper from responsible sources
FSC® C105338
FSC
www.fsc.org

If you have any concerns about our products,
you can contact us on
ProductSafety@springernature.com

In case Publisher is established outside the EU,
the EU authorized representative is:
Springer Nature Customer Service Center GmbH
Europaplatz 3, 69115 Heidelberg, Germany

Printed by Libri Plureos GmbH
in Hamburg, Germany